Nanotechnology-Assisted Recycling of Textile Waste

Scrivener Publishing
100 Cummings Center, Suite 541J
Beverly, MA 01915-6106

Publishers at Scrivener
Martin Scrivener (martin@scrivenerpublishing.com)
Phillip Carmical (pcarmical@scrivenerpublishing.com)

Nanotechnology-Assisted Recycling of Textile Waste

Sustainable Tools for Textiles

Edited by

Prashansa Sharma

Dept. of Home Science, Banaras Hindu University, Varanasi, India

and

Shilpi Shree Sahay

Department of Home Science, MMV, Banaras Hindu University, Varanasi, Uttar Pradesh, India

Scrivener Publishing

WILEY

This edition first published 2025 by John Wiley & Sons, Inc., 111 River Street, Hoboken, NJ 07030, USA and Scrivener Publishing LLC, 100 Cummings Center, Suite 541J, Beverly, MA 01915, USA

For more information about Scrivener publications please visit www.scrivenerpublishing.com.

Wiley Global Headquarters
111 River Street, Hoboken, NJ 07030, USA

For details of our global editorial offices, customer services, and more information about Wiley products visit us at www.wiley.com.

Library of Congress Cataloging-in-Publication Data

ISBN 978-1-394-17449-2

Cover image: Adobe Firefly
Cover design by Russell Richardson

Set in size of 11pt and Minion Pro by Manila Typesetting Company, Makati, Philippines

Printed in the USA

10 9 8 7 6 5 4 3 2 1

Contents

Preface

In an era where innovation drives progress, the realm of textiles stands at the forefront of transformation, propelled by the convergence of nanotechnology and recycling, we are delighted to present this significant volume, "Nanotechnology-Assisted Recycling of Textile Waste: Sustainable Tools for Textiles," edited by Prashansa Sharma and Shilpi Shree Sahay. This book represents a collaborative effort to explore the intersection of nanotechnology, recycling, and sustainable textile practices.

Divided into four comprehensive parts, comprising a total of 16 chapters, this book provides insights into the potential of nanotechnology in revolutionizing textile recycling and shaping the future of sustainable textiles.

Part I sets the stage with an insightful overview of textile waste and management, delving into the conceptual dimensions and challenges in handling and organizing textile waste. It also explores the innovative realm of textile recycling.

In Part II, the spotlight shines on comprehensive, sustainable, and productive recycling of waste using nanotechnology. Here, readers are invited to explore the transformative contributions of nanotechnology in shaping sustainable textile design and characterizing functional properties of novel recycled nano-textiles. Future perspectives of nanotechnology in textile applications, particularly concerning waste recycling, are also examined.

Part III explores deeper into the advanced application of recycled and nano-assisted novel textiles generated through waste. From sports textiles to technical textiles, this section explores the diverse applications of recycled waste, bolstered by nano-engineered innovations.

Finally, Part IV addresses the critical aspects of quality control and regulatory compliance in the realm of advanced nano-textile materials. Through an exploration of global legislation, schemes, and standards.

To our readers, we extend a warm welcome. Within these chapters, you'll find a wealth of information waiting to be uncovered.

As editors, we express our heartfelt gratitude to each author who has shared their expertise and insights, enriching this book with a diverse array of perspectives and knowledge. Special acknowledgment is owed to Martin Scrivener, the publisher, whose efforts have ensured the realization of our vision. In closing, we express our heartfelt appreciation to all those who have contributed to this endeavor, whether through their writing, support, or encouragement. We hope this book sparks inspiration and helps pave the way for a more sustainable and responsible future.

Prashansa Sharma
Shilpi Shree Sahay
December 2024

Part I

OVERVIEW OF TEXTILE WASTE AND MANAGEMENT

1

Overview of Textile Waste

Amisha Singh, Shilpi Shree Sahay and Prashansa Sharma*

Department of Home Science (Clothing and Textile), MMV, Banaras Hindu University, Varanasi, U.P., India

Abstract

The textile sector is one of the oldest sectors. After food, clothing comes as the next primary human necessity. Rapid population growth encourages the textile sector to boost production to meet the demand for clothing and apparel, which leads to a significant amount of textile waste. The resulting waste is handled via landfilling or incineration procedures, both of which have a detrimental environmental impact. A significant volume of textile effluent is produced through wet processing because it uses a lot of water and chemicals. When textile waste is dumped on the ground or in water sources, it contaminates both the land and the water bodies. The improper disposal of textile waste leads to soil and water pollution.

By using effective waste management techniques, it is possible to reduce environmental contamination and thus establish a circular economy by turning waste into valuable goods. The present paper provides a thorough description of the various types and categories of textile waste and effluents, as well as the sources that produce them. This paper also discusses the major causes of textile waste and also highlights the impact of textile waste on the environment and human health. This study examines diverse textile processing steps along with a wide range of pollutants produced during textile processing and their toxic effects on weavers, employees, and the environment (air pollution, soil pollution, water, and noise pollution).

Keywords: Textiles, waste, fiber, production, pollution, environmental hazard, clothing

**Corresponding author*: prashansa.sharma@bhu.ac.in

Prashansa Sharma and Shilpi Shree Sahay (eds.) Nanotechnology-Assisted Recycling of Textile Waste: Sustainable Tools for Textiles, (3–28) © 2025 Scrivener Publishing LLC

List of Abbreviations

CAGR	Compound annual growth rate
CO_2	Carbon dioxide
PET	Polyethylene terephthalate
BOD	Biochemical oxygen demand
COD	Chemical oxygen demand
PVA	Polyvinyl alcohol
VOCs	Volatile organic compounds
TDS	Total dissolved solids
GHG	Greenhouse gases
CMW	COVID-19 medical waste
PPE	Personal protective equipment
OEF	Organization environmental footprint
PEF	Product environmental footprint
REACH	Registration, evaluation, authorization, and restriction of chemical
5R	Refuse, reduce, reuse, recycle, and recover

1.1 Introduction

Fabrics and fibers have been integral to human existence since ancient time, serving as the fundamental elements of clothing. Today, textiles play diverse roles from shielding the body against weather conditions to symbolizing social status and expressing individual style all while keeping pace with evolving trends. With market values of almost USD 1.3 trillion and employment opportunities for over 300 million people along the whole value chain, the textile industry is one of the largest and oldest industries in existence [1]. The dominant industry for clothing and apparel is the textile sector, which is regarded as a key sector for consumer goods. Over the past few decades, fast population expansion, rising global incomes, and improved living standards have all enabled an increase in both clothing manufacturing and demand. The average yearly consumption of textiles has increased by two times in the last two decades, from 7 to 13 kg per person, and reached the maximum amount of 100 million tons worldwide [2].

In the present era, Clothes are designed and produced for rapid trend transitions via depreciation, early disposal, and discard, which facilitates fast income, instead of considering how design and production might accommodate consumer needs and lead toward sustainability. By 2025, the textile industry's demand for textile fibers is anticipated to grow, by around 400%, to reach 130 million tons, with an annual growth rate of 4.3% [3].

Between 2025 and 2026, the Indian textile and apparel market is forecasted to expand by a Compound annual growth rate (CAGR) of 10%, reaching over US$190 billion. India holds a 4% share in the global textile and clothing trade, and it is the leading producer of cotton worldwide. In the 2021–2022 cotton season, approximately 362.18 lakh bales were harvested, with an anticipated domestic consumption of 338 lakh bales. By 2030, India's cotton production is expected to rise to 7.2 million tons (around 43 million bales of 170 kg each) due to increasing consumer demand. [4] As can be shown from statistics, as demand grew, the manufacturing sector developed, and mechanized manufacturing techniques acquired the place of manual labor incentive systems. This enabled to manufacture textiles at affordable prices along with higher productivity and a wide range of variety. This has led to an excessive stock of economical, mass-produced products that tend to low quality as well as a significant amount of textile fiber waste, having restricted end uses.

One of the most intricate and harmful industries is the textile sector. This is challenging because it entails a very extensive and diverse procedure, including raw material production, fabrication of textiles, apparel development, transportation, and waste disposal. It is polluting because a significant amount of waste is generated while manufacturing and consuming it. The processing and manufacturing procedure of textile materials such as the harvesting, extraction; agriculture; process to make fiber, fiber to yarn, and yarn to fabric; wet-processing of fiber or fabric; blow room; carding; draw frame; combing; roving frame; ring frame; spinning process sequences; weaving/knitting, dyeing/printing, and finishing steps; and transportation are additionally required an excessive number of natural resources and non-renewable resources. Aspects related to the environmental impact of fiber production and the subsequent disposal activities have also become more prominent as the demand for fabrics has increased. Several recent studies highlighted the threat and hazard the textile industry generates due to the use of large volumes of industrially dangerous and toxic chemicals during the manufacturing process and the discharge of pollutants during the lifecycle of a textile product in aquatic systems and the atmospheric micro-system. The textiles industry emits over 1.2 billion tons of greenhouse gas (GHG) emissions (more than all of the emissions from international flights and maritime shipping combined). Sixty-three percent of textile fibers are composed of petrochemicals, whose manufacture and disposal result in significant carbon dioxide (CO_2) emissions [5]. One of the top three industries for water waste is the textile industry, which severely degrades freshwater supplies. For the production of one pair of jeans and one single t-shirt, respectively, approximately 8,500 and 2,600 liters of water are needed [6]. The mean lifespan of a piece of fabric is approximately 3 years, and more than one million tons of textile materials are discarded annually, the majority of which comes from domestic sources. About 3% of the total volume of residential waste is made up of textiles. Currently, about 80%–90% of textile waste produced is no longer biodegradable because it consists of polyester, which contains PET (polyethylene terephthalate). Compared to natural fibers, synthetic materials are non–eco-friendly and non-biodegradable. Developing nations produce more than 60% of the world's apparel. The world's leading producer of apparel today, accounting for over 32% of global exports, is Asia. The global supply is generated by arranging raw materials and finished products in shipping containers and then transported by rail and vehicles. The health and well-being of people living in coastal and rural regions around the world are commencing to be impacted by marine shipping pollution, which has expanded dramatically over the preceding 20 years. Because there is insufficient information and details available regarding the quality of the discarded textiles, it is difficult to determine whether they can be recycled, transformed, reduced, and recovered or should simply be disposed of and discarded as waste materials. Therefore, it becomes essential to inquire into how textile waste is managed in the most environment-friendly manner [7].

Waste

Waste is defined as any material or goods that has reached the end of its shelf life. They are considered a challenge because they degrade and diminish precious resources, the environment, and human health; occupy landfill space; and increase the expense of using current landfills and building new ones.

1.2 Textile Waste

Textiles perform a variety of purposes and are constructed from a variety of fiber types, blended in a range of ratios. Three classes—apparel, home furnishings, and industrial—can be used for classifying the applications of fiber. Growing product consumption is an indication of rising waste production globally, which prompted greater social responsibility and environmental awareness, reinforced by stringent legal restrictions in developed nations, and led to the development of more effective waste management techniques. Green consumerism has encouraged industries to portray a positive environmental picture to the general public. The producers of clothing and apparel have employed recycling initiatives and environment-friendly techniques of manufacturing.

1.3 Classification of Textile Waste

The textile sector emits a lot of pollutants into the environment, including air emissions, noise pollution, solid waste, and textile effluent (liquid discharge). The waste from the textile sector is generated via numerous industries, including (the step from the production of fiber to fabric and its commercialization and maintenance of the clothing) textile industry, the apparel manufacturing industry (including loom, wet-processing, and finishing), dyeing and printing industries, consumers, and the retail and service sectors. Hence, textile industry is a complex entity. Industrial manufacturing operations are one cause of the production of toxic or harmful waste, which may exist in a variety of forms such as liquids, solids, gases, and sludges. Textile waste can be majorly classified as (a) pre-consumer textile waste, (b) post-consumer textile waste, (c) factory waste, and (d) post-consumer technical textile waste as shown in Figure 1.1.

1.4 Pre-Consumer Textile Waste

Every stage of textile production and manufacturing that contributes to the production of finalized textile material yields some waste known as "pre-consumer textile waste." It is also known as "production waste" and is eliminated from the initial phase of the supply chain. Pre-consumer textile wastes are waste materials that originated straight from the initial steps of the manufacturing process, i.e., all the steps involved before the final product reaches actual customers. It includes all waste associated with processing steps for spinning, weaving, knitting, and non-woven, as well as dyeing, printing, and finishing as illustrated in Table 1.1. It consists of remnants of fabric from cutting, damaged or flawed material samples, fabric selvages, and scraps of other materials. In other words, it counts from fiber to fabric manufacturing procedures, including their maintenance. The wastes produced during spinning, weaving, dyeing, printing, and finishing processes are present in the form of fibers, yarns, and fabric that have been combined with chemicals (finishing agents, coloring agents, and chemical auxiliaries), correspondingly [8]. In one of the studies, it was found that 15% of the fabric needed for the creation of clothing was wasted, whereas,

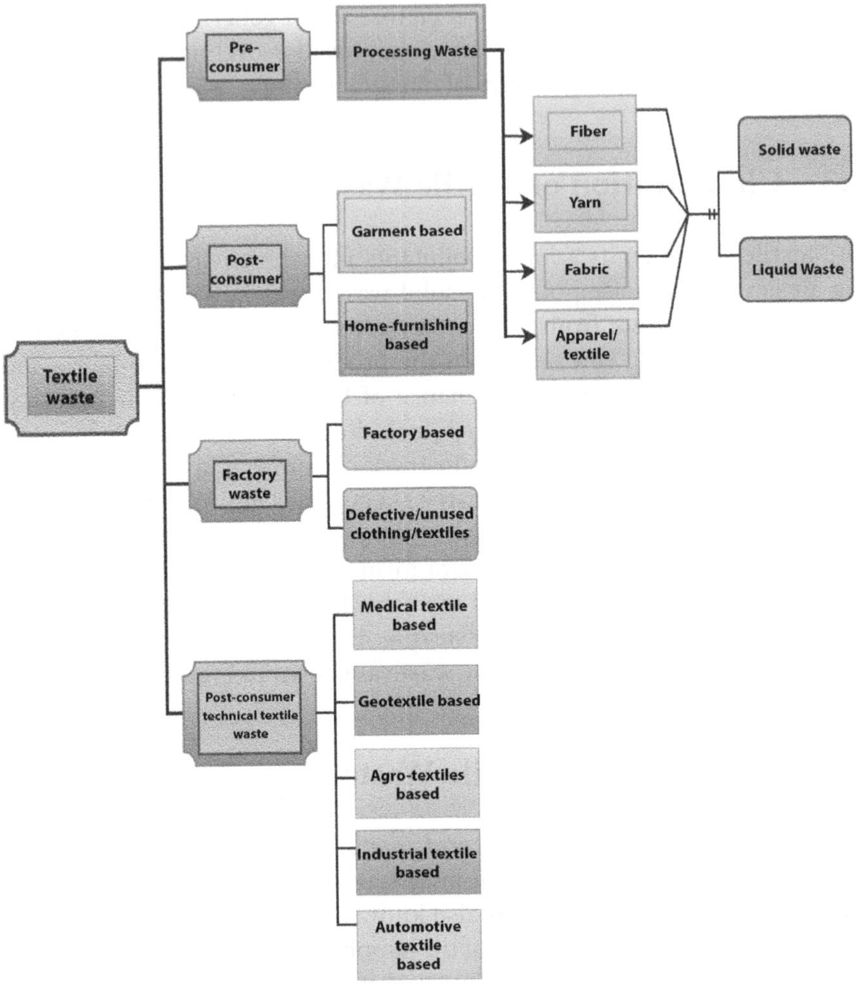

Figure 1.1 Classification of textile waste.

in another investigation, the figure was 10% or more for shirts, jackets, jeans, and undergarments sometimes reaching as high as 25% or 30% [9]. The various kinds of pre-consumer textile waste are examined in particular. Hence, the waste percentage depends on many factors such as garment type, design, and print of the surface of the textile.

Production Chain

Fabric production requires yarn, and yarns are made by fibers. The figure below (Figure 1.2) is a diagrammatic representation of this process.

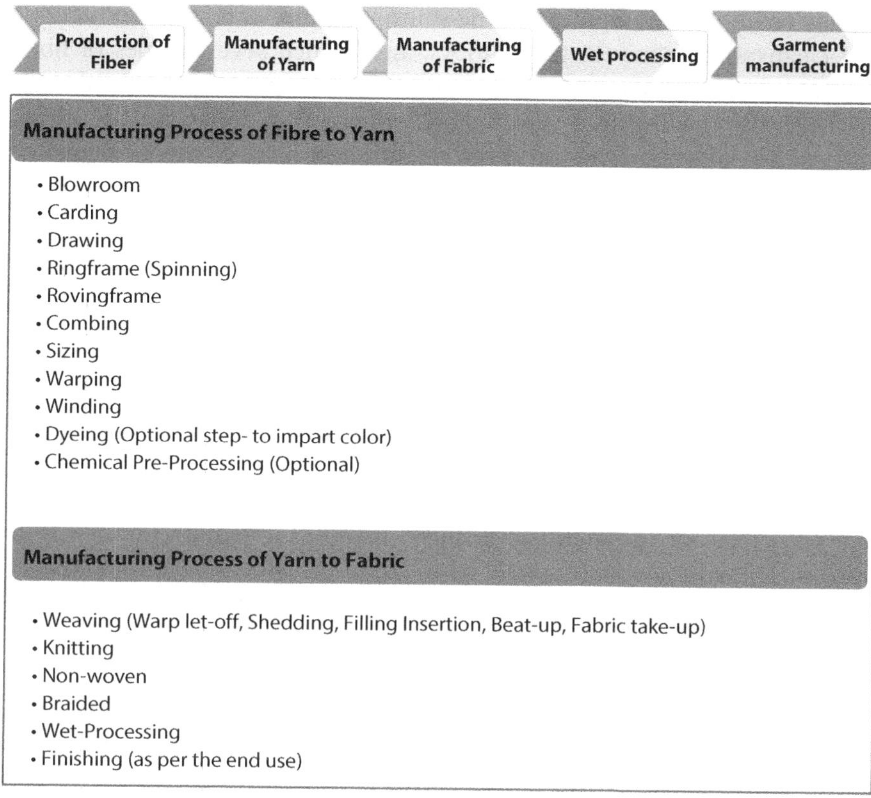

Figure 1.2 Production chain of textile manufacturing.

1.4.1 Fiber Production–Based Waste

Different types of fiber require different types of chemicals or fertilizers/pesticides for production purposes; it depends on the nature of the fiber to be produced. For instance, pesticides, insecticides, and fertilizers are the most frequently applied chemicals to produce cellulosic fiber, fertilizers are used when the natural resources alone are insufficient to support in cellulosic plant growth; whereas, as for the protein fiber, chemicals are essential to suppress parasites in animals and also in the scouring process after the fiber had been shredded. In the case of synthetic fiber (polyester, polypropylene, nylon, etc.), petrochemicals serve as a basic raw material for the processing of synthetic fibers; moreover, dye and pigment can be easily applied to textile materials to obtain desired yarn or fabric. The ginning, blow room, card, draw frame, comber, roving frame, and ring frame machine processes are linked with several kinds of textile spinning wastes.

A blow room is an equipment in the spinning phase of textile production, that cleans and opens the fiber flocks, after which several types of fiber waste, such as filters, lap-cut waste, and floor sweep waste, are produced. Equivalently, carding is a mechanical process that separates the fibers to form a continuous web or sliver, resulting in wastes such as taker-in, licker-in, and sliver-cut waste. The ring frame is the final step used for inserting a required amount of twist to impart strength into yarn, resulting in wastes like bonda waste, pneumafil waste, and waste thread [8, 10].

1.4.2 Yarn Production–Based Waste

In the conversion of fiber to yarn, spinning oil is required to enable the strengthening of the fiber and limit spinning friction. The majority of yarn-based waste is produced during the warp processing process, in the winding operation, and as well as the weaving operation, where yarn is prepared and processed for the manufacturing of woven, knitted, and non-woven textiles. These leftover yarns may be recycled or thrown away as garbage. This is determined by several variables, including the quality and state of the yarn, the desired fiber size, strength, and cleanliness, and has the potential for harmonious mixing and blending [8, 10].

The three main categories of yarn waste are natural yarn (cotton, jute, linen, silk, wool, pina, etc.), synthetic/manmade yarn (polyester, polypropylene, nylon, rayon, etc.), and blended or mixed yarn (polyester-viscose blend, a polyester-cotton blend, cotton-wool, etc).

1.4.3 Fabric Production–Based Waste

Fabric construction processes primarily consist of three types: weaving, which involves the intertwining of warp and weft yarns; knitting, where yarns are inter-looped; and nonwoven, which is produced through the felting process. All the waste generated during these operations falls under fabric-related waste generation; moreover, during their coloration, printing, and finishing procedure, various types of quality defects in the fabric are also generated. Sometimes, fabric waste occurs as a result of weaving machinery malfunctions, including issues with primary motions (such as shedding, picking, and beating up), secondary motions (like take-up and let-off), and auxiliary motions (such as warp stop motion, weft stop motion, warp protector motion, and temple), or due to friction of yarn [8].

1.4.4 Wet Processing–Based Textile Waste

Wet processing is a process that consists of pre-treatment, coloration, and finishing stages. It is a process in which dyes and auxiliaries or other chemicals are applied to the textile (fibers, yarn, woven, and knitted) to eliminate fabric impurities such as fats, oils, waxes, dirt, lubricants, protruding fiber ends, sizing agents, and cotton seed husks and to enhance some attributes like improved absorbency and/or whiteness of the fabric; minimum fiber damage; and uniform residual size, pH, alkalinity, and moisture content. This step involves a variety of pre-treatment, coloration, and finishing procedures, including sizing, singeing, de-sizing, pre-treatment, printing (including digital printing), finishing, and laundering, scouring, bleaching, and mercerizing processes as shown in Figure 1.3.

The de-sizing procedure is responsible for approximately 50% of the wastewater's substantially elevated biochemical oxygen demand (BOD) content [11]. Through the scouring process, the volatile organic compounds from glycol ethers and scouring solvents are released into the air. The disinfectant and insecticide residues, surfactant, oil, lubricants, etc., contaminate water bodies. The chemical used during the bleaching process such as organic stabilizers and sodium silicate is alkali or base in nature, which leads to high pH, in water bodies. Chlorine is an extensively utilized bleaching agent that is extremely harmful to both ecosystems and humans.

It has been reported that the textile sector utilized between 700,000 and 1,000,000 tons of dyes, of which 280,000 tons ended up being released into textile wastewater [10]. The effluent produced by the textile industry

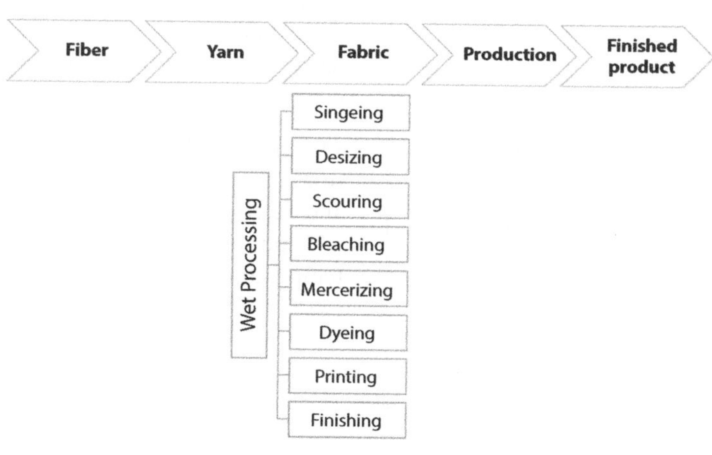

Figure 1.3 Diagrammatic representation of the wet-processing process.

Table 1.1 Waste production during various textile manufacturing processes.

Waste generated during textile manufacturing process			
Process	**Emission**	**Water effluent**	**Solid waste**
Fiber preparation	Minimal	Negligible	Fiber and packaging waste
Yarn spinning	Minimal	Minimal or negligible	Packaging waste, sized yarn, fiber waste, processing waste
Sizing	Starch by-products, synthetic and semisynthetic sizing (PVA, urea, waxes, and oils) VOCs	BOD, COD, metals, and temperature	Fiber lint; yarn waste, unused starch, and sizes
Singeing	Minimal	none	Fabric waste by an error in processing
Desizing	VOCs from glycol ethers	BOD, COD, temperature (70°C–80°C) lubricants, and biocides	Yarn waste: maintenance materials and fiber lint
Scouring	Scouring solvent and VOCS from glycol ethers	NaOH, insecticides residues, oily fats and detergents, and spin finishes; temperature (70°C–80°C), BOD, and high pH	None

(*Continued*)

Table 1.1 Waste production during various textile manufacturing processes. (*Continued*)

Waste generated during textile manufacturing process			
Process	**Emission**	**Water effluent**	**Solid waste**
Bleaching	None	High pH, H_2O_2, NaOCl, stabilizers, wetting agents, and TDS	Little
Mercerizing	Minimal or negligible	High pH, NaOH, BOD, and dissolved matters	Minimal or negligible
Dyeing	VOCs	Toxicity, BOD, high pH, organohalogens (toxic), heavy metals, sulfide, formaldehyde, and surfactant	None
Printing	Acetic acid, steaming and curing oven emission, and gases	High pH, BOD, COD, urea, solvent, metals, foam, pigments	None
Finishing	VOCs, formaldehyde vapors, and combustion gasses	COD, toxic materials, chlorinated compounds, resins, and waxes	Fabric scraps, trimming wastes, and packaging waste

is extremely difficult to decompose, is toxic, has a substantial chemical oxygen demand (COD), contains a significant number of surfactants, and has strong color along a pH and temperature range that are extremely changeable.

Common heavy metals associated with contrasting dyes include nickel, chromium, copper, zinc, cobalt, and lead. After getting into bodies of water and soil, heavy metals can be hazardous to both human beings and the environment as a whole. On average, the level of total nitrogen in the effluent is 300 mg/L. Phosphorous in the effluent is a by-product of the phosphor of the cleaning agent. Due to its low cost and efficacy as a dye, sulfide is another substance that is frequently present in water effluent. Consequently, it has been prohibited in a few nations because it is extremely hazardous to the environment and water bodies. The effluent from the textile sector is typically distinguished by its high BOD (200 mg/L), COD (700 mg/L), and pH (11) values [10]. Aquatic life is harmed by the highly concentrated of this chemical.

1.5 Post-Consumer Textile Waste

This includes clothing, home textiles (such as bedding and towels), and other fabric-based items that are no longer wanted or usable by their owners. They are thrown away in the waste bin and ultimately end up in municipal landfills. These wastes are produced at the consumer level as a result of low fabric performance, changing fashion or design, ill-fitting issues, or other factors. There are two types of post-consumer textile waste.

1.5.1 Garment-Based Waste

Garment-based waste refers to the discarded materials generated during the production, use, and disposal of clothing items. This type of waste includes fabric scraps, offcuts, trimmings, and damaged or unsold garments. Garment-based waste can occur at various stages of the garment lifecycle, from manufacturing to consumer use and disposal.

1.5.2 Home-Furnishing Textile Waste

These types of textile waste include used household goods (including cloth, bags, bedding, carpet, and draperies) and packaging, and masks.

Any textile is supposed to have an end-of-life cycle of 3 to 4 years. According to a research study, approximately 10.5 million tons of post-consumer textile waste are disposed of in landfills annually in the United States, 350,000 tons in the United Kingdom, and 287,000 tons in Turkey [12].

Traditionally, old clothing materials were recycled, to be used again as a mop or washcloth for various household tasks such as a duster or handmade quilted matt, but, regarding the present invasion of disposable textiles and due to fast fashion, the use-and-throw philosophy is highly popular, leading to the old textile's being discarded off. The depressing truth is that most people no longer know how to fix or mend clothing and accessories, and this evolution in consumer behavior has made the world a dangerous place.

Even though textile garbage also consists of organic substances, it takes a long time for them to decay in landfills (6 months to 20 years) and requires ample space. The degradation rate depends on the fabric compositions and landfill habitat. Wool requires half a year to break down while cotton textiles lose 50%–70% of their weight after 3 months in the trash [10]. Polyester is a synthetic fiber and is non-biodegradable in nature. Additionally, landfilling has the potential to release harmful compounds and produce methane, a gas that enhances global warming. Methane and carbon dioxide, two GHGs that play a significant role in climate change, are produced as post-consumer textile waste breaks down in landfills or outdoor dumping sites. So, it is safe to say that this mass of waste contributes a major portion to the problem of climate change.

1.5.3 Factory-Based Textile Waste

Factory waste can be categorized into two parts: one of them is factorial trash is the type of waste, which originated from commercial and industrial textile utilization of such as filtration, conveyor belting, and the use of textiles for things like carpets and durries. Factorial textile waste is typically referred to as "dirty waste" because this group has the lowest likelihood of the recovery process due to accumulation and chemical contamination difficulties. However, other organizations and companies, especially the carpet industry, are actively researching to find effective ways of waste management. The second type of waste is defective pieces of clothing or unused textile materials that originated from industry or factory. Toxic chemicals and GHGs are emitted into the environment as a consequence of the decomposition of textile garbage during landfilling and incineration process. The dependence of South Asian nations like India, China, and Bangladesh on petroleum and coal for power supply has led to an upsurge in the carbon footprint of every garment manufactured [13].

1.5.4 Post-Consumer Technical Textile Waste

Recently, the need for technical textiles in today's society is steadily rising because the market for them was recently established. Textiles for medical applications, agro-textiles, automotive textiles, geotextiles, and industrial textiles are the sources of post-consumer technical textile waste.

For instance, one of the technical textile domains, medical textile, is expanding more quickly as a result of advancements in polymer science and technology as well as innovation in the creation of novel textile structures.

Medical textile includes post-consumer textile waste that has been used by the consumer like dressing, implants, surgical sutures, health care textiles, diapers, surgical gowns, menstrual pads, wipes, gloves, PPE kits, and barrier fabrics, which have been used by the consumer. Globally, a significant amount of medical textile waste was produced during COVID-19 and is specifically referred to as COVID-19 medical waste (CMW).

Geotextile waste generated from textiles has been used for geotechnical applications such as road and separation of the soil layer, drainage, and reinforcement.

Agro-textile waste is produced from textiles that have been used in agriculture, including for growing crops, raising livestock, and other activities. Similar to this, the textiles used for industrial purposes and the textiles used for automotive purposes (such as tire cords, seat covers, and belt materials) are used and discarded, resulting in industrial and automotive textile waste.

Because they are technical textiles and contain various chemicals and dangerous compounds, their disposal requires particular consideration to ensure environmental safety. Due to environmental safety concerns, landfilling and incineration are given less importance than reuse and recycling. Before final disposal, there must be some type of physical and/or chemical modification through the degradation and removal of harmful materials from the post-consumer technical textile wastes.

1.6 Soft and Hard Waste

Production wastes or pre-consumer textile waste can also be classified based on recovery span into two categories:

(a) Soft Waste and (b) Hard Waste

(a) Soft Waste: This type of waste has a more porous web of fiber and can be recycled. These are generally obtained from earlier stages of production. Low-grade yarn is produced using these soft wastes. It is produced from the blow room to ring frame such as during carding, combing, and drawing [14].
(b) Hard Waste: It is non-recyclable trash. The fibers in this trash are compressed densely and require a further approach to be recycled. It is produced through the processes of spinning and twisting, weaving preparation, weaving, and knitting [14].

1.7 Major Reasons for Textile Waste Production

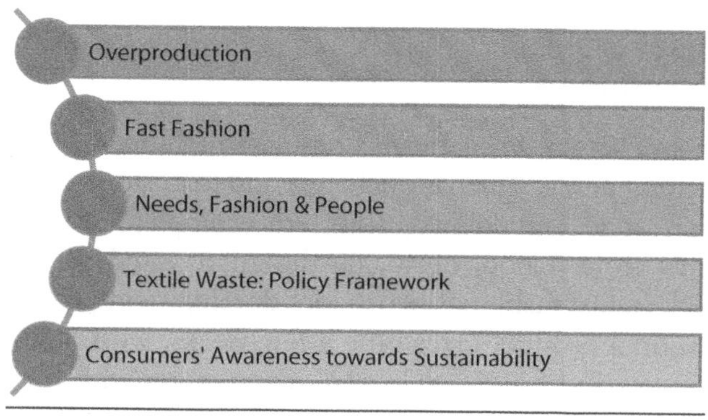

1.7.1 Overproduction

One of the major reasons for textile waste is overproduction. The population across the world increasing and the demand for textile goods is also rising day by day. The demand increases, and, then, the production rate also increases to fulfill the demand of consumers. With the advent of revolutionary industrial and technological developments in the 18th century, machines could produce thousands of apparel in a single minute with minimal effort and expense. These machines could generate desired textile materials far more quickly than humans could. For instance, John Kay developed the flying shuttle in 1733, which employed cords with selecting pegs. The loom could be operated by a weaver with just one hand. James Hargreaves developed the spinning jenny in 1764 to spin by numerous threads. A tool was made such that eight threads may be spun at once by rotating a single wheel at a time. Richard Arkwright designed the water frame, which worked with a water wheel in which spindles impart the twist in the fibers and rollers generate yarn. These are the major advantage of industrialization.

The major technological advances in spinning and loom, for instance, air jet loom, water jet loom, rapier loom, projectile loom, and multiphase loom increased the manufacturing rate of all types of textiles and apparel goods that enabled manufacturers to export inexpensive cloth and other items worldwide.

After the arrival of synthetic textiles in the market, textiles became cheaper and easily accessible due to enhancements in the production rate and supply chain. But these machines require additional energy and utilized fossil fuels. Synthetic materials are non-biodegradable and release GHGs (CO_2, CH_4, N_2O, HFCs, and O_3), which leads to climate change.

1.7.2 Fast Fashion

Fast fashion is characterized by the rapid availability of the latest and trendiest clothing items, which quickly respond to customers' demands for various design, style, and color options. For example, Zara selects 12,000 styles annually from a pool of 40,000 designs created by 200 in-house designers [12]. In fast fashion, the product development process from initial design to final product takes only a few weeks, with H&M's marketing span being as short as 3 weeks [12], contrasting with the traditional six-month timeframe for apparel sector commercialization. Fast fashion is presently influencing the apparel and textile market, which leads to excessive consumption where customers purchase more than their actual needs. Consequently, besides basic needs, the demand for fashionable and attractive articles encourages a greater extent of consumption.

Numerous renowned fast fashion enterprises, for instance, H&M and Zara, supply products that were swiftly replaced while they proceed quickly across the cupboards or shelves and lastly end up in the trash can.

The globe still purchases 80 billion fresh pieces of apparel annually. According to statistics, a typical American produces 82 pounds of textile trash annually, and, in 2018, the USA disposed of around 14 million tons of textile waste, of which, only 17.8% material was recycled. On a worldwide scale, 92 million tons of textile waste are produced yearly, and, by the end of 2030, approximately 134 million tons of waste will be anticipated annually [15]. This statistical finding indicates the consequences of fast fashion.

1.7.3 Needs, Fashion, and People

The standard of living is high in developed countries. Most people can easily fulfill their daily needs, and they have enough income to spend on fashion. Clothing and apparel are a tool to maintain status, so people, in general, want to be up-to-date and follow the latest style. Eventually, this leads to an enhancement in the demand for fashion and hence textile waste. Shopping addiction is also one of the causes of waste generation. Currently, numerous apps and websites for fashion stores facilitate easy shopping and are accessible for consumers to expend more capital on fashion and accessories. The manufacturers and retailers design the apparel as per the trend but use low-quality fabric and materials to reduce the cost of the product. Along with this, many people in the name of brands try to sell low-quality products to middle and lower-income groups. Consequently, almost all groups of society can afford "fashion" in their own ways, and, thus, unnecessary textile consumption will be increased. Although this may be beneficial to the chain of industries involved but the youth need to understand the need of the hour. These days' quick fashion trends are based on an inexpensive, "buy in bulk" mindset that leads to the mass production of clothing and a consequent rise in carbon emissions and global warming. If each and every citizen of "the planet Earth," no matter a celebrity or a college student, could adopt sustainability in their own life, then the severity of the problem might get reduced.

1.7.4 Textile Waste: Policy Framework

For the textile industry to thrive sustainably, the legislative and regulatory framework is of utmost importance. There is a dire need for regulation and laws at every stage of manufacturing textiles. Textile and fashion industries are very complex and hard to manage due to various production chains. The textile and fashion industries develop a worldwide market link, within which multiple production phases are carried out in a variety of nation-states; these intricate links, which are closely linked to one another, include the designing, the production of raw materials and fashion goods (such as textiles and clothing, footwear, leather products, technical and health and hygiene textile, and fur products), along with their transportation and sale at retailers till they reach to consumers. Currently, Sri Lanka has no regulations for controlling textile waste regulation [16]. There are developing countries that are not enough attentive toward textile waste such as India, which has strict environmental laws but insufficient enforcement of those laws [6]. Although the policies are effective, they are not consistently followed. The EU entity has approved several regulations that are necessary for the textile and apparel industry and assist to manage value chains in this industry sustainably, even having an impact on third-party nations that are a part of these systems. To provide customers with adequate information and safety concerns, the EU issued a regulation in 2011 setting standardized norms on the labeling and markings of textile commodities. Additionally, the EU introduces the EU Ecolabel, an opt-in label for commodities having minimal negative environmental effects across their entire lifespan, from raw material extraction to manufacturing and consumption to disposal. The adoption and implementation of the Ecolabel standards, such as the usage of biocides, water utilization, water releases, emissions into the atmosphere, and the employment of harmful compounds, must be thoroughly examined and validated. Furthermore, the EU adopted a package to measure the environmental performance of the product and organization, such as both the Organization Environmental Footprint (OEF) and the Product Environmental Footprint (PEF) pilot particularly focused on leather, footwear, and t-shirt; besides, this OEF also handles retail enterprises. The REACH Regulation, which controls the production, importation, sales, marketing, and final consumption of chemicals, is an especially significant rule for the industries. The rule concerns enterprises and industries that use colorants, auxiliaries, adhesives, and related products and additional materials to modify raw materials into commercial final products, such as textile and apparel producers, and shoemakers [17].

1.7.5 Consumer Awareness Toward Sustainability

Consumers have to be responsible for their waste and what they have created, because it contributes to carbon footprint and global warming. Consumers should be aware of the materials and chemicals that have been used for the manufacturing of textiles and apparel goods. Advertisement is another factor that drives the buying decision of consumers; as a result, this action leads to unwanted textile waste. Consumer disposal habits and environmental consciousness are very important in the life cycle of any textile product in terms of minimizing waste. If customers choose to keep utilizing an item till the end of its life and recycle it rather than putting it in a landfill, then the amount of textile waste will undoubtedly be significantly reduced.

1.8 Textile Waste–Associated Environmental Hazards

The textile sector produces around 1.2 billion tons of CO_2, approximately 10% of the world's total emissions of GHGs. Natural resources are required due to the rising demand for apparel, which becomes worse with the expansion of the rapid fashion business model. By 2050, it is projected that the fashion industry will use more than a quarter world's carbon budget. Synthetic fibers emit a higher amount of carbon dioxide as compared to cotton and contribute a considerable portion to climate change. To fabricate one polyester t-shirt, approximately 5.5 kg of CO_2 is released, in contrast to 2.1 kg CO_2 for a cotton t-shirt [13]. Due to faulty dyeing procedures, 10%–15% of the synthetic dyes are discharged into industrial waste, which poses severe environmental risks. For instance, contaminating surface waters results in sunlight penetration, decreased activity for photosynthesis, and oxygen deprivation, all of which can destabilize ecosystems.

1.8.1 Water Pollution

There are numerous types of auxiliaries such as surfactants, binders, starch, dyes, and other chemical materials that are used, of which some are very toxic to humans and as well as environment. They can endanger aquatic life, diminish dissolved oxygen levels in the water, and degrade the overall quality of the water. Short-term (acute) or long-term (chronic) impacts of these substances on living beings in the ecosystem are both possible.

There are primarily four possible tracks for colorants reaching the environment that must be taken into account:

a) By regular procedures discharge or emission
b) By disposing of surplus supplies and processing waste
c) By accidental emission
d) By discarding plastic packets and other toxic materials

Water pollution carried out by the emission of untreated wastewater is often one of the environmental issues affiliated with the textile sector as shown in Figure 1.4. These discharges have a mostly strong odor and are

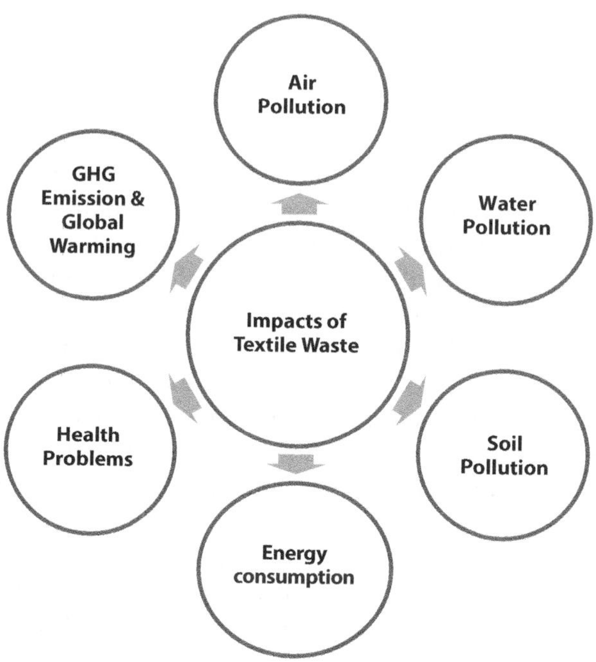

Figure 1.4 Major impact of textile waste.

extremely hot and alkaline in nature. Some dyes have been identified to be mutagenic and carcinogenic and include poisonous compounds like aromatic amines. The carcinogenicity is related to the development of nitremium or carbonium. It also persists in the food chain and is resistant to biodegradation and decomposition. The known hazardous consequences of dyes in both humans and animals include cancer, immune system diseases, and dermal impacts, and these have all been frequently recorded in employees who have encountered benzidine.

Primarily, zinc, cadmium, lead, mercury, chromium, arsenic, iron, cobalt, aluminium, and copper are the heavy metals that have been known to cause a variety of health issues. If heavy metals come into contact with the human body, then they may result in several health issues. These heavy metals have been responsible for memory issues, neurotoxicity, abnormalities, kidney failure, brain damage, neutral damage, bone disorders, numerous malignancies, kidney failure, brain damage, and other issues [18].

It has been claimed that the Remazol Red Brilliant and Indigo dye discharges disrupt the food web in the water cycle, resulting in instability in the ecology. In areas where dyes are discharged, aquatic life often suffers due to two main reasons: the depletion of algae, which serves as a crucial part of the ecosystem, and the direct harmful effects of the dyes themselves. The harmful impacts of dyes emitted in textile effluents can be quite dangerous to humans [19].

In underdeveloped nations, dye-carrying textile effluent is commonly used for soil irrigation. As a result, a significant quantity of dyes assembles on the outermost layers of the soil of fields near textile industries and factories. Humans may be exposed to these dyes either through direct contact via the consumption of tainted food such as fish or indirectly through the consumption of crops grown in soil that has absorbed harmful colorants or degraded forms. Thus, these toxins can affect human organs due to reaching the food web, leading to several illnesses. Certain azo dyes can partially or completely deteriorate into chemicals like aromatic amines; in the creation of synthetic colors, by-products like benzidine and 2-naphthylamine are produced as well, and these have been suspected of bladder cancer. Studies have generally shown that health issues resulting from textile dyes vary from dermatitis to ailments of the neurological system. Employees who produce or handle reactive colors are more prone to suffer from health issues like dermatitis, allergic rhinitis, allergic conjunctivitis, occupational asthma, and other allergic reactions. Reactive dyes were discovered to cause occupational asthma because their chromogen aids in adhering dye to hydroxyl groups on cellulose fibers, which, in turn, triggers an immunological response by functioning as haptens [19, 20].

In the textile production sector, chlorine-based solvents are employed for a variety of processes, hazardous to both human and environmental systems. They are poisonous and have a history of leading to cancer diseases or are detrimental to marine life.

1.8.2 Land Pollution

The land is contaminated by textile effluent. The essential medium for cultivating plants, vegetation crops, etc., is the soil. The quality of the soil affects the crops' quality. Therefore, the quantity and quality of crops also drop when soil quality declines as a result of industrial effluent pollution. As effluents are eventually dumped in the lower lands, it is also observed that the lower lands end up more contaminated than the upper fields. Metals are important to life in small amounts, but their overabundance is a serious problem. One study discovered that the abundance of earthworms in the field of Northern France is negatively impacted by heavy metal-enriched soil [21]. Pollutants that are extremely concentrated in the soil diminish the soil's fertility and its overall worth. Therefore, their impact on the human race, whether direct or indirect, requires considerable thought. The concentration and length of time that a living being has been subjected to dyes determine the severity of the consequences. Because of the substantial number of hazardous metals and organic substances found in dye effluent, various crops, including maize, kidney beans, Bengal gram, and soybeans, suffer from poor germination of the seeds and growth in plants [20].

1.8.3 Air Pollution

Generally, the environment is exposed to gaseous textile manufacturing wastes consisting of solvent gases such as ammonia and formaldehyde. Boilers are the source of an additional type of air waste. The majority of textile factories utilize fossil fuels like gas or coal, and, as a result, significant amounts of gases are released into the environment, leading to toxicity in the air. Approximately 60% of India's electricity is produced by burning fossil fuels, primarily coal, which accounts for around 16% of the country's air pollution, with industry coming in at around 12% [22].

Air emissions can be produced in various forms and sources:

Point sources: boilers, ovens, and storage tanks.
Diffusive: solvent-based, water effluent treatment, warehouses, and spills.

The textile industry releases a variety of air pollutants, e.g., nitrous oxides and sulfur dioxide (energy generation phases), volatile organic compounds (during coating, curing, and drying), aniline vapors, carrier hydrogen sulfide, chlorine, and chlorine dioxide (dyeing and bleaching steps).

A health risk can result from spinning processes that discharge particles of cotton dust into the air. It leads to acute respiratory illnesses such as byssinosis and chronic bronchitis among weavers and employees.

1.9 Future Prospect

The textile and apparel industry is extremely complex. The value chain of the procedures is linked to each other, and it contributes to economic growth and provides huge employment. Along with the pros, the textile and apparel industries have numerous cons too; it contributes to GHG emission and leads to global warming. It is challenging to adopt sustainable business practices because the fashion sector continues to face intense cost inflation and competitiveness. Nevertheless, it is crucial that this sector as an entire (from the manufacturing of fiber through retailing), sooner or later, will be made accountable for its hazardous environmental effects notably water resource depletion, high energy consumption, employment of toxic chemicals, hazardous gaseous emission, and generation of pollutants. Cooperation between stakeholders, designers, manufacturers, innovations in technology, supportive infrastructure, and legislative action and regulation are all essential for systemic change to occur in industries. The solutions include technological developments, sustainable agricultural practices, recycling technologies, biomaterials, reuse, recycling, and resale operations, which encourage the concept of a circular economy. The industry needs to convert from its existing linear business model to a circular (closed loop) one, in order to reduce its ecological footprint. This requires sustainable methods of textile production and the application of the 5R concept (refuse, reduce, reuse, recycle, and recover) throughout the life of textile materials. Consumers need to be mindful of their decisions and actively contribute to sustainable development. The rise of green consumerism has transformed the textile industry, highlighting its commitment to waste reduction and the adoption of eco-friendly practices. However, on a global scale, the pursuit of economic growth and the lack of comprehensive international policies pose challenges to establishing a sustainable fashion system. Green consumerism is now recognized for its commendable efforts in waste reduction and the adoption of sustainable practices, which have the potential to drive transformation within the textile sector.

1.10 Conclusion

The processing of textiles comprises multiple complicated steps and has detrimental impacts on the environment at each phase of processing. Textile industries required extensive consumption of water, and it also generates a lot of toxic effluents that contain inorganic salts, colorants, and toxic metals as well as organic pollutants. Pollution of land and water results from improper dumping of ill-defined or poorly processed wastewater. Additionally, the production of textiles contributes to noise and air pollution. The complicated and varied textile waste comprised a variety of chemicals and auxiliaries including hazardous metals as well as colors, wetting agents, surfactants, bleaching agents, and detergents. The aquatic ecosystem and human health are negatively impacted by textile effluents, which are extremely poisonous and dangerous. The organic chemicals that are present in effluent enhance the water's biological oxygen demand, inorganic minerals, and pH-balancing substances change the water's pH, and toxic metals get into a food cycle through bio-amplification. This wastewater discharge needs to undergo the appropriate treatment with systematic procedures. Chemicals such as chlorinated solvents, hydrocarbon solvents, oxygenated solvents, oils, waste fabrics covered in grease and oil, and dyestuffs and pigments are all present in the textile wastewater discharge. A possible solution to deal with this kind of waste is the adoption of sustainable choices so as to utilize textile effluent as a source of renewable energy and create electricity from the effluents. The existing clothing chain proceeds in a primarily linear fashion in terms of apparel production, distribution, and consumption. At present, in order to produce clothes, more than 68% of fibers are made from resources that are obtained from non-renewable resources, such as fossil fuels, that are used for a very brief period of time before being burned or disposed of in landfills. Annually, cloth about \$183 million is discarded into trash dumps. The linear framework system leads to excessive waste and depletes the environment, it leaves economic opportunities unexplored, puts strains on scarce resources, deteriorates the natural habitat, and has a substantial detrimental influence on society at the local, regional, and worldwide levels. To tackle the issues of the linear economy, it is a need of the hour to shift from such a linear economy toward a circular one. Three primary actions and approaches primarily operate the circular economy, i.e., reduce, reuse, and recycle—are common conventional strategies for managing traditional trash. The next factor is the most important factor, that is, consumers should be responsible and aware of textile waste and try to adopt a habit of reducing, reusing, and recycling.

References

1. Ry, M.A., *O F F I N D I N, A New textiles economy: Redesigning fashion's future.*
2. Shirvanimoghaddam, K., Motamed, B., Ramakrishna, S., Naebe, M., Death by waste: Fashion and textile circular economy case. *Sci. Total Environ.*, 718, pp. 137317, May 2020, doi: 10.1016/j.scitotenv.2020.137317.
3. Geyer, R., Production, use, and fate of synthetic polymers. *Plastic Waste and Recycling: Environmental Impact, Societal Issues, Prevention, and Solutions*, 13–32, 2020, doi: 10.1016/B978-0-12-817880-5.00002-5.
4. Textile Industry in India, Leading Yarn Manufacturers in India, IBEF, https://www.ibef.org/industry/textiles (accessed May 26, 2023).
5. Sandin, G. and Peters, G.M., Environmental impact of textile reuse and recycling – A review. *J. Cleaner Prod.*, 184, 353–365, May 20, 2018, doi: 10.1016/j.jclepro.2018.02.266, Elsevier Ltd.
6. Prerna, J. and Charu, G., The story of textile waste-Reasons and solutions. *Int. J. Appl. Home Sci.*, 5, 4, 871–890, 2018.
7. Nørup, N., Pihl, K., Damgaard, A., Scheutz, C., Development and testing of a sorting and quality assessment method for textile waste. *Waste Manage. (Oxford)*, 79, 8–21, Sep. 2018, doi: 10.1016/j.wasman.2018.07.008.
8. Mishra, P.K., Izrayeel, A.M.D., Mahur, B.K., Ahuja, A., Rastogi, V.K., A comprehensive review on textile waste valorization techniques and their applications. *Environ. Sci. Pollut. Res.*, 29, 44, 65962–65977, Sep. 01, 2022, doi: 10.1007/s11356-022-22222-6, Springer Science and Business Media Deutschland GmbH.
9. Niinimäki, K., Peters, G., Dahlbo, H., Perry, P., Rissanen, T., Gwilt, A., The environmental price of fast fashion. *Nat. Rev. Earth Environ.*, 1, 4, 189–200, Apr. 01, 2020, doi: 10.1038/s43017-020-0039-9, Springer Nature.
10. Pensupa, N. *et al.*, Recent Trends in Sustainable Textile Waste Recycling Methods: Current Situation and Future Prospects. *Top. Curr. Chem.*, 375, pp. 76, 5, Oct. 01, 2017, doi: 10.1007/s41061-017-0165-0, Springer International Publishing.
11. Khan, W.U., Ahmed, S., Dhoble, Y., Madhav, S., A critical review of hazardous waste generation from textile industries and associated ecological impacts. *J. Indian Chem. Soc.*, 100, 1, pp. 100829, Jan. 01, 2023, doi: 10.1016/j.jics.2022.100829, Elsevier B.V.
12. Yalcin-Enis, I., Kucukali-Ozturk, M., Sezgin, H., Risks and Management of Textile Waste. *Nanoscience and Biotechnology for Environmental Applications*, 29–53, 2019, doi: 10.1007/978-3-319-97922-9_2.
13. Leal Filho, W. *et al.*, An overview of the contribution of the textiles sector to climate change. *Front. Environ. Sci.*, 10, pp. 973102, Frontiers Media S.A., Sep. 05, 2022, doi: 10.3389/fenvs.2022.973102.
14. Doctor, S.J. and Khatwani, P.A., 'Solid Textile Wastes: Different Types and their Causes of Generation', 83/1. *J. Text. Assoc.*, 83, 1, 422–426, May 2022.

15. Chen, X., Memon, H.A., Wang, Y., Marriam, I., Tebyetekerwa, M., Circular Economy and Sustainability of the Clothing and Textile Industry. *Mater. Circ. Econ.*, 3, 1, pp. 12, Dec. 2021, doi: 10.1007/s42824-021-00026-2.
16. Sinha, P., Dissanayke, D.G.K., Abeysooriya, R.P., Bulathgama, B.H.N., Addressing post-consumer textile waste in developing economies. *J. Text. Inst.*, 113, 9, 1887–1907, 2022, doi: 10.1080/00405000.2021.1954428.
17. Jacometti, V., Circular Economy and Waste in the Fashion Industry. *Laws*, 8, 4, 27, Oct. 2019, doi: 10.3390/laws8040027.
18. Islam, T., Repon, M.R., Islam, T., Sarwar, Z., Rahman, M.M., Impact of textile dyes on health and ecosystem: a review of structure, causes, and potential solutions. *Environ. Sci. Pollut. Res.*, 30, 4, pp. 9207–9242, Springer Science and Business Media, Deutschland GmbH, Jan. 01, 2023, doi: 10.1007/s11356-022-24398-3.
19. Mehra, S., Singh, M., Chadha, P., Adverse impact of textile dyes on the aquatic environment as well as on human beings. *Toxicol. Int.*, 28, 2, 165–176, Apr. 01, 2021, doi: 10.18311/ti/2021/v28i2/26798, Informatics Publishing Limited.
20. Singh, K., Kumar, P., Srivastava, R., An overview of textile dyes and their removal techniques: Indian perspective Impact assessment of pollution generated in Varanasi city on river Varuna-A tributary of River Ganga View project, [Online], Available: https://www.researchgate.net/publication/322699631.
21. Markandeya, Mohan, D., Shukla, S.P., Hazardous consequences of textile mill effluents on soil and their remediation approaches. *Cleaner Eng. Technol.*, 7, pp. 100434, Elsevier Ltd, Apr. 01, 2022, doi: 10.1016/j.clet.2022.100434.
22. Aishwariya, S. and Jaisri, M.J., Harmful effects of textile waste Understanding Sustainability in Textiles and Fashion View project Exploring Fashion Accessories View project Article published in fiber2fashion Harmful effects of textile wastes, 2020, [Online], Available: https://www.fibre2fashion.com/industry-article/8696/harmful-effects-of-textile-waste.

2

Challenges in Handling and Systematically Organizing Textile Waste

A.K. Choudhary* and Prashant Kumar

Department of Textile Technology, Dr. B.R. Ambedkar National Institute of Technology, Jalandhar, Punjab, India

Abstract

The necessity of clothes for human civilization is demonstrated by Maslow's 1943 placement of clothing at the base of the hierarchy of requirements, along with food and shelter. The amount of time that clothing is worn before being discarded has significantly decreased in many industrialized nations, nevertheless, as a result of fashion and style. The amount of time until disposal for these textile products is gradually decreasing as they transition from being necessities for humans to fashion accessories. In reality, there are noticeably more things being thrown away, and the need of managing end-of-life textiles properly is growing. A further difficulty is that increased recycling may increase the availability of fibers globally, lowering the cost and raising demand for fibers. Recycling is definitely the exception rather than the standard in the textile business, where it is far less developed.

***Keywords*:** Fashion, textile recycling, fibers, disposal

List of Abbreviations

NGO	Non-governmental organization
EU	European Union
NCR	National Capital Region
MSW	Municipal solid waste
HDPE	High-density polyethylene
PET	Polyethylene terephthalate

**Corresponding author*: choudharyak@nitj.ac.in

Prashansa Sharma and Shilpi Shree Sahay (eds.) Nanotechnology-Assisted Recycling of Textile Waste: Sustainable Tools for Textiles, (29–58) © 2025 Scrivener Publishing LLC

AD	Anaerobic digestion
WWTP	Waste water treatment plant
TRC	Textile-reinforced concrete composite
FRP	Fiber reinforced polymers

2.1 Introduction

The linear model of manufacturing, in which raw materials are sourced from countries with abundant resources and new technologies are used to make commodities, is being dominated the industrial revolution for the past 150 years. Customers use, drop, and eventually replace these mass-produced goods with brand-new ones. Essentially, the current financial system is one of "take-make-dispose," where businesses benefit by producing goods that eventually end up in landfills [1 , 2]. Figure 2.1 shows the different disposed textile cloths.

If we look at the big picture, changes in land usage, climate variations, and deterioration of the biosphere's ethics, together with an excess of toxic gases like N and P in chemical and biological cycles, are already making a bigger contribution to the global danger zone [3]. The textile and apparel industry might be seen as one of the most socially divisive and damaging industries on Earth. The main problem to how to maintain the essence of

Figure 2.1 Disposed textile cloths.

the fibers in parallel is looking at the enormous number of trashes created by the company and recycling textile waste [4–7]. At the intermediate level, a number of firms have begun to notice that the intrinsic financial mechanism increases their susceptibility to risks, particularly through resource cost volatility and exposure to supply constraints [8, 9].

Many fabrics nowadays are "down," which means they are being transformed into things like carpet underlay or rugs rather than recycled. Nowadays, most clothes are a blend of many textile types (e.g., cotton and polyester). Although blends give fabrics desired properties like breathability, current technology cannot accurately determine the actual material composition of blended cloth piles, rendering the separation of fibers an impossible task. For users, there are numerous alternatives. When people want to donate or sell their worn clothing to NGOs, collectors (who may be professionals or merchants), or governments, for instance, most often in Europe the waste product is destroyed or burned as shown in Figure 2.2 and the soil is filled with municipal trash. The donated or purchased textiles are screened according to quality prior to being sent to facilities for recycling or reusing. The majority of apparel and textiles that still have a high enough rating for reusing are exported to countries in East Africa or Europe [10] and the remaining inventory is then distributed to recycling facilities. Due to the fact that there are not many systems in place for recycling materials, almost fabric waste is converted into rugs, wipes, and insulation fabrics. The remaining used fabrics that have gathered are either filled or incinerated. The clothes that are no longer in use are occasionally accumulated in closets or unofficially traded by family members or acquaintances of those who work

Figure 2.2 Disposed cloths collection by NGOs.

for their company [11]. In the EU region, only 25% of the fabric waste is now collected by NGOs or other businesses with the sole intention of recovering and reusing. More garbage than what is being disposed in a landfill or burned as municipal waste perishes due to a lack of knowledge about and access to technological recycling processes, and many recycling techniques for synthetic and natural fiber take time to develop [12].

The consumer apparel item typically lands around NGOs, manufacturers, or even the small scale market in India as well—the tale is the same everywhere. The commodities are recycled into fabrics and fibers by both small firms and the brands above. To determine the level of waste creation at various stages of garment production, a survey has been undertaken among the various garment manufacturers situated in Himachal Pradesh and the Delhi NCR region. The final study was primarily descriptive in nature. Eight factories with a sample of 50 respondents working in a range of roles were questioned. The selection of responders was done using a subjective and easy sampling procedure. Primary data was collected by interviewing them and observing their work processes at the plants. One of the preliminary conclusions from the analysis of the data received from the respondents revealed that the main offender and contributor to trash generation were in the top section. The weight ratio between the final output product and the input material proved the same thing [13].

According to a survey on the fashion sector, which accounts for an astounding 92 million tons of waste yearly, the business generates 4% of all global waste each year. Inorganic and synthetic materials make up the majority of fast textile and fashion clothing. These fabrics include chemicals that contaminate water because they cannot break down correctly. Adopting different marketing techniques that take into account both environmental and socioeconomic factors will help resolve this redundancy problem. It suggests morality and product reuse in the emerging field of "sustainable fashion," which can raise the value and price of local products and even play a key role in extending the lifespan of clothing, textiles, and other allied materials.

2.2 Types of Textile Waste

According to function in the production cycle, there are three categories of textile waste that can be distinguished in Figure 2.3.

i. Material created by manipulating tissues, such as after a cut, is considered a post-industrial waste.

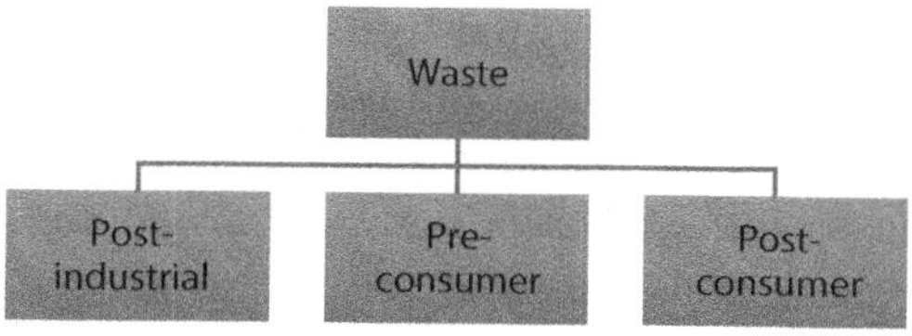

Figure 2.3 Type of textile waste.

ii. A pre-consumer waste (pre-consumer) is a garment that is not sold because it is either deemed to be defective or a manufacturer's trash, for example. This trash is removed from the production process but is not used by the end user. Most often, it is burned.
iii. After being utilized and having reached the end of its useful life, a post-consumer waste is created.

Additionally, there is a classification of the goods produced from these wastes that are made in accordance with the substance employed and whether it was a residue before or after consumption:

i. A post-consumer product is one that is manufactured from recyclable garbage that has previously been thrown away by a consumer.
ii. A pre-consumer product was created using the manufacturing-related trash, or the so-called post-industrial and pre-consumer waste. Its foundation is in the recycling of resources and the minimization of trash.

The products must be appropriately labeled in order to accurately track the traceability of the fabric. This helps the buyer make a responsible purchase decision by letting them know what kind of textile they are getting.

According to the economic principles, the decision-making process for mass clothes consumption is a little quicker than it is for conscious clothing purchases. Timing is thus one of those crucial elements. Whenever people dress, all communicate something about the world around us, but when people are aware of the garment's origin, we also become its source. As a result, we can respect the circumstances, origin, and making of this clothing, which makes it a unique item and makes the selection process more conscientious. On the other hand, in the case of a mass clothing purchase, the process of clothing acquisition, usage, and also drop occurs over a very brief time period, similar to its manufacturing process. Even less is known about the origin, use, and final destination of clothing. Due to this, new ideas advocate not only product reuse but also upcycling, which adds value to the final product as opposed to recycling, which often involves reusing materials but does not always strive to enhance the quality of the new item.

2.3 Textile Waste Management

According to the most recent report by the Worldwide Fashion Agenda and the Consulting Firm, waste production will increase by roughly 60% between 2015 and 2030 due to an additional 57 million tons of waste being produced annually, moving at the exact same rate as current waste production rates from various pattern or manufacturing procedures and final processes. As a result, by 2030, there will be 148 million tons of total fashion trash or 17.5 kg per person per year on Earth. The top 10 producers of fabric waste from the European Union in 2014 will be Italy, Germany, the U.K., Germany, Russia, Italy, Dutch, the Czech Republic, and Portugal (based on Euro stat data). With the exception of Poland, Belgium, and Germany, where the amount of fabric waste increased between 2004 and 2014, most of those countries were able to reduce pollution to below the I line by 2004. The problem that wastes create is dependent not only on how much of each type there is but also on how it is handled. Only 20% of the world's clothing waste is collected for recycling or reusing over time, whereas the other 80% is dumped or filled with soil, resulting in a significant decrease in raw resources and further tenderness energies.

The ability to access the necessary infrastructure, the type of textile product and its physiologic state, the threshold of using it, the fiber composition, the end, the structure of the garment, icons and symbols, accessories, and the manner of prototype are just a few of the important considerations. Especially over the past but again not least, the way the clothing has already been made is another important consideration. From the very beginning of the design process, the item performance must be described. The majority of research focuses on material finishes and fiber composition. However, nothing significant was accomplished. You will succeed, but they are not economically beneficial. Cleaning up the most effective recycling options and sustainable solutions, together with sustainable design, is a difficult task for engineers and designers. Entirely renewable and biodegradable products tend to be less appealing to some consumers, who then have little needs and eventually lose their effectiveness [14]. The decision on how cloth waste should be built and assembled is another crucial and challenging step in the development of a sustainable method. Thus, the most important factors that must be carefully considered in order to solve this problem, in particular successful recovery, reprocessing of infrastructure, and successful communication across the distribution chain.

2.4 Major Challenges in Managing Textile Waste

Various types of barriers prevent the opened and closed-circuit supply chain in the clothing and apparel industry from operating:

- Because clothes are left after being worn for a comparatively short time, the current fashion and textile production and consumption practices follow a rhythm that adds to a significant amount of fabric waste.
- Another issue with overproduction is that, of the 100% of clothes manufactured, only 30% are sold at the suggested retail price, another 30% are sold from sales, and the remaining 40% are either left over or do not get to their destination.
- Dealing with the burning or territory-filling problems will be the main obstacle for the aforementioned three types of wastes. Reducing the amount of these three waste types will also reduce the amount of waste that is now being burned or dumped in landfills. The streams of textile waste that must be reduced or eliminated in order to carry out the necessary disposal activities may be identified in the value string as significant actions.
- Consumer entry patterns, including education level and behavioral traits.
- Clinics for grouping, sorting, and later disposal: necessary processes and infrastructure including trash collection and categorization.
- The rise in worldwide clothing use and production has led to a large rise in the production of textile waste, creating grave problems in many developed nations. Globally, textile waste is acknowledged as the MSW waste source with the quickest rate of growth.
- Infrastructure for garbage collection and profitable sorting, however, continues to be a problem. Sorting out textile waste

requires a lot of time and effort, and problems brought on by different fiber blends are a big difficulty.

- Growing interests include sorting automation and textile recycling advances [15]. Some countries' ban on the import of used clothing, the market for textile reuse, which is the favored choice, is dwindling.

Customers appear to be less aware of how to deal with end-of-life clothing and textiles than they are with newspaper, glass, or plastics, for example. The most significant problem with this might be your improved efficiency while collecting and sorting used clothing and clothing trash. The recycling industry is therefore heavily dominated by low-quality fabric materials and their combinations. Due to the quick creation of low-grade textiles and combinations, several trial-and-error processes are being used to reduce the tremendous demand on economically feasible recycling technology. Moreover, only 15% to 20% of textiles (depending on the region) are recycled, with the remaining materials being disposed of in landfills or burned. The most recent EU prescription is that the speed for 27 recycled or soiled clothing is only 18%, which is significantly worse than it is in the US. They make obvious comparisons between the costs of various goods (the cost of packing is 98% in Europe and 79% in France).

Textile product marketing nowadays is a business that makes money. Customers that are on a tight budget are looking for affordable cloth products. The industrial sector can use recycled products for packaging, cleaning, protection, and agriculture, namely, for greenhouses and animal husbandry. Many middlemen (collectors) focus on products made of reusable fabric because recycling cloth is a very expensive process, and this eliminates the need for such products. They are less aware of things and lack regional infrastructure. Commercially, a number of factors affect how much recycling technologies cost. First off, there are not enough recycling processes available for non-grade fabric materials. The notion of separating fiber from mixed mixtures and composite building is the next to be overlooked. Despite the fact that the first recycling procedure is highly expensive, low-quality recycling materials predominate on the market because the final goods are so inexpensive. Transportation costs and the limited accessibility of local or regional fabric recycling enterprises are two important cost-effective factors. There are still just a few common recycling technologies and important solutions that might remove obstacles from the textile industry's blueprint of a worldwide closed loop. The solution to this issue would not materialize until all relevant industries,

NGOs, academic institutions, and research figures work together on a regular basis to address this urgent issue as swiftly as feasible. The recycling process maintains ecological and environmental balance. All of these recycling enterprises use fibers of both the natural and synthetic kind. The synthetic fibers made from primordial oil and atoms include a variety of toxic elements. In our talk, a few textile firms are included under.

a) G-Star: G-Star is a finance company for the production of denim fabric. New denim materials being developed by G-Star might compete with virgin cotton lace in terms of price, quality, and appeal. They used to make an effort to start a manufacturing business and promote environmental responsibility by recycling lace. The percentage of recycled fibers that may be used to make yarn is 30%. However, it is made sure that the yarn retains its ability for finishing and weaving. About 12% of denim fabric is made from recycled materials. As a result, it will have a far less environmental effect than fresh materials. About 9.8% more water may be consumed, but energy consumption falls by 4.2% and CO_2 emissions drop by 3.8%. The Circle Economy is Wieland Textiles' pilot initiative to reintroduce denim products through trash recycling and gathering. G-marketing Star's strategy calls for the use of recycled materials in the production of their lace textiles. The objective was to broaden the project's potential impact beyond just one set of capsules and incorporate recycled denim into their long-term procurement strategy. Engineered lace fabric is 12.5% more expensive than the virgin equivalent.
b) BIONIC: Using marine and coastal waste plastic from the sea, some of the best enterprises in BIONIC produced high-quality textiles. Recycled artificial materials are converted by cutting-edge technology into reusable and sophisticated fabric materials. This process meets new standards for aesthetics and fabric materials to satisfy customer demand. They produced higher-density polyethylene (HDPE) and polyethylene terephthalate (PET) textile fiber from 40% to 100% of marine and coastal vinyl using blow mould injection technology. The BIONIC company is promoting the preservation of marine environments by using the recycling

process. They state that, over the past three decades, 7,000 plastic bottles have been gathered from seashores all around the world to make yarn.

The fashion sector gains enormous environmental advantages from the transition from the current linear economy to a circular economy while reducing the impacts of increased demand for clothing brought on by a growing global population [16].

2.5 Challenges in Textile Recycling (Solid Waste)

2.5.1 Challenges in Recycling

Additionally, recycling is a component of their 3Rs (Figure 2.4), which also include reducing, recycling, and reusing materials. Reducing intake or utilization of resources is a band-aid solution [17] and would not result in lasting change. In contrast, it is critical to create goods from components that may be repeatedly modified and reused, raising the quality of the finished product every time. Sustainable clothing brands with circular economy-based business practices are presently facing a wave because they promote ideas of much more mindful use to change customers' spending patterns simultaneously striving to sell more goods [18].

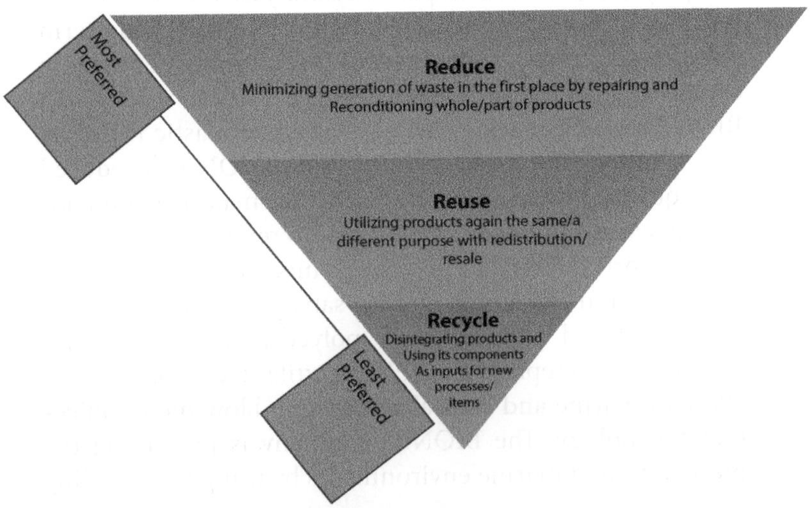

Figure 2.4 Importance of 3Rs in textile recycling.

In general, the manufacturing of virgin textile fibers might be decreased, and certain operations later down the textile product lifespan, which could be avoided if more textiles were reused and recycled. In addition, recycling and reusing textiles are now more sustainable than incineration and open dumping. Reuse is still preferred over recycling, especially whenever the utilizing stage is adequately extended [19]. Textile reuse includes a variety of techniques for expanding the lifespan of textile items from the actual purchaser to a subsequent one [20]. This is often done by renting, trading, exchanging, borrowing, and receiving, made possible through thrift shops, yard sales, internet retail sites, flea markets, and NGOs. The term "textile recycling" describes the regenerating of pre- and post-consumer recyclables in order to make innovative textile or other than textile goods.

2.5.2 Challenges in Textile Recycling

Fiber use has increased globally due to factors such as population expansion, rising living standards, a wider variety of textile materials, and shorter product lifespan [21]. This has resulted in a substantial volume in post-industrial and post-consumer fiber waste. Because of globalization, the garment industry is able to make more clothes at cheaper prices, and, as a result, many customers have adopted a tendency known as "fast fashion," which views apparel as a commodity that can be discarded after just a single use. A rise in the number of garments purchased may be attributed to the rise in popularity of fast fashion, which is characterized by mass manufacturing, a wide selection of options, flexibility, and competitive pricing.

2.5.2.1 Why in Textile Recycling?

Businesses all around the world are being pressured by the growing costs of electricity, resources, and waste management related to the textile manufacturing industry. About 10% of all carbon emissions [22] come from the textile sector, placing it as the fifth greatest source of carbon emissions [23, 24]. Under this context, it is necessary to ascertain that attempting to satisfy 21st-century expectations using 20th-century methods is impractical for sustainable growth (Figure 2.5) [25].

It is crucial to take into account the effective usage management of natural resources by lowering the feedstock intake via the reuse and recycling of textile goods that are considered to be trash, as this would provide a sustainable method for the management of textile waste. A long-term national policy should be devised to reduce the existing consumption of clothes and waste production [26] in an ecologically and commercially

Figure 2.5 Textile recycling.

responsible manner. Under 1% including all textiles are recycled back into apparel globally, with 75% of textile waste ending up in landfills and 25% being reused or recycled [27]. This is why developing reuse and recycling solutions for textile waste is essential for preventing wastage from going to landfills.

2.5.3 Basic Terminology Related to Textile Recycling

"Circular economy" refers to a civilization that never creates waste but instead constantly repairs and regenerates goods, parts, and resources to maximize their usability and worth. Several studies have provided numerous terminologies for recycling textiles, among which are mentioned in Table 2.1 [28].

2.5.4 Major Challenges in Systematically Organizing Classification of Textile Waste Recycling

There are so many ways to segregate the textile recycling so it is very difficult for anyone to identify the proper channel for textile recycling. In this study, the textile recycling is segregate in systematically organizing way.

Table 2.1 Definitions of terminology related to textile recycling.

Terminology	Definition
Closed-loop recycling	Recycling in a closed loop refers to the process through which a product's material is recycled and utilized to create a (almost) similar product.
Open-loop recycling	This refers to procedures that allow the components of one product to be recycled and utilized in another.
Fiber recycling	Fiber recycling is the practice of disassembling a cloth while retaining the original fibers.
Monomer/oligomer/ polymer recycling	Recycling of polymer occurs if fibers are disassembled although polymers are kept intact.
Textile recycling	This refers to recycling of prior or subsequent consumer textile waste for utilization in brand-new textile or other than textile items, such PET bottles, to create textile products.

2.5.5 General Classification of Textile Recycling

Typically, the three types of textile recycling methods are mechanical, chemical, and, less commonly, thermal. In many situations, this oversimplifies the situation because recycling pathways often include mechanical, chemical, and thermal processes. Figure 2.6 shows the different classification of the textile recycling methods including it's general classification.

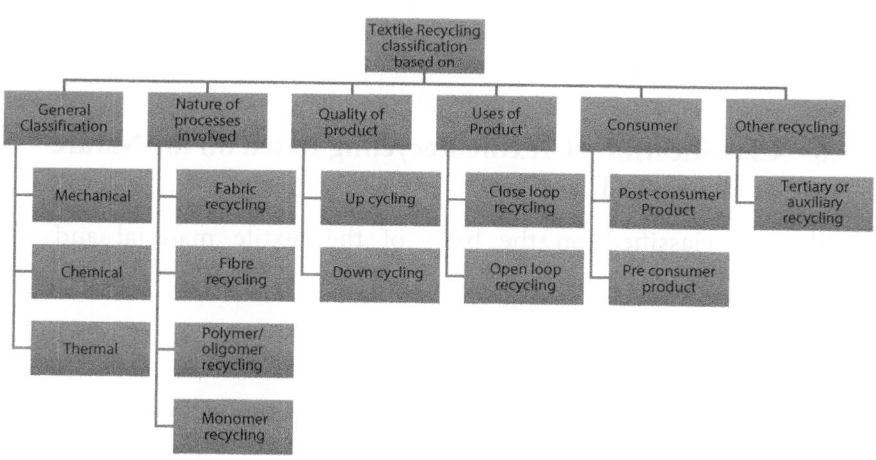

Figure 2.6 Classification of textile recycling.

2.5.5.1 Mechanical Recycling

In order to recycle mechanically, one of the two methods may be utilized: melting of synthetic fibers that are being used for the spinning of new fibers and breaking textiles to retrieve the fibers (mechanical processing). In the second instance, the fabric is divided into tiny bits and feed to tearing machine in which tearing of fabric occurs and fibers are come out. Waste may be used for gardening, agriculture, building, and decorating thanks to mechanical recycling. Polymers such as polyester are dispersed or depolymerized through the chemical recycling process. Chemically recycling may provide fibers that are on par with those made from virgin resources [29]. Cellulosic fibers for the manufacturing of bioethanol and protein-based fibers for wood panel adhesives may both be extracted chemically from the sorting textile waste.

2.5.5.2 Chemically Recycling

Chemically recycling process sometimes delivered the fibers, which may be equivalent to virgin resourced one. Cellulosic fibers for the manufacturing of bioethanol and protein-based fibers for wood panel adhesives may both be extracted chemically from the sorting textile waste. Chemical recycling by depolymerization is a practical method for certain synthetic fibers. In order to create new polymers, the polymer chains are divided into monomer, which are then isolated from one another and cleaned. During the purification (typically distillation) process, additives are eliminated. Today, nylon 6 and PET are chemically recycled on a small-scale commercial basis. By pulping and solution spinning, some cellulosic fibers, like cotton, may be chemically recycled to create regenerated cellulose fibers.

2.5.6 Classification of Textile Recycling Based on the Nature of Processes Involved

Textiles are classified on the basis of the textile material and its disintegration.

2.5.6.1 Fabric Recycling

By the recycling and recovering of fabric, the new product can be created with the utilization of recycled fabric.

2.5.6.2 Fiber Recycling

Fabric is disassembled in fiber recycling, but the original fibers are kept.

2.5.6.3 Polymer/Oligomer Recycling

Polymer recycling necessitates the deconstructing fibers without damaging the polymer.

2.5.6.4 Monomer Recycling

Recycling of monomer comprises the disintegration of polymer without any change in monomer.

2.5.7 Classification of Textile Recycling Based on the Product Quality

Textile recycling is of upcycling and downcycling type.

2.5.7.1 Upcycling

Upcycling refers to the process of creating goods with recycled resources and that goods are better than previous one.

2.5.7.2 Downcycling

"Downcycling" is the reverse of upcycling. Downcycling is the word that refers to the manufacturing goods from the recycled resources, but the quality of those goods is lower than the original one.

2.5.8 Classification of Textile Recycling Based on the Product Use

Textile recycling, based on the product, was mainly divided into two categories: closed-loop and open-loop.

2.5.8.1 Closed-Loop Recycling

Closed-loop recycling is recycling a product's material and reusing it in a product that is essentially similar to the original. Closed-loop recycling refers to the recycling of goods material and reuses it which are equally reliable to the original one [30].

2.5.8.2 Open-Loop Recycling

Open loop recycling refers to the recycling of material, but it can utilize as different goods.

2.5.9 Classification of Textile Recycling Based on Consumer

The goods made from these wastes are also classified in accordance with the material utilized and whether it was a leftover before or after consumption:

2.5.9.1 Post-Consumer Product

A post-consumer product is one that is created from garbage that was recycled after a customer threw it away.

2.5.9.2 Pre-Consumer Product

With the help of the production waste—the so-called post-industrial and pre-consumer waste—a pre-consumer product has been created. Its foundation is the reuse of resources and the minimization of waste.

The items must be appropriately labeled in order to accurately track the fabric's traceability. This helps the customer make a responsible purchase choice by informing them of the sort of textile they are about to buy.

2.5.10 Other Classification of Textile Recycling

Primarily, it is necessary to recycle the industrial waste. Product of post-consumer is processed mechanically which is the intermediate recycling phase. In tertiary or auxiliary recycling the hydrolysis and pyrolysis of plastics takes place to convert plastic into useful chemicals and fuels. Burning of waste is categorized under tetra recycling method [31].

Examples:

- Patagonia is one of the chief manufacturers of sustainable products, and it has a global market. It urges its customers to return the clothing they no longer use so that they may be repaired and sold to new customers. Patagonia is conscious

about the impression of human responses, particularly in the fabric sector.
- Pratibha Syntex, an Indian fabric factory that started recycling textiles and converting trash into recyclable yarns and clothes, is another excellent example.

By doing this, the company has not only reduced waste while increasing the manufacturing of original goods, but even more importantly, its reprocessing and intrinsic worth proposals have significantly altered the nature of thinking of their workers regarding sustainability so that they now contribute their thoughts of creativity during service.

2.6 Challenges in Different Methods for Recovering Waste from the Textile Industry

Today, a variety of technologies are available to support the recycling and recovery of textile waste. Biotechnology methods feasible for the recycling of textile waste include anaerobic digestion of textile waste, fermentation, and composting. Textile wastes and its thermal recovery and its transformation into insulation and construction materials are also covered in the sections.

2.6.1 In Textile Waste Anaerobic Digestion

Biodegradable portion of organic waste is often treated using anaerobic digestion (AD) in order to produce biogas. More than 50% of the cellulose in cotton made it a suitable medium for biomass conversion (Table 2.2). Studies on AD producing methane-rich biogas from cotton waste have been done throughout the last 10 years. Cotton waste anaerobic processing, consisting of cotton stem and cotton kernel, may result in the production of biogas. Potential substrate for AD is cotton waste from spinning machines [32]. An enhanced biogas output of almost 92% was shown in the AD of medical cotton industrial wastes under thermophilic conditions using cattle manure as inoculums [33]. Pre-treatment techniques of AD promote the biodegradation of complicated morphological natural substances, increase the quality and generation of biogas, and improve the quality of bio solids while reducing output [34, 35]. For the improvement of the digestion process, it is better to select primarily mechanical, chemical and thermal pre-treatment methods [36]. Cotton/polyester jeans waste

Table 2.2 Cotton waste contents [32].

Content	Level in percentage
Cellulosic content	54.5
Non-cellulosic	16–18
Ether extractive	11
water	9
Ash	7
Metal or others	3.4

and cellulosic waste materials may provide biggest methane production when pre-treated chemically with Na_2CO_3 at approximately 150°C for 2 h before digestion process [37]. Additionally, operating both single- and double-batching operations with textiles made of blend of cotton/polyester or polyester/viscose and loaded with cellulosic contents of approximately 20–22 g/L, a maximum methane generation rate of 80% is achieved [38]. From the examined literature, Table 2.3 highlights the ideal operating conditions for the batch AD process.

Table 2.3 Reliable conditions for biogas production by cotton waste [32, 33, 37, 39].

Cotton waste web	Pre-treatment	Inoculum	Processing temperature (°C)	Digestion time (Days)	CH_4 yield (mL/g VS)	CH_4 (%)
Cotton waste (cotton stalks, cottonseed hull, and cotton oil cake)		Effluent from WWTP anaerobic digester	35	23	65 (cotton stalks); 86 (cotton seed hull), and 78 (cotton oil cake)	60
Cotton waste from spinning mills		5%–7.5% cow dung/pig dung	30	50		77

(Continued)

Table 2.3 Reliable conditions for biogas production by cotton waste [32, 33, 37, 39]. (*Continued*)

Cotton waste web	Pre-treatment	Inoculum	Processing temperature (°C)	Digestion time (Days)	CH_4 yield (mL/g VS)	CH_4 (%)
Medical cotton waste	Alkaline (Na_2CO_3)	Cattle manure	55	90	37.57	60–70
Waste jeans (60% cotton) and cotton waste (100%)	0.5 M Na_2CO_3 at 150°C for 120 min	Effluent from municipal WWTP anaerobic digester	37	40	328.9 (60% cotton) and 361.08 (pure cotton)	-
Cotton textile waste (100% cotton)	0.5 M Na_2CO_3 at 150°C for 3 h	Digested sludge from municipal WWTP anaerobic digester	37	15	306.73	>50

2.6.2 In Textile Waste Fermentation for Ethanol Production

Texas Tech University began researching the utilization of cotton seed waste as a raw stock for production of ethanol in mid-1990s; however, only a few studies looked at the effectiveness of textile waste for ethanol production [40]. Using polyester/cotton mix textile, the impact of alkali pre-treatment to improve ethanol production was assessed. Chemical processing with sodium hydroxide or urea at 20°C, which was regarded as the most ideal resulted in a maximum ethanol production with fermentation process [41]. Additionally, the possibility of producing ethanol via the processes of fermentation and enzymatic hydrolysis of the cellulosic portion of jeans waste (60/40 cotton polyester blend) has been examined. Cellulose is hydrolyzed by enzymes to produce fermentable sugars. Cotton fabric showed promise as a different feedstock for the manufacture of bioethanol [42]. Based on the literature assessment, Table 2.4 lists the ideal parameters for ethanol production.

Table 2.4 Optimum operating conditions for ethanol production using cotton wastes [41–43].

Cotton waste stream	Pre-treatment	Enzymatic hydrolysis	Fermentation condition	Glucose yield	Ethanol yield
Cotton part from polyester cotton textile	NaOH/urea, −20°C, 72 h	Cellulose and β-glucosidase enzyme, pH 4.8, 45°C, 72 h	S. cerevisiae, 36°C, 72 h	91%	70%
Bleached and mercerized cotton fabric (100% cellulose)	Corona pre-treatment	Celluclast enzyme, 50°C, 8 days	S. cerevisiaevarellipsoideus, pH 5, 30°C, 100 rpm	0.94 g/g	0.9 g/L·h
Waste jeans (60% cotton) Cotton (pure)	1 M Na_2CO_3, 150°C, 120 min	Cellulose and β-glucosidase, 45°C, 72 h, 120 rpm	S. cerevisiae, 36°C, 72 h	81.7% (60% cotton) and 88% (pure cotton)	59.5% (60% cotton) and 69.4% (pure cotton)

2.6.3 In Textile Waste Composting

Composting is a biological process that turns organic waste, like cotton slack, into a beneficial soil amendment. The amount of organic waste is reduced by up to 50% during the composting process' active phase thanks to the low-tech, bio-oxidative method of composting [40]. In the process of composting, different microorganisms, such as bacteria and fungus, break down complex organic waste into simpler elements in the presence of air. Composting was seen as a solution to the problem of cotton garbage being directly disposed of in landfills, which is a serious waste disposal issue today. Cotton waste that has been digested and vermicomposted may be a fantastic long-term supply of nutrients [43].

Earthworms are used in the biotechnological vermicomposting process, which produces compost with increased soil fertility that is much superior than ordinary compost [44]. The amount of bacterial communities in compost and organic manure samples using cotton waste substrate was comparable. In contrast to compost samples, which provide greater humus, vermicomposting samples had a higher concentration of bacterial isolates [45].

Investigations on the vermicomposting of cotton textile waste from ginning companies took the form of willow waste. Willow trash is simply thrown of in landfills because it is not suitable for textile applications. Cow dung slurry, cellulose and amylase enzymes (derived from cow dung), and an efficient microorganism solution were combined with the collected willow waste. Water was regularly sprayed over the mixture while it was rotated. The trash completely broke down after 20 days, at which point earthworms were added. After 14 days, the waste mixture became light brown or dark brown, marking the conclusion of the vermicomposting process. The plants were then cultivated in pots using the resultant vermicomposting, and it was discovered that they had a great rate of growth in stem diameter, stem length, and index of leaf area when compared to a standard pot [46].

Additionally, there are concerns associated with directly reusing cotton gin waste on farms; therefore, composting is a recognized alternative. In two distinct ratios of 4:3 and 3:4, pig slurry and cotton gin waste, respectively, cotton gin waste was employed as a reinforcing agent for composting pig manure [47]. This investigation came to the conclusion that the temperature development and aeration need were caused by the bulking agent's thermal characteristics. The decomposition of organic substances was a factor in the gaseous emissions. Compost with a higher pig slurry content (4:3) exhibited increased humification of organic matter and higher nutrient contents.

The substrate made from discarded cotton has also been used to grow oyster mushrooms since the 1980s. Over 90% of producers of oyster mushrooms used discarded cotton as substrate [48]. The best production on oyster mushroom fruit bodies was seen when cotton waste and fermented poplar sawdust were combined, yielding 742 g per 4 kg of substrate. When compared to a conventional natural fermentation technique with a yield of 43.6%, a novel cotton waste composting system for growing oyster mushrooms produces a greater output of 65.1% over substrate dry weight. Before drying of cotton waste to the moisture level of 65%, it is being pre composted and soaked in lime solution for 2–3 days. Previously produced Ctec2 enzyme was then sprayed over the cotton substrate under ideal synthetically conditions (pH 4–53, 600 min, 50°C, and 0.45% ratio between enzyme and substrate) before being infected from cultural fungus. After then, harvesting was done [49].

2.6.4 In Regeneration of Fiber from Textile Waste

The reuse alternative to export as further utilization in apparel in many developing nations is no longer a viable choice; virgin cotton fiber manufacturing necessitates the utilization of significant resources. Alternative solution for cotton waste via closed-loop upcycling method by which the regeneration of fiber is achieved [50]. Regenerating fibers entails pulping used cotton garments; in the solvent, the pulp is dissolved; and further going for the spinning of the pulp until the fibers form. Cellulosic content is totally dissolved in N-methylmorpholine N-oxide, and it is harmless to the environment. To create fibers resembling lyocell, wood pulp may be combined with pulp recovered from discarded cotton-based clothing [51].

Additionally, phosphoric acid is used for the pre-treatment of textiles waste to retrieve glucose and PET. There are some parameters of pre-treatment have to examine which are such as concentration of phosphoric acid, temperature, time period of pre-treatment, and proportion of textile/phosphoric acid. It can generally be observed, that under the optimal circumstances of about 85.55% phosphoric acid at 55°C, 7 hours, and a ratio of 0.066% between textiles and phosphoric acid, about almost all polyester has recovered and accomplished highest level of recovery of sugar 80% [52]. Some researchers observed that, by utilization of fungal cellulose, recovery of glucose yields up to 42% and 45%. In order to handle textile waste and recover glucose and polyester for use in value-added goods, the suggested technique offers a lot of promise.

2.6.5 In Textile Waste as a Raw Material for Building Construction

Textile waste can be of source as raw materials in the making of buildings. Utilization of textile wastage as fillers/reinforcements for concrete, material for noise, and as a temperature insulation [53]. Additionally, textile-reinforced concrete composite (TRC) is employed in a variety of applications, such as precast constructions, repairs, and alignment of structured buildings. It is the innovation from the building sector that encourages the use of sustainable building materials by reusing textile industry waste. It blends multi-axially oriented textiles with fine-grained concrete to provide benefits such as optimum size, high breaking load strength, lightweight, anti-corrosion, and excellent ductility [54, 55]. Additionally, strong ropes used for protecting slopes from erosion and slide are made from textile waste. Ropes were created using leftover nonwoven fabric made from a mixture of recycled fibers and leftover insulation made from low-quality wool. The outcomes supported the technology's value in protecting steep slopes [56].

2.6.6 In Recovery of Thermal Energy

An effective substitute for land filling is incinerating unwanted textiles that cannot be recycled (such rugs or textiles with unidentified fibers). High calorific carpet fibers may minimize the requirement for power resources (like diesel and fuel), in which the results may be utilized as cement's primary raw ingredient. The benefit of incineration is that the majority of unused textile waste may be handled, and power can be recovered through burning. In the igniting propagation stage, burning textiles on their own might, however, result in unpredictable temperature behavior, ignition rate, and weight reduction %. To ensure that textile waste burns uniformly, it should be combined with cardboard trash during incineration [57]. One ton of domestic textile waste may be burned to produce 27 kg of ash and 15,800 MJ of energy [58, 59].

2.6.7 In Regeneration

Wiping cloth and yarns that have been untwisted and spun again to create a different variety of thread or yarn, are used to make cushions, and wadding products. It may be created by recycled fibers for the regeneration of many product varieties. Regeneration is a different method that uses heat and chemicals to recreate fiber from a natural source. For instance, some of the well-known companies that produce textile fibers from wood are Tencel, Lyocell, and Seacell. The wood is chopped into little pieces when the trees are felled, and these pieces are then treated chemically, heated to a high temperature, and put under pressure before the conversion of it into the filament form for textile application. Further, it employed to create eco-friendly textiles [60].

2.6.8 In Nonwoven Technology

Non-woven refers to the method of producing textiles that employs extremely short staple fibers, which are bonded together by means of heat application, binders, or by chemically producing a material that resembles textiles. Because a structured fiber length and fabric creation may not be achievable due to the heterogeneity of the recycled fibers, non-woven materials may be creative. The composites used in this technique combine a fiber and polymer matrices that are bonded together under pressure and heat to create a compressive medium that may be particularly well suited for textiles used in agriculture, construction, geo, acoustics, and filtration.

With the development of fiber reinforced polymers (FRPs), which starts by the recycling resources, textile reinforced composite engineering is also expanding. Composites may also be made using natural fibers. Pineapple fibers are used for reinforcing. Most of the well-liked non-traditional fiber varieties making their way into for the merchandising are jute, bamboo, flax, kenaf, hemp, sisal, coir, and banana. By the recycling of these fibers, it can be further utilized as a matrix of non-woven material. The possibility for using such recycled materials in insulation is excellent [61–63].

2.6.9 In Technical Textiles

When textiles were first created, they were evaluated to see whether they were suitable for the intended use, and, if they passed, they were then turned into the necessary product. In contrast, parameters of yarn, fiber criteria, and finishing parameters are determined and carried out depending on the attributes needed in the final product. Because all of the application areas united in the main bark, today, the technical textile conceptualized likewise massive banyan tree. Textiles from previous uses are employed as filters [64]. Fiber recycling is majorly contributed in food packaging, upholstery, textiles in agriculture, geotextile, soundproofing, and car interiors.

2.6.10 In Alternatives of Paper Making

It is intriguing to learn that old textiles are used to make paper while talking about the potential for recycling old textiles. This is a typical method of producing high-quality paper. The American money and bond documents are thought to be the cause [65]. An emerging firm that makes environmentally friendly paper discovers that used clothes make wonderful building materials. By the agriculture refuses, biomass, and cotton fabric waste, the handmade paper is to be made. Deforestation is responsible for up to 25% of carbon emissions. Tea bag, tote bag, envelope, and book sheets are all manufactured from the recycled fibers used to make paper. Downcycling is the term used for items that are produced with less value than traditional. Compared to normal paper manufacture, it requires less energy. Chlorine is never used in an effluent that is dangerous. The paper quality is also significantly higher, reaching up to 70%, than the other categories of papers, which add up to 35%. Because recovered materials are already coloured, we may reduce the amount of dye used by using recycling technology. There is a significant portion of post-consumer textiles that have been underutilized but have a great deal of promise for recycling and use in various technological textile products.

References

1. Maslow, A.H., A theory of human motivation. *Psychol. Rev.*, 50, 370–396, July, 1943.
2. Koszewska, M., Circular Economy — Challenges for the Textile and Clothing Industry. *AUTEX Res. J.*, 18, 4, 337–347, December 2018, DOI: 10.1515/aut-2018-0023, © AUTEX.
3. Danigelis, A retailers bank on environmentally friendly clothing for increased sales. Retailers Hope Environmentally-friendly clothing boost sales, July 18, 2017, https://www.environmentalleader.com/2017/07/retailersbank-environmentally-friendly-clothing-increase.
4. Claudio, L., Waste couture: Environmental impact of the clothing industry. *Environ. Health Perspect.*, Focus, Waste Couture Left, 115, A 454, September 2007.
5. BlackburnR.S., *Sustainable Apparel: Production, Processing and Recycling*, 1st edition, p. 288, Woodhead Publishing Ltd, Cambridge, United Kingdom, 21 August 2015, ISBN-10: 1782423397, ISBN-13 : 978-1782423393.
6. Dahlbo, H., Aalto, K., Eskelinen, H., Salmenperä, H., Increasing textile circulation—Consequences and Requirements. *Sustain. Prod. Consump.*, 9, 44–57, June 2016, http://dx.doi.org/10.1016/j.spc.2016.06.005.

7. Black, S., The sustainable fashion handbook sustainable fashion, in: *Sustainable fashion?*, p. 259, 2012.
8. The Ellen MacArthur Foundation, Towards the circular economy., Economic and business rationale for an accelerated transition, pp. 23–44, 2012.
9. The Ellen MacArthur Foundation, Circular Fibres Initiative, A new textiles economy:redesigning fashion's future, 2017, http://www.ellenmacarthurfoundation.org/publications.
10. Resta, B., Gaiardelli, P., Pinto, R., Dotti, S., Enhancing environmental management in the textile sector: An organisational-Life Cycle Assessment approach. *J. Cleaner Prod.*, 135, 620–632, 2016, http://dx.doi.org/10.1016/j.jclepro.2016.06.135.
11. Prieto-Sandoval, V., Jaca, C., Ormazabal, M., Towards a consensus on the circular economy. *J. Cleaner Prod.*, 179, 605–615, 2018, https://doi.org/10.1016/j.jclepro.2017.12.224.
12. De Paoli, A., Towards the Circular Economy, April, 2016, https://vancouvereconomic.com/research/towards-the-circular-economy/.
13. Weetman, C., *A Circular Economy Handbook*, 2nd Edition, 2 Kogan Page Ltd, London, November 2020, EAN: 9781789665314, https://www.koganpage.com/product/a-circular-economy-handbook-9781789665314.
14. Payne, A., Open- and closed-loop recycling of textile and apparel products, in: *Handbook of Life Cycle Assessment (LCA) of Textiles and Clothing*, pp. 103–123, Woodhead Publishing, Cambridge, United Kingdom, 2015, http://dx.doi.org/10.1016/B978-0-08-100169-1.00006-X.
15. Bukhari, M.A., Carrasco-Gallego, R., Ponce-Cueto, E., Developing a national programme for textiles and clothing recovery. *Waste Manage. Res.*, 36, 4, 321–331, March 2018, DOI: 10.1177/0734242X18759190.
16. Lehmann, M., Arici, G., Boger, S., Martinez-Pardo, C., Krueger, F., Schneider, M., Carrière-Pradal, B., Schou, D., Pulse of the Fashion Industry, Global Fashion Agenda, Boston Consulting Group, and Sustainable Apparel Coalition, 2019, https://www.globalfashionagenda.com/publications/#pulseofthefashionindustryreport.
17. Gullingsrud, A. and Perkkins, L., Designing for the circular economy: Cradle to Cradle®design, in: *Sustainable Fashion What's Next?*, J. Hethorn and C. Ulasewicz (Eds.), p. 293, Bloomsbury, New York, 2015.
18. Hussain, T., Re-fashioning the garment industry: Exploring innovations for a circular economy. *Clothing Cult.*, 5, 1, 61–86, 2018, doi: 10.1386/cc.5.1.61_1.
19. Sandin, G. and Peters, G.M., Environmental impact of textile reuse and recycling—A review. *J. Cleaner Prod.*, 184, 353–365, 2018, https://doi.org/10.1016/j.jclepro.2018.02.266.
20. Fortuna, L.M. and Diyamandoglu, V., Optimization of greenhouse gas emissions in second-hand consumer product recovery through reuse

platforms. *Waste Manage. (Oxford)*, 66, 178–189, April, 2017, http://dx.doi.org/10.1016/j.wasman.2017.04.032.

21. Wang, Y., Fiber and textile waste utilization. *Waste Biomass Valor.*, 1, 135–143, Feb, 2010, DOI 10.1007/s12649-009-9005-y.
22. Chao Chi Chen, J., Low Carbon Intelligent Operations for Textile Industry in APEC Economies, Taiwan Textile Research Institute (TTRI) APEC#213-PP-01.3, August, 2013, https://www.apec.org/docs/default-source/Publications/2013/8/Low-Carbon-Intelligent-Operations-forTextile-Industry-in-APEC-Economies—Project/TOC/Main-Report.pdf.
23. Athalye, A., Carbon footprint in textile processing, Colourage, Technical Service, Atul Ltd (Colors Division), Atul – 396 020, 2012.
24. Rana, S., Pichandi, S., Moorthy, S., Bhattacharyya, A., Carbon footprint of textile and clothing products, in: *Handbook of Sustainable Apparel Production*, pp. 141–165, CRC Press, Boca Raton, FL, USA, April 2015, DOI: 10.1201/b18428-10.
25. Chavan, R.B., Environmental sustainability through textile recycling. *J. Text. Sci. Eng.*, 2(2), S2–007, 2014, DOI: 10.4172/2165-8064.S2-007.
26. Ekstrom, K.M. and Salomonson, N., Reuse and recycling of clothing and textiles: A Network Approach. *J. Macromarketing*, 34, 3, 383–399, 2014, DOI: 10.1177/0276146714529658.
27. Textile Exchange, Preferred Fiber& Materials, Market Report, 2020, Available online: https://textileexchange.org/wp-content/uploads/2020/06/Textile-Exchange_Preferred-Fiber-Material-Market-Report_2020.pdf, (accessed on 8 January 2023).
28. Sandin, G. and Peters, G.M., Environmental impact of textile reuse and recycling – a review. *J. Cleaner Prod.*, 184, 353–365, 2018, https://doi.org/10.1016/j.jclepro.2018.02.266.
29. Collective Responsibility, Unspoken Crisis: Mounting Textile Waste in China, July 2018, Available online: https://www.coresponsibility.com/unspoken-crisis-mounting-textile-waste-in-china, (accessed on 8 January 2023).
30. Wang, H., Kaur, G., Pensupa, N., Uisan, K., Du, C., Yang, X., Sze Ki Lin, C., Textile waste valorization using submerged filamentous fungal fermentation. *Process Saf. Environ. Prot.*, 118, 143–151, June 2018, https://doi.org/10.1016/j.psep.2018.06.038.
31. Ignatyev, A.I., Thielemans, W., Vander Beke, B., Recycling of polymers: A review. *ChemSusChem*, 00, 1 – 16, 2014, DOI: 10.1002/cssc.201300898.
32. Raj, C.S., Arul, S., Sendilvelan, S., Saravanan, C.G., Biogas from Textile Cotton Waste—An Alternate Fuel for Diesel Engines. *Open Waste Manage. J.*, 2, 1–5, 2009.
33. Giakoumakis, G.E. and Sidiras, D.K., Recycled Medical Cotton Waste Modified via Torrefaction to be Used as an Enhanced Material for Energy Production. *Wseas Trans. Environ. Dev.*, 14, 69–75, 2018, E-ISSN: 2224-3496.
34. Shrestha, B., Hernandez, R., Fortela, D.L.B., Sharp, W., Chistoserdov, A., Gang, D., Revellame, E., Holmes, W., Zappi, M.E., A review of pretreatment

methods to enhance solids reduction during anaerobic digestion of municipal wastewater sludges and theresulting digester performance: Implications to Future Urban Biorefineries. *Appl. Sci.*, 10, 9141, December 2020.
35. Ariunbaatar, J., Panico, A., Esposito, G., Pirozzi, F., Lens, P.N.L., Pretreatment methods to enhance anaerobic digestion of organicsolid waste. *Appl. Energy*, 123, 143–156, 2014, http://dx.doi.org/10.1016/j.apenergy.2014.02.035.
36. Pilli, S., Pandey, A.K., Katiyar, A., Pandey, K., Tyagi, R.D., Pre-treatment Technologies to Enhance Anaerobic Digestion, in: *Sustainable Sewage Sludge Management and Resource Efficiency*, IntechOpen, London, UK, 2020, DOI: http://dx.doi.org/10.5772/intechopen.93236.
37. Hasanzadeha, E., Mirmohamadsadeghia, S., Karimi, K., Enhancing energy production from waste textile by hydrolysis of syntheticparts. *Fuel*, 218, 41–48, 2018, https://doi.org/10.1016/j.fuel.2018.01.035.
38. Jeihanipour, A., Aslanzadeh, S., Rajendran, K., Balasubramanian, G., Taherzadeh, M.J., High-rate biogas production from wastetextiles using a two-stage process. *Renew. Energy*, 52, 128–135, 2013, http://dx.doi.org/10.1016/j.renene.2012.10.042.
39. Dixon, P.J., Ergas, S.J., Mihelcic, J.R., Hobbs, S.R., Effect of substrate to inoculum ratio on bioenergy recovery from food waste, yard waste, and biosolids by high solids anaerobic digestion. *Environ. Eng. Sci.*, 36, 1459–1465, 2019, DOI: 10.1089/ees.2019.0078.
40. Hamawand, I., Sandell, G., Pittaway, P., Chakrabarty, S., Yusaf, T., Chen, G., Seneweera, S., Al-Lwayzy, S., Bennett, J., Hopf, J., Bioenergy from Cotton Industry Wastes: A Review and Potential. *Renew. Sustain. Energy Rev.*, 66, 435–448, 2016, http://dx.doi.org/10.1016/j.rser.2016.08.033.
41. Gholamzad, E., Karimi, K., Masoomi, M., Effective conversion of waste polyester–cotton textile to ethanol and recovery of polyester by alkaline pretreatment. *Chem. Eng. J.*, 253, 40–45, April 2014, S1385-8947(14)00551-8, DOI: http://dx.doi.org/10.1016/j.cej.2014.04.109.
42. Nikoli, S., Lazi, V., Veljovi, D., Mojovi, L., Production of bioethanol from pretreated cotton fabrics and waste cotton materials. *Carbohydr. Polym.*, 164, 136–144, 2017, http://dx.doi.org/10.1016/j.carbpol.2017.01.090.
43. Mahitha, U., Dhaarini Devi, G., AktherSabeena, M., Shankar, C., Kirubakaran, V., Fast Biodegradation of Waste Cotton Fibersfrom Yarn Industry using Microbes. *Procedia Environ. Sci.*, 35, 925 – 929, 2016, http://creativecommons.org/licenses/by-nc-nd/4.0/.
44. Selvi, C.P. and Koilraj, A.J., Bacterial Diversity in Compost and Vermicompost of Cotton Waste at Courtallam, Nellai District in Tamilnadu, India. *Int. J. Curr. Microbiol. Appl. Sci.*, 4, 582–585, 2015, ISSN: 2319-7706.
45. Aishwariya, S. and Amsamani, S., Evaluating the efficacy of compost evolved from bio-managing cotton textile waste. *J. Environ. Res. Dev.*, 6, 4, 941–952, June 2012.

46. Santos, A., Bustamante, M.A., Tortosa, G., Moral, R., Bernal, M.P., Gaseous emissions and process development during composting of pig slurry: The Influence of the Proportion of Cotton Gin Waste. *J. Cleaner Prod.*, 112, 81–90, 2016, http://dx.doi.org/10.1016/j.jclepro.2015.08.084.
47. Oh, S.J., Park, S.J., Shin, P.G., Yoo, Y.B., Jhune, C.S., An improved compost using cotton waste and fermented sawdust forcultivation of oyster mushroom. *Mycobiology*, 32, 115–118, June 2004, https://doi.org/10.4489/MYCO.2004.32.3.115.
48. Phuong, T.T., Development of a New Cotton Waste Composting Technology for Cultivation of Oyster Mushroom (Pleurotus Ostreatus). *ARPN J. Eng. Appl. Sci.*, 11, 21, 12607–12613, November 2016, ISSN 1819-6608.
49. Haule, L.V., Carr, C.M., Rigout, M., Preparation and physical properties of regenerated cellulose fibers from cotton waste garments. *J. Cleaner Prod.*, 112, 4445–4451, 2016, http://dx.doi.org/10.1016/j.jclepro.2015.08.086.
50. Woodings, C.R., The development of advanced cellulosic fibers. *Int. J. Biol. Macromol.*, 17, 305–309, November 1995.
51. Shen, F., Xiao, W., Lin, L., Yang, G., Zhang, Y., Deng, S., Enzymatic saccharification coupling with polyester recovery from cotton-based waste textiles by phosphoric acid pretreatment. *Bioresour. Technol.*, 130, 248–255, 2013, http://dx.doi.org/10.1016/j.biortech.2012.12.025.
52. Pichardo, P.P., Martínez-Barrera, G., Martínez-López, M., Ureña-Núñez, F., Ávila-Córdoba, L.I., Waste and Recycled Textiles asReinforcements of Building Material, in: *Natural and Artificial Fiber-Reinforced Composites as Renewable Sources*, pp. 89–105, 2017, http://dx.doi.org/10.5772/intechopen.70620.
53. Ghiassian, H., Poorebrahim, G., Gray, D.H., Soil Reinforcement with Recycled Carpet Wastes. *Waste Manage. Res.*, 22, 108–114, 2004.
54. Williams, N., Lundgren, K., Wallbaum, H., Malaga, K., Sustainable potential of textile-reinforced concrete. *J. Mater. Civil. Eng.*, 27, 1–12, 2015, http://dx.doi.org/10.1061/(ASCE)MT.1943-5533.0001160.
55. Anjana, E.A. and Shashikala, A.P., Sustainability of construction with textile reinforced concrete—A state of the art. *Mater. Sci. Eng.*, 936, 012006, 2020, doi:10.1088/1757-899X/936/1/012006.
56. Broda, J., Przybyło, S., Gawłowski, A., Grzybowska-Pietras, J., Sarna, E., Rom, M., Laszczak, R., Utilisation of textile wastes for theproduction of geotextiles designed for erosion protection. *J. Text. Inst.*, 110, 435–444, 2018, http://www.tandfonline.com/action/showCitFormats?doi=10.1080/00405000.2018.1486684.
57. Ryu, C., Phan, A.N., Yang, Y., Sharifi, V.N., Swithenbank, J., Ignition and burning rates of segregated waste combustion in packedbed. *Waste Manage. (Oxford)*, 27, 802–810, 2007, doi:10.1016/j.wasman.2006.04.013.
58. Swedish, E.P.A., Förbränningsanläggningar för Energiproduction Inklusive rökgaskondensering, Swedish EnvironmentalProtection Agency, s.l, 2005,

Available online: https://www.naturvardsverket.se/globalassets/media/publikationer-pdf/8100/91-620-8196-9.pdf (accessed on 8 January 2023).

59. Youhanan, L., Environmental Assessment of Textile Material Recovery Techniques, Examining Textile Flows in Sweden, Master's Thesis, Royal Institute of Technology, Stockholm, Sweden, 2013, Available online: https://www.diva-portal.org/smash/get/diva2:630028/FULLTEXT01.pdf (accessed on 8 January 2023).
60. Hugill, R., Ley, K., Rademan, K., Coming Full Circle: Innovating towards Sustainable Man-Made Cellulosic Fibres, Available online: https://reports.fashionforgood.com/report/coming-full-circle-innovating-towards-sustainable-man-made-cellulosicfibres/ (accessed on 8 January 2023).
61. Sülar, V., Oner, E., Devrim, G., Aslan, M., Eser, B., A comparative study on performance properties of yarns and knitted fabrics made of biodegradable and conventional fibers. *Fibers Polym.*, 17, 12, 2085–2094, 2016, DOI 10.1007/s12221-016-6808-x.
62. Bongarde, U.S. and Shinde, V.D., Review on natural fiber reinforcement polymer composites. *Int. J. Eng. Sci. Innov. Technol. (IJESIT)*, 3, 2, 431–436, March 2014.
63. Leão, A.L., Cherian, B.M., Narine, S., Souza, S.F., Sain, M. *et al.*, The use of pineapple leaf fibers (PALFs) as reinforcements in composites. *Biofiber Reinforcements in Composite Materials,* 7(1) 211–235, 2015, https://doi.org/10.1533/9781782421276.2.211
64. Zander, N.E., Gillan, M., Sweetser, D., Composite Fibers from Recycled Plastics Using Melt Centrifugal Spinning. *Materials*, 10, 1044, 2017, doi:10.3390/ma10091044.
65. Effects of Recycled Fiber Use, Metafore, (The Fiber Cycle Technical Document, available at http://postcom.org/eco/sls.docs/Metafore-Paper%20Fiber%20Life%20Cycle.pdf. (accessed on 8 January 2023).

3

Overview of Textile Recycling (Solid Waste)

Debarati Maity, Akash Borkar, Suvojit Ghosh and Anagha S. Sabnis*

Department of Polymer and Surface Engineering, Institute of Chemical Technology, Mumbai, India

Abstract

Globally, apparel consumption has increased 400% since the past two decades by introducing at least one new "collection" each week. With growing frequency and quantity of garment purchases, the amount of clothes being discarded is increasing. Three different types of solid textile waste are produced by the apparel industries: production waste, pre-consumer waste, and post-consumer waste. These wastes are dumped as landfills at an astonishing rate of one garbage truck every second, which poses a serious risk to the environment. Around 75% of solid textile waste is landfilled, whereas only 25% is reused or recycled.

The existing clothing system operates in a linear approach mostly in terms of production, distribution and usage, which misses out on a lot of economic opportunities, pollutes the environment, exerts pressure on insufficient resources, and has an impact on the economics of societies. To address these concerns, the urgent need for transition from linear to circular economy is necessary, which is highly influenced by six major conventional waste management approaches (6 R's): reduce, reuse, recycle, refuse, rethink, and repair. A circular system would promote the development and adoption of innovative recycling techniques, which could lead to the growth of high-tech enterprises. Governments and relevant institutions are appealing to various stakeholders to transform the textile industry toward a circular economy.

The majority of the waste produced and recycled by the apparel industry comprises cotton followed by cotton/polyester blend, polyester and other waste like viscose, rayon, and nylon. The approach toward recycling techniques is classified into four different categories named as primary, secondary or mechanical, tertiary or chemical/biochemical, and quaternary recycling or thermal recovery.

**Corresponding author*: as.sabnis@ictmumbai.edu.in

Prashansa Sharma and Shilpi Shree Sahay (eds.) Nanotechnology-Assisted Recycling of Textile Waste: Sustainable Tools for Textiles, (59–96) © 2025 Scrivener Publishing LLC

Apart from the benefits and drawbacks of these recycling techniques, several nations have started recycling strategies throughout the world.

Keywords: Recycling, solid textile waste, cotton waste, reuse, circular economy

List of Abbreviations

GDP	Gross domestic product
EU	European Union
MT	Million tons
NDRC	National Development and Reform Commission
MSW	Municipal solid waste
GRS	Global Recycled Standard
PET	Poly (ethylene terephthalate)
PA	Polyamide
PU	Polyurethane
BHET	Bis(hydroxyethyl) terephthalate
DMT	Dimethyl terephthalate
EG	Ethylene glycol
TPA	Terephthalic acid
TAD	Terephthalamide
LiOH	Lithium hydroxide
NaOH	Sodium hydroxide
NMMO	N-methylmorpholine N-oxide
[AMIM][Cl]	1-Allyl-3-methylimidazolium chloride
PP	Polypropylene
PE	Polyethylene
$FeCl_3$	Ferric (III) chloride
CNC	Cellulose nanocrystals
H_2O_2	Hydrogen peroxide
$Cr(NO_3)_3$	Chromium (III) nitrate
[Emim][OAc]	1-Ethyl-3-methylimidazolium acetate
DMSO	Dimethyl sulfoxide
SDBS	Sodium dodecylbenzene sulfonate
ZnO	Zinc oxide
(CH_4)	Methane
Fe_2O_3	Ferric (III) oxide
CSTR	Continuous stirred tank reactor
TRFC	Textile-reinforced fiber composites
ACF	Activated charcoal fiber

BHET	Bis(2-Hydroxyethyl) terephthalate
BHETA	Bis(2-hydroxyethyl) terephthalamide
AI	Artificial intelligence
IoT	Internet of Things
ICT	Information and communication technology
RFID	Radio frequency identification
AIHIF	Artificial intelligence-based hybridized intelligent framework
WEEE	Waste electrical and electronic equipment
6R's	Reduce, reuse, recycle, refuse, rethink, and repair

3.1 Introduction

"Fast fashion" has made the apparel industry produce more clothing at cheaper rates within a very short span of time to satisfy consumer fashion demand. The dominating linear model of the apparel business is currently producing greater waste owing to the rise of fast fashion, absence of government restrictions, unaware customers, and a disposable culture [1].

The textile sector is in charge of consuming a lot of resources and as a result polluting the environment. Textile industry is the second highest industry in terms of land consumption, the fourth highest in terms of water usage, and the fifth highest in terms of greenhouse gas emissions. According to Haslinger *et al.* (2019), the dyeing process can use up to 150 liters of water per kilogram of fabric and release harmful chemicals and heavy metals in the environment causing water contamination [2]. This contaminated water can go into the food chain causing severe health problems [3]. Figure 3.1 shows textile waste dumped at various places.

Figure 3.1 Textile waste dumped in places.

The growth in the manufacture and consumption of clothing has led to a sharp rise in textile waste on a global scale. The global fashion apparel sector contributes almost 2% of the gross domestic product (GDP) [4]. On an average, 75% of the global fashion market is concentrated in China, the EU, the USA and Canada [5]. China is the largest exporter of apparel and textiles in the world producing nearly 65% of clothing and roughly 31% of synthetic fibers required by the modern textile industry [6]. According to the NDRC (People's Republic of China), about 22 million tons of textile waste was generated by China in 2020 with a recycling rate of 20% [7]. The textile industry in the EU produces about 16 million tons of waste each year where only 26% is recycled [8]. In Europe, just 18% of clothing is recycled and reused, 30% is incinerated, and a substantial amount of the remaining 70% is disposed of in landfills [9]. In France, 40% of the waste textiles collected are exported to African countries for reuse [10]. In the USA, discarded clothing accounts for the maximum of the textile waste in the municipal waste stream (MSW). About 16 million tons of textile waste are produced annually in the United States of America (USA), of which approximately 14% is recycled, 71% is disposed off as landfills, and the remaining 15% are incinerated for energy recovery [11]. In India, only 40% of one million tons of clothing waste produced is recycled [12]. In Canada, an estimated 500,000 tons of clothing waste are disposed of each year where only 20% is recycled [13]. Figure 3.2 depicts the waste generated and recycled across various countries.

Recycling has some difficulties, despite being vital. Because the waste produced is largely mixed and hence not fully recyclable, it is either dumped in landfills or incinerated. Thus, appropriate waste management is required to categorize the waste and then separate and segregate them.

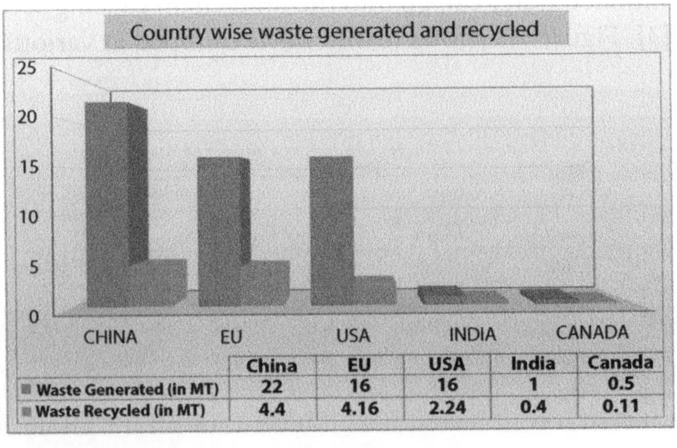

Figure 3.2 Waste generated and recycled country wise.

3.2 Overview of Textile Recycling

In the modern world, recycling of textiles is an essential part of sustainable and environment-friendly practices. This chapter provides an overview about types of solid textile waste generated, importance of circular economy (CE) over linear economy, its importance in waste management, various methodologies of textile recycling, and their application in different fields. This study provides readers with a thorough understanding of the textile recycling procedures utilized for various types of waste, the difficulties encountered during recycling, and the most recent technological advancements used.

3.2.1 Types of Solid Textile Waste

Three different types of textile waste are discussed in the following.

Production Waste

Fibers, yarns, fabric scraps, and cloth cuttings are the by-products generated during the manufacturing of fibrous materials [14]. This sort of waste can be formed at any stage of the manufacturing process including spinning, weaving, and cut-make-sew processes. Production waste falls into two categories: (i) soft waste: produced from carding, combing, drawing, and spinning; and (ii) hard waste: produced after spinning, twisting, weaving, and knitting [15]. Soft waste can also be categorized into two parts: clean waste and trashy waste. Clean waste does not need to be cleaned further, whereas trashy waste needs cleaning before reprocessing. Hard waste or yarn waste is obtained from the ring frame and winding department and is not recyclable [16]. Other commercial and industrial waste are produced by the carpet and curtain industries, conveyor belts, automotive industry, home furnishings industry, paper industry, etc. [17].

Pre-Consumer Waste

These waste are composed of clothing waste that result from unsold and damaged apparels, which are the result of poor design, poor fabric quality, or incorrect color choices [18]. Companies offer a variety of recommendations for these products. Marks & Spencer donates these things to charities, whereas H&M sells these items in their own stores [19]. Around 65% of the total estimate is supplied to customers in the form of brand-new clothing, whereas 35% is discarded during the production and pre-consumer stages [20].

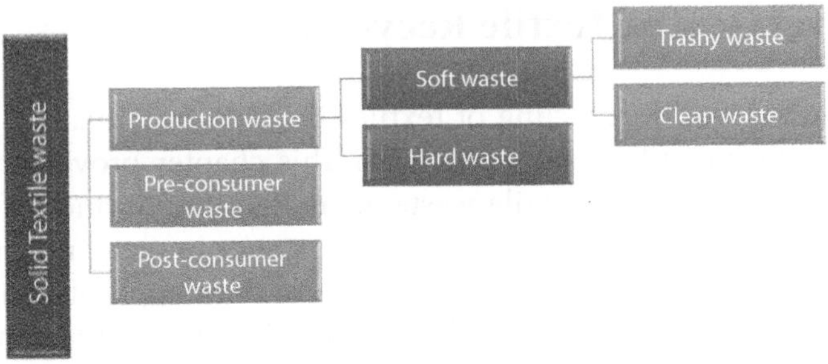

Figure 3.3 Types of various solid textile waste.

Post-Consumer Waste

Any type of clothing or household items that are worn out, damaged, outdated, or no longer in style are considered as post-consumer waste. Whereas some of this post-consumer waste is donated to non-profit organizations or given to friends and family, the majority is discarded in the waste and ends up in landfills [21]. An estimated 10.5 MT of post-consumer textile waste is discarded in landfills each year in the USA, 0.35 MT in the United Kingdom (UK), and approximately 0.29 MT in Turkey [20]. Figure 3.3 illustrates types of various solid textile waste.

3.3 Circular Economy

Circular economic growth, consumption, and production are all expanding linearly, but the sources of this system are depleting resulting in market volatility, unpredictability, and financial crises. This linear pattern of production and consumption is harming ecosystems owing to weather change and environment pollution [22]. Numerous theories have been proposed as solutions to this problem. Concepts like carbon footprint [23], life cycle assessment (LCA) [24], zero emissions [25], and eco-efficiency [26] built the ideas that tackle the ecological deterioration linked to business operations.

In contrast to eco-efficiency, the concept of the CE offers another approach to sustainability [27]. The CE, a form of industrial economy, seeks to improve sustainability through regenerative purpose a recycling processes nd design [28]. Figure 3.4 describes the pictorial difference between linear and circular economy. When products and manufacturing processes are carefully designed and handled in closed loops (industrial metabolism), the concept of waste is avoided [27]. Reduce, reuse, recycle, refuse, rethink, and repair (6R's) are the six main activities and strategies governed by the CE to manage waste [29].

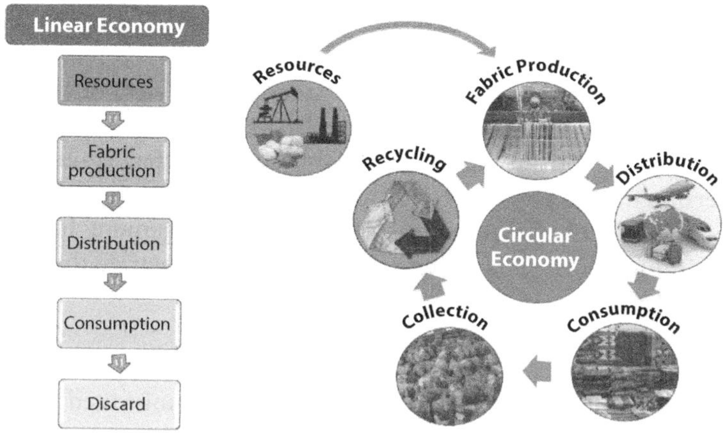

Figure 3.4 Difference between linear and circular economy.

- Reduce: Consumers could try to buy fewer products or think about purchasing products that consume less energy.
- Reuse: Consumers who no longer need things for their own use might sell or donate them so that others can use.
- Recycle – Products are converted back to their constituent parts and then made into new ones.
- Refuse: Consumers are free to decide whether or not to purchase a thing.
- Rethink: Both designers and manufacturers can produce items more effectively to accomplish the same task.
- Repair: It is the role and responsibility of designers to develop items that are easier to fix.

Textile exchange is a non-profit establishment that has been working since 2002 to make CE feasible and long-lasting for companies to valorize textile waste by promoting strategies for reducing the usage of feedstocks and increasing the standard of recycled products. The Global Recycled Standard (GRS) was created to accredit businesses that develop new textile products using pre- or post-consumer waste. The GRS certification must attest that the products comprise more than 20% recycled material along with additional requirements for transparency, such as those pertaining to the entire manufacturing process, the types of chemicals used, water management, harmful emissions, and waste-to-energy recovery [30].

3.4 Waste Management

Although managing waste mostly falls under the purview of the government, especially local governments, it also falls within the purview of numerous private businesses. It is a viable strategy that applies to both the government and private sectors at various levels to comprehend what is meant by effective governance. The waste management hierarchy is depicted in Figure 3.5. It is necessary to roughly summarize the following principles [31]:

- To gain regulatory authority; in other words, this method will regulate unlimited waste disposal and unlimited reuse of usable materials.
- To have a permitted release; in other words, this approach will manage the reuse of usable resources and the permitted disposal of waste to the environment.
- To ensure that waste is properly disposed off and that useful resources are properly transferred to other procedures.

3.4.1 Prevention (Reduction)

According to Nielsen and Schmidt (2014), waste prevention (reduction) can be defined as the process of raising public awareness about the harm that waste generation causes to the environment and to people, as well as the value of waste minimization and product reuse [32].

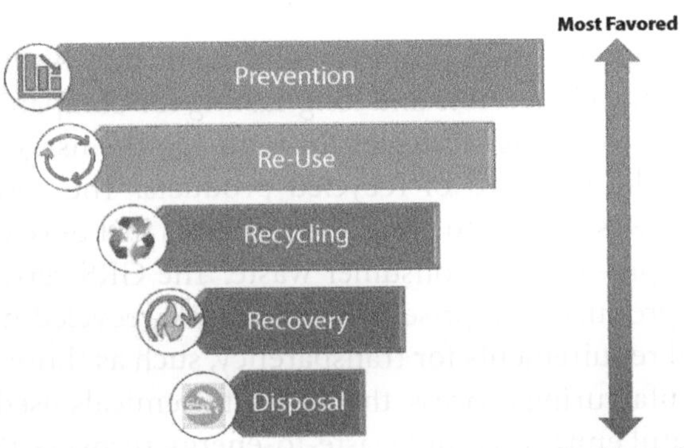

Figure 3.5 Waste management hierarchy.

3.4.2 Reuse

Reuse of products helps to control waste because it saves resources, minimizes the impact on the environment, and relieves the stress on waste management [33]. Reuse of apparels (apart from producing the same kind from waste materials) can decrease energy usage by 90%–95%, even after the pre-stages of collection, sorting, and reselling of worn clothing [34]. Additionally, consumption of 6,000 L of H_2O, 3.6 kg of CO_2, 0.3 kg of chemical fertilizer, and 0.2 kg of pesticides can be avoided by reusing 1 kg of a textile rather than fabricating a new one [35]. Therefore, it is important to encourage customers to reuse things in order to increase their environmental consciousness.

3.4.2.1 Second-Hand Clothing

Products that have reached a new stage of usage without alteration in the product development or possibly just after minor refurbishment are referred to as second-hand products. The time between purchasing a new product and the point at which its quality is deemed to be substandard is referred to as the useful life of the product. From the perspective of a consumer, the life cycle of the product can be defined as the time between the time of purchase and the time the product is tossed aside or replaced. Because people frequently buy new things instead of used ones, the life cycle is typically shorter than that of its useful life [36].

3.4.2.2 Vintage Clothing Trend

The age of the clothing can be used to identify vintage clothing (generally manufactured in 1920–1980s). According to Strähle and Matthaei (2017), clothing made during the 1980s is considered modern, whereas textiles made before 1920 are regarded as antiques [36].

3.4.3 Recycling

Four main approaches have been proposed for the recycling of solid textile waste.

3.4.3.1 Primary Recycling

This recycling process involves reusing off-grade and scrap materials by recycling plants themselves directly [37].

3.4.3.2 Secondary Recycling (Mechanical Recycling)

It entails separating, cleaning, and reusing the product as a blend or as individual polymers. Depending on the recycling method used, the mechanical recycling process can be categorized into two different directions. The first route is melt-extrusion process which involves shredding, crushing, carding, and grinding of waste before melting, re-extruding, and respinning fibers into yarns. In this manner, synthetic fabrics like nylon or polyester can be transformed back into fibers either via extrusion or by first turning flasks into pellets and then melting them. The second route involves defibrillation of textile materials to generate fibers. The basic goals of the carding method are to separate small tufts into individual fibers, to start the parallelization process and deliver the fibers as a web. Crushed or shredded materials are compression-molded to create sheets for a variety of uses including thermal and acoustic insulation [38, 39].

3.4.3.3 Tertiary Recycling (Chemical Recycling)

Chemical recycling involves complete or partial conversion of polymers into monomers, oligomers, and other chemicals or fuels. It can be carried out in two ways. The first involves depolymerization and repolymerization route, which depolymerizes waste material back into its component monomers. Plastic polymers, like polyester, can be broken down into monomers or oligomers during the depolymerization process and then re-polymerized to create new fibers or materials. Condensation-type textile wastes [polyester (PET), polyamide (PA), polyurethanes (PU), etc.] are the easiest ones to depolymerize. The molecules in these materials link in the presence of a chemically reactive agent in such a way that they break the molecules into shorter chains in controllable ways on application of appropriate amount of heat and pressure. Technologies for breaking down such polymers, primarily PET, have already been demonstrated and are regarded as relatively cost-effective.

PET recycling processes involves the following (as shown in Figure 3.6):

- **Glycolysis**

In the glycolysis reaction, PET is molecularly degraded by glycols while being exposed to trans-esterification catalysts, usually metal acetates, where the ester linkages are broken down and substituted with hydroxyl functional groups. Bis(hydroxyethyl) terephthalate (BHET) and its oligomers are produced from PET waste. The glycolysis process can be carried out for a period of 0.5 h to 8 h at temperatures ranging between 180°C and 250°C [40].

- **Hydrolysis**

PET waste can be hydrolyzed in acidic, alkaline, or neutral media. Acid hydrolysis can be performed using concentrated acids such as sulfuric, nitric, or phosphoric acid under low pressure or by using diluted acids at higher pressure. Alkaline sodium or potassium hydroxide solutions between 2 and 20 wt% are used to perform alkaline hydrolysis [41]. Depending on pressure and concentration of the alkaline solution, reaction time may be 2–7 h, which can be shortened to 40 min by adding methanol. Ethylene glycol (EG) and salt of disodium terephthalate are the byproducts of the process. Neutral hydrolysis is carried out in autoclaves in the pressure range of 1–4 MPa with excess hot water at 200°C–300°C. Undoubtedly, the high ecological purity of neutral hydrolysis is an advantage, but the main drawback, i.e., mechanical impurities in the polymer retained in TPA, leads to a product with a much lesser purity than that of acid or alkaline hydrolysis.

- **Aminolysis**

Aminolysis is basically a bond cleavage reaction that is carried out by using an amine. Bis (2-hydroxy ethylene) terephthalamide (BHETA) and EG are often produced in aminolysis process. BHETA is not a chemical that is particularly profitable to produce, but it can be turned into something with higher added value. The products recovered from these chemical reactions can either be a combination of oligomers with reactive end groups or monomers with a purity level acceptable for repolymerization [42].

The second approach, namely, dissolution process, involves separation, filtration, and regeneration of fibers. For cellulosic fibers like cotton or viscose, ionic liquids are typically utilized as solvents to participate in cellulose and obtain regenerated cellulose [43]. In a similar manner, polyester can be broken down using chemicals like dimethyl isophthalate, dimethyl terephthalate, or methyl-p-toluate, then retrieved, and spun back into the material. Instead of using a single fiber, the majority of textile items are made of fiber blends. Therefore, numerous researchers have also looked at the separation of mixes utilizing solvents.

- **Biochemical Recycling**

The environment-friendly method of recycling textiles is a biochemical recycling that decomposes polymers into monomers. Bio-catalysts called enzymes make chemical reactions more effective. Typically, biochemical recycling starts with a pre-treatment step that breaks down the macrostructure of the fiber using an acid or alkali. The most popular pre-treatment method is acid pre-treatment, which exposes the crystalline part of the

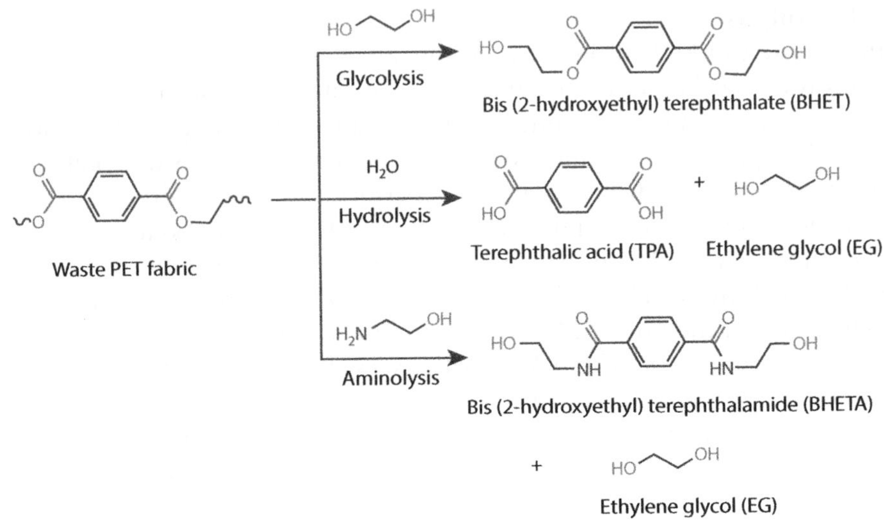

Figure 3.6 Various recycling methods of PET.

fiber to enzymatic destruction while the amorphous region of the fiber can be hydrolyzed. The intra- and inter-chain links of cellulose molecules can be broken down by pre-swelling treatment in alkali creating more room for the enzymatic treatment. In the enzymatic hydrolysis of cellulose, glycosidic linkages are broken down by cellulosic enzymes [44]. Utilizing biochemical recycling techniques, synthetic materials like polyester can be hydrolyzed, causing the ester linkages in the polymer chain to be broken and converted to monomers.

3.4.3.4 *Quaternary Recycling/Energy Recovery*

The process of burning waste materials to generate energy is referred to as "thermal recovery." Thermal recovery is employed when none of these recycling methods are suitable though it is not the most advised alternative with regards to recovery of resources. The main thermal recycling methods are combustion, pyrolysis, incineration, gasification, and anaerobic digestion. Combustion is a sequence of exothermic interactions between a fuel source and an oxidizer that releases heat. Pyrolysis is a heat degradation process without oxygen. Pyrolysis can be carried out on a variety of textile materials without prior sorting of the waste resulting in production of gas or oil that can be utilized to fulfill the energy requirements for different applications like wastewater treatment and soil restoration [45].

During the controlled burning procedure known as incineration, waste is turned into CO_2 and other gases which can reduce waste by up to 90% [34]. Gasification is the process when organic molecules are partially oxidized to produce synthetic gas at extreme temperatures (500°C–1800°C) [46]. There are two types of gasification techniques: direct gasification (using a gasification agent) and indirect gasification (without gasification agent). The main benefit of gasification over incineration is its better efficiency for generating energy. Anaerobic digestion manages to convert waste into biogas (mainly methane) by using bacteria [47].

Mechanical recycling has attracted the most researchers (43%) compared to other recycling techniques. Chemical recycling accounts for 38% of the demand, suggesting that it is not much behind mechanical recycling. Fourteen percent of the contribution of biochemical recycling indicates a far lower demand than chemical or mechanical recycling. Moreover, fabrics can be recycled by combining mechanical, chemical, and thermal processes.

- **Merits and Demerits of Recycling Processes**

While each of these technologies for recycling fabrics has its own benefits and drawbacks, they are all highly beneficial in keeping waste out of landfills and retrieving its fundamental qualities. The spinning drums and metal pins employed in mechanical recycling machines exert a considerable mechanical pressure on the machine resulting in shorter fibers, which are not beneficial for recycled products. However, chemical recycling has gained popularity without compromising quality though it needs to be operated on a bigger scale in order to be economically viable. Additional barriers to chemical recycling involve the use of mixed fibers, various dyeing agents, finishing and processing chemicals, the risk of fiber deterioration during the process, and the environmental impact of using toxic chemicals as solvents. Biochemical recycling process using biocatalysts in the depolymerization process is thought to be an environmentally benign recycling method as it eliminates the need for hazardous chemicals. Because the components of thermal recovery cannot be recovered for reuse, it is the least desirable technique of recycling solid textile waste [48].

Based on the type of methods utilized or the degree of disassembly of the recycled material, the textile recycling pathway can be categorized. Recovery and reuse of a fabric in new products is fabric recycling. Fiber recycling, on the other hand, includes disassembling a fabric while retaining the original fibers. Polymer/oligomer recycling involves disintegrating fibers while keeping the polymers/oligomers intact. Similarly, monomer recycling involves disassembling of polymers/oligomers without affecting monomers [49].

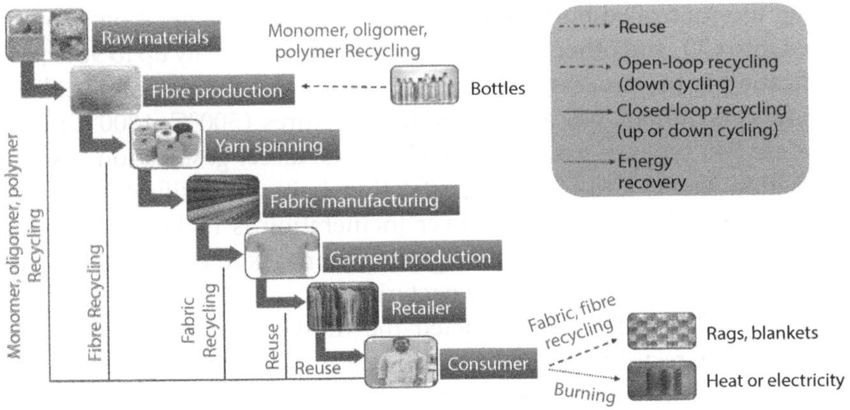

Figure 3.7 Classification of textile recycling routes.

However, depending on the quality of the finished product, fabric recycling can also be divided as upcycling and downcycling. Upcycling refers to the process of creating something that is of higher quality from the recycled materials. Upcycling can be demonstrated by converting old curtains into new clothes or damaged jeans into new bags. Downcycling is a restoration technique in which a waste material is transformed into a raw material that has a lesser quality than the original material. Downcycling can reduce the demand for new resources and decrease air and water pollution [35].

Depending on the final product desired, fabric recycling methods can be categorized into open-loop and closed-loop recycling. Closed-loop recycling a technique in which a product is recycled and reused in a product that is essentially identical to the original. On the other hand, open-loop recycling involves recycling of a product and using it again in a different product/form not essentially identical to the original one. Therefore, closed-loop recycling is beneficial in reducing the negative environmental effects caused by the textile industry. Figure 3.7 illustrates the classification of various solid textile recycling routes elaborately. It can be accomplished once a used item of recycling worn-out clothing and then repurposing it to make a product that is identical to it [38].

3.4.4 Recovery (Energy from Waste)

Energy recovery is an activity of extracting energy from waste in the form of heat or electricity. Inflammable elements such as methane (CH_4), methanol (CH_3OH), and ethanol (C_2H_5OH) generate energy in this process.

Energy recovery can be accomplished through a variety of methods, the most common of which are incineration, gasification, and anaerobic digestion which is discussed in details in Section 2.4.3.4. Quaternary recycling/ energy recovery section [14].

3.4.5 Disposal (Landfilling)

Disposal of solid textile waste is the least desirable way of waste management as landfills serve as the final destination of this waste [50]. Due to its characteristics, waste made of synthetic textiles does not degrade in landfills, but clothing made of wool decomposes emitting CH_4 and CO_2, which contributes to global warming. A lot of waste is thrown into landfills due to rising population and lack of awareness of textile recycling. By adopting the right recycling techniques, about 40%–45% of the post-consumer textile waste can be reused as secondhand clothes, 25-30% can be reused as industrial rags, whereas 20%–25% can be bio-degraded after being dumped, whereas only 5% of it will be unusable [51].

3.5 Applications of Recycled Fabrics

3.5.1 Cotton Waste

Waste textile cottons are either mechanically shredded or dissolved in various solvents and ionic liquids to produce regenerated cellulose fibers, and their characteristics were studied. To test the spinnability, pre-consumer knitted 100% cotton fabric wastes were sorted and shred into short fibers [52]. Waste fibers were retrieved from spinning mill, separated into hard and soft wastes, and made into a web of fibers that were further rotor-spun to make yarns in order to observe the impact of waste fibers on yarn quality [16]. Utilizing a Shirley analyzer, dyed yarn wastes were recycled to generate fibers that were blended with virgin cotton fibers and compared to those of virgin fibers [53].

Liu *et al.* (2019) recycled 100% cotton T-shirts (white/blue) into regenerated cellulose fibers using alkali (LiOH or NaOH)/urea aqueous solutions as solvents [54]. The lyocell process was taken into consideration to regenerate cellulose fibers from waste cotton [55] and three distinct types of solid cotton waste (waste denim clothing, easy-care finished cotton fabrics, and a mixture of cotton-based waste clothing and wood pulps) [56]. The discarded clothes were cleaned, soaked in NMMO solution, and spun into fibers whose characteristics were compared to those of regular lyocell fibers spun from wood pulps. Waste cotton lints (>92% cellulose, DP3133) were dissolved in 1-Allyl-3-methylimidazolium chloride ([AMIM][Cl]) ionic liquid after alkali treatment and then extruded into regenerate fibers [43].

Cotton-based textile waste can be reused as reinforcement for composite materials. To create high value-added composite materials with higher thermal and acoustic qualities, cotton fibers derived from recycled denim fabrics were employed as the reinforcing material, whereas PP and PE granules made from waste food and cleaning containers and water bottle caps, respectively, were used as the matrix material [57]. Short cotton dyed fibers that were produced as waste during various phases of production were used as reinforcing materials in PP matrix, and the composites created using the Brabender Plastograph intense melt mixer had superior qualities to those made using other natural fiber and glass fiber reinforced composites [58]. Cotton with short fibers that are based from old jeans fabric flocks have been injection molded in PP matrix using compression molding, dry mixing, extruding using a twin screw extruder, and finally injection-molding processes to preserve the possible properties without further degrading [59].

Activated carbon was produced using $FeCl_3$ from discarded cotton textiles, which is extensively used as an effective adsorbent for air and water purification due to its huge surface area and great porosity [60, 61]. Cotton textile waste was employed as a precursor to create hydrochars, a type of solid fuel by hydrothermal carbonization (HTC) process catalyzed by $FeCl_3$ [62] and a mixture of three surfactants—Span 80, SDBS, and Tween 80 [63].

Cellulose nanocrystal (CNC) formation is one of the best ways to recycle waste cotton clothes. Wang *et al.* created CNCs from old, printed bed sheets with a nearly 100% cotton content after extracting cellulose by alkali and bleaching pre-treatments followed by hydrolysis in the presence of sulfuric and hydrochloric acids and lastly ultrasonicating the same [64]. CNC, on the other hand, was synthesized using $H_2O_2/Cr(NO_3)_3$ solution as a bleaching agent in a simple one-pot oxidative hydrolysis process [65]. The non-dyed bed sheet labeled as 100% cotton were dissolved overnight in a mixture of ionic liquid ([Emim][OAc]) and DMSO solution and casted at room temperature to prepare ultra-filtration membrane [66].

One of the most intriguing alternatives to fossil fuels is bioethanol, a liquid renewable biofuel. The oxygen content of bioethanol (35%) provides the potential to reduce automobile emission [67]. Bioethanol was produced from cotton gin trash, dust, waste, etc. by pretreatment with dilute sulfuric acid [68], dilute sodium hydroxide [69], and corona [70] and then enzymatic hydrolysis producing glucose and finally after ethanol fermentation. Glucose can be produced from discarded bed linens by directly hydrolyzing with sulfuric acid without pretreatment [71].

Woven waste cotton fiber cloths were carbonized using controlled pyrolysis process at argon atmosphere at a heating rate of 10 up to 800 to obtain the 2D carbon fibers and to prepare flexible Li-ion battery negative electrodes [72]. Ismail *et al.* investigated the anaerobic digestion method for producing biogas (methane) from medical cotton industry waste, which were primarily made up of cotton ball waste, roll and pad gauge, crepe bandage, and cotton threads [73]. Various recycling processes of cotton waste and their final products are mentioned in Table 3.1.

3.5.2 Mixed Waste

Mixed textiles are the second most abundantly generated waste that need novel sustainable routes for their disposal. Pyrolysis, co pyrolysis, HTC, anaerobic digestion, fungal fermentation, dissolution, etc.., are some of the chemical routes that can be incorporated, whereas the mechanical way includes compression molding of the waste.

Waste jeans, consisting of cotton rich material that acts as a sustainable cellulosic feedstock that gives bio products such as ethanol and bio gas, are a significant component of the millions of tons of clothing waste generated each year [74]. Combating the shortage of produced cotton can also be accomplished by recovering cotton from textile waste [75]. Pyrolysis of waste jeans collected from MSW is used to obtain char, gas, oil, porous carbon, and activated charcoal fibers (ACFs) [76, 77]. The obtained char can be further functionalized by chemical treatment and used as single and hybrid filler in cement industry [45]. Types of metal oxides play an important role in the recycling of waste. During pyrolysis of mixed waste, ZnO shows the highest catalytic effect, whereas Fe_2O_3 shows the highest for char gasification process [78]. Temperature and kinetics also play a significant role in the pyrolysis of waste.

Biomass-derived waste such as wood, paper, and textile have much lower pyrolysis temperature which is in the range of 220°C–380°C as compared to the PET waste, which is in the range of 360°C–480°C [79, 80]. Behavior of mixed waste consisting of waste textile, wood, paper, etc., with HTC

Table 3.1 Waste cotton recycling methods and final products.

Sr. no.	Cotton waste recycled	Method of recycling	Final product	References
1	Pre-consumer knitted cotton textile waste	Sorting and shredding	Recycled cotton fiber	[52]
2	Waste fiber generated during spinning	Mechanical recycling	Recycled cotton fiber	[16]
3	Waste cotton yarn	Mechanical recycling	Reclaimed fibers	[53]
4	Post-consumer cotton waste	Acid hydrolysis	Regenerated cellulose fibers	[54]
5	Cotton waste garment	Dissolution	Fiber regeneration	[55]
6	Easy care finished cotton fabrics	Dissolution	Regenerated cellulose fibers	[56]
7	Waste cotton lint	Dissolution	Regenerated cellulose fibers	[43]
8	Cotton fibers from waste denim fabric	Hot press	Composite material	[57]
9	Cotton residues	Brabender Plastograph	Composite in PP matrix	[58]
10	Jeans short fiber–based flocks	Compression molding, dry mixing, extrusion, and injection molding	PP composite	[59]
11	Waste cotton textile	Pyrolysis	Activated carbon	[60]

(Continued)

Table 3.1 Waste cotton recycling methods and final products. (*Continued*)

Sr. no.	Cotton waste recycled	Method of recycling	Final product	References
12	Waste cotton woven	Pyrolysis	Activated carbon	[61]
13	Cotton textile waste	Hydrothermal carbonization	Hydrochars solid fuel	[62]
14	Cotton textile waste	Hydrothermal carbonization	Clean solid fuel	[63]
15	Old and printed bed sheet	Acid hydrolysis	Cellulose nanocrystals	[64]
16	Waste cotton cloth	Oxidative hydrolysis	nanocellulose	[65]
17	Non dyed bed sheet	Casting and membrane fabrication	Cellulose membrane	[66]
18	Cotton gin trash	Enzymatic hydrolysis and fermentation	Bioethanol	[68]
19	Cotton gin waste and cotton gin dust	Enzymatic hydrolysis and fermentation	Cellulosic ethanol	[69]
20	Desized and bleached cotton fabric	Enzymatic hydrolysis and fermentation	Bioethanol as renewable biofuel	[70]
21	Discarded bed linen	Acid hydrolysis	Glucose	[71]
22	Woven waste cotton fiber cloth	Controlled pyrolysis	Self-standing and flexible Li-ion battery	[72]
23	Medical cotton industry waste	Anaerobic digestion	Biogas production	[73]

and co-HTC showed more stable hydrochars [81, 82]. With the increasing demand for renewable energy, sustainable routes to formulate bio-based gas and oil are on a rise. Waste textiles are majorly made up of synthetic and cellulose polymers and are considered as a treat for the environment as their cellulosic part bears a high potential for biofuel production. Treated waste jeans consisting of cotton and polyester are subjected to anaerobic digestion, enzymatic hydrolysis, and fermentation for the production of bio gas, ethanol, and sugar [44, 47, 82]. A single stage as well as two stage batch CSTR reactors can be incorporated for their production [83].

Mixed waste can also be recycled into fiber reinforced composites for building applications, sound insulation materials, etc. Natural rubber when compressed with mixed cotton/polyester waste gives a composite having better acoustic properties [84]. Textile-reinforced fiber composites (TRFCs) show enhanced performance for moisture resistance as they inherently contain a high percentage of synthetic polymer fibers [85]. Nowadays, recycling waste to get ethanol is on a rise due to the increase in its demand. Pretreatment of textile waste with alkali solutions leads to the improvement of enzymatic hydrolysis, thereby increasing the yield of ethanol, whereas that of treated with acid solution leads to the recovery of polyester and cellulose [86–88]. Nylon/cotton-blended fabrics when dissolved in an ionic liquid with subsequent filtration can give recycled cellulose and nylon 6 as a product [89]. Textile waste when subjected to fermentation gives cellulase as a by-product, which then, by further hydrolysis, gives PET fiber. PET fibers when re-spun gives new textile [74, 90]. Recycling of mixed waste also finds its applications in children's mattresses, etc. [91]. Table 3.2 lists all possible recycling methods of mixed solid textile waste.

3.5.3 Polyester Waste

Polyester waste, also known as poly (ethylene terephthalate), is the third widely produced waste in the textile industry, which is mostly recycled through glycolysis, aminolysis, hydrolysis, and pyrolysis. De-dyeing of PET fibers is very important as it can incorporate coloration into the final depolymerized product. Sodium formaldehyde sulfoxylate was employed to decolorize dispersed dyes developed for PET [92].

One of the most important chemical recycling methods of post-consumed PET waste clothes is glycolysis. PET fabric wastes were glycolyzed with EG in the presence of various metal acetate catalysts such as Zn, Pb, Co, Mn [40, 93], glacial acetic acid, lithium hydroxide, sodium sulfate, and potassium sulfate [94] to produce bis(2-Hydroxyethyl) terephthalate (BHET). Zinc acetate is one of the most explored catalysts in glycolytic

Table 3.2 Mixed waste recycling methods and final products.

Sr. no.	Mixed waste recycled	Method of recycling	Final product	References
1	Cotton/PET-blended textile waste	Fermentation	Cellulase	[74]
2	Waste jeans	Separation of cotton and polyester	Cellulose fiber and polyester	[75]
3	Denim fabric waste	Slow pyrolysis	Mesoporous activated carbon fibers	[76]
4	Cotton/polyester-based textile waste	Pyrolysis	Porous carbon	[77]
5	Waste jeans	Pyrolysis	Char formation, hybrid filler, and CNT/char/cement	[45]
6	Polyester fiber (PF), cotton, and woolen	Catalytic pyrolysis and gasification under CO_2 atm	Textile char, CO, and H_2	[78]
7	Municipal solid waste	Co-pyrolysis	Bio mass	[79]
8	Technical solid waste (polyethylene, paper towel and textiles, and their mixture with a blending ratio of 4:1:1)	Isothermal pyrolysis	Solid, liquid, and gas	[80]

(Continued)

Table 3.2 Mixed waste recycling methods and final products. (*Continued*)

Sr. no.	Mixed waste recycled	Method of recycling	Final product	References
9	Municipal solid waste	Hydrothermal carbonization	Hydro char	[81]
10	Waste jeans (70/30 cotton/ polyester)	Catalyst supported pyrolysis	Bio oil, bio gas, and char	[82]
11	Wool textile waste	Anaerobic digestion	Bio gas and methane	[47]
12	Waste jeans	Anaerobic digestion, enzymatic hydrolysis, and fermentation	Bio gas and ethanol	[44]
13	Viscose and cotton fibers, both blended with polyester fibers	Anaerobic digestion	Bio gas	[83]
14	Post-industrial textile waste (C/P)	Compression molding	Composite with natural rubber	[84]
15	Local municipal solid waste	Compression Molding	Fiber reinforced composites	[85]
16	Waste polyester/ cotton-blended fabrics (WBFs)	Separation of cotton and polyester	Cellulose acetate and polyester	[86]
17	C/P textile waste	Enzymatic hydrolysis and fermentation	Ethanol and polyester fiber	[87]

(*Continued*)

Table 3.2 Mixed waste recycling methods and final products. (*Continued*)

Sr. no.	Mixed waste recycled	Method of recycling	Final product	References
18	Cotton-based waste textiles	Regeneration and enzymatic hydrolysis	Polyester and glucose	[88]
19	Waste nylon/ cotton-blended fabrics	Dissolution and filtration	Regenerated cellulose and nylon 6 fibers	[89]
20	Municipal solid waste	Fungal fermentation	Fungal cellulase	[90]
21	Fashion waste material	Chemical modification	Flame-retardant mattress	[91]

depolymerization of PET fabric waste. BHET obtained can be decolorized using Electro-Fenton process (EF process) with magnetic iron oxide particles (Fe_3O_4) [95] or chitosan/Fe_3O_4 nanoparticles [96]. BHET can be used as a raw material to synthesize various fatty amides [97] and water soluble polymers [98] or to regenerate PET [99, 100] comparable to virgin grade. BHET was further hydrolyzed to form terephthalic acid (TPA) to obtain azo disperse dyestuffs [101].

Another important method of depolymerization of PET fiber waste is aminolysis. PET fiber waste are depolymerized with ethanolamine in the presence of various catalysts like glacial acetic acid, sodium acetate, and potassium sulphate to produce BHETA [42]. BHETA can also be reacted with isocyanates to form polyurethanes used in coatings and adhesives [102].

PET waste fibers were also depolymerized by alkaline hydrolysis in the presence of sodium hydroxide (NaOH) [41] and chemo-enzymatic hydrolysis in the presence of Humicola insolens Cutinase (HiC) enzyme [103] to recover TPA and EG.

In order to produce, activated carbon by pyrolysis method, mangnesium acetate ($Mg(Ac)_2$), magnesium oxalate (MgC_2O_4) [104], magnesium chloride ($MgCl_2$) as precursor of magnesia (MgO) [105], ferric chloride ($FeCl_3$), ferrous chloride ($FeCl_2$), and ferrous sulphate ($FeSO_4$) [106] were selected as the activating agents from polyester fabric waste. Table 3.3 tabulates all recycling methods of waste PET.

Table 3.3 Waste polyester recycling methods and final products.

Sr. no.	Polyester waste recycled	Method of recycling	Final product	References
1	Dyed PET fibers	Decolorization	Decolorized PET fibers	[92]
2	Polyester clothes	Glycolysis with EG (zinc acetate catalyst)	BHET	[40]
3	PET fiber waste	Glycolysis with EG (metal acetate catalysts)	BHET	[93]
4	PET waste fiber	Glycolysis with EG (CH_3COOH, LiOH, Na_2SO_4, and K_2SO_4)	BHET	[94]
5	PET fabric waste	Glycolysis and electrochemical decolorization (chitosan/Fe_3O_4)	Decolorized BHET	[95]
6	Waste PET fibers	Glycolysis and electrochemical decolorization (Fe_3O_4)	Decolorized BHET	[96]
7	PET waste fiber	Glycolysis (sodium sulfate)	Fatty amides from BHET	[97]
8	PET fabric waste	Glycolysis (zinc acetate)	Water soluble polymer from BHET	[98]
9	Polyester textile waste	Glycolysis (Mg-Al double oxide pellets)	Regenerated PET from BHET	[99]
10	PET fiber waste	Glycolysis (zinc acetate)	Regenerated PET	[100]
11	PET fiber waste	Glycolysis (zinc acetate)	Azo disperse dye	[101]

(Continued)

Table 3.3 Waste polyester recycling methods and final products. (*Continued*)

Sr. no.	Polyester waste recycled	Method of recycling	Final product	References
12	PET waste fiber	Aminolysis with ethanolamine (CH_3COOH, $NaCOOCH_3$, and K_2SO_4)	BHETA	[42]
13	PET staple waste fiber	Aminolysis with ethanolamine (sodium acetate)	Polyurethane	[102]
14	PVC-coated PET woven fibers	Alkaline hydrolysis	PVC, EG	[41]
15	Wellman PET fiber	Chemo-enzymatic hydrolysis	Terephthalic acid (TA) and ethylene glycol	[103]
16	Waste polyester textile	Pyrolysis	Mesoporous activated carbon	[104]
17	Waste polyester textile	Pyrolysis	Activated carbon	[105]
18	Waste polyester textile	Pyrolysis	Activated carbon	[106]

3.5.4 Other Waste

Apart from waste cotton, polyester, and cotton-polyester mixed fibers, other types of waste fibers and fabrics like viscose, rayon, and polyamide can be subjected to recycling for obtaining various products. Short viscose yarn waste can be acid hydrolyzed in the presence of sulfuric acid to extract nanocellulose (NC) [107, 108]. Waste viscose fibers were dissolved in various concentrations of NaOH to regenerate cellulose fibers. Blended nylon 6/spandex waste can be recycled by heat treatment, washing, and melt mixing [109]. A large amount of post-consumer nylon waste is being generated from fishing nets, paint brushes etc., and can be recycled into fibers to be used in mortar reinforcement, construction industry, etc. [110, 111].

3.6 Challenges in Fabric Recycling

Fabric-to-fabric recycling in a closed-loop system has proven to be a severe challenge to date. Complexity of materials is the most challenging issue, as they are made up of irregular compositions of natural and synthetic fibers in a variety of blended textiles and products. However, a number of intricate problems limit the ability of fabric waste to be recycled, and these are as follows [112]:

- Material complexity
- Lack of technological advancement
- Color and chemical complexity
- Quality issues
- Challenges in the reverse logistics process
- Complexities in garment recycling

3.7 Technologies and Innovation Involved in Textile Recycling

Several recent technologies and innovations aimed at improving recycling outcomes across various industries are developed. Industry and research enthusiasts are developing more sophisticated advanced sorting technologies that use artificial intelligence (AI), machine learning, computer vision and robotics in order to identify and separate different types of wastes which can help to improve the efficiency of recycling process [113, 114]. Digital innovations are becoming more practicable, dependable, clear, efficient, and optimal due to the application of IoT, digitalization, and information and communication technology (ICT). Following mentioned technologies and innovation are being employed for better recycling outcomes:

3.7.1 Radio Frequency Identification (RFID) Tags

RFID tags are used to sort clothes at fabric level. They are added to fabric hems during garment production and are capable of remaining active during the whole process of clothing use. Sorting with the help of RFID tags uses the same approach employed by the infrared sensors for automated systems in plastic sorting industries [115]. Beyond, sorting capabilities, with RFID tags, the manufacturer can track the clothing within and out of the store and are sources of big data to the manufacturer to monitor product use and later collection after use. These tags facilitate fast self-checkouts in stores and are capable of communicating with smart washing machines for optimum washing settings which is beneficial for customers as well [116].

3.7.2 Artificial Intelligence–Based Hybridized Intelligent Framework (AIHIF) Recognition Technology

In order to improve waste management, Yu *et al.* designed AI-based hybridized intelligent framework (AIHIF). The technology improves collection of waste over a brief period of time by combining machine learning and graph theory. Furthermore, a variety of strategies can be used to support AI design technology that can be customized for particular interest groups, gathering their data, and enhancing ecological productivity, accuracy, and effectiveness in waste management [113].

3.7.3 Importance of Computing Technology for Sorting Waste

Cloud computing technology is a data management solution that has resulted in a massive scale of data sharing among several users and has become the latest data management standard [117]. Cloud manufacturing technology intelligently handles processes and resources throughout the product life cycle, utilizing IoT technologies for control and management, simplifying digital sharing, and complementing the service-oriented framework [118]. Enevo One software, featuring ultrasonic wireless sensors, optimizes waste collection for over 100 container types, including mixed, textile, glass, metal, and waste electrical and electronic equipment (WEEE). In addition to the cloud service operating in the background, the Enevo One web interface and the sensors themselves have contributed to the optimization of waste collection [119].

3.7.4 Implementation of Internet of Things (IoT)

IoT technologies are evolving into an entire ecosystem which includes computer vision, fuzzy logic, machine learning, and natural language interfacing. This convergence assists in integrating diverse data sources with distributed computing, decreasing data dimension, and managing exponential data expansion in the IoT environment [120]. The design approach for the IoT-based categorization and detection of textile waste is illustrated in Figure 3.8. Waste is separated using a funnel and a single metallic sensor. The metallic sensor detects any metal compounds that move through the funnel. The metal components are preserved in metal storage. Then, the rest of the wastes are placed on the sensing platform, which detects objects by the capacitive sensors based on their moisture content. The materials that have passed through capacitive sensors are then divided into two groups: textiles and plastic bottles. The sorted textile waste is passed through numerous steps of sorting and identification due to the complex mixture of constituents [121].

3.7.5 Chemical Innovation

Advances in chemical recycling such as irradiating the waste with gamma radiation [122], electron beam [123], etc., can help to breakdown complex materials by generating free radicals, which are capable of inducing the molecular changes by degradation of macromolecules into the molecules with small chain length, hence increasing the efficiency by reduction in depolymerization time. Also, the degradation in

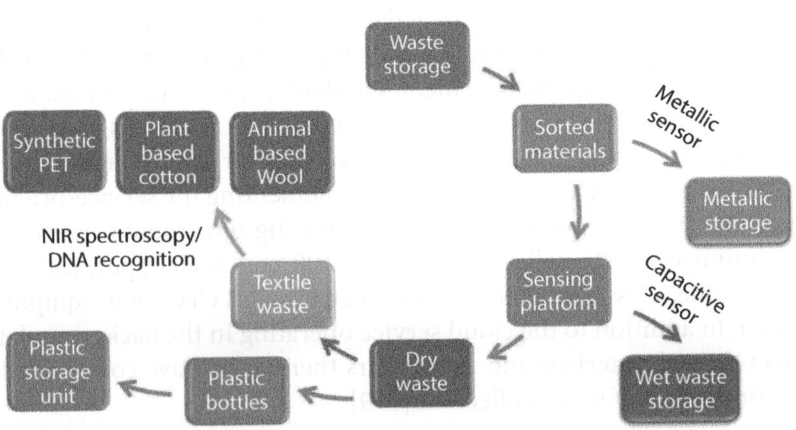

Figure 3.8 Implementation of IoT in sorting waste.

molecular weight has its subsequent effect on the physical and chemical properties. Unexplored recycling routes such as acidolysis [124] and ammonolysis [125] are being explored in order to develop new recycling routes.

3.8 Conclusions

Increase in the global population, industrial development, and higher fashion standards have all contributed to an expansion of the textile sector, which has, in turn, increased the production of textile waste. Overexposure to fast fashion trends has caused the apparel sector to produce millions of tons of solid, liquid, and gaseous waste. Compared to production and pre-consumer textile waste, post-consumer wastes generated by consumers are increasing day by day, resulting in dumping as landfills or incinerating. Hence, proper waste management hierarchy is required to minimize waste by increasing public awareness of environmental issues. Additionally, recycling and reuse of waste are favored to disposal or energy recovery.

In this chapter, we have emphasized the amount of waste produced and recycled globally, followed by the requirement for different types of textile waste (particularly solid waste) to be sorted and segregated before recycling. Each stage of the waste management hierarchy is carefully detailed in order to concentrate on the CE. There is a detailed discussion of various reuse and recycling techniques for solid textile waste. A thorough analysis of the literature focused on the techniques for recycling the most commonly produced textile waste, such as cotton, polyester, viscose, nylon, and their blends, is discussed. Challenges encountered during recycling are properly investigated to determine their root causes for the sake of the environment.

References

1. Niinimäki, K., Peters, G., Dahlbo, H., Perry, P., Rissanen, T., Gwilt, A., The environmental price of fast fashion. *Nat. Rev. Earth Environ.*, 1, 4, 189–200, 2020.
2. Haslinger, S., Wang, Y., Rissanen, M., Lossa, M.B., Tanttu, M., Ilen, E., Määttänen, M., Harlin, A., Hummel, M., Sixta, H., Recycling of vat and reactive dyed textile waste to new colored man-made cellulose fibers. *Green Chem.*, 21, 20, 5598–5610, 2019.

3. Bediako, J.K., Wei, W., Yun, Y.S., Low-cost renewable adsorbent developed from waste textile fabric and its application to heavy metal adsorption. *J. Taiwan Inst. Chem. Eng.*, 63, 250–258, 2016.
4. Fashion united, Global Fashion Industry statistics, 2022, https://fashionunited.com/global-fashion-industry-statistics.
5. Business 2 Community, 30 Shocking Figures and Facts in Global Textile and Apparel Industry, 2015, https://www.business2community.com/fashion-beauty/30-shocking-figures-facts-global-textile-apparel-industry-01222057.
6. Dudin, M.N., Lyasnikov, N.V., Dzhurabaeva, G.K., Kuznecov, A.V., Chinese textile industry: Sustainable development challenges and competitiveness issues in economic environment dynamics. *Fibres Text. East. Eur.*, 23, 4, 14–18, 2015.
7. Textile Technology, Recycling of 25 % of all textile waste by 2025, 2022, https://www.textiletechnology.net/technology/news/china-recycling-of-25–of-all-textile-waste-by-2025-32004#:~:text=China%20aims%20to%20recycle%2025%20%25%20of%20all,South%20China%20Morning%20Post%20on%20April%2013%2C%202022.
8. European Commission, Questions and Answers: A European Strategy for Plastics, 2018, https://ec.europa.eu/commission/presscorner/detail/en/MEMO_18_6.
9. Recovery recycling technology worldwide, The limitations of textile recycling, The limitations of textile recycling - recovery (recovery-worldwide.com), 2019.
10. Bukhari, M.A., Carrasco-Gallego, R., Ponce-Cueto, E., Developing a national programme for textiles and clothing recovery. *Waste Manag. Res.*, 36, 4, 321–331, 2018.
11. United States Environmental Protection Agency, Textiles : Material-Specific Data, Textiles: Material-Specific Data | US EPA, 2022.
12. Vogue, What actually happens to your clothes when you recycle them?, 2019, https://www.vogue.in/fashion/content/what-happens-to-recycled-clothes-sustainable-fashion-environmental-impact.
13. Storry, K. and McKenzie, A., Unravelling the Problem of Apparel Waste in the Greater Vancouver Area, Metro Vancouver and City of Vancouver, pp. 11–63, 2018.
14. Yalcin-Enis, I., Kucukali-Ozturk, M., Sezgin, H., Risks and Management of Textile Waste, in: *Nanoscience and Biotechnology for Environmental Applications. Environmental Chemistry for a Sustainable World*, K.M. Gothandam (Eds.), p. 29–53, Springer, Cham, 2019.
15. Doctor, S.J. and Khatwani, P.A., Solid Textile Wastes : Different Types and their Causes of Generation. *J. Text. Assoc.*, 83, 1, 422–426, 2022.
16. Jamshaid, H., Hussain, U., Mishra, R., Tichy, M., Muller, M., Turning textile waste into valuable yarn. *Clean. Eng. Technol.*, 5, 100341, 2021.
17. Responsible Global Fashion LLC, Tackling textile waste, 2016, http://www.responsibleglobalfashion.com/blog/2016/7/tackling-textile-waste.

18. Türe, M. (Ed.), *Waste management and sustainable consumption: reflections on consumer waste*, Routledge, London, Taylor & Francis Group, 2015.
19. Gardetti, M.A. and Torres, A.L. (Eds.), *Sustainability in Fashion and Textiles – Values, Design, Production and Consumption*, Routledge, London, Taylor & Francis Group, 2013.
20. Karaosman, H., Brun, A., Morales-Alonso, G., Vogue or Vague: Sustainability Performance Appraisal in Luxury Fashion Supply Chains, in: *Sustainable Management of Luxury*, M.A. Gardetti (Ed.), pp. 301–330, Springer, Singapore, 2017.
21. Stanescu, M.D., State of the art of post-consumer textile waste upcycling to reach the zero waste milestone. *Environ. Sci. Pollut. Res.*, 28, 12, 14253–14270, 2021.
22. Leonas, K.K., The Use of Recycled Fibers in Fashion and Home Products, in: *Textiles and Clothing Sustainability Recycled and Upcycled Textiles and Fashion*, S.S. Muthu (Ed.), pp. 55–77, Springer, Singapore, 2017.
23. Zamani, B., Svanström, M., Peters, G., Rydberg, T., A Carbon Footprint of Textile Recycling: A Case Study in Sweden. *J. Ind. Ecol.*, 19, 4, 676–687, 2014.
24. Subramanian, K., Chopra, S.S., Cakin, E., Li, X., Lin, C.S.K., Environmental life cycle assessment of textile bio-recycling – valorizing cotton-polyester textile waste to pet fiber and glucose syrup. *Resour. Conserv. Recycl.*, 161, 104989, 2020.
25. Braungart, M., Mcdonough, W., Bollinger, A., Cradle-to-cradle design: creating healthy emissions- a strategy for eco-effective product and system design. *J. Clean. Prod.*, 15, 1337–1348, 2007.
26. Ozturk, E., Koseoglu, H., Karaboyaci, M., Yigit, N.O., Yetis, U., Kitis, M., Sustainable textile production: cleaner production assessment/eco-efficiency analysis study in a textile mill. *J. Clean. Prod.*, 138, 2, 248–263, 2016.
27. Ellen Macarthur Foundation, What is a circular economy?, 2022, https://ellenmacarthurfoundation.org/topics/circular-economy-introduction/overview.
28. Ghisellini, P., Cialani, C., Ulgiati, S., A review on circular economy: The expected transition to a balanced interplay of environmental and economic systems. *J. Clean. Prod.*, 114, 11–32, 2016.
29. The 6Rs, 2014, https://penyrheol-comp.net/technology/wp-content/uploads/sites/2/2014/06/6-Rs1.pdf.
30. Textile Exchange, Textile Exchange Standards, 2021, https://textileexchange.org/standards/.
31. Islam, M.S. and Islam, J.M.M., Sources and Fates of Textile Solid Waste and Their Sustainable Management, in: *Handbook of Solid Waste Management*, C. Baskar (Eds.), pp. 1–21, Springer, Singapore, 2021.
32. Nielsen, R. and Schmidt, A., Changing consumer behaviour towards increased prevention of textile waste : Background report, Nordic Council of Ministers, pp. 1–28, 2014.

33. Fortuna, L.M. and Diyamandoglu, V., Disposal and acquisition trends in second-hand products. *J. Clean. Prod.*, 142, 4, 2454–2462, 2017.
34. Zamani, B., Towards Understanding Sustainable Textile Waste Management: Environmental impacts and social indicators, Chalmers University of Technology, pp. 1–41, 2014.
35. Vats, S. and Rissanen, M., Parameters affecting the upcycling of waste cotton and PES/CO textiles. *Recycling*, 1, 1, 166–177, 2016.
36. Kinyua, N.G., Effects of Imported Second Hand Clothes on the Performance of Textile Industry in Kenya (A Survey of Selected Market in Nairobi County), Management University of Africa, pp. 1–56, 2018.
37. Gadkari, R. and Burji, M.C., Textiles Waste Recycling, 2014, https://textilevaluechain.in/in-depth-analysis/articles/textile-articles/textiles-waste-recycling/.
38. Payne, A., Open-and closed-loop recycling of textile and apparel products, in: *Handbook of Life Cycle Assessment (LCA) of Textiles and Clothing*, S. Muthu (Ed.), pp. 103–123, Woodhead Publishing, United Kingdom, 2015.
39. Peña-Pichardo, P., Martínez-Barrera, G., Martínez-López, M., Ureña-Núñez, F., dos Reis, J.M.L., Recovery of cotton fibers from waste Blue-Jeans and its use in polyester concrete. *Constr. Build. Mater.*, 177, 409–416, 2018.
40. Barot, A.A. and Sinha, V.K., Chemical scavenging of post-consumed clothes. *Waste Manag.*, 46, 86–93, 2015.
41. Kumagai, S., Hirahashi, S., Grause, G., Kameda, T., Toyoda, H., Yoshioka, T., Alkaline hydrolysis of PVC-coated PET fibers for simultaneous recycling of PET and PVC. *J. Mater. Cycles Waste Manag.*, 20, 439–449, 2017.
42. Shukla, S.R. and Harad, A.M., Aminolysis of polyethylene terephthalate waste. *Polym. Degrad. Stab.*, 91, 1850–1854, 2006.
43. De Silva, R. and Byrne, N., Utilization of cotton waste for regenerated cellulose fibres : Influence of degree of polymerization on mechanical properties. *Carbohydr. Polym.*, 174, 89–94, 2017.
44. Hasanzadeh, E., Mirmohamadsadeghi, S., Karimi, K., Enhancing energy production from waste textile by hydrolysis of synthetic parts. *Fuel*, 218, 41–48, 2018.
45. Yousef, S., Kalpokaitė-Dičkuvienė, R., Baltušnikas, A., Pitak, I., Lukošiūtė, S.I., A new strategy for functionalization of char derived from pyrolysis of textile waste and its application as hybrid fillers (CNTs/char and graphene/char) in cement industry. *J. Clean. Prod.*, 314, 128058, 2021.
46. Bosmans, A., Vanderreydt, I., Geysen, D., Helsen, L., The crucial role of Waste-to-Energy technologies in enhanced landfill mining: a technology review. *J. Clean. Prod.*, 55, 10–23, 2013.
47. Kabir, M.M., Forgács, G., Horváth, I.S., Enhanced methane production from wool textile residues by thermal and enzymatic pretreatment. *Process Biochem.*, 48, 4, 575–580, 2013.

48. Piribauer, B. and Bartl, A., Textile recycling processes, state of the art and current developments : A mini review. *Waste Manag. Res.*, 37, 2, 112–119, 2019.
49. Sandin, G. and Peters, G.M., Environmental impact of textile reuse and recycling- A review. *J. Clean. Prod.*, 184, 353–365, 2018.
50. Strähle, J., *Green Fashion Retail*, pp. 1–291, Springer, Singapore, 2017.
51. Lee, K.E., Environmental Sustainability in the Textile Industry, in: S.S. Muthu (Ed.), Textile Science and Clothing Technology, pp. 17–55, Springer, Singapore, 2016.
52. Ütebay, B., Çelik, P., Çay, A., Effects of cotton textile waste properties on recycled fibre quality. *J. Clean. Prod.*, 222, 29–35, 2019.
53. Wanassi, B., Azzouz, B., Hassen, M.B., Value-added waste cotton yarn: Optimization of recycling process and spinning of reclaimed fibers. *Ind. Crop. Prod.*, 87, 27–32, 2016.
54. Liu, W., Liu, S., Liu, T., Liu, T., Zhang, J., Liu, H., Eco-friendly post-consumer cotton waste recycling for regenerated cellulose fibers. *Carbohydr. Polym.*, 206, 141–148, 2019.
55. Haule, L.V., Carr, C.M., Rigout, M., Preparation and physical properties of regenerated cellulose fibres from cotton waste garments. *J. Clean. Prod.*, 112, 4445–4451, 2016.
56. Haule, L.V., Carr, C.M., Rigout, M., Investigation into the supramolecular properties of fibres regenerated from cotton based waste garments. *Carbohydr. Polym.*, 144, 131–139, 2016.
57. Sezgin, H., Kucukali-ozturk, M., Berkalp, O.B., Yalcin-Enis, I., Design of composite insulation panels containing 100% recycled cotton fibers and polyethylene/polypropylene packaging wastes. *J. Clean. Prod.*, 304, 127132, 2021.
58. Serra, A., Tarrés, Q., Llop, M., Reixach, R., Mutjé, P., Espinach, F.X., Recycling dyed cotton textile byproduct fibers as polypropylene reinforcement. *Text. Res. J.*, 89, 11, 1–13, 2018.
59. Petrucci, R., Nisini, E., Puglia, D., Sarasini, F., Rallini, M., Santulli, C., Minak, G., Kenny, J.M., Tensile and fatigue characterisation of textile cotton waste/polypropylene laminates. *Compos. Part B*, 81, 84–90, 2015.
60. Xu, Z., Sun, Z., Zhou, Y., Chen, W., Zhang, T., Huang, Y., Zhang, D., Insights into the pyrolysis behavior and adsorption properties of activated carbon from waste cotton textiles by $FeCl_3$-activation. *Colloids Surf., A*, 582, 123934, 2019.
61. Xu, Z., Zhou, Y., Sun, Z., Zhang, D., Huang, Y., Gu, S., Chen, W., Understanding reactions and pore-forming mechanisms between waste cotton woven and $FeCl_3$ during the synthesis of magnetic activated carbon. *Chemosphere*, 241, 125120, 2020.
62. Qi, R., Xu, Z., Zhou, Y., Zhang, D., Sun, Z., Chen, W., Xiong, M., Clean solid fuel produced from cotton textiles waste through hydrothermal

carbonization with $FeCl_3$: Upgrading the fuel quality and combustion characteristics. *Energy*, 214, 118926, 2021.

63. Xu, Z., Qi, R., Xiong, M., Zhang, D., Gu, H., Chen, W., Conversion of cotton textile waste to clean solid fuel via surfactant-assisted hydrothermal carbonization: Mechanisms and combustion behaviors. *Bioresour. Technol.*, 321, 124450, 2021.
64. Wang, Z., Yao, Z., Zhou, J., Zhang, Y., Reuse of waste cotton cloth for the extraction of cellulose nanocrystals. *Carbohydr. Polym.*, 157, 945–952, 2017.
65. Chen, Y.W. and Lee, H.V., Revalorization of selected municipal solid waste as new precursors of "green" nanocellulose via a novel one-pot isolation system: A source perspective. *Int. J. Biol. Macromol.*, 107, 1, 78–92, 2018.
66. Lopatina, A., Anugwom, I., Blot, H., Corde, A.S., Mäntäri, M., Kallioinen, M., Re-use of waste cotton textile as ultrafiltration membrane. *J. Environ. Chem. Eng.*, 9, 4, 105705, 2021.
67. Dahman, Y., Syed, K., Begum, S., Roy, P., Mohtasebi, B., Biofuels: Their characteristics and analysis, in: *Biomass, Biopolymer- Based Materials, and Bioenergy*, D. Verma, E. Fortunati, S. Jain, X. Zhang (Eds.), pp. 277–325, Woodhead Publishing, United Kingdom, 2019.
68. Mcintosh, S., Vancov, T., Palmer, J., Morris, S., Ethanol production from cotton gin trash using optimised dilute acid pretreatment and whole slurry fermentation processes. *Bioresour. Technol.*, 173, 42–51, 2014.
69. Fockink, D.H., Maceno, M.A.C., Ramos, L.P., Production of cellulosic ethanol from cotton processing residues after pretreatment with dilute sodium hydroxide and enzymatic hydrolysis. *Bioresour. Technol.*, 187, 91–96, 2015.
70. Nikolić, S., Lazić, V., Veljović, Đ., Mojović, L., Production of bioethanol from pre-treated cotton fabrics and waste cotton materials. *Carbohydr. Polym.*, 164, 136–144, 2017.
71. Sanchis-sebastiá, M., Ruuth, E., Stigsson, L., Galbe, M., Wallberg, O., Novel sustainable alternatives for the fashion industry: A method of chemically recycling waste textiles via acid hydrolysis. *Waste Manage.*, 121, 248–254, 2021.
72. Jagdale, P., Nair, J.R., Khan, A., Armandi, M., Meligrana, G., Hernandez, F.R., Rusakova, I., Piatti, E., Rovere, M., Tagliaferro, A., Winter, M., Gerbaldi, C., Waste to life: Low-cost , self-standing , 2D carbon fiber green Li-ion battery anode made from end-of-life cotton textile. *Electrochim. Acta*, 368, 137644, 2021.
73. Ismail, Z.Z. and Talib, A.R., Recycled medical cotton industry waste as a source of biogas recovery. *J. Clean. Prod.*, 112, 5, 4413–4418, 2016.
74. Hu, Y., Du, C., Pensupa, N., Lin, C.S.K., Optimisation of fungal cellulase production from textile waste using experimental design. *Process Saf. Environ. Prot.*, 118, 133–142, 2018.
75. Yousef, S., Tatariants, M., Tichonovas, M., Kliucininkas, L., Lukošiūtė, S., Yan, L., Sustainable green technology for recovery of cotton fibers and polyester from textile waste. *J. Clean. Prod.*, 254, 120078, 2020.

76. Silva, T.L., Cazetta, A.L., Souza, P.S.C., Zhang, T., Asefa, T., Almeida, V.C., Mesoporous activated carbon fibers synthesized from denim fabric waste: Efficient adsorbents for removal of textile dye from aqueous solutions. *J. Clean. Prod.*, 171, 482–490, 2017.
77. Gu, S., Zhang, D., Gao, Y., Qi, R., Chen, W., Xu, Z., Fabrication of porous carbon derived from cotton/polyester waste mixed with oyster shells: Pore-forming process and application for tetracycline removal. *Chemosphere*, 270, 129483, 2021.
78. Wu, Y., Wen, C., Chen, X., Jiang, G., Liu, G., Liu, D., Catalytic pyrolysis and gasification of waste textile under carbon dioxide atmosphere with composite Zn-Fe catalyst. *Fuel Process. Technol.*, 166, 115–123, 2017.
79. Ansah, E., Wang, L., Shahbazi, A., Thermogravimetric and calorimetric characteristics during co-pyrolysis of municipal solid waste components. *Waste Manag.*, 56, 196–206, 2016.
80. Hu, D.H., Chen, M.Q., Huang, Y.W., Wei, S.H., Zhong, X.B., Evaluation on isothermal pyrolysis characteristics of typical technical solid waste. *Thermochim. Acta*, 688, 178604, 2020.
81. Lin, Y., Ge, Y., Xiao, H., He, Q., Wang, W., Chen, B., Investigation of hydrothermal co-carbonization of waste textile with waste wood , waste paper and waste food from typical municipal solid waste. *Energy*, 210, 118606, 2020.
82. Yousef, S., Eimontas, J., Striūgas, N., Tatariants, M., Abdelnaby, M.A., Tuckute, S., Kliucininkas, L., A sustainable bioenergy conversion strategy for textile waste with self- catalysts using mini-pyrolysis plant. *Energy Convers. Manag.*, 196, 688–704, 2019.
83. Jeihanipour, A., Aslanzadeh, S., Rajendran, K., Balasubramanian, G., Taherzadeh, M.J., High-rate biogas production from waste textiles using a two-stage process. *Renew. Energy*, 52, 128–135, 2013.
84. Dissanayake, D.G.K., Weerasinghe, D.U., Thebuwanage, L.M., Bandara, U.A.A.N., An environmentally friendly sound insulation material from post-industrial textile waste and natural rubber. *J. Build. Eng.*, 33, 101606, 2021.
85. Echeverria, C.A., Handoko, W., Pahlevani, F., Sahajwalla, V., Cascading use of textile waste for the advancement of fibre reinforced composites for building applications. *J. Clean. Prod.*, 208, 1524–1536, 2019.
86. Sun, X., Lu, C., Zhang, W., Tian, D., Zhang, X., Acetone-soluble cellulose acetate extracted from waste blended fabrics via ionic liquid catalyzed acetylation. *Carbohydr. Polym.*, 98, 1, 405–411, 2013.
87. Gholamzad, E., Karimi, K., Masoomi, M., Effective conversion of waste polyester – cotton textile to ethanol and recovery of polyester by alkaline pretreatment. *Chem. Eng. J.*, 253, 40–45, 2014.
88. Shen, F., Xiao, W., Lin, L., Yang, G., Zhang, Y., Deng, S., Enzymatic saccharification coupling with polyester recovery from cotton-based waste textiles by phosphoric acid pretreatment. *Bioresour. Technol.*, 130, 248–255, 2013.

89. Lv, F., Wang, C., Zhu, P., Zhang, C., Isolation and recovery of cellulose from waste nylon/cotton blended fabrics by 1-allyl-3-methylimidazolium chloride. *Carbohydr. Polym.*, 123, 424–431, 2015.
90. Wang, H., Kaur, G., Pensupa, N., Uisan, K., Du, C., Yang, X., Lin, C.S.K., Textile waste valorization using submerged filamentous fungal fermentation. *Process Saf. Environ. Prot.*, 118, 143–151, 2018.
91. Nayak, R., Houshayar, S., Patnaik, A., Nguyen, L.T.V., Shanks, R.A., Padhye, R., Fegusson, M., Sustainable reuse of fashion waste as flame-retardant mattress filing with ecofriendly chemicals. *J. Clean. Prod.*, 251, 19620, 2020.
92. Fei, X., Freeman, H.S., Hinks, D., Toward closed loop recycling of polyester fabric: Step 1. decolorization using sodium formaldehyde sulfoxylate. *J. Clean. Prod.*, 254, 120027, 2020.
93. Ghaemy, M. and Mossaddegh, K., Depolymerisation of poly (ethylene terephthalate) fibre waste using ethylene glycol. *Polym. Degrad. Stab.*, 90, 570–576, 2005.
94. Shukla, S.R. and Harad, A.M., Glycolysis of Polyethylene Terephthalate Waste Fibers. *J. Appl. Polym. Sci.*, 97, 513–517, 2005.
95. Li, Y., Li, M., Lu, J., Li, X., Ge, M., Decoloration of waste PET alcoholysis liquid by electrochemical method. *Water Sci. Technol.*, 77, 9-10, 2463–2473, 2018.
96. Li, M., Li, Y., Lu, J., Li, X., Lu, Y., Decolorization and reusing of PET depolymerization waste liquid by electrochemical method with magnetic nanoelectrodes. *Environ. Sci. Pollut. Res.*, 25, 34531–34539, 2018.
97. Shukla, S.R., Harad, A.M., Jawale, L.S., Recycling of waste PET into useful textile auxiliaries. *Waste Manage.*, 28, 51–56, 2008.
98. Lu, J., Li, M., Li, Y., Li, X., Gao, Q., Ge, M., Synthesis and sizing performances of water-soluble polyester based on derived from depolymerized waste poly (ethylene terephthalate) fabrics. *Text. Res. J.*, 89, 572–579, 2019.
99. Guo, Z., Eriksson, M., De La Motte, H., Adolfsson, E., Circular recycling of polyester textile waste using a sustainable catalyst. *J. Clean. Prod.*, 283, 124579, 2021.
100. Hu, Y., Wang, Y., Zhang, X., Qian, J., Xing, X., Wang, X., Synthesis of poly (ethylene terephthalate) based on glycolysis of waste PET fiber. *J. Macromol. Sci. Part A*, 57, 430–438, 2020.
101. Li, M., Huang, Y., Yu, T., Chen, S., Ju, A., Ge, M., Chemical recycling of waste poly (ethylene terephthalate) fibers into azo disperse dyestuffs. *RSC Adv.*, 87, 46476–46480, 2014.
102. Shamsi, R., Abdouss, M., Sadeghi, G.M.M., Taromi, F.A., Synthesis and characterization of novel polyurethanes based on aminolysis of poly (ethylene terephthalate) waste , and evaluation of their thermal and mechanical properties. *Polym. Int.*, 58, 1, 22–30, 2009.
103. Quartinello, F., Vajnhandl, S., Valh, J.V., Farmer, T.J., Vončina, B., Lobnik, A., Acero, E.H., Pellis, A., Guebitz, G.M., Synergistic chemo-enzymatic

hydrolysis of poly (ethylene terephthalate) from textile waste. *Microb. Biotechnol.*, 10, 6, 1376–1383, 2017.

104. Yuan, Z., Xu, Z., Zhang, D., Chen, W., Huang, Y., Zhang, T., Tian, D., Deng, H., Zhou, Y., Sun, Z., Mesoporous activated carbons synthesized by pyrolysis of waste polyester textiles mixed with Mg-containing compounds and their Cr (VI) adsorption. *Colloids Surf., A*, 549, 86–93, 2018.
105. Yuan, Z., Xu, Z., Zhang, D., Chen, W., Zhang, T., Huang, Y., Gu, L., Deng, H., Tian, D., Box-Behnken design approach towards optimization of activated carbon synthesized by co-pyrolysis of waste polyester textiles and MgCl2. *Appl. Surf. Sci.*, 427, 340–348, 2018.
106. Xu, Z., Tian, D., Sun, Z., Zhang, D., Zhou, Y., Chen, W., Deng, H., Highly porous activated carbon synthesized by pyrolysis of polyester fabric wastes with different iron salts: Pore development and adsorption behavior. *Colloids Surf., A*, 565, 180–187, 2019.
107. Li, Y., Peng, J., Liu, X., Song, D., Xu, W., Zhu, K., Dissolving waste viscose to spin cellulose fibers. *Polymer*, 237, 124349, 2021.
108. Prado, K.S., Gonzales, D., Spinacé, M.A.S., Recycling of viscose yarn waste through one-step extraction of nanocellulose. *Int. J. Biol. Macromol.*, 136, 729–737, 2019.
109. Lv, F., Yao, D., Wang, Y., Wang, C., Zhu, P., Hong, Y., Recycling of waste nylon 6/spandex blended fabrics by melt processing. *Compos. Part B*, 77, 232–237, 2015.
110. Spadea, S., Farina, I., Carrafiello, A., Fraternali, F., Recycled nylon fibers as cement mortar reinforcement. *Constr. Build. Mater.*, 80, 200–209, 2015.
111. Farooq, M.A., Fahad, M., Ali, B., Ullah, S., Ouni, M.H.E., Elhag, A.B., Influence of nylon fibers recycled from the scrap brushes on the properties of concerte: Valorization of plastic waste in concrete. *Case Stud. Constr. Mater.*, 16, e01089, 2022.
112. Dissanayake, D.G.K. and Weerasinghe, D.U., Fabric Waste Recycling: a Systematic Review of Methods, Applications, and Challenges. *Mater. Circ. Econ.*, 3, 24, 1–20, 2021.
113. Yu, K.H., Zhang, Y., Li, D., Montenegro-Marin, C.E., Kumar, P.M., Environmental planning based on reduce, reuse, recycle and recover using artificial intelligence. *Environ. Impact Assess. Rev.*, 86, 106492, 2021.
114. Damayanti, D., Wulandari, L.A., Bagaskoro, A., Rianjanu, A., Wu, H.S., Possibility Routes for Textile Recycling Technology. *Polymers*, 13, 21, 3834, 2021.
115. Denuwara, N., Maijala, J., Hakovirta, M., Sustainability Benefits of RFID Technology in the Apparel Industry. *Sustainability*, 11, 22, 6477, 2019.
116. Nayak, R., Singh, A., Padhye, R., Wang, L., RFID in textile and clothing manufacturing: technology and challenges. *Fash. Text.*, 2, 1, 9, 2015.
117. Zhong, R.Y., Xu, X., Klotz, E., Newman, S.T., Intelligent Manufacturing in the Context of Industry 4.0: A Review. *Engineering*, 3, 5, 616–630, 2017.

118. Wu, D., Greer, M.J., Rosen, D.W., Schaefer, D., Cloud manufacturing: Strategic vision and state-of-the-art. *J. Manuf. Syst.*, 32, 4, 564–579, 2013.
119. Giacobbe, M., Puliafito, C., Scarpa, M., The Big Bucket: An IoT Cloud Solution for Smart Waste Management in Smart Cities, in: *Advances in Service-Oriented and Cloud Computing*, ESOCC 2016, Communications in Computer and Information Science, vol. 707, Lazovik, A. and Schulte, S. (Eds.), p. 43–58, 2018.
120. Larsson, J.K.J., Digital innovation for sustainable apparel systems. *Res. J. Text. Appar.*, 22, 4, 370–389, 2018.
121. Lopes, S. and Machado, S., IoT based Automatic Waste segregator, in: *2019 International Conference on Advances in Computing, Communication and Control (ICAC3)*, p. 1–5, 2019.
122. Bhuiyan, M.A.R., Rahman, M.M., Shaid, A., Bashar, M.M., Khan, M.A., Scope of reusing and recycling the textile wastewater after treatment with gamma radiation. *J. Clean. Prod.*, 112, 3063–3071, 2016.
123. Jamdar, V., Kathalewar, M., Dubey, K.A., Sabnis, A., Recycling of PET wastes using Electron beam radiations and preparation of polyurethane coatings using recycled material. *Prog. Org. Coat.*, 107, 54–63, 2017.
124. Grdadolnik, M., Drinčić, A., Oreški, A., Onder, O.C., Utroša, P., Pahovnik, D., Žagar, E., Insight into Chemical Recycling of Flexible Polyurethane Foams by Acidolysis. *ACS Sustain. Chem. Eng.*, 10, 3, 1323–1332, 2022.
125. Coeck, R., De Bruyne, A., Borremans, T., Stuyck, W., De Vos, D.E., Ammonolytic Hydrogenation of Secondary Amides: An Efficient Method for the Recycling of Long-Chain Polyamides. *ACS Sustain. Chem. Eng.*, 10, 9, 3048–3056, 2022.

4

Nanoparticle-Facilitated Treatment of Wastewater Containing Textile Dye

Anirban Roy[1] and Sampa Chakrabarti[2]*

[1]Department of Materials Science and Technology, Maulana Abul Kalam Azad University of Technology, Simhat, Haringhata, Nadia, West Bengal, India
[2]Department of Chemical Engineering, University of Calcutta, Kolkata, India

Abstract

Textile industry is a water-intensive industry where its wastewater contains different types of dye and metal salts. A few dyes are carcinogenic and recalcitrant. Sometimes, pollutants are mutually interfering and resist removal. Conventional wastewater treatment technologies generate either large quantity of sludge or spent filtering media, which may cause secondary pollution.

Nanoparticles and nanotechnology can be useful for treatment of textile wastewater. Nano-adsorbents can remove pollutants more effectively, whereas semiconductor nano-catalysts can oxidize refractory dyes into comparatively less-toxic products with the help of photocatalysis using sunlight or artificial light. Nanosized zerovalent iron or iron oxide particles may help oxidizing organic pollutants by Fenton-type reaction. Nanostructured membranes may also be helpful for such purification. However, separation of nanoparticles after treatment of wastewater is difficult and costly; hence, nanomaterials may be immobilized on supports. There are various methods for the synthesis of nanoparticles including sol-gel, sonochemical, microwave, biosynthesis, pulse-laser, and sputtering. Other than the specific surface area, the characteristics related to the performance of nanoparticles as adsorbent or photocatalysts are surface morphology, zeta-potential, band-gap, and x-ray crystallography.

This chapter reviews various techniques for the treatment of textile wastewater involving nanoparticles. Synthesis, characterization, and experimental details are also compiled. Performances of the mentioned processes, current standing, and future prospects of nanoparticle-driven technologies for the treatment of textile

**Corresponding author*: scchemengg@caluniv.ac.in

Prashansa Sharma and Shilpi Shree Sahay (eds.) Nanotechnology-Assisted Recycling of Textile Waste: Sustainable Tools for Textiles, (97–128) © 2025 Scrivener Publishing LLC

wastewater are discussed. Ecotoxicological aspects and treatment of real wastewater are also discussed.

Keywords: Nanoparticles, characterization, textile effluent, dyes, metal salts, adsorption, photocatalysis, ecotoxicity

Abbreviations

AOP	Advanced oxidation processes
BET	Braunner, Emmet, Teller
BOD	Biological oxygen demand/biochemical oxygen demand
CNT	Carbon nanotubes
CuO	Cupric oxide
COD	Chemical oxygen demand
EDX	Energy-dispersive x-ray
FESEM	Field emission scanning electron microscopy
FTIR	Fourier transformed infrared spectroscopy
GCS	Graphene-coated silica
GO	Graphene oxide
GSH	Glutathione
MBR	Membrane bioreactor
MFC	Microbial fuel cell
MW-CNT	Multi-walled carbon nanotubes
NF	Nanofiber
NM	Nanomaterials
NP	Nanoparticles
NS	Nanosheets
NW	Nanowires
nZVI	Nano zerovalent iron
PU	Polyurethane
PVC	Polyvinyl chloride
UV	Ultraviolet
TEM	Transmission electron microscopy
TBARS	Thiobarbituric acid reactive substance
SW-CNT	Single-walled carbon nanotubes
rGO	Reduced graphene oxide
XPS	X-ray photoelectron spectroscopy

4.1 Introduction

Every day, due to rapid increase in population and industrialization, water is getting scarce. To meet the ever increasing demand of water, there is no other way than to use treated industrial wastewater. Textile is a water-intensive industry requiring 100–200 liters of water per kg of fabric processed per day, on an average. Textile wastewater contributes about 20% of the total industrial wastewater load [1]. In India average rate of generation of textile wastewater is 11.4 million liters per day [2].

Textile industries discharge a complex mixture of chemicals with low biodegradability and high salinity as its wastewater [3]. The various steps of textile sizing, dyeing and finishing process *vis-a-vis* the waste generated from them are given in Figure 4.1. In the final effluent, pollutants from all the stages are accumulated.

Treatment technologies are largely dependent on the composition and characteristics of the wastewater. Yaseen and Scholz (2019) made an extensive review on the characteristics of textile wastewater [1]. Based on their report, the important characteristics and the major constituents with the average quantities are given in Table 4.1. However, the compositions vary with the type of fiber, speciality of processing, and the country of origin.

Besides the above, a few researchers have noted presence of heavy metals including chromium (Cr), cadmium (Cd), lead (Pb), zinc (Zn), copper (Cu), and arsenic (As) in textile wastewater depending upon the source of water and chemicals used. More than 70 kinds of hazardous and toxic pollutants may be present in the textile wastewater of which about 30 are

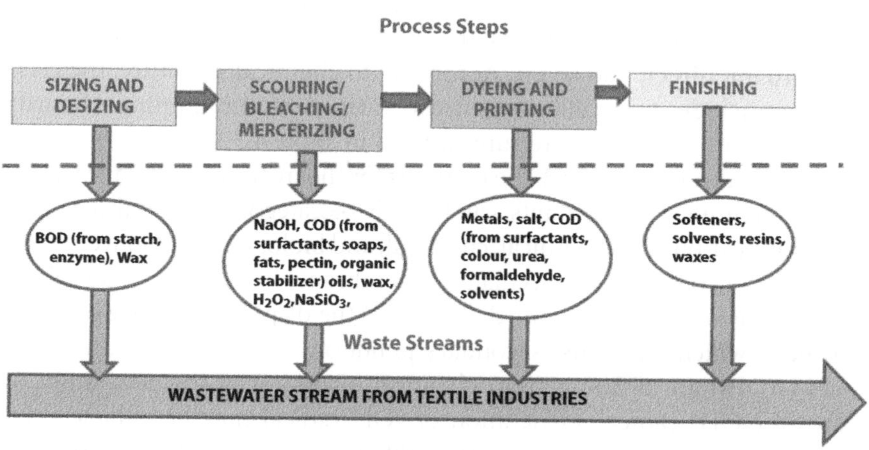

Figure 4.1 Steps of process as sources of pollutants in textile wastewater.

Table 4.1 Typical composition of textile wastewater.

Characteristics/constituent	Average value/range
Temperature, °C	35–45
pH	6–10
BOD, mg/L	100–6,000
COD, mg/L	150–20,000
Color, Pt-Co	50–2,500
TSS, mg/L	15–8,000
TDS, mg/L	1,500–8,000
Chloride, mg/L	200–6,000
Sulphate, mg/L	600–1,000
Sodium, mg/L	400–5,000
Total alkalinity, mg/L, as $CaCO_3$	500–800
Oil and grease, mg/L	10–50

non-biodegradable and cause severe health hazards. The various chemicals other than dye are used in several steps of textile finishing processes; some of them with their corresponding health hazards are given in Table 4.2 [4].

The major pollutant in the textile wastewater is dye; dyes can be classified according to various criteria like chemical structure, color, or applications [5, 6]. According to modes of applications, dyes can be classified as vat, mordant, direct, and so on. According to chemical structure, the classification includes types like azo, nitroso, nitro, or anthraquinine. Hazards associated with a few dyes are summarized in Table 4.3 [5, 7].

As evident from Figure 4.2, the methods for the treatment of textile wastewater, containing dye pollutants, can broadly be categorized into three main types: physical/physico-chemical, chemical, and biological [7]. Conventional methods like coagulation-flocculation, membrane, or adsorption are physical methods suffering from the drawback of the disposal of sludge or spent adsorbent, which may cause secondary pollution. Advanced oxidation processes (AOPs) are different from chemical oxidation in the generation of hydroxyl (•OH) radical. Generation of •OH radical may be enhanced by external irradiations like UV-visible light or ultrasound. These processes

Table 4.2 Hazards of a few chemicals used in textile industries.

Chemical	Applied for	Health hazard
Chlorinated solvents liketrichloroethane and chlorobenzenes like hexachlorobenzene	Used in scouring of fabric. Chlorobenzenes are used as dye carriers or for the preparation of azo dyes.	Can cause depletion of ozone and trigger the allied issues; can influence function of liver, kidneys, or central nervous system. Chlorobenzenes are known endocrine disruptor and carcinogens. It may also affect liver and thyroid.
Phenols and its derivatives like alkyl phenols, nonylphenol and ethoxylates	Used as cleaning detergents, emulsifiers or dye-dispersing agents; also used for scouring of wool and to stabilize polymeric fibers.	Dangerous for aquatic life, carcinogenic, endocrine-disruptor, and neurotoxic to humans.
Heavy metals like arsenic, antimony, lead, mercury, nickel and chromium	Used as mordants in dyeing of textiles to stabilize the colors; also, in production of dyes and pigments.	Causes environmental pollution; may be genotoxic, carcinogens and mutagens; may cause reproductive disorders, damage of DNA, and blood cells; can influence function of kidney, and liver.
Formaldehyde	Used in textile finishing, as softener; to make water-resistant, crease-resistant and shrink-proof.	Human carcinogen, causes allergies and dermatitis, and respiratory problems.
Volatile organic compound (VOC) s like methyl ethyl ketone, toluene, xylene, dichloromethane, and 1,1,1-trichloroethane	Used for the preparation of solvent-based colorants for dyeing and printing.	On short exposure, may cause nausea, dizziness, drowsiness, headaches, tremors, and weakness or skin/eye irritation; may affect function of kidney and liver damage on exposure at longer duration.

Table 4.3 Hazard/toxicity of a few common organic dyes.

Dye	Textile substrate	Hazard
Methyl orange and Sunset yellow (Azo)	Wool and silk	Carcinogenic
Methylene Blue (Thiazine) and Rhodamine 6G (Xanthene)	Polyester	Carcinogenic and mutagenic
Disperse orange 33 and Disperse red (Azo)	Nylon and acrylic fiber	Allergenic and carcinogenic
Congo Red and Direct Red 23 (Azo)	Cotton	Carcinogenic
Reactive Red (Azo)	Nylon, cotton, and wool	Dermatitis and allergic conjunctivitis

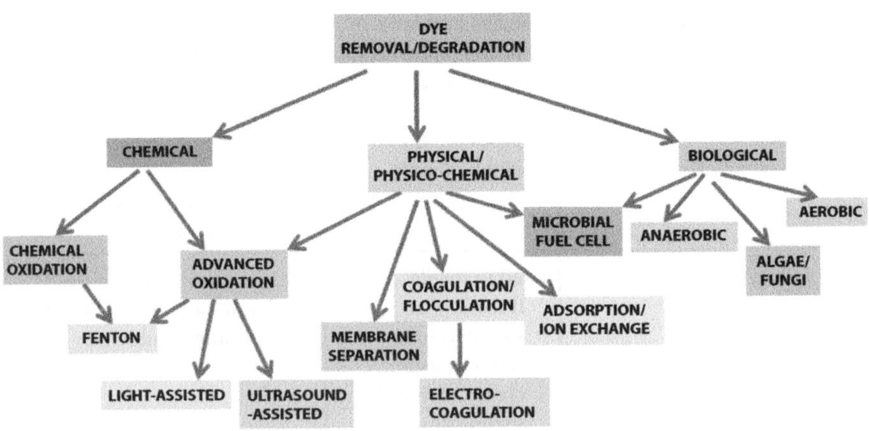

Figure 4.2 Various techniques for degradation dye-polluted textile wastewater.

may be homogeneous like photo-or sono-Fenton processes or heterogeneous like photocatalytic processes. Hence, these processes may be considered as a hybrid process of physical and chemical processes. Similarly, microbial fuel cell (MFC) method may be considered a hybrid of physical and biological processes.

Among the various techniques, there are a few where nanomaterials can take part and enhance their efficacies. Adsorption and catalysis are heterogeneous surface phenomena; the more is the active surface, the more will be the efficiency of the processes. Nanoparticles can contribute in these areas. Nano-membrane is another area where nanotechnology can be effective.

4.2 Aim, Objectives, and Scope

This chapter aims at compilation of various technologies of treating textile wastewater containing dye pollutants where nanotechnology can play an important role. It has been observed that the efficiency of the adsorption, catalysis, bioremediation, or membrane separation processes involving nanoparticles increases by several orders of magnitude from the same that use their bulk counterparts. Therefore, literatures on the nano-adsorption, nano-bio-remediation, nano-catalysis, and nano-membrane technology, applied for treatment of textile wastewater, are reviewed in this chapter. As it is known that nanoparticles have some inherent problems like toxicity, ecotoxicological impact of using nanomaterial for water treatment is also discussed. Because reports with the real industrial wastewater are scarce, majority of the reviews are based on the studies with simulated wastewater. Very few reports are available on application of nanotechnology for the treatment of actual wastewater from textile industries in laboratory scale; some of them are reviewed. However, full-scale continuous treatment with nanotechnology is rarely available and not covered in this chapter.

4.3 Nanoparticle-Based Adsorption Processes for Treatment of Textile Wastewater

4.3.1 Basics of Adsorption

Adsorption is a physico-chemical surface phenomenon involving interaction between the adsorbate and the adsorbent. There are several advantages of the process including ease of operation, ability to remove non-biodegradable contaminants, economic regeneration of the adsorbent, and relatively small quantity of spend adsorbents compared to sludges in the chemical process. Different factors are affecting the adsorption process of dye including pH, initial concentration of the dye solution, quantity of adsorbent dosed, interactive forces between the adsorbent and adsorbate, temperature, and also presence of other materials [8]. Different isotherm equations and kinetic models have been proposed and validated with experimental data to understand the mechanistic pathway of the adsorption process.

Based on the adsorbate–adsorbent interactions, the process of adsorption can be classified into two types: chemisorption (chemical adsorption) and physisorption (physical adsorption). The chemisorption or chemical adsorption is usually accomplished by some kind of chemical interaction of the adsorbate with the adsorbent. Physisorption or physical adsorption becomes effective by different types of interactive forces like intermolecular van der Waals force, hydrogen bonding, or dipole–dipole interaction.

After Gupta *et al.* (2016), we may summarize that nanoparticles as adsorbents should possess the following criterion to make it as an efficient and ideal adsorbent material [9].

- They should be non-toxic.
- The adsorbent should have high sorption capacity, high affinity, and good selectivity for removing contaminants of low concentration.
- They should be easily reactivated or regenerated.

Active carbon is the most popular adsorbent both in the laboratory and commercial scales, but the size was generally not below micrometre size before. Moreover, commercial activated carbon is costly. Nanomaterial-based adsorbents can be broadly classified into carbon-based adsorbents (including those derived from low-cost materials and agricultural wastes), metal and metal oxide nano-adsorbents, natural adsorbents, and nano-composite adsorbents [10].

4.3.2 Nanoparticles as Adsorbents

Many researchers have worked on the preparation, characterization, and performances of nanosized activated carbon adsorbents derived from waste materials like shells of different nuts, peels of orange, banana or potato, husks of rice or millet, stalks of corn or cotton, straws of wheat or rice, and cobs of corn [10]. A few such reports of using straw [11], oil palm fiber [12], sunflower seed hull [13], brasiliensis seed coat [14], coir pith [15], palm kernel shell [16–18], and date-pit [19] are also available. Crushing, heating, pyrolysis, and solvent exaction are among the physical activation processes, whereas acid or base treatment is among the chemical activation processes. Because adsorption is a surface phenomenon, adsorbents are characterized by their specific surface area, pore size and its distribution, surface morphology, particle size distribution, and zeta potential.

Carbon nanotubes (CNT), fullerenes, and graphenes are different forms of nanosized carbon. Carbon nanotubes may be single-walled (SW-CNT) or multi-walled (MW-CNT). Ultrasonic irradiation, vapor deposition, and electric arc method are some of the techniques of synthesizing CNTs. The exceptionally high sorption capacity of CNTs is due to controlled pore size distribution and high specific surface-active site. Wu *et al.* (2007) worked with multi-walled (CNT) for adsorption of Procion Red MX-5B dye. They found that the adsorption was endothermic and followed Langmuir isotherm equation [20]. Time-adsorption data were modeled by pseudo-second order equation. At 321 K, maximum 30 mg/g adsorption capacity was achieved within 3 h [21]. Multi-walled CNTs (MW-CNT) impregnated with chitosan were used for adsorption of Congo Red dye by Chatterjee *et al.* (2010) [22]. Equilibrium data fitted well into the Langmuir isotherm equation; maximum adsorption capacity obtained was 450 mg/g. Experimental data for the kinetic study fitted well with the pseudo-second order kinetic equation. Single-walled CNTs were also used to remove Reactive dye 120 from textile wastewater by Bazrafshan *et al.* (2013) [23]. They found that the adsorption capacity was dependent on pH and adsorbent dosage.

However, CNTs are costly and less preferred for bulk water treatment. Graphene, a two-dimentional array of carbon, along with its oxides, is a substitute for CNT, and their use for water treatment is now being explored. Due to presence of oxygen atoms on the surface, graphene oxide is a good adsorbent that can bind pollutants, especially metal ions. Modified Hummer's method is generally used for synthesizing graphene-oxide from graphite and then reduction can be done by different methods. Graphene-based materials are excellent support for other adsorbents like metal oxides [21].

Zero-valent metals like iron, copper, or aluminium have large specific surface area and high surface reactivity; therefore, they are good adsorbents. Marcelo *et al.* (2018) used zerovalent copper to remove Reactive Blue 4 dye [24]. Methyl Orange dye was removed by Ebrahiminezhad *et al.* (2018) using zerovalent iron of 19-nm size prepared by a green route [25].

Nanosized oxides of various metals like iron, magnesium, aluminium, titanium, or zinc are utilized as adsorbents for dye and metal-ion pollutants present in wastewater. These oxides may be supported or bare. TiO_2 nanocomposite impregnated with functionalized CNTs selectively and efficiently adsorbed anionic dyes; it adsorbed 94% and 100% of the initial concentration of Methyl Orange dye in 10 min and 30 min, respectively. Titania nanotubes, functionalized with a surfactant, hexadecyl-trimethylammonium chloride, accomplished adsorption of two anionic dyes:

Acid Red 1 and Acid Blue 9 [26]. Doping of TiO_2 nanoparticles with different transition metals like Cu, Fe, or Zn can enhance the adsorption capacity to remove different dyes. Patra *et al.* (2022) studied the adsorption removal of Reactive Blue 21 (RB-21) dye through manganese oxide nanoparticles prepared by sol-gel method using Acacia Concinna fruit extract. The removal efficiency was found to be 98% at pH 3.0 [27].

Wang *et al.* (2012) synthesized zinc oxide–reduced graphene oxide (ZnO-rGO) nanocomposite with novel structural features and explored its adsorption efficiency for Rhodamine B dye [28]. Another graphene-based nanocomposite, functionalized by $CoFe_2O_4$, was synthesized by hydrothermal method and appeared to be an effective adsorbent to remove Methyl Orange dye from wastewater [29]. Several organophosphorus pesticides were also removed efficiently from polluted agricultural wastewater using graphene-coated silica (GCS) nanocomposite [30].

Separation of spent absorbents from treated wastewater is a challenge especially for nano-adsorbents. Magnetic nanoparticles of metals and metal oxides may be used as a remedy. The spent magnetic adsorbents may easily be separated by using external magnetic force; in addition, iron is biocompatible; hence, treated wastewater may be less toxic. Aragaw *et al.* (2021) reviewed synthesis and application of iron-based adsorbents [31]. Application of magnetite (Fe_3O_4), maghemite (γ-Fe_2O_3), hematite (α-Fe_2O_3), iron oxyhydroxide (FeOOH), and metallic zerovalent iron were reported. Absalan *et al.* (2011) modified Fe_3O_4 nanoparticles with imidazolium-based ionic liquids and utilized them for removing Reactive Red 120 dye from water [32]. Grafting of magnetic iron oxide nanoparticles by β-cyclodextrin improved its adsorption capacity for adsorbing azo dyes like Direct blue 15, Evans blue, and Chicago sky blue from wastewater, compared to the ungrafted one [33].

4.4 Nano-Photocatalysts and Their Application for Treatment of Textile Wastewater

4.4.1 Basics of Photocatalysis

While adsorption transfers pollutants from one phase to another, photocatalysis degrades the pollutants into less toxic forms. In this process, the pollutants are degraded with the help of a catalytic reaction enhanced by light. When light, having energy more than the bandgap energy of a semiconductor, falls on the surface of the same, pair(s) of hole and electron is generated. These hole-electron pairs subsequently generate hydroxyl (•OH) and superoxide ($\bullet O_2^-$) radicals that can oxidize most of the organic and microbial pollutants. Photogenerated electrons may be used for reduction of the reducible pollutants. Figure 4.3 indicates the scheme of photocatalytic oxidation and reduction. Some of the photocatalysts, most commonly used for the degradation of dyes, detergents, pesticides, and volatile organic pollutants present in wastewater are TiO_2, ZnO, Fe_2O_3, ZnS, CdS, SnO_2, ZrO_2, WO_3, SnO_2, and CdS.

Because photocatalysis is a surface phenomenon and nanoparticles have a large specific surface area, nano-photocatalysts are more efficient than their bulk counterpart. Moreover, nano-photocatalysts can be engineered as required; for example, bandgap can be decreased by doping, shape can be varied, and so on. There are many techniques of synthesizing doped and undoped semiconductor photocatalysts: sol-gel, ultrasonication, evaporation, mechanical grinding, microwave-irradiation, and so on. Performances of the photocatalysts are dependent on the morphology and crystal structure along with several experimental conditions. Major characterization includes surface morphology by FESEM/TEM, specific surface area by BET method, crystal details by x-ray crystallography, and bandgap analysis by UV-spectroscopy. Composition of pre-cursor, nature of solvent, and other experimental conditions determine the characteristics of the nanoparticles. Banerjee *et al.* (2012) sonochemically synthesized ZnO nanoparticles using different solvents to get particles of different morphologies [34]. They used these nanoparticles to oxidize Trypan Blue dye [35] and to reduce hexavalent chromium [34, 35] under sunlight.

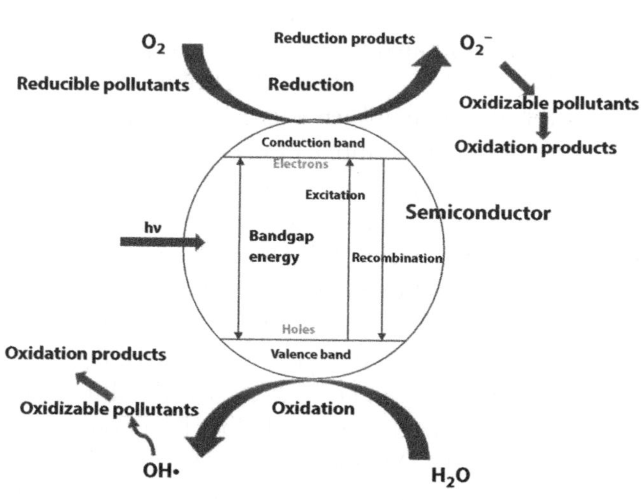

Figure 4.3 Schematics of photocatalytic process.

Light sources used for photo-excitation may be ultraviolet or visible. Use of sunlight makes the process less energy-consuming.

Among many semiconductors, anatase-Titanium dioxide or Titania (TiO_2) is the most popular due to its corrosion resistance and 3.2-eV bandgap. TiO_2 photocatalyst has been used widely under UV light to degrade a number of organic dyes such as Orange G, Malachite Green, Rhodamine 6G, Alizarin Green, Amido Black 10B, Safranine O'azure II, Indigo-carmine, Solvent red 23, and Sudan III [36]. Chen *et al.* (2007) used TiO_2 nanoparticles for degrading 99.9% of Malachite Green dye under UV radiation [37]. TiO_2 is generally more photoactive under UV radiation than under solar radiation; it absorbs a small fraction (4%) of the solar spectrum [38]. To enhance its photocatalytic activity in the visible light zone, researchers attempted functionalization and/or doping the TiO_2-based nanostructures. Temam *et al.* (2022) in their work prepared Al- and Ni-doped TiO_2 thin films via sol-gel dip-coating method. The photocatalytic efficiency of these thin films was then tested for the photo-degradation of Methylene Blue dye under sunlight and about 93% degradation of the dye was achieved [39]. TiO_2 nanocrystals, doped with nitrogen, were prepared and used by Liu *et al.* (2005) for photodegradation of azo dyes (acid orange 7, Procion red MX-5B, and reactive black 5) both under UV light and sunlight. N-doped TiO_2 performed better as a photocatalyst than the commercially available TiO_2 catalyst under both the light sources. About 95% and 70% of acid orange 7 dye was degraded under UV and sunlight within 1 h and 3 h, respectively [40].

Pure or doped zinc oxide (ZnO) nanomaterials are nowadays considered as an alternative to titania-based nano-photocatalysts for the decontamination of organic pollutants in wastewater [41]. Among most of the semiconductor photocatalysts studied, ZnO performed better under sunlight as it has the ability to absorb a larger fraction of solar spectrum. Mydeen *et al.* (2019) adopted the application of nanosized ZnO for removal of Methylene Blue with sunlight and artificial UV-visible light. The result shows 85% degradation of the dye in UV and 95% for sunlight in 460 min and 360 min period, respectively [42]. Dilaeyenna *et al.* (2018) reported photocatalytic degradation of dye-containing industrial wastewater using nanosized ZnO prepared by precipitation method in presence of polyvinyl pyrrolidone as capping agent. The degradation rate of dye was 90.61% at pH 7 within 30-min time [43]. In another study, Alhadhrami *et al.* (2018) applied microwave-synthesized nanosized-ZnO for photo-catalytic degradation of Methyl Red (MR), Methyl Orange (MO), and Methylene Blue (MB). The results showed that the synthesized nano-ZnO could degrade more than 93% of the mentioned dyes [44]. The specific surface area and

Figure 4.4 FESEM image of iron-doped ZnO nanoparticle.

crystal defects influenced the photocatalytic activities of metal oxides and that has been enhanced by doping with metals, transition metals, and lanthanides [45].

Fe-doped ZnO has been prepared sonochemically [46] by Roy *et al.* (2016), and their photocatalytic activity has been checked for degradation of composite film made of PVC under both UV and sunlight [47]. Figure 4.4 shows a FESEM image of Fe-doped ZnO nanoparticle. The average crystal size and bandgap of the nanoparticles calculated was 33.67 nm and 2.69 eV, respectively. Under sunlight, 12% of the initial weight of the PVC-nanocomposite film was degraded in 210 min, whereas under UV light the degradation was 10%.

4.4.2 Oxide Nanoparticles as Photocatalysts

Spent nano-adsorbents or nano-photocatalysts are difficult as well as costly to remove from the treated wastewater. To address this issue, immobilization of the adsorbents or photocatalysts on a suitable support has been tried. Activated carbon- and carbon-based compounds, fiberglass, fiber optic cables, glass-beads, quartz sand, clay, zeolites, polymers silica gel, natural fibers, and stainless steel are some of the matrices used for immobilizing nanoparticles [48]. In another study, Roy *et al.* (2022) described simultaneous photocatalytic degradation of PVC-nano-Fe-ZnO composite film and Methylene Blue dye in water. Within 3 h and 45 min, 28% of dye and 5% of the PVC-nanocomposite film were degraded under sunlight [49].

A few other semiconductor oxide nanoparticles are also used for photocatalytic treatment of dye pollutants than the main players, TiO_2 and ZnO. Oxides of copper oxide is generally obtained in two forms: cupric oxide (CuO) and cuprous oxide (Cu_2O), and they are used as photocatalyst. CuO is found to be more efficient than Cu_2O as photocatalyst. Ramesh (2021) prepared CuO as a good photocatalyst for decoloration of wastewater polluted with different azo dyes (Reactive Blue RB-5, Auramine Yellow AY-23, and Methylene Blue MB) [50]. It was prepared through a chemical precipitation route using copper (II) nitrate trihydrate precursor; the bandgap energy of CuO-NPs was determined as about 1.76 eV. The results showed that compared to MB and AY-23, RB5 decolorized more, which was around 67.8% at 5-h illumination. Dulta *et al.* (2022) prepared cupric oxide nanoparticles (CuO-NPs) following a green route of synthesis where Rhizome extract of Bergenia ciliata was used as a bio-functional reducing agent. As-prepared spherical-shaped CuO nanoparticles were then used as photocatalyst to degrade Methylene Blue and Methyl Red dye. About 85% degradation was achieved within 135 min [51].

An efficient and visible-light-active CuO/TiO_2-zeolite heterogeneous nanostructures, with p-n junction, was synthesized by Zhao *et al.* (2015). Nanoparticles prepared by standard impregnation method exhibited excellent photocatalytic activity by achieving 90% degradation of Methylene Blue dye in 1-h exposure to visible light. Under identical experimental conditions, zeolite and its composites with TiO_2 and CuO nanoparticles separately removed 79.1%, 84.1%, and 85.2% of Methylene Blue dye from wastewater, respectively [52]. SnO_2 is an n-type semiconductor having a comparatively larger bandgap energy of ~3.6 eV and applied for various photocatalytic applications [53]. SnO_2-MgO composite nanoparticle prepared via sol-gel method showed enhanced photocatalytic degradation ability toward Methylene Blue dye degradation than pure SnO_2. MgO functionalized the surface of SnO_2 that could eliminate the recombination of photo-generated electron-hole pairs; this resulted in better photocatalytic efficiency. SnO_2, functionalized by chitosan (CS-SnO_2), displayed good photocatalytic efficacy under UV light activation for the degradation of Methyl Orange and Rhodamine B dye. The adsorption and subsequent photocatalysis was enhanced by the electrostatic interaction between the (CS-SnO_2) nano-composite and Methyl Orange dye molecules. On the contrary, the repulsion between the same composite nanoparticles and Rhodamine B dye molecules led to comparatively lower photodegradation [54]. Tungsten oxide (WO_3) was proved as a reasonably good photocatalyst for its high absorption in the visible light range (up to 480 nm), its tuneable bandgap, and good optical properties. Surakashi *et al.* (2022) synthesized WO_3 by hydrothermal process. The nanoparticles were characterized by XRD, UV-spectroscopy, FESEM-EDX, FTIR, and XPS. Photocatalytic activity was examined by Methylene Blue dye under visible light from a Xenon lamp; within 150 min, about 90% of the dye was degraded [55].

Bismuth oxide (Bi_2O_3), a p-type semiconductor has also attracted the attention of researchers nowadays for degrading organic dyes due to its non-toxic nature and having a bandgap of 2.8 eV, which is comparatively lower than that of TiO_2 or ZnO [56]. For this low bandgap energy and the consequent fast recombination of the photogenerated hole-electron pairs, pure Bi_2O_3 has a low photocatalytic efficiency under visible light. This problem could be overcome by doping with a transition metal or its oxide which increases the bandgap as well as its capacity to degrade organic pollutants by of photocatalysis [57]. Matsumura *et al.* (2015) studied the photocatalytic effectivity of Ce and Nd-doped Bi_2O_3 nanorods for degrading organic dyes like Coomassie brilliant blue G250, acid yellow 29, and Acid green 25. The doped atoms scavenged the photo-excited electrons from the conduction band of Bi_2O_3, thereby prevented the hole-electron recombination, which, in turn, enhanced its photocatalytic efficiency [58].

4.4.3 Carbon-Based Nano-Photocatalysts

Graphene, graphene oxide (GO), and reduced graphene oxide (RGO) are nowadays being widely explored as promising supports for various graphene-based nanocomposites. Graphene-based matrices can help stopping the recombination of the photo-excited hole-electron pairs in the electron bands of the metal oxides, and, consequently, the photocatalytic activity of the graphene-nanocomposites is enhanced [59]. However, in such cases, determination of the optimum quantity of graphene-based material to be added is critical for the desired photocatalytic activity [60].

Phan *et al.* (2011) studied the role of the quantity of GO used in the GO/TiO_2 composite for the degradation of Methylene Blue by photocatalysis. The amount of GO was increased from 1% to 10%, and a red shift in the higher wavelength region and a stronger absorption intensity in the visible region were observed after incorporation of GO with TiO_2. The shift of the absorption edge indicates the formation of Ti-O-C chemical bonds, which helps in charge separation and reduction of the bandgap energy. Consequently, the photocatalytic properties increased under both UV and visible light irradiation. The maximum photocatalytic activity was obtained with 10% GO [61].

4.4.4 Other Nano-Photocatalysts

Homogeneous Fenton reaction involving Fe^{2+} solution and hydrogen peroxide (H_2O_2) is an efficient catalytic process for degrading dyes. It suffers from disadvantages like formation of sludge, requirement of acidic pH and large chemical consumption. To eliminate such problems, heterogeneous Fenton-like reaction involving nano zerovalent iron (nZVI) has been introduced. nZVI have specific characteristics that enhance the catalytic activities. It typically has a core-shell structure with the core of iron and shell of the iron oxides. Chemical reduction, sol-gel method, and ball milling are the common methods of synthesizing nZVI. Recently, a trend of using natural products or plant extracts for synthesis of nZVI is observed [62]. nZVI synthesized using green tea extract, along with H_2O_2, was used for degradation of Bromothymol Blue dye by heterogeneous Fenton process. About 85% degradation was achieved within 15 min at pH 6 [63]. In another report, nZVI, prepared through the same route, could degrade 80% of Malachite Green dye within a few minutes [64]. nZVI is relatively unstable and prone to aggregate in water; to avoid this, it is immobilized onto a suitable support like zeolite or nanostructured silica [65, 66]. Liu *et al.* (2014) synthesized kaolin-supported nZVI and used it for degradation of Direct Black G dye. More than 87% dye was removed within a small period at pH 7 [67].

4.5 Membrane Processes Involving Nanotechnology for Treatment of Textile Wastewater

For removing the undesired pollutants from the textile wastewater, a membrane acts as a barrier depending on their sizes and affinity [68]. Membrane processes like microfiltration, ultrafiltration and reverse osmosis can be used to eliminate color, BOD, and COD from textile wastewater. These processes may need application of high differential pressure. The relative sizes of the pores in the three types of processes are shown in Figure 4.5. In some reports, it was shown that a combination of more than one of the techniques may be more effective [69].

The structure of membrane is the primary consideration for its efficacy. The membrane technology is advantageous due to smaller environmental footprint, less chemical requirement, modular installation, and easy operation. The challenge is the selectivity and fouling of the membrane as well as requirement of high pressure. Nanotechnology is generally used to increase selectivity of the membrane and to reduce fouling. Incorporation

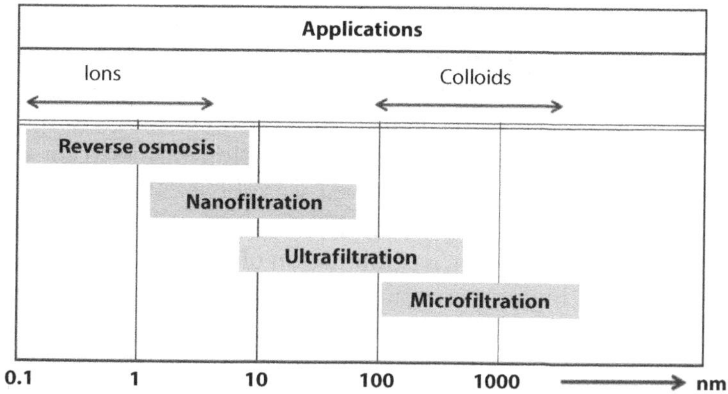

Figure 4.5 Pore sizes and application of several nano-membrane processes.

of engineered and functional nanomaterials in the form of nanocomposites or nano-coating can impart hydrophilicity, selectivity, efficiency and anti-fouling nature. Nanocomposite membranes are divided into two types: mixed matrix and thin-film. In case of mixed matrix type, nanoparticles are dispersed before the membrane was cast, whereas the thin-film membranes have nanoparticles embedded or coated on a polymer membrane [70]. The term nano-filter or nano-filtration was first proposed by Peter Ericsson in 1985 to describe a filtration that can reject particles of 1–20 nm. At that time, the focus was on fabricating finer membranes rather than enhancement of the properties. Afterward, various nanoparticles like metals and metal oxides nanoparticles were used for membrane fabrication with polymer matrix. Engineered nanoparticles like TiO_2, silica, alumina, or zeolite can enhance mechanical stability, hydrophilicity, permeability, and fouling resistance when incorporated in the polymeric ultrafiltration membrane in addition to increasing the thermal and mechanical stability [68, 71].

Recently, in research scale studies, incorporation of CNTs and graphene of different morphologies in polymer or ceramic matrices are reported. Silver (Ag) nanoparticles are known for their antimicrobial activities. These particles are impregnated in the membrane matrix for disinfection as well [72]. CNTs and nano-Ag can reduce membrane biofouling by inhibiting microbial attachment and biofilm formation on the membranes [73]. Composite nanomaterials comprising reduced graphene oxide (rGO) and titanium dioxide (TiO_2) could remove more than 97% of organic dyes like Rhodamine B, Methyl Orange, Methylene Blue, and Congo Red at a flow rate of 9.36 $Lm^{-2}\ h^{-1}\ bar^{-1}$.

Sahinkiya *et al.* (2018) employed reverse osmosis to treat textile wastewater and achieved 94% decolorization [74]. Jager *et al.* (2018) showed that nanofiltration is more cost-effective than reverse osmosis, because it needs lower pressure [75]. There are reports on removal of different dyes from synthetic wastewater [76, 77]. When ultrafiltration was used for removal of Reactive dye 5, the rejection observed was above 70% with initial dye concentration of 500 mg/L, showing that ultrafiltration was not suitable for dye removal [78]. However, a combination of UF and RO showed a better result regarding higher permeate flux, higher retention of multivalent salts along with relatively lower capital investment, and cost of operation and maintenance [79]. Methylene blue, Victoria blue, and Methyl violet dyes could be effectively removed by nanofiltration membrane with cellulose nanocrystals. While most of the studies were batch type, a continuous flow experiment was conducted by Ruan *et al.* (2019). About 99% of the Reactive dye 5 was removed at a flow rate of 480 L/h, a pressure of 15 bar, and a temperature of 35°C [5].

4.6 Bioremediation Involving Nanomaterials

4.6.1 Microbial Biodegradation of Textile Wastewater

As evident from Figure 4.1 and Table 4.1, water-soluble unadsorbed free dyes, heavy metals, acids, alkalis, inorganic salts, and many organic compounds are present in textile wastewater, and, as shown in Figure 4.1, bioremediation is one of the methods to treat textile wastewater. However, some dyes are non-biodegradable and presence of inorganic salts and heavy metals in high concentrations generally inhibit growth of microbes. Hence, a combination of nanotechnology and biotechnology was found suitable and eco-friendly for degradation of the pollutants in textile wastewater. Nanoparticles are highly reactive with the pollutants, and they can adsorb the pollutants or enzymes required for biodegradation. They may also act as carrier for the enzymes or nutrients and help immobilizing the microbial cells. Plant extracts and microbial cultures may also facilitate sustainable synthesis of nanoparticles, which, in turn, may be used as adsorbents or catalysts for treatment of textile wastewater. Membrane bioreactor and MFCs are the other two areas of bioremediation of textile wastewater where nanoparticles may take part.

Because bioremediation involves pollutants, nanoparticles and living cells, the interaction of all the three players are critical for the efficacy of

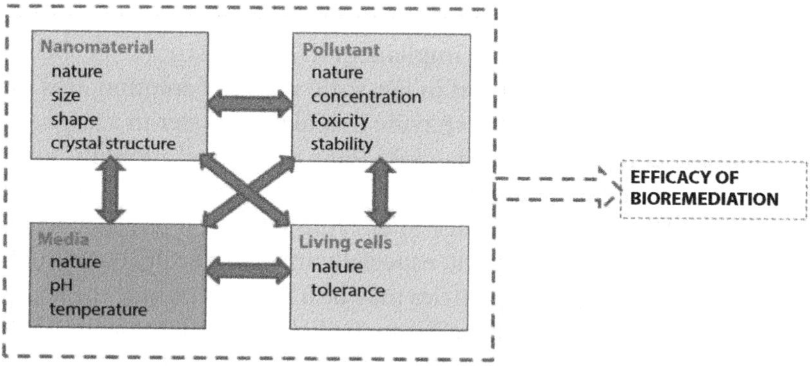

Figure 4.6 Different players and their interactions in bioremediation.

the process (Figure 4.6). For example, the temperature of pH suitable for the stability of the nanoparticle may not be suitable for survival of the living cells [80].

4.6.2 Role of Nanoparticles in Enzyme Facilitated Bioremediation

Enzyme-assisted remediation of dye pollutants is found effective with the help of nanoparticles. Yogalakshmi *et al.* (2020) reviewed efficacy of nanoparticles-supported enzymes for dye degradation. The generic enzymes used generally for the treatment of the pollutants present in the textile wastewater include peroxidase, cytochrome reductase (Fe III), and oxido-reductase. The enzyme Laccase immobilized on a composite nanoparticle comprising copper phosphate [$Cu_3(PO_4)_2$, abbreviated as CuP], GO, and CNTs could remove 100% of Neutral red and Crystal violet dye, whereas free enzyme could remove only 20%. Immobilization helped preserving the activity of the enzyme; stored under the identical conditions, free Laccase retained 30% of its activity; whereas immobilized enzyme retained 80% of it after 6 days [81]. Ginger peroxidase, immobilized on amino functionalized TiO_2 nanoparticles, showed 80% removal of acid yellow 42 dye in both acidic and basic medium at 60°C, but the free enzyme could remove only 10% of the dye [82]. Laccase enzyme retained its efficacy up to 10 cycles to remove 79% of Procion Red MX-5B dye when immobilized on Fe_3O_4-SiO_2 composite nanoparticles. Laccase, when immobilized on chitosan nanoparticles encapsulated glass beads, could withstand a temperature of 90°C during removal of Congo Red dye, whereas the free enzyme lost its activity [83].

Textile wastewater treatment with nanoparticle-immobilized enzymes has some associated problems to implement in an industrial scale. Because most of the studies are conducted in lab-scale with dye solution, it is difficult to envisage the behavior the enzyme in real wastewater in a large scale. After immobilization, enzyme may lose their diffusional mobility, and, at the same time, the conformational change may also take place. It is difficult to segregate such causes for change of enzyme activity, if any. Recovery of enzyme and its support is difficult, especially in a large scale. For eliminating this issue, magnetic nanoparticles may be a solution, where the support particles may be recovered using a magnetic field [84]. All these limitations should be taken care of before implementing enzyme technology at commercial scale.

Bacteria, fungi, and algae can help removing dye pollutants from wastewater in a sustainable way. Ihsanullah *et al.* (2020) presented an exhaustive review on this [85]. However, most of them do not use nanoparticles or nanotechnology and, hence, are not included in this chapter.

4.6.3 Synthesis of Nanoparticles by Biological Routes

Nanoparticles used for adsorption, catalysis, or membrane-impregnation may be synthesized in a green way by using plant extracts or via microbial route. This may also be considered as nano-bioremediation in an indirect sense. Noman *et al.* (2020) synthesized Cu-nanoparticles from E. sp. SINT7; the nanoparticles thus produced could remove azo dyes present in textile effluent. More than 80% of the Reactive Black, Congo Red, Direct Blue and Malachite green dyes were removed by the biogenic nanoparticles at low (25 mg/L) concentration; at higher concentration of about 100 mg/L, the degradation percentages were lower. When real industrial wastewater was treated, concentrations of suspended solid, phosphate, and chloride also reduced [86]. Al-Buriahi *et al.* (2022) published a review on ZnO nanoparticles as photocatalyst, synthesized using plant extract, for degradation of Rhodamine B dye under UV light. One hundred percent of the dye was removed after 210 min when watermelon rind extract was used for synthesis of ZnO [7].

4.6.4 Combination of Biological System with Other Systems

In membrane bioreactors (MBRs), wastewater is treated via a combination of biological and membrane systems for biodegradation and separation respectively. The MBR technology is about 30 years old and has the advantages of smaller ecological footprint, less sludge production, and lower energy demand in addition to high-quality treated water. MBR may also be combined with other systems like photocatalytic or adsorption for a better result. Fouling is a major limitation of membrane bioreactor technology. Nanomaterials may be used for remediation of the issue; the technology is nanomaterial–membrane bioreactor (NM-MBR). Based on the structural characteristics and applications, nanomaterial-based MBRs are classified as nanofiber (NF)–MBR, nanoparticle (NP)–MBR, nanotube (NT)–MBR, nanocrystal (NC)–MBR, nanowire (NW)–MBR, and nanosheet (NS-MBR) [87]. A composite membrane with NFs of polyerathane (PU) and graphene oxide (GO) was synthesized by Sundaran *et al.* (2019) that removed 95% methylene blue and 92% Rhodamine B [88]. An area of the potential use of NM-MBR is treatment of landfill leachates that contain organic and inorganic pollutants. Dabaghian*et al.* (2018) reviewed potentiality of leachate treatment by NM-MBR [89].

Microbial fuel cells (MFC) are recently becoming popular for wastewater treatment because simultaneous generation of energy is possible. The components are micro-organisms, electrodes and proton-selective membrane. Electrogenic bacteria like G. sulfur reduces or E. coli consume the substrate anerobically and generate proton and electron. Electrons flow through the external circuit showing current generation. On the cathode surface, a chemical reaction takes place with oxygen and proton [90, 91]. In the case of textile dye removal, proton travels to cathode via ion-permeable-membrane where reduction of dye takes place. Slow process and incomplete degradation results in small current—these are the major shortcomings of MFC. Nanotechnology can play important role to eliminate such shortcomings. Nanomaterials can accelerate biocatalytic activities of the micro-organisms by providing large specific surface area for biofilm production. Use of nanosized carbon or metal can enhance oxygen reduction. Metal oxides can increase air-cathode life. Modifications of electrodes can achieve better performance of electricity generation, wastewater treatment and bio-sensing. The efficiency of MFCs was improved by coating its electrodes with iron(II)-molybdate nanocomposites so that the maximum columbic efficiency of about 21.3% was achieved; the power density was about 106 mW/m^2 and about 80% of the initial COD could be removed [92].

4.7 Ecotoxicological Impact of Using Nanoparticles for Textile Wastewater Treatment

As discussed in the above sections, nanomaterials can enhance the efficiency of conventional treatment techniques of the textile wastewater.

However, scientists are yet to have the unconditional confidence in using nanomaterials for wastewater treatment, mainly due to the intrinsic toxicity of nanomaterials by virtue of its size and configuration. Due to the small size, nanomaterials can penetrate through the cell walls, can reach into various organs of living bodies, and may damage them [93, 94]. Nanomaterials may leach into treated water when used for water treatment. Size and shape-dependent toxicities in the end point are reported for silver and other metal nanoparticles and CNTs. Khan *et al.* (2017) reported increase in oxidative stress and genotoxicity in freshwater fish in water treated by Ag-nanoparticles [95]. Banerjee *et al.* (2014) used ZnO nanoparticles for photocatalytic degradation of wastewater comprising dye and hexavalent chromium. They have studied the toxicity of treated water possibly caused by nanoparticles using freshwater fish Channa Punctatus as biomarker. As indicated by lipid peroxidation (TBARS) and reduced glutathione (GSH), they observed toxicity in the fish body [35]. Hence, required precautions should be taken and nanoparticles should be removed from treated water before releasing the same to surface water bodies. However, separation of nanoparticles is a costly and tedious process.

4.8 Real-Life Textile Wastewater Treatment Using Nanoparticles

Although full-scale continuous treatment of actual textile wastewater using nanoparticles or nanotechnology is not easily available, a few researchers reported laboratory scale treatment of real-life textile wastewater with nanotechnology. Maroufi *et al.* (2021) reported treatment of actual textile wastewater by adsorption with graphene oxide [96]. The wastewater was from a carpet industry that contained residual dyes, benzene derivatives, aromatic amines, and long chain alkanes. Initial COD was 2,568 mg/L, BOD was 2,181 mg/L, and pH was 7.5. Before subjecting the wastewater to adsorption process, it was centrifuged. A 200-ml sample of this wastewater was subjected to batch adsorption process with graphene oxide particles of 153.6 m^2/g BET surface area and average pore diameter of 2.58 nm. Central composite design yielded optimum conditions of 125 mg/L adsorbent dosing at pH 8, whereby, after 25 min, 62% of the initial COD was removed. Amirian *et al.* (2018) used CuO nanoparticles (size less than 50 nm) as photocatalyst to treat actual textile wastewater having BOD of 96 mg/L and COD of 1,226 mg/L. They have treated raw wastewater by poly aluminium chloride (PACL) and reduced the BOD and COD to 66 mg/L and 532 mg/L, respectively [97]. Optimum conditions for about 83% color removal were pH 8 and CuO dosage of 0.05 g/L, whereas the same for about 99% COD removal were pH 6.8 and CuO dosage of 0.02 g/L; both at 60-min time interval and 60-W UV light. Jorfi *et al.* (2016) sequenced the treatment of actual wastewater with coagulation by $AlCl_3$ and UV light–assisted photocatalysis with MgO nanoparticles. Before working with the real-life wastewater, they optimized the process by using model dye AR 73. Initial COD of the actual wastewater was 5,400 mg/L, and, after coagulation, it came down to 2,250 mg/L. When this wastewater was subjected to the photocatalytic treatment, 96% of the COD was removed within 300 min. UV lamp was of 20W. Without the pre-treatment by coagulation, the COD removal was 49%, due to the turbidity through which UV light could not penetrate [98].

4.9 Conclusion

Textile dyeing and finishing is a water-intensive industry, and, in view of the current water scarcity, textile wastewater should be treated, recycled, and reused. This wastewater contains organic dyes, suspended solids, and dissolved salts of metals, which are often difficult to manage by conventional treatment techniques. Nanoparticles and nanotechnology may render help to enhance the efficiencies of the wastewater treatment processes like adsorption, membrane separation, and catalytic degradation by several orders of magnitude. Nanoparticles generally used for treatment of wastewater treatment are metals like iron, silver and copper as well as metal oxides like titanium oxide, zinc oxide, and iron oxide. These are used as such or as composites or in supported forms. Supports may be natural zeolites, activated carbon or graphene, and its derivatives. Bioremediation of textile wastewater may also be facilitated by nanotechnology; nanostructured materials play an important role in the enzyme-based treatment technologies. Most of the studies reviewed here are of laboratory scale and with simulated wastewater. Nanotechnology-assisted technologies are still rarely used in case of real life textile wastewater in a full scale at a factory-setting. This may be due to the lack of funds as well as lack awareness among the medium scale textile dyeing industries surviving with less trained workforce. Efficacy of nanoparticle-based techniques with actual industrial wastewater is yet to be explored completely. From a few laboratory scale studies with actual wastewater, it appears that a pre-treatment like coagulation-flocculation may be necessary before applying nanoparticle-enhanced treatment technologies to it, because raw industrial wastewater is generally turbid, which can inhibit lights for photocatalysis, or the suspended solids may clog the nanosized pores of nano-adsorbents.

Moreover, interaction of nanoparticles with living organisms is also not yet completely explored. Generally, nanoparticles are inherently toxic by virtue of its small size that can penetrate cell membranes; hence, precautions should be taken before wide use of nanoparticles and nanotechnology for water treatment.

References

1. Yaseen, D.A. and Scholz, M., Textile dye wastewater characteristics and constituents of synthetic effuents: a critical review. *Int. J. Environ. Sci. Technol.*, 16, 1193–1226, 2019.
2. https://indianinfrastructure.com/2020/02/01/wastewater-generation-and-treatment-present-status-in-india/ accessed in July 2022.
3. Bidu, J.M., Bruggen, B.V.d., Rwiza, M.J., Njau, K.N., Current status of textile wastewater management practices and effluent characteristics in Tanzania. *Water Sci. Technol.*, 83, 2363–2376, 2021.
4. Kishor, R., Purchase, D., Saratale, G.D., Saratale, R.G., Ferreira, L.F.R., Bilal, M., Chandra, R., Bharagava, R.N., Ecotoxicological and health concerns of persistent coloring pollutants of textile industry wastewater and treatment approaches for environmental safety. *J. Environ. Chem. Eng.*, 9, 105012, 2021.
5. Ruan, W., Hu, J., Qi, J., Hou, Y., Zhou, C., Wei, X., Removal of dyes from wastewater by nanomaterials: A review. *Adv. Mater. Lett.*, 10, 09–20, 2019.
6. Velusamy, S., Roy, A., Sundaram, S., Mallick, T.K., A review on heavy metal ions and containing dyes removal through graphene oxide-based adsorption strategies for textile wastewater treatment. *Chem. Rec.*, 21, 1570–1610, 2021.
7. Al-Buriahi, A.K., Al-Gheethi, A.A., Kumar, P.S., Mohamed, R.M.S.R., Yusof, H., Alshalif, A.F., Khalifa, N.A., Elimination of rhodamine B from textile wastewater using nanoparticle photocatalysts: A review for sustainable approaches. *Chemosphere*, 287, 132162, 2022.
8. Yagub, M.T., Sen, T.K., Afroze, S., Ang, H.M., Dye and its removal from aqueous solution by adsorption: a review. *Adv. Colloid Interface Sci.*, 209, 172–184, 2014.
9. Gupta, V.K., Moradi, O., Tyagi, I., Agarwal, S., Sadegh, H., Shahryari-Ghoshekandi, R., Makhlouf, A.S.H., Goodarzi, M., Garshasbi, A., Study on the removal of heavy metal ions from industry waste by carbon nanotubes: effect of the surface modification: a review. *Crit. Rev. Environ. Sci. Technol.*, 46, 93 –118, 2016.
10. Badawi, A.K., Elkodous, M.A., Ali, G.A.M., Recent advances in dye and metal ion removal using efficient adsorbents and novel nano-based materials: an overview. *RSC Adv.*, 11, 36528–36553, 2021.

11. Kannan, N. and Sundaram, M.M., Kinetics and mechanism of removal of methylene blue by adsorption on various carbons—a comparative study. *Dyes Pigm.*, 51, 25–40, 2001.
12. Tan, I.A.W., Ahmad, A.L., Hameed, B.H., Adsorption of basic dye using activated carbon prepared from oil palm shell: batch and fixed bed studies. *Desalination*, 225, 13–28, 2008.
13. Thinakaran, N., Panneerselvam, P., Baskaralingam, P., Elango, D., Sivanesan, S., Equilibrium and Kinetic Studies on the Removal of Acid Red 114 from Aqueous Solutions Using Activated Carbons Prepared from Seed Shells. *J. Hazard. Mater.*, 158, 142–150, 2008.
14. Hameed, B.H. and Daud, F.B.M., Adsorption studies of basic dye on activated carbon derived from agricultural waste: Hevea brasiliensis seed coat. *Chem. Eng. J.*, 139, 48–55, 2008.
15. Namasivayam, C. and Kavitha, D., Removal of congo red from water by adsorption onto activated carbon prepared from coir pith, an agricultural solid waste. *Dyes Pigm.*, 54, 47–58, 2002.
16. Habeeb, O.A., Ramesh, K.R., Ali, G.A.M., Yunus, R.b.M., Low-cost and eco-friendly activated carbon from modified palm kernel shell for hydrogen sulfide removal from wastewater: adsorption and kinetic studies. *Desalin. Water Treat.*, 84, 205–214, 2017.
17. Omar, A.H., Ramesh, K., Gomaa, A.M.A., Rosli, b.M.Y., Experimental design technique on removal of hydrogen sulfide using CaO-eggshells dispersed onto palm kernel shell activated carbon: Experiment, optimization, equilibrium and kinetic studies. *J. Wuhan Univ. Technol.-Mat. Sci. Edit.*, 32, 305–320, 2017.
18. Gomaa, A.M.A., Omar, A.H., Algarni, H., Chong, K.F., CaO impregnated highly porous honeycomb activated carbon from agriculture waste: symmetrical supercapacitor study. *J. Mater. Sci.*, 54, 683–692, 2019.
19. El-Naas, M.H., Al-Zuhair, S., Alhaija, M.A., Reduction of COD in refinery wastewater through adsorption on date-pit activated carbon. *J. Hazard. Mater.*, 173, 750–757, 2010.
20. Wu, C.-H., Adsorption of reactive dye onto carbon nanotubes: Equilibrium, kinetics and thermodynamics. *J. Hazard. Mater.*, 144, 93–100, 2007.
21. Madhura, L., Singh, S., Kanchi, S., Sabela, M., Bisetty, K., Inamuddin, Nanotechnologybased water quality management for wastewater treatment. *Environ. Chem. Lett.*, 17, 65–121, 2019.
22. Chatterjee, S., Lee, M.W., Woo, S.H., Adsorption of congo red bychitosan hydrogel beads impregnated with carbon nanotubes. *Biores. Technol.*, 101, 1800–1806, 2010.
23. Bazrafshan, E., Mostafapour, F.K., Hosseini, A.R., Khorshid, A.R., Mahvi, A.H., Decolorisation of reactive red 120 dye byusing single-walled carbon nanotubes in aqueous solutions. *J. Chem.*, Volume 2013, Article ID 938374, 8 pages, Hindawi Publishing Corporation, 2013.

24. Marcelo, C.R., Puiatti, G.A., Nascimento, M.A., Oliveira, A.F., Degradation of the Reactive Blue 4 dye in aqueous solution using zero-valent copper nanoparticles. *J. Nanomater.*, Volume 2018, Article ID 4642038, 9 pages, 2018.
25. Ebrahiminezhad, A., Taghizadeh, S., Ghasemi, Y., Berenjian, A., Green synthesized nanoclusters of ultra-small zero valent iron nanoparticles as a novel dye removing material. *Sci. Total Environ.*, 621, 1527–1532, 2018.
26. Lee, C.-K., Liu, S.-S., Juang, L.-C., Wang, C.-C., Lyu, M.-D., Hung, S.-H., Application of titanate nanotubes for dyes adsorptive removal from aqueous solution. *J. Hazard. Mater.*, 148, 756–760, 2007.
27. Patra, T., Mohanty, A., Singh, L., Muduli, S., Parhi, P.K., Sahoo, T.R., Effect of calcination temperature on morphology and phase transformation of MnO_2 nanoparticles: A step toward green synthesis for reactive dye adsorption. *Chemosphere*, 288, 132472, 2022.
28. Wang, J., Tsuzuki, T., Tang, B., Hou, X., Sun, L., Wang, X., Reduced graphene oxide/ZnO composite: Reusable adsorbent for pollutant management. *ACS Appl. Mater. Interfaces*, 4, 3084–3090, 2012.
29. Li, N., Zheng, M., Chang, X., Ji, G., Lu, H., Xue, L., Pan, L., Cao, J., Preparation of magnetic $CoFe_2O_4$-functionalized graphene sheets via a facile hydrothermal method and their adsorption properties. *J. Solid State Chem.*, 184, 953–958, 2011.
30. Liu, X., Zhang, H., M,a, Y., Wu, X., Meng, L., Guo, Y., Yu, G., Liu, Y., Graphene-coated silica as a highly efficient sorbent for residual organophosphorus pesticides in water. *J. Mater. Chem. A*, 1, 1875–1884, 2013.
31. Aragaw, T.A., Bogale, F.M., Aragaw, B.A., Iron-based nanoparticles in wastewater treatment: A review on synthesis methods, applications, andremoval mechanisms. *J. Saudi Chem. Soc.*, 25, 101280, 2021.
32. Absalan, G., Asadi, M., Kamran, S., Sheikhian, L., Goltz, D.M., Removal of reactive red-120 and 4-(2- pyridylazo) resorcinol from aqueous samples by Fe_3O_4 magnetic nanoparticles using ionic liquid as modifier. *J. Hazard. Mater.*, 192, 476–484, 2011.
33. Arslan, M., Sayin, S., Yilmaz, M., Removal of carcinogenic azo dyes from water by new cyclodextrin-immobilized iron oxide magnetic nanoparticles. *Water Air Soil Pollut.*, 224, 1527, 2013.
34. Banerjee, P., Chakrabarti, S., Maitra, S., Dutta, B.K., Zinc oxide nanoparticles – Sonochemical synthesis, characterization and application for photo-remediation of heavy metal. *Ultrason. Sonochem.*, 19, 85–93, 2012.
35. Banerjee, P., Das, D., Mitra, P., Sinha, M., Dey, S., Chakrabarti, S., Solar photocatalytic treatment of wastewater with zinc oxide nanoparticles and its ecotoxicological impact on Channa punctatus –a freshwater fish. *J. Mater. Environ. Sci.*, 5, 1206–1213, 2014.
36. Aarthi, T., Narahari, P., Madras, G., Photocatalytic degradation of Azure and Sudan dyes using nano TiO_2. *J. Hazard. Mater.*, 149, 725–734, 2007.

37. Chen, C.C., Lu, C.S., Chung, Y.C., Jan, J.L., UV light induced photodegradation of malachite green on TiO_2 nanoparticles. *J. Hazard. Mater.*, 141, 520–528, 2007.
38. Kim, H.-S., Lee, J.-W., Yantara, N., Boix, P.P., Kulkarni, S.A., Mhaisalkar, S., Grätzel§, M., Park, N.-G., High efficiency solid-state sensitized solar cell-based on submicrometer rutile TiO_2 nanorod and CH3NH3PbI3 perovskite sensitizer. *Nano Lett.*, 13, 2412–2417, 2013.
39. Temam, E.G., Djani, F., Rahmane, S., Temam, H.B., Gasmi, B., Photocatalytic activity of Al/Ni doped TiO_2 films synthesized by sol-gel method: Dependence on thickness and crystal growth of photocatalysts. *Surf. Interfaces*, 31, 102077, 2022.
40. Liu, Y., Chen, X., Li, J., Burda, C., Photocatalytic degradation of azo dyes by nitrogen-doped TiO_2nanocatalysts. *Chemosphere*, 61, 11–18, 2005.
41. Daneshvar, N., Salari, D., Khataee, A., Photocatalytic degradation of azo dye acid red 14 in water on ZnO as an alternative catalyst to TiO_2. *J. Photochem. Photobiol. A*, 162, 317–322, 2004.
42. Mydeen, S.S., Kottaisamy, M., Vasantha, V.S., Photocatalytic performances and antibacterial activities of nano-ZnO derived by cetrimide-based co-precipitation method by varying solvents. *Int. J. Innov. Technol. Explor. Eng.*, 9, 930–938, 2019.
43. Sidik, D.A.B., Hairom, N.H.H., Zainuri, N.Z., Desa, A.L., Misdan, N., Yusof, N., Ong, C.B., Mohammad, A.W., Aripen, N.S.M., Photocatalytic degradation of industrial dye wastewater using zinc oxide–polyvinylpyrrolidone nanoparticles. *Malays. J. Anal. Sci.*, 22, 693–701, 2018.
44. Alhadhrami, A., Almalki, A.S.A., Adam, A.M.A., Refat, M.S., Preparation of Semiconductor Zinc Oxide Nanoparticles as a Photocatalyst to Get Rid of Organic Dyes Existing Factories in Exchange for Reuse in Suitable Purpose. *Int. J. Electrochem. Sci.*, 13, 6503 – 6521, 2018.
45. Cojocaru, B., Avram, D., Kessler, V., Parvulescu, V., Seisenbaeva, G., Tiseanu, C., Nanoscale insights into doping behavior, particle size and surface effects in trivalent metal doped SnO2. *Sci. Rep.*, 7, 9598, 2017.
46. Roy, A., Maitra, S., Ghosh, S., Chakrabarti, S., Sonochemically synthesized iron-doped zinc oxide nanoparticles: Influence of precursor composition on characteristics. *Mater. Res. Bull.*, 74, 414–420, 2016.
47. Das, P., Roy, A., Chakrabarti, S., Photocatalytic degradation of the nanocomposite film comprising polyvinyl chloride (PVC) and sonochemically synthesized iron-doped zinc oxide: A comparative study of performances between sunlight and UV radiation. *J. Polym. Environ.*, 25, 1231–1241, 2017.
48. Chawla, H., Garg, S., Ingole, P.P., Chandra, A., Immobilization of photocatalytic material on the suitable substrate, in: *Green Photocatalytic Semiconductors-Green Chemistry and Sustainable Technology*, S. Garg and A. Chandra (Eds.), Springer, Cham, 2022, https://doi.org/10.1007/978-3-030-77371-7_15.

49. Roy, A., Chakrabarti, S., Maitra, S., Simultaneous solar photo-degradation of PVC-Fe-Doped ZnO-nanocomposite flakes and methylene blue dye in water. *J. Environ. Eng. Landsc. Manag.*, 30, 268–275, 2022.
50. Ramesh, M., CuO as efficient photo catalyst for photocatalytic decoloration of wastewater containing Azo dyes. *Water Pract. Technol.*, 16, 1078–1090, 2021.
51. Dulta, K., Ağçeli, G.K., Chauhan, P., Jasrotia, R., Chauhan, P.K., Ighalo, J.O., Multifunctional CuO nanoparticles with enhanced photocatalytic dye degradation and antibacterial activity. *Sustain. Environ. Res.*, 32, 2, 2022.
52. Zhao, L., Cui, T., Li, Y., Wang, B., Han, J., Han, L., Liu, Z., Efficient visible light photocatalytic activity of p–n junction CuO/TiO_2 loaded on natural zeolite. *RSC Adv.*, 5, 64495–64502, 2015.
53. Long, J., Xue, W., Xie, X., Gu, Q., Zhou, Y., Chi, Y., Chen, W., Ding, Z., Wang, X., Sn^{2+} dopant induced visible-light activity of SnO_2 nanoparticles for H_2 production. *Catal. Commun.*, 16, 215–219, 2011.
54. Gupta, V.K., Saravanan, R., Agarwal, S., Gracia, F., Khan, M.M., Qin, J., Mangalaraja, R.V., Degradation of azo dyes under different wavelengths of UV light with chitosan-SnO_2 nanocomposites. *J. Mol. Liq.*, 232, 423–430, 2017.
55. Surakasi, R., Alimuddin, Rao, Y.S., Sanuj, A.K., Patil, P.P., Jayaganthan, A., Hechhu, R., Methylene blue dye photodegradation during synthesis and characterization of WO_3 nanoparticles. *Adsorp. Sci. Technol.*, Volume 2023, Article ID 2882048, 1 page, 2023.
56. Xie, T., Liu, C., Xu, L., Yang, J., Zhou, W., Novel heterojunction Bi_2O_3/$SrFe_{12}O_{19}$ magnetic photocatalyst with highly enhanced photocatalytic activity. *J. Phys. Chem. C*, 117, 24601–24610, 2013.
57. Xiao, X., Hu, R., Liu, C., Xing, C., Qian, C., Zuo, X., Nan, J., Wang, L., Facile large-scale synthesis of β-Bi_2O_3 nanospheres as a highly efficient photocatalyst for the degradation of acetaminophen under visible light irradiation. *Appl. Catal., B*, 140–141, 433–443, 2013.
58. Raza, W., Haque, M.M., Muneer, M., Harada, T., Matsumura, M., Synthesis, characterization and photocatalytic performance of visible light induced bismuth oxide nanoparticle. *J. Alloys Compd.*, 648, 641–650, 2015.
59. Gusain, R., Kumar, P., Sharma, O.P., Jain, S.L., Khatri, O.P., Reduced graphene oxide–CuO nanocomposites for photocatalytic conversion of CO_2 into methanol under visible light irradiation. *Appl. Catal., B*, 181, 352–362, 2016.
60. Chen, F., Jia, D., Jin, X., Cao, Y., Liu, A., A general method for the synthesis of graphene oxide-metal sulfide composites with improved photocatalytic activities. *Dyes Pigm.*, 125, 142–150, 2016.
61. Nguyen-Phan, T.-D., Pham, V.H., Shin, E.W., Pham, H.-D., Kim, S., Chung, J.S., Kim, E.J., Hur, S.H., The role of graphene oxide content on the adsorption-enhanced photocatalysis of titanium dioxide/graphene oxide composites. *Chem. Eng. J.*, 170, 226–232, 2011.

62. Litter, M.I. and Slodowicz, M., An overview on heterogeneous Fenton and photoFenton reactions using zerovalent iron materials. *J. Adv. Oxid. Technol.*, 20, 20160164, 2017.
63. Hoag, G.E., Collins, J.B., Holcomb, J.L., Hoag, J.R., Nadagouda, M.N., Varma, R.S., Degradation of bromothymol blue by 'greener' nano-scale zero-valent iron synthesized using tea polyphenols. *J. Mater. Chem.*, 19, 8671–8677, 2009.
64. Wu, Y., Zeng, S., Wang, F., Megharaj, M., Naidu, R., Chen, Z., Heterogeneous Fenton-like oxidation of malachite green by iron-based nanoparticles synthesized by tea extract as a catalyst. *Sep. Purif. Techn.*, 154, 161–167, 2015.
65. Wang, W., Zhou, M., Mao, Q., Yue, J., Wang, X., Novel NaY zeolite-supported nano scale zero-valent iron as an efficient heterogeneous Fenton catalyst. *Catal. Commun.*, 11, 937–941, 2010.
66. Saad, R., Thibutot, S., Ampleman, G., Hawari, J., Sorptive removal of trinitroglycerin (TNG) from water using nanostructured silica-based materials. *J. Environ. Qual.*, 39, 580–586, 2010.
67. Liu, X., Wang, F., Chen, Z., Megharaj, M., Naidu, R., Heterogeneous Fenton oxidation of Direct Black G in dye effluent using functional kaolin supported nanoscale zero iron. *Environ. Sci. Pollut. Res.*, 21, 1936–1943, 2014.
68. Qu, X., Alvarez, P.J.J., Li, Q., Applications of nanotechnology in water and wastewater treatment. *Water Res.*, 47, 3931–3946, 2013.
69. Ahmed, S.F., Mofijur, M., Nuzhat, S., Chowdhury, A.T., Rafa, N., Uddin, M.A., Inayat, A., Mahlia, T.M.I., Ong, H.C., Chia, W.Y., Show, P.L., Recent developments in physical, biological, chemical, and hybrid treatment techniques for removing emerging contaminants from wastewater. *J. Hazard. Mater.*, 416, 125912, 2021.
70. Ahmada, N.A., Goh, P.S., Zulhairuna, A.K., Wonga, T.W., Ismaila, A.F., The Role of Functional Nanomaterials for Wastewater Remediation (Chapter 1), in: *Chemistry in the Environment Series No. 3 Functional Hybrid Nanomaterials for Environmental Remediation*, A.F. Ismail and P.S. Goh (Eds.), The Royal Society of Chemistry, London, United Kingdom, 2021.
71. Hasbullah, H., Sabri Md., N.S., Said, N., Rosid, S.M., Roslan, I., Ismail, A.F., Jye, L.W., Yusof, N., Nanoengineered materials for water and wastewater treatments, in: *Nanotechnology in water and wastewater treatment*, pp. 303–335, Amsterdam, Elsevier, 2019.
72. Poornima, S., Manikandan, S., Karthik, V., Balachandar, R., Subbaiya, R., Saravanan, M., Chi, N.T.L., Pugazhendhi, A., Emerging nanotechnology based advanced techniques for wastewater treatment. *Chemosphere*, 303, 135050, 2022.
73. De Gusseme, B., Hennebel, T., Christiaens, E., Saveyn, H., Verbeken, K., Fitts, J.P., Boon, N., Verstraete, W., Virus disinfection in water by biogenic silver immobilized in polyvinylidene fluoride membranes. *Water Res.*, 45, 1856–1864, 2011.

74. Sahinkaya, E., Sahin, A., Yurtsever, A., Kitis, M., Concentrate minimization and water recovery enhancement using pellet precipitator in a reverse osmosis process treating textile wastewater. *J. Environ. Manage.*, 15, 420–427, 2018.
75. Jager, D., Kupka, D., Vaclavikova, M., Ivanicova, L., Gallios, G., Degradation of Reactive Black 5 by electrochemical oxidation. *Chemosphere*, 190, 405–416, 2018.
76. Dasgupta, J., Sikder, J., Chakraborty, S., Curcio, S., Drioli, E., Remediation of textile effluents by membrane based treatment techniques: a state of the art review. *J. Environ. Manage.*, 147, 55–72, 2015.
77. Laqbaqbi, M., García-Payo, M.C., Khayet, M., El Kharraz, J., Chaouch, M., Application of direct contact membrane distillation for textile wastewater treatment and fouling study. *Sep. Purif. Technol.*, 209, 815–825, 2018.
78. Alventosa-deLara, E., Barredo-Damas, S., Alcaina-Miranda, M.I., Iborra-Clar, M.I., Ultrafiltration technology with a ceramic membrane for reactive dye removal: optimization of membrane performance. *J. Hazard. Mater.*, 30, 492–500, 2012.
79. Cheng, S., Oatley, D.L., Williams, P.M., Wright, C.J., Characterisation and application of a novel positively charged nanofiltration membrane for the treatment of textile industry wastewaters. *Water Res.*, 46, 33–42, 2012.
80. Edgar Vázquez-Núñez, E., Molina-Guerrero, C.D., Peña-Castro, J.M., Fernández-Luqueño, F., Rosa-Álvarez, M.G.d.l., Use of Nanotechnology for the Bioremediation of Contaminants: A Review. *Processes*, 8, 826–842, 2020.
81. Yogalakshmi, K.N., Das, A., Rani, G., Jaswal, V., Randhawa, J.S., Nano-bioremediation: A New Age Technology for the Treatment of Dyes in Textile Effluents, in: *Bioremediation of Industrial Waste for Environmental Safety*, G. Saxena and R. Bharagava (Eds.), Springer, Singapore, 2020.
82. Ali, M., Husain, Q., Alam, N., Ahmad, M., Enhanced catalytic activity and stability of ginger peroxidase immobilized on amino-functionalized silica-coated titanium dioxide nanocomposite: a cost-effective tool for bioremediation. *Water Air Soil Pollut.*, 228, 22, 2017.
83. Vedhantham, K., Girigoswami, A., Harini, A., Girigoswami, K., Waste water remediation using nanotechnology-A review. *Biointerface Res. Appl. Chem.*, 12, 4476–4495, 2022.
84. Wong, J.K.H., Tan, H.K., Lau, S.Y., Yap, P.-S., Danquah, M.K., Potential and challenges of enzyme incorporated nanotechnology in dye wastewater treatment: A review. *J. Environ. Chem. Eng.*, 7, 103261, 2019.
85. Ihsanullah, I., Jamal, A., Ilyas, M., Zubair, M., Khan, G., Atieh, M.A., Bioremediation of dyes: Current status and prospects. *J. Water Process Eng.*, 38, 101680, 2020.
86. Noman, M., Shahid, M., Ahmed, T., Niazi, M.B.K., Hussain, S., Song, F., Manzoor, I., Use of biogenic copper nanoparticles synthesized from a native Escherichia sp. as photocatalysts for azo dye degradation and treatment of textile effluents. *Environ. Pollut.*, 257, 113514, 2020.

87. Pervez, M.N., Balakrishnan, M., Hasan, S.W., Choo, K.-H., Zhao, Y., Cai, Y., Zarra, T., Belgiorno, V., Naddeo, V., A critical review on nanomaterials membrane bioreactor (NMs-MBR) for wastewater treatment. *npj Clean Water*, 3, 43, 2020.
88. Sundaran, S.P., Reshmi, C.R., Sagitha, P., Manaf, O., Sujith, A., Multifunctional graphene oxide loaded nanofibrous membrane for removal of dyes and coliform from water. *J. Environ. Manage.*, 240, 494–503, 2019.
89. Dabaghian, Z., Peyravi, M., Jahanshahi, M., Rad, A.S., Potential of advanced nano-structured membranes for landfill leachate treatment: A review. *ChemBioEng Rev.*, 5, 119–138, 2018.
90. Masoudi, M., Rahimnejad, M., Mashkour, M., Enhancing operating capacity of microbial fuel cells by using low-cost electrodes and multi anode-cathode connections in a membrane-less configuration. *Int. J. Hydrog. Energy*, 46, 8226–8238, 2021.
91. Lu, L., Xing, D., Ren, Z.J., Microbial community structure accompanied with electricity production in a constructed wetland plant microbial fuel cell. *Bioresour. Technol.*, 95, 115–121, 2015.
92. Mohamed, S.N., Thomas, N., Tamilmani, J., Boobalan, T., Matheswaran, M., Kalaichelvi, P., Alagarsamy, A., Pugazhendhi, A., Bioelectricity generation using iron (II) molybdate nanocatalyst coated anode during treatment of sugar wastewater in microbial fuel cell. *Fuel*, 277, 118119, 2020.
93. Jain, K., Patel, A.S., Pardhi, V.P., Flora, S.J.S., Nanotechnology in wastewater management: A new paradigm towards wastewater treatment. *Molecules*, 26, 1797, 2021.
94. Sukhanova, A., Bozrova, S., Sokolov, P., Berestovoy, M., Karaulov, A., Nabiev, I., Dependence of nanoparticle toxicity on their physical and chemical properties. *Nanoscale Res. Lett.*, 13, 1–21, 2018.
95. Khan, M.S., Qureshi, N.A., Jabeen, F., Assessment of toxicity in fresh water fish Labeorohita treated with silver nanoparticles. *Appl. Nanosci.*, 7, 167–179, 2017.
96. Maroufi, E.M., Amirkhani, L., Zakaryazadeh, H., Removal of real multicomponent textile wastewater by adsorption onto graphene oxide nanoparticles: optimization of operating parameters. *Desalin. Water Treat.*, 226, 104–112, 2021.
97. Amirian, P., Bazrafshan, E., Payandeh, A., Textile Wastewater Treatment Using Photonanocatalytic Process (UV/CuO Nanoparticles): Optimization of Experiments by Response Surface Methodology. *Healthscope*, 7, 3, e57689, 2018.
98. Jorfi, S., Barzegar, G., Ahmadi, M., Soltani, R.D.C., Haghighifard, N.A.J., Takdastan, A., Saeedi, R., Abtahi, M., Enhanced coagulation-photocatalytic treatment of Acid red 73 dye and real textile wastewater using UVA/synthesized MgO nanoparticles. *J. Environ. Manage.*, 177, 111–118, 2016.

5

Innovative Approaches to Recover Waste in the Textile Sector

Dipak Kumar Sahu[1]* **and Goutam Rath**[2]

[1]School of Pharmacy, Amity University, Raipur, India
[2]School of Pharmacy, Siksha O Anusandhan (Deemed to be University), Bhubaneswar, India

Abstract

The bulk of textiles are composed of a blend of processed plant-derived polymers and synthetic components, making recovery and recycling difficult and expensive. The wasted or undesired textile materials such as clothes, fabric, and other related products generated are called solid waste. Waste generated in liquid form throughout various stages of textile manufacture, processing, and finishing is referred to as liquid textile waste. The solid waste form can be managed with various programs like extended producer responsibility, waste bin tracking, take-back schemes, and automation in retracing as well as sorting. Furthermore, for effective waste management, advanced robotics, and artificial intelligence can be integrated into these programs. Dyes, metals, and other contaminants in liquid waste can be handled with nanotechnology as a dependable and safer solution. Hence, this book chapter discusses several novel concepts such as mechanical recycling, chemical recycling, upcycling, and waste-to-energy conversion of both solid and liquid textile waste. In addition. few case studies are highlighted as the application of these innovative concepts. Finally, regulatory collaboration is highlighted as a crucial instrument for the efficient management of textile waste.

***Keywords*:** Solid waste recovery, innovative approaches, EPR program, retracing and sorting, liquid waste, nanoparticle, nanomaterials, waste recycling to devices

**Corresponding author*: dipaksahu_il1@hotmail.com

Prashansa Sharma and Shilpi Shree Sahay (eds.) Nanotechnology-Assisted Recycling of Textile Waste: Sustainable Tools for Textiles, (129–170) © 2025 Scrivener Publishing LLC

List of Abbreviations

EPR	Extended producer responsibility
GDP	Gross domestic product
NGO	Non-governmental organizations
GPS	Global Positioning System
RFID	Radio-frequency identification detection
AI	Artificial intelligence
CNN	Convolutional neural networks
QR	Quick response
NFC	Near-field communication
PET	Polyethylene terephthalate
PACI	Polyaluminium chloride
EPS	Expanded polystyrene
ROS	Reactive oxygen species
GO	Graphene oxide
CNT	Carbon nanotube
ZDHC	Zero discharge of hazardous chemicals
CA	Cellulose aerogels

5.1 Introduction

5.1.1 Importance of Waste Recovery in the Textile Sector

The worldwide textile industry has a substantial economic and environmental impact. It is estimated to be around $3,000 billion, demonstrating its pervasive impact on the global economy [1]. A major share of more than 2% of the global gross domestic product (GDP) underscores its importance as a key economic contributor [2]. The doubling of textile production over the last two decade emphasizes the industry's tremendous growth and expansion. This rise can be ascribed to reasons such as fast fashion, worldwide supply chains, and increased clothing demand. With annual textile consumption per person increasing from 7 kg to 13 kg, it is apparent that worldwide clothing consumption is on the rise. However, this growing consumption has had severe environmental repercussions [3, 4]. Massive amounts of waste, which includes fabric scraps, dyes, chemicals, and wasted clothes, if not effectively managed, might end up in landfills. This contributes to pollution and consumes valuable land space [5, 6].

The textile industry is a resource-intensive business that uses a lot of water, energy, and raw materials at various stages of production [7].

Recovery procedures, such as recycling and upcycling, aid in the conservation of these precious resources. As a result, the overall sustainability of the sector increases by reusing materials and lowering the requirement for virgin resources [8]. Recovery efforts take textiles from landfills, lowering environmental effects and the pressure on waste management systems. Thus, waste recovery in the textile industry is critical due to the considerable environmental, economic, and social ramifications. Moreover, recycling and recovering textiles can help to reduce the energy required to create new materials from scratch. Carbon emissions are reduced and reliance on fossil fuels becomes diminished. In addition, this process generates new economic opportunities [9, 10]. Businesses that recycle and upcycle can arise, creating jobs and contributing to local economies. By reusing materials, businesses can minimize procurement costs and improve overall cost-efficiency [11]. Waste recovery stimulates creative approaches to design and manufacturing. Designers can experiment with novel techniques to repurpose textile waste into unique products. This results in the creation of novel and sustainable fashion pieces or utilitarian materials [12, 13].

5.1.2 Significance of Innovative Approaches

Innovative approaches to waste recovery in the textile industry are critical. They can address major environmental, economic, and social issues. These strategies provide unique solutions that go beyond typical waste management procedures. The advantages are the conservation of natural resources, reduced energy usage, reduced greenhouse gas emissions, and reduced pollution connected with raw material extraction and production [14]. Innovative waste recovery methods are consistent with the concepts of the circular economy, which seeks to reduce waste generation while increasing the value derived from resources [15]. These ideas contribute to a more sustainable and regenerative economic model by utilizing closed-loop systems and repurposing materials. In order to pursue innovative waste recovery technologies, research and development efforts are required. This push for innovation may result in the development of new technologies, techniques, and materials with uses beyond the textile industry [16].

A circular economy seeks to reduce waste and maximize resource utilization. This is possible by keeping products and materials in use for as long as possible. Waste recovery adheres to circular economy principles by prolonging the lifespan of textiles by recycling and reusing [17]. As a result, the demand for continual manufacturing of new materials is decreased [18]. Environmental and ethical considerations are becoming increasingly important to modern consumers. As the textile business expands, implementing waste recovery practices becomes not only an ethical option but also a strategic decision for long-term viability and competitiveness. Textile companies that prioritize waste reduction and recycling can attract environmentally concerned customers. They may gain a market advantage in the end [19, 20].

Sustainability has become an increasingly important component of corporate operations. Companies should be capable of adapting to shifting consumer preferences, regulatory needs, and market trends. Collaborative environments are typically conducive to innovation. Hence, companies who invest in creative waste recovery technologies are likely to be more competitive in the long run [21, 22]. Adopting innovative waste recovery practices fosters collaboration among various stakeholders. This includes corporations, research institutions, governments, and non-governmental organizations (NGOs). Further, it boosts knowledge sharing and cross-sector cooperation. The ability of the textile industry to successfully manage waste is inextricably tied to its growth and survival. By addressing resource limitations, environmental concerns, and societal expectations, innovative waste recovery technologies provide a road to long-term profitability. Finally, innovative waste recovery technologies in the textile sector are critical for them to remain responsible, and resilient. These essential components help the industries move to a greener future [23, 24].

5.1.3 Overview of the Current Waste Management Practices in the Textile Industry

Waste management practices in the textile industry have evolved recently. This is to meet both environmental issues and promote sustainability. Textile manufacturers have been concentrating on lowering waste creation at the source station. They are achieving it, by optimizing production processes, minimizing material waste, and boosting efficiency in cutting, stitching, and other manufacturing stages [25, 26].

Recycling is the collecting and processing of post-consumer or post-industrial textile waste to create new materials. Fabrics can be mechanically or chemically recycled to recover fibers for use in new textiles [27]. Upcycling is the process of recycling textile waste into higher-value items while preserving the original material. This technique stimulates design originality and frequently results in novel objects [28]. Unsold or unsellable textiles are frequently donated to charitable organizations or reused. Some businesses form alliances with organizations that donate textiles to individuals in need. Non-recyclable textile waste can be used as a fuel source in waste-to-energy operations in some situations. This helps to recover energy while reducing landfill consumption [29]. This practice, however, is not as environmentally friendly as recycling. Some textile firms are developing closed-loop methods. Through this process, they collect old garments or items, recycle them, and reuse the recovered materials in new products. This method promotes the cyclic movement of materials [30, 31].

Significant amounts of wastewater and chemical waste are produced during the dyeing and finishing operations in textiles. To lessen the environmental effects, the industry has been researching more sustainable dyeing procedures and eco-friendly chemicals. Biodegradable nanoparticles, nanomaterials, and recycling procedures for liquid waste are being explored continuously [32, 33]. Figure 5.1 illustrates various procedures for the current management of solid and liquid waste.

Some countries have enacted legislation and certifications to encourage or compel textile companies to follow specified waste management standards. These regulations frequently seek to limit the use of hazardous chemicals while also encouraging safe waste disposal practices. It is crucial to note that adoption of these practices varies by company and location. Finally, the textile sector is always looking for new ways to enhance waste management and minimize its environmental imprint [34, 35].

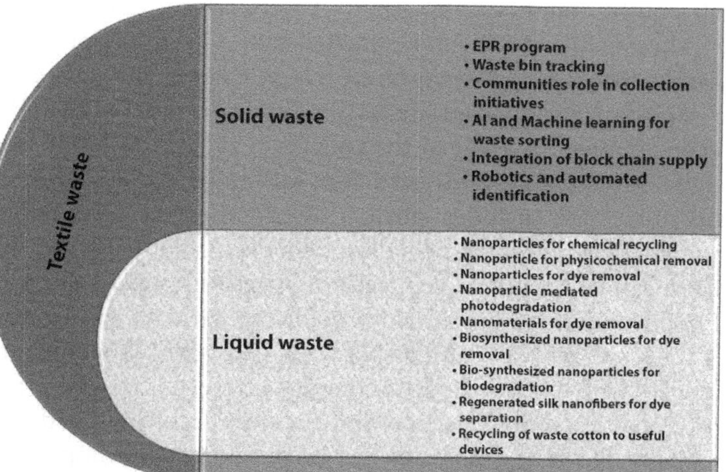

Figure 5.1 Recovery procedures for both solid and liquid textile waste.

5.2 Innovative Approaches for Solid Waste Management

5.2.1 Collection of Textile Waste

5.2.1.1 Implementing EPR Programs

Implementing an extended producer responsibility (EPR) program for the collection of textile waste encourages the textile industry to take greater responsibility for the entire lifecycle of its products. EPR programs shift the responsibility for waste management from the end-consumer to the producers, motivating them to design products that are more sustainable, recyclable, and easier to manage. In consultation with government agencies, industry associations, environmental organizations, and other stakeholders, insights, perspectives, and support can be gained by them. By doing this, a comprehensive policy framework for the EPR program can be developed [36, 37].

Textile makers can be mandated to assume responsibility for the collecting, recycling, or safe disposal of their products. This can be done at the end of their useful life through legislation or regulations. It is possible to set specific goals, timetables, and responsibilities. These goals should be ambitious but attainable in order to promote enhanced sustainability and resource recovery. Products could be developed to facilitate recycling and material separation. Collection mechanisms, such as collection centers or drop-off points, could be established. A network for collecting, sorting, and recycling textile waste can be created [38]. Collaboration boosts efficiency and encourages information exchange. Consumers can be educated on the importance of appropriate textile disposal and recycling. They also may be encouraged to assist in the collection process through awareness campaigns. Investment in infrastructure for textile waste collection, sorting, and processing includes efficient development of collection points, recycling facilities, and transportation systems. A reporting and monitoring system can be set up to keep track of progress toward collection goals [39]. Textile manufacturers should provide regular data on the amount of textiles collected, recycled, and disposed of. Incentives should be provided to producers who exceed collection targets or create innovative recycling systems. Penalties should be imposed for noncompliance with program requirements. Novel technologies can be developed and implemented to facilitate textile recycling and recovery. This could involve improvements in sorting methods, material identification, and recycling operations [40]. Reviewing and improving the EPR program on a regular basis helps to correct any issues, adapt to changing market conditions, and align with developing best practices [41]. Program implementation transparency can be maintained by sharing progress updates, triumphs, and issues with stakeholders as well as the general public. All parties must work together to implement EPR programs for textile waste collection. Thus, this program holds manufacturers accountable for their goods' whole lifecycle. Finally, it can result in positive environmental, economic, and social effects in the textile sector [42].

5.2.1.2 Leveraging Technology for Smart Textile Waste Bins and Tracking

Using smart textile waste bins and tracking technologies can considerably improve the efficiency and efficacy of textile waste collection and handling procedures. This may include technology adaptation into the design and execution of smart textile waste bins and tracking systems. Integration and installation of sensors in textile waste bins help to determine when the container is full. These sensors can provide signals to waste management workers or directly to collection trucks, allowing collection routes to be optimized and wasteful collections reduced. Compaction mechanisms can be included in the bins to enhance capacity, reducing the frequency of emptying and lowering collection costs [43].

Smart bins may feature devices for sorting various forms of textile waste, making recycling more efficient. Bins, for example, could include divisions for various materials such as cotton, polyester, or wool. It may gather information about bin usage, fill levels, and collection frequency. Predictive analytics can determine when bins are likely to be full by analyzing historical data. This allows waste management businesses to organize collections in advance, lowering the risk of overflowing bins. GPS (Global Positioning System) and RFID (radio-frequency identification) technologies can be used to track the location and movement of bins [44]. This helps to guarantee that dumpsters are placed in the proper locations and collected on time. Remote monitoring systems allow waste management personnel to monitor bin fill levels, status, and position in real-time from a central control center. Automated notifications set up help to warn waste collection crews when bins reach particular fill levels. This lowers the need for manual inspections and optimizes collection routes [45].

Smartphone apps for waste collection staff provide real-time information regarding bin locations, fill levels, and collection schedules. Further, the app allows citizens to report full or overflowing bins, encouraging a sense of community participation in waste management. This app can be used to teach users about proper waste disposal, recycling, and the environmental benefits of smart textile waste containers. Smart bin data connection to waste management software helps for easy collection activity coordination and data analysis. This data provides information on the types and quantities of textile waste they collect [46, 47]. Sharing data with textile recycling facilities to help them streamline the sorting and recycling process. Technology should respect user privacy and adhere to data protection standards. Using technology in this way can lead to more effective waste collection procedures, lower operational costs, and higher recycling rates. It also helps the textile industry migrate to a more data-driven and innovative waste management approach [48].

5.2.1.3 Collaboration with Fashion Brands for Take-Back Schemes

Collaborating with fashion labels on take-back programs is a strategic technique that can considerably boost waste recovery efforts in the textile industry. By collaborating with businesses, a win-win situation is created that promotes viability, decreases waste, and increases customer engagement. Efficient take-back programs in partnership with apparel brands can be set up. Out of them, fashion brands that are compatible with waste recovery goals and principles can be selected. Noticingly, these brands should also value sustainability, ethics, and corporate social responsibility [49, 50].

Fashion brand's participation in take-back schemes can enhance their brand image. This can be demonstrated by commitment to sustainability and engaging customers in meaningful ways. Different brands have different operational capacities and priorities. The approach can customized on the size, resources, and specific objectives of each brand. Collaborative discussions can be held with brands to co-design the take-back scheme. These could be logistical considerations, such as collection methods, transportation, and sorting processes. Incentives or rewards can be offered to brands for their participation [51]. This could include marketing exposure, co-branded campaigns, or other value-added benefits. Instructional materials, social media campaigns, engaging consumers, and communicating the take-back program with brands inspire people to participate. A system can be created for tracking and reporting the amount of textiles gathered from each brand. This information can be used to illustrate the effectiveness of the take-back program and to promote continual improvement. Collaboration with brands is needed, which have circular economy plans such as repurposing materials, recycling textiles, or partnering with textile recycling partners. Regular communication fosters trust and ensures that both sides are working in unison. Based on feedback, statistics, and changing market dynamics, the take-back system should be adjusted. It can be through programs to change consumer preferences and technological improvements in waste recovery [52].

5.2.1.4 Engaging Local Communities and NGOs in Waste Collection Initiatives

Engaging local communities and NGOs in textile waste collection activities is critical for developing a comprehensive and meaningful waste recovery strategy. It could be through insights, relationships, and co-design collecting initiatives, workshops, focus groups, or meetings [53].

NGOs with experience in waste management, environmental protection, or community participation can be selected. These can be used to provide training and capacity-building workshops for local communities in textile waste collection projects [54]. The initiatives for each community's unique needs and capabilities may include infrastructure, resources, and cultural sensitivities. Decision-making processes, thoughts, and feedback help the projects be relevant and effective. Further, collaboration with NGOs helps to arrange funds, grants, or sponsorships for the projects [55]. NGOs frequently have established networks and financial support skills. Making use of their existing communication channels and networks smoothens the process of collection campaigns. In addition. incentives or rewards to encourage active engagement help for high levels of participation. Data collection, involvement rates, and impact can be used to evaluate the projects' progress and make adjustments. Initiatives' long-term viability and continuance include developing community leaders for forming local waste management. Through this, it promotes a sense of community ownership and responsibility. Thus, the beneficial change by engaging these players helps in the textile waste management landscape by harnessing knowledge, resources, and passion [56, 57].

5.2.2 Retracing and Sorting Textile Waste

5.2.2.1 Utilizing AI and Machine Learning for Efficient Waste Sorting

AI (artificial intelligence) and machine learning for effective textile waste sorting have the potential to revolutionize the textile recycling process.

This can be achieved through increasing accuracy, speed, and resource utilization. A broad range of photos or samples can be collected that illustrate various sorts of textiles, colors, patterns, and situations. Data has to be labeled correctly in order to properly train the machine learning algorithms [58, 59]. Machine learning models can recognize and categorize textiles based on visual features, such as convolutional neural networks (CNNs). Fabric kinds (cotton, polyester, etc.), colors, patterns, and any damage can be identified by these models. To identify textiles based on their material composition, spectroscopy or hyperspectral imaging methods can be used in conjunction with machine learning. This method goes beyond aesthetic characteristics to sort mixed-material textiles [60, 61]. AI technologies can be employed in sorting and recycling operations. AI-powered sorting eliminates the need for human sorting, which can be time-consuming, labor-intensive, and error-prone. AI-powered sorting systems have the potential to greatly boost the throughput of textile waste sorting plants, therefore enhancing overall efficiency. As textiles travel over conveyor belts, cameras, and sensors can recognize and categorize them using pre-trained algorithms. Further, AI algorithms can be used to evaluate textile quality and condition. This can aid in distinguishing textiles suited for recycling, upcycling, or reusing, reducing waste, and maximizing resource recovery. AI systems can make split-second decisions regarding the best sorting destination for each item, increasing sorting accuracy while decreasing human error. Adaptive machine learning algorithms can learn from fresh data and feedback constantly. This ensures that the system evolves over time and is capable of dealing with new cloth types and trends [62, 63]. The recycling process becomes more effective by properly detecting and classifying textiles, decreasing the environmental impact associated with waste generation and disposal. AI models can be fine-tuned to meet the individual needs of various textile waste sorting facilities, taking into account differences in material types, sizes, and sorting criteria. The use of AI and machine learning in textile waste sorting fosters innovation and research in the industry. This potentially leads to the creation of new technologies and processes. To effectively develop and execute these solutions, AI and machine learning experts, technology companies, and recycling industry players can collaborate. When working with sensitive data, privacy requirements, and ethical issues are made sure to be followed. Thus, using AI and machine learning to sort textile waste can lead to more effective recycling operations. Further, it lowers environmental impact, and higher resource recovery rates, ultimately supporting a more sustainable textile business [64, 65].

5.2.2.2 Robotics and Automated Textile Identification Techniques

Robotics and automated textile waste identification systems have the potential to dramatically improve the efficiency and accuracy of textile waste sorting and processing in recycling plants [66]. These technologies use robotics, sensors, and artificial intelligence-driven systems to automate the identification, sorting, and processing of textile materials [67]. Textiles can be handled on sorting lines by robots equipped with robotic arms and conveyor systems. It decreases the requirement for manual labor and enhances operational efficiency. Robot-mounted cameras and sensors may collect photos of textiles and utilize AI algorithms. This is to classify them based on attributes such as material type, color, pattern, and condition [68, 69].

Hyperspectral cameras can detect materials based on their unique spectral fingerprints. This allows accurate textile sorting even when they appear superficially similar. Real-time material analysis can be performed by robots equipped with integrated spectroscopic instruments. It enables precise identification of textile types, blends, and compositions [70]. Textiles can be felt and manipulated by robots equipped with powerful touch sensors, allowing them to categorize items. This is based on texture, thickness, and other physical attributes. AI-powered systems can evaluate textile quality by detecting wear and tear, stains, and defects. Robots can then evaluate whether an object may be recycled, upcycled, or reused. Textile disassembly robots can be trained to separate buttons, zippers, and other components for effective recycling [71].

Cobots can aid human operators in performing tasks that demand precision and strength. This collaborative strategy improves productivity while lowering the danger of occupational injuries. Automated systems make decisions regarding each item's sorting location in real-time, decreasing processing time and human error. It may learn from new data indefinitely, enhancing its accuracy and adaptability over time. Sorting routes and sequences can be optimized by automated systems to reduce transportation and boost overall efficiency. Automation and robotics can be scaled up or down to match the size and capacity of various recycling plants [72, 73].

Hence, automated systems can optimize energy use, lowering the environmental impact of waste sorting. Automation lessens the possibility of contamination due to human errors in sorting. Using robots and automated textile waste identification techniques optimizes waste sorting, and increases resource recovery. Further, it helps to a more sustainable and efficient textile recycling sector [74].

5.2.2.3 *Integration of Blockchain for Supply Chain Transparency*

Integrating blockchain technology for supply chain transparency in textile waste management can provide the process with a new degree of responsibility, traceability, and confidence. The decentralized and unchangeable structure of blockchain makes it ideal for tracking and validating transactions, materials, and processes [75]. This can be done throughout the textile waste supply chain. From collecting and sorting through recycling and final product development, blockchain can track every stage of the textile waste journey. Each transaction is chronologically and tamper-proof connected, resulting in a transparent and auditable trail. By creating a tamper-proof record of each material's origin and route, blockchain can assist in preventing counterfeit or fraudulent practices [76, 77]. The technology can be used to authenticate and verify vendors along the textile waste supply chain. This guarantees that waste is generated legally and responsibly. Blockchain records can contain precise information on the sources of textile materials, such as fiber kinds, manufacturing methods, and locations. Smart contracts, which are self-executing codes on the blockchain, can automate specific portions of the supply chain. This is by requiring participants to meet established conditions before progressing. It can be used by brands and businesses to provide consumers with information about the route of their products [78]. Blockchain can trace the recycling process, recording each step from deconstruction through the development of a new product. This ensures that recycling is carried out in accordance with industry standards. Blockchain can record the carbon footprint of materials, processes, and transportation when integrated with IoT (Internet of Things) devices [79, 80]. Blockchain ensures that data is securely stored and shared only with authorized parties, while sensitive information remains private. The shared data can be accessed by all stakeholders in the textile waste supply chain. This enhances boosting collaboration and mutual understanding. Thus, transparency can help organizations demonstrate compliance with rules and industry standards by simplifying auditing processes. Blockchain technology can be used by brands to authenticate their sustainability claims, giving consumers proof of their commitment to ethical and responsible practices [81, 82]. Consumers can be educated about the importance of proper waste management and recycling practices through blockchain-based platforms. Blockchain records are permanent, allowing stakeholders to follow the history of resources and goods for years to come. The use of blockchain in textile waste management can promote research and innovation by allowing for the exploration of novel approaches to optimize the waste recovery process.

5.2.2.4 *Adopting Sustainable Labels for Easy Categorization*

Adopting sustainable labels for easy categorization of textile waste is a practical strategy to encourage responsible waste management. It empowers customers to make informed decisions. Sustainable labels inform consumers about a product's materials, manufacturing techniques, and environmental impact. By adding these labels, both consumers and waste management systems will be able to quickly recognize textiles with comparable properties, hence streamlining sorting and recycling operations [83, 84].

A standardized set of sustainable labels can be created with clear and straightforward information. This could be on the composition, care instructions, and environmental features of the textile. The composition includes information on the label about the principal material composition of the cloth (e.g., cotton, polyester, wool). This aids in the classification of textiles based on their material type [85]. Further, a color-coded system can help in classifying textiles according to their material makeup. Sustainable textile production practices, such as organic farming, low-impact dyeing, or water-efficient processes can be highlighted [86]. Whether the cloth is recyclable or contains recycled materials can be indicated. Advice for customers can be included on how to properly dispose of or recycle the cloth [87]. This promotes appropriate disposal and recycling. QR codes or near-field communication (NFC) technology and mobile apps on labels can be created to provide consumers with more information about the product's life cycle and recycling procedures [88]. This visual indication assists both consumers and waste management employees in sorting. Generally recognized symbols and images can be used to communicate sustainability features like "organic," "recycled," or "low impact" [89, 90].

Educational programs can be used to educate customers on the importance of labeling for responsible waste management. Training for employees can be included in the categorization of textiles. It helps to speed up the sorting process. Collaboration with organizations and certification programs ensures label accuracy and alignment with industry standards. Imposing restrictions or rewarding brands that use these labels can be done. Feedback from customers, waste management systems, and recycling can facilitate, modify, and improve the labeling system [91].

5.2.2.5 *Engaging Stakeholders for Effective Waste Management*

Stakeholder engagement is critical for effective textile waste management. It promotes teamwork, shared accountability, and a comprehensive approach to addressing waste-related concerns [92]. A variety of stakeholders from many sectors can be used to develop a comprehensive and impactful waste management strategy [93]. Important stakeholders with an interest in textile waste management can be determined. This includes government agencies, fashion brands, manufacturers, merchants, waste management firms, NGOs, consumers, and local communities. Alliances can be formed with stakeholders who have similar goals and objectives. A stakeholder map can be created that depicts the many stakeholders' responsibilities, impact, and interests in textile waste management [94]. Engagement strategies can be tailored to each stakeholder group's requirements and expectations. Different approaches to participation may be required by different stakeholders. Stakeholders should be involved in the planning and decision-making processes. Ideas, opinions, and viewpoints can be collected to develop a more inclusive and effective waste management strategy [95]. Open and transparent avenues of contact should be maintained with stakeholders with updates on advances, progress, and issues in textile waste management. Stakeholders can be educated through workshops, seminars, webinars, and awareness campaigns. Stakeholders should be apprised of participation advantages, such as lower environmental impact, economic savings, improved brand reputation, and improved community well-being [96]. Sharing and collaborating on data can be executed to better understand waste generation, collection, and recycling patterns. Data sharing can help to inform more successful strategies [97]. For decision-making processes, they can be involved, allowing them to contribute to the development of waste management policies, guidelines, and initiatives. They should be updated with the status and impact of waste management programs regularly. Channels for stakeholders can be created to share feedback, suggestions, and concerns. Their feedback can be utilized to continually improve waste management practices. Recognition events, awards, or partnerships can be done to emphasize their commitment to waste reduction [98]. Pilot projects, research efforts, and innovative solutions that harness the skills and resources of various stakeholders are encouraged. Long-term connections can be developed with stakeholders to guarantee that waste management goals are met on an ongoing basis. Thus, stakeholder participation in textile waste management fosters a sense of ownership and shared responsibility. Furthermore, comprehensive and successful solutions can be established by using the pooled knowledge, resources, and efforts of multiple stakeholders [99, 100]. Figure 5.2 illustrates the process of treatment of both solid and liquid waste.

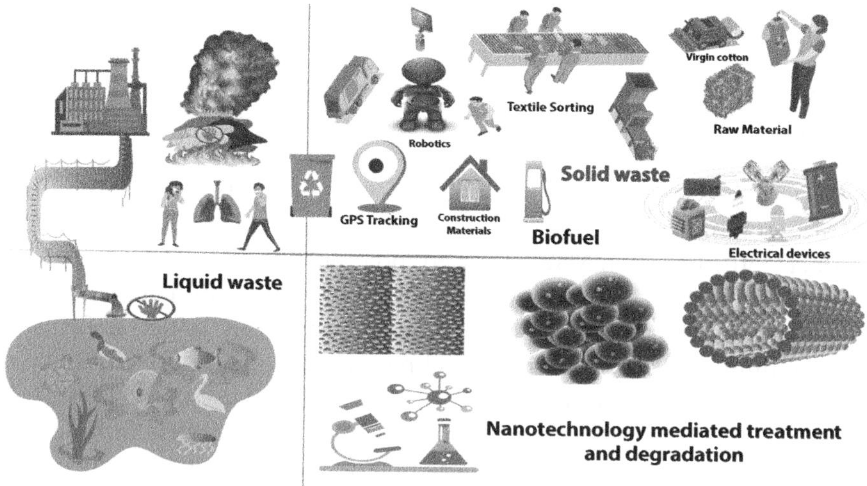

Figure 5.2 Treatment of both solid and liquid waste out of the textile factory.

5.3 Case Studies of Successful Innovations

H&M, a multinational fashion retailer, launched a garment collecting program in which customers can drop off unwanted items (of any brand) at any of their stores. The gathered textiles are then sorted for reuse, recycling, or energy recovery. H&M has also made investments in the development of novel recycling technologies to convert textile waste into new materials. Efforts like this encourage textile circularity and ethical disposal [101, 102]. Renewcell, a Swedish firm, invented a method for converting worn-out cotton and other cellulosic fabrics into Circulose®, a dissolving pulp that can be used to make new fibers. This closed-loop approach minimizes the requirement for virgin resources. This further lowers the environmental effect of textile production dramatically [103, 104]. They collaborated with denim companies such as Levi's to recycle old denim into new fibers. Their novel approach degrades cotton fibers and regenerates them into cellulosic pulp, which can then be spun into new yarns. This effort lessens the demand for virgin cotton while also reducing waste [105]. Patagonia, an outdoor clothing company, introduced the Worn Wear program to encourage customers to mend, reuse, and recycle their gear. They provide Patagonia gear repairs, sell used things on their website, and recycle worn-out garments to generate new products. This effort fosters a long-term and sustainable culture [106]. The R Collective is a fashion firm that specializes in upcycling abandoned apparel and textile waste into high-quality garments. They work with established designers to transform textile waste into fashionable collections, demonstrating that waste can be a significant resource for the creation of new products [107]. Adidas teamed with Parley for the Oceans to manufacture sporting products from ocean plastic waste. Waste plastic is converted into yarn, which is then utilized to make performance apparel. This project not only fights plastic pollution but also emphasizes the value of recycled materials in high-performance products [50]. The Ellen MacArthur Foundation's Circular Fibres Initiative brings together stakeholders from across the fashion industry to accelerate the transition to a circular textile economy. This initiative focuses on designing waste, promoting recycling technologies, and encouraging new business models [108]. These case studies showcase how innovative approaches to textile waste management can create positive environmental and social impacts, demonstrating the potential for the textile industry to become more responsible.

5.4 Innovative Approaches for Liquid Textile Waste Management

The wastewater produced by numerous processes in the textile industry is referred to as liquid textile waste. If liquid waste is not properly handled before being discharged into water bodies or sewage systems, then it can have a negative impact on the environment. The contamination of various water sources can harm aquatic life, disrupt ecosystems, and potentially affect human health [109]. The textile industry effluent contains a variety of pollutants like dyes, chemicals, suspended particles, heavy metals, and other contaminants. They are discharged during the dyeing, printing, finishing, and other textile manufacturing operations [110]. Particulate matter, fibers, and other solid materials can be suspended in the wastewater, causing turbidity and reducing water clarity [111]. Unabsorbed dyes and various chemicals, including detergents, acids, alkalis, and other processing chemicals, can cause water discoloration and potential toxicity. Some dyes and chemicals used in textile processes contain heavy metals, which can be carcinogenic or mutagenic to ecosystems [112]. To mitigate the negative impact of liquid textile waste, various wastewater treatment methods can be employed, including coagulation, flocculation, sedimentation, and chemical precipitation. Biological processes like aerobic and anaerobic treatment, as well as constructed wetlands, can help break down organic compounds and enhance the overall treatment efficiency. Techniques like ozonation, UV/H_2O_2 treatment, and other advanced oxidation methods can be effective in breaking down complex organic pollutants. Polyaluminum chloride (PACl), Alum, etc., are being used as catalysts in the catalytic ozonation process after their use as coagulants [113, 114]. Nanotechnology can be pivotal in handling liquid textile waste. Nano-sized pigments and dyes adhere more effectively to the textile fiber reducing the amount of dye runoff into waste water. Some other domains of research are the nanoencapsulation of dyes, nanofiltration membranes, nanoadsorbents, photocatalytic nanoparticles, biosynthesized nanoparticles, nanosensors, and smart textiles.

5.4.1 Nanoparticle for Physicochemical Removal

Nanoparticles with high surface area and adsorption capacity can be used to remove heavy metals present in textile wastewater. The nanoparticles can adsorb these metals, reducing their concentration and potential environmental impact. Coagulation and flocculation are common processes in wastewater treatment where suspended solids, colloidal particles, and other impurities are aggregated and settled out of the water, resulting in improved water clarity and quality. They can be used as coagulants or flocculants to aggregate and settle suspended solids and colloidal particles present in textile wastewater. A large surface area and high surface reactivity, allow them to interact with suspended solids and colloidal particles in wastewater more effectively. This interaction leads to the aggregation of these particles into larger clusters. The clusters settle more rapidly under the influence of gravity. Traditional coagulation and flocculation processes often rely on the use of chemical coagulants and flocculants. The properties of bio-synthesized nanoparticles can be tuned to optimize their interaction with specific contaminants in textile wastewater. This customization can lead to more efficient aggregation and settling of the targeted pollutants. Nanoparticle-enhanced coagulation and flocculation can be integrated with other treatment methods, such as sedimentation, filtration, and biological processes, to create a comprehensive and efficient wastewater treatment system. These agents possess the ability to interact with and reduce metal ions or other precursor materials, resulting in the formation of nanoparticles. The biological agents facilitate the reduction of metal ions from their soluble forms to nanoparticles through enzymatic or non-enzymatic processes. In many cases, these agents also play a role in stabilizing the nanoparticles to prevent agglomeration.

5.4.2 Nanoparticles for Textile Polyester Recycling

Traditional methods of recycling textile polyester involve melting down the material and reforming it into new products. Chemical recycling, on the other hand, focuses on breaking down the polymer chains into their monomer units. Polyester textiles are made from polymers known as polyethylene terephthalate (PET). Over time, these textiles can become worn or outdated, leading to textile waste. PET polymer chains can be broken into ethylene glycol and terephthalic acid [115]. Nanoparticles can offer higher efficiency and selectivity for breaking as they target specific chemical bonds in structure. Further, nanoparticles help to accelerate the breakdown of the polymer chains under milder conditions. Silver-doped zinc oxide nanoparticles present an innovative approach to discarded polyester textiles. These nanoparticles are believed to catalyze the hydrolysis of the ester bonds in the polyester chains. This hydrolysis breaks the polymer into smaller segments, ultimately leading to the release of the monomers. They can then be used to synthesize new polyester polymers. The addition of silver to zinc oxide nanoparticles can enhance their catalytic activity. Moreover, silver is known for its antimicrobial and catalytic properties, which can help improve the efficiency of the depolymerization process [116]. This reduces the need for incineration or landfill disposal, which are less environmentally friendly options. It also decreases the demand for virgin polyester production, contributing to resource conservation. Thus, nanoparticles showcase the potential of nanotechnology to revolutionize waste management and contribute to a more circular and sustainable textile industry [117].

5.4.3 Nanoparticles for Wastewater Dye Removal

Nanoparticles with high surface area and specific surface properties can adsorb dye molecules onto their surfaces. This process relies on the interaction between the dye molecules and the nanoparticles' surface functional groups. Nanoparticles can be used in combination with adsorbent materials to enhance dye removal. For instance, they can be immobilized onto the surface of a porous adsorbent material. Thus, creating a composite material benefits both adsorption and catalytic degradation mechanisms. Certain nanoparticles can facilitate redox reactions with dye molecules. For example, zero-valent iron nanoparticles can donate electrons to dye molecules. It leads to reduction and subsequent removal from the solution. Electrochemical methods involving nanoparticles can be employed for dye removal. Nanoparticles can act as catalysts in electrochemical reactions that degrade dye molecules through processes like electro-oxidation. Nanoparticles can be incorporated into membrane materials used in filtration processes. These nanocomposite membranes can selectively adsorb or degrade dye molecules as wastewater passes through them. Membranes with enhanced permeability, fouling resistance, stability, and removal capabilities are needed. The effectiveness of nanoparticle-based dye removal can be influenced by factors such as pH and temperature. Optimal conditions need to be determined for each specific dye and nanoparticle combination. Depending on the mechanism used, nanoparticles can be regenerated and reused. Adsorbent nanoparticles can be desorbed from the dye molecules, allowing the nanoparticles to be reused multiple times. Their catalytic and adsorption properties make them promising candidates for advanced wastewater treatment methods. Common nanoparticle materials used for adsorption include gold, silver, iron, cobalt, zinc, titanium, and molybdenum nanoparticles. The different forms such as metal oxides, carbon-based nanoparticles [e.g., graphene oxide (GO)], and activated carbon are also efficient in the removal of dyes [118]. The adsorption capacity can be enhanced by modifying the nanoparticle surface chemistry to increase dye binding affinity. Some nanoparticles have catalytic properties that can initiate chemical reactions, leading to the breakdown of dye molecules. Sulfonated and gamma-irradiated waste expanded polystyrene (EPS) with iron oxide nanoparticles is a material that has been developed for the removal of indigo carmine dye from textile wastewater. This material combines EPS with iron oxide nanoparticles, resulting in a multifunctional material havimg enhanced dye removal capabilities. Gamma irradiation can further modify the EPS structure, potentially increasing its surface area and reactivity. Iron oxide nanoparticles, which can possess magnetic properties due to their iron content, are incorporated into the modified EPS. These nanoparticles provide additional adsorption capacity and can facilitate the magnetic separation of the material from the wastewater. Electrostatic interactions, hydrogen bonding, and the combination of gamma irradiation, and iron oxide nanoparticles can increase the material's surface area and reactivity This enhances its overall adsorption capacity for dye molecules. The magnetic separation simplifies the removal process and facilitates material recovery for reuse [119]. After adsorbing dye molecules, the material can potentially be regenerated through desorption methods, such as changing pH or using appropriate solvents. This regeneration allows for multiple cycles of dye removal and material reuse. But factors like material stability, optimization of sulfonation and irradiation processes, and scalability need to be considered for practical application.

Similarly, phytogenic magnetic nanoparticles functionalized with 3-mercaptopropanic acid can effectively sorb and remove malachite green from wastewater. This is through a combination of magnetic attraction and chemical interactions. The nanoparticles can be tailored to have specific sizes and surface properties. 3-Mercaptopropanic acid is a thiol-based compound with a sulfur-containing functional group. Functionalizing the magnetic nanoparticles with this acid introduces thiol groups on their surfaces, which can form chemical interactions with various types of pollutants, including malachite green. The interaction between the thiol-functionalized nanoparticles and malachite green involves chemical bonding, such as thiol-malachite green interactions, as well as electrostatic interactions. These interactions facilitate the adsorption of the dye molecules onto the nanoparticle surfaces. The magnetic nature of the nanoparticles allows them to be easily separated from the wastewater using an external magnetic field. This magnetic separation simplifies the separation process, making it more efficient and effective. The combination of chemical interactions and magnetic attraction enhances the overall removal efficiency of malachite green from wastewater. The chemical interactions provide a strong binding mechanism, while the magnetic separation ensures easy recovery of the nanoparticles for reuse. After adsorbing malachite green, the nanoparticles can be regenerated by desorption methods, such as changing pH or using appropriate solvents. This regeneration allows for multiple cycles of dye removal and nanoparticle reuse. The use of phytogenic materials and the efficient removal of malachite green contribute to a more sustainable and eco-friendly approach to treating dye-contaminated wastewater. In summary, the use of phytogenic magnetic nanoparticles functionalized with 3-mercaptopropanic acid presents an innovative and efficient solution for the removal of malachite green dye from wastewater. The combination of magnetic attraction and chemical interactions offers an effective and sustainable approach to addressing dye pollution challenges [120, 121].

5.4.4 Nanomaterials for Dye Removal

Nanomaterials possess unique properties that make them efficient adsorbents and catalysts for dye removal and degradation processes [122]. Graphene is a single layer of carbon atoms arranged in a two-dimensional honeycomb lattice structure. Its large surface area, high electrical conductivity, and excellent mechanical strength make it an attractive material for adsorption and catalysis [123]. Graphene-based materials, such as GO and reduced GO (rGO), have functional groups on their surfaces that can interact with dye molecules through π-π stacking and electrostatic interactions. Graphene can be used as an adsorbent to remove dyes through physical interactions or as a support for catalysts to enhance degradation reactions [124]. Carbon nanotubes (CNTs) are cylindrical nanostructures composed of carbon atoms. They have high aspect ratios, mechanical strength, and large surface areas, which make them effective adsorbents. Functionalized CNTs can interact with dye molecules through van der Waals forces, π-π stacking, and electrostatic interactions. CNTs can also serve as support materials for catalysts, enhancing degradation reactions through increased surface area and reactivity. The utilization of these nanomaterials for dye removal and degradation offers several benefits. The large surface areas and unique properties of nanomaterials enhance their efficiency in adsorption and catalysis, leading to rapid dye removal and degradation. Functionalized nanomaterials can be engineered to selectively interact with specific dye molecules, enabling targeted removal [125]. Many nanomaterials can be regenerated and reused, making the process economically viable and environmentally friendly. The combination of different nanomaterials can create synergistic effects, boosting overall dye removal and degradation performance [126].

5.4.5 Nanoparticle-Mediated Photodegradation

Photocatalytic nanoparticles like titanium dioxide (TiO2) or zinc oxide (ZnO) can be activated by UV light to generate reactive oxygen species (ROS) that degrade the dye molecules into simpler compounds. This process is known as photocatalytic degradation [127]. Metal oxide nanoparticles, including titanium dioxide (TiO_2) and iron oxide (Fe_2O_3), are often used as photocatalysts for dye degradation. Under UV or visible light irradiation, metal oxide nanoparticles break down dye molecules into simpler compounds. Activated mesoporous ZnO/g-$C_{(3)}N_{(4)}$ catalyst under sunlight irradiation absorbs solar energy and initiates the generation of ROS, which oxidizes and breaks down textile pollutants into non-toxic compounds [128, 129]. The use of an activated mesoporous ZnO/g-$C_{(3)}N_{(4)}$ catalyst under sunlight irradiation is an advanced and sustainable approach for the degradation of textile pollutants in wastewater. This method takes advantage of photocatalysis, a process in which a catalyst absorbs solar energy and uses it to generate ROS that can break down organic pollutants into non-toxic compounds. The catalyst consists of two main components: mesoporous ZnO (zinc oxide) and g-$C_{(3)}N_{(4)}$ (graphitic carbon nitride). Both materials are semiconductors that have photocatalytic properties, meaning they can absorb light and use its energy to drive chemical reactions. The mesoporous structure of the catalyst provides a large surface area and abundant active sites for pollutant adsorption and catalytic reactions to occur. The g-$C_{(3)}N_{(4)}$ component of the catalyst is capable of absorbing sunlight across a broad range of wavelengths, including visible light. This absorption generates electron-hole pairs (excited states) within the material. When the g-$C_{(3)}N_{(4)}$ absorbs sunlight, the excited electrons and holes can migrate to the ZnO component, where they initiate redox reactions with water and oxygen from the surrounding environment [128, 130]. These reactions produce ROS such as hydroxyl radicals (•OH) and superoxide radicals (•O2–). The ROS generated by the catalyst is highly reactive and can oxidize the organic pollutants present in the wastewater. This oxidation breaks down complex organic compounds into simpler, non-toxic molecules such as carbon dioxide and water. The process offers an eco-friendly way to treat textile wastewater as it directly utilizes solar energy to drive the degradation of pollutants. Additionally, the end products of the oxidation reactions are harmless and can be safely released into the environment [131]. The selectivity and efficiency of the catalyst depend on factors like its composition, morphology, and the characteristics of the pollutants being treated. Optimization is crucial to achieving the best results. The catalyst operates as a heterogeneous system, meaning it exists as a distinct phase from the wastewater. This allows for easy separation and recycling of the catalyst for multiple cycles. Nanocellulose composites are evaluated as photocatalysts for the absorption of textile black dye and methylene blue dye. The results show a significant decrease in dye concentration after each hour, highlighting the bio-adsorbent property of the composites [132].

5.4.6 Biosynthesized Nanoparticles for Dye Removal

Bio-synthesized nanoparticles offer a promising avenue for improving the efficiency of biodegradation processes, especially in the context of treating textile dyes and other pollutants [133]. This approach combines the advantages of both nanotechnology and bioremediation to create a more effective and sustainable solution for wastewater pollution. They are often derived from bacteria, fungi, plants, or algae. Nanoparticles, which are extremely small particles with unique properties due to their size, can be bio-synthesized using various biological methods. Bio-synthesized nanoparticles, particularly those with adsorption properties, can effectively remove dyes from textile wastewater [134]. These nanoparticles can bind to the dye molecules through electrostatic reactions, hydrogen bonding, and van der Waals forces, leading to their removal from the solution and subsequent decolorization. The properties of nanoparticles, such as size, shape, and surface chemistry, can be tailored to optimize their interactions with specific enzymes or microorganisms, further enhancing their performance in pollutant degradation. By incorporating bio-synthesized nanoparticles into existing treatment processes such as activated sludge systems or membrane filtration, the overall efficiency of wastewater treatment can be improved. Bio-synthesized nanoparticles are often biodegradable and have low toxicity, making them a more environmentally friendly option compared to conventional chemical treatments. Bio-synthesized nanoparticles can be used in conjunction with other wastewater treatment technologies like membrane filtration, adsorption columns, and biological treatment to achieve a comprehensive and effective treatment process [135, 136]. Ismali *et al.* bio-synthesized silver nanoparticles through two species of cyanobacteria, Anabaena variabilis, and *Spirulina platensis*. They experimented with Malachite green dye removal and found the synthesized nanoparticles to be highly efficient in the removal of dyes [137]. Other biosynthesized nanoparticles of iron oxide, ZnO with Citrus Sinensis, and MnO with Ficus leaves have given significant results in the dye removal process [138].

5.4.7 Bio-Synthesized Nanoparticles for Biodegradation

Biosynthesized nanoparticles have a high surface area-to-volume ratio, which provides a larger area for the attachment of enzymes or microorganisms. This increases the availability of active sites for the biodegradation process to occur, leading to faster and more efficient pollutant breakdown. By using bio-synthesized nanoparticles, it is possible to reduce the reliance on synthetic chemicals, thereby minimizing the potential negative environmental impacts [139]. Bio-synthesized nanoparticles can serve as carriers for enzymes or microorganisms that are effective in degrading textile dyes and other pollutants. This approach can enhance the biodegradation process and increase treatment efficiency [140]. Nanoparticle carriers can provide protection and stability to these biomolecules, allowing them to remain active for longer periods and under challenging conditions. Nanoparticle carriers can be designed to deliver enzymes or microorganisms directly to the polluted site. This targeted delivery minimizes the dispersion of these biodegrading agents and ensures that they are concentrated where they are needed the most. Also, the nanoparticles can help control the release and distribution of the biodegrading agents, preventing unintended ecological consequences. Some nanoparticles can be recovered and reused after the biodegradation process [141]. This adds to the sustainability of the approach by reducing waste and resource consumption. The properties of biosynthesized nanoparticles, such as size, shape, and composition, can be influenced by the type of biological agent used and the specific conditions of the synthesis process [142]. Dhaka *et al.* studied the effectiveness of *Balanites aegyptiaca*-biosynthesized silver nanoparticles for both methylene blue and congo red dye degradation. The results demonstrate effective dye removal and provide a green solution in the dye treatment process [143].

5.5 Nanofibers for Dye Separation

Nanofibers can be easily separated from the treated water due to their larger size compared to nanoparticles. They possess a high surface area due to their nanoscale dimensions and porous structure. It provides ample active sites for the adsorption of anionic dye molecules [144, 145]. Factors such as nanofiber structure, surface modification methods, regeneration techniques, and specific dye characteristics need to be optimized for maximum efficiency [146]. After adsorbing dye molecules, the nanofibers can be regenerated through desorption techniques, such as changing pH or using appropriate solvents. This allows for multiple cycles of dye removal and nanofiber regeneration, making the process economically and environmentally efficient [147]. This simplifies the post-treatment separation process. Silk is a natural biopolymer, making regenerated silk nanofibers biocompatible and environmentally friendly. Silk nanofibers are known for their mechanical strength and durability. Regenerated silk nanofibers offer the advantage of being regenerable and reusable. It offers an effective approach to wastewater treatment for robust and cyclic adsorption-desorption of anionic dyes. The surface of silk nanofibers can be modified to introduce functional groups that enhance the interaction with anionic dye molecules. These modifications can improve the affinity of the nanofibers for the target dyes. Regenerated silk nanofibers can be used in batch systems, continuous-flow columns, or even integrated into filtration membranes, offering versatility in application. Regenerated silk nanofibers can be combined with other materials to create composite materials with enhanced dye adsorption properties or additional functionalities. Further, the use of regenerated silk nanofibers for the adsorption-desorption of anionic dyes aligns with the principles of green chemistry and sustainable wastewater treatment.

5.6 Waste Cotton Recycling

Waste cotton accounts for 24% of the total textile waste generated. One viable solution presented is the production of thermal insulating products using carding-folding technology, which involves using natural fibers, as well as recycled PET (rPET) and polyester (rPES) waste converted into fibers. Waste cotton can be modified to serve as adsorbents for pollutants in wastewater treatment. Their porous structure and surface chemistry make them effective at removing contaminants from water. Waste cotton can contribute to the creation of closed-loop recycling systems in the textile industry. It is through end-of-life textiles that are being collected, processed, and converted into new products. The use of a superbase-based ionic liquid [DBNH] [OAc] for selectively dissolving cellulose from cotton-polyester blended waste garments is a noteworthy innovation. This separation technique could contribute to the efficient recovery of valuable cellulose fibers [148]. Waste cotton is recycled into yarns, composite reinforcements, regenerated cellulose fibers, cellulose nanocrystals (CNCs), adsorptive materials, flexible electronic devices, and biofuels [149]. It can be mechanically or chemically processed to break down the fibers, creating recycled cotton fibers that can be spun into yarns. These yarns can be used for new textile products, reducing the demand for virgin cotton. Regenerated cellulose fibers, like viscose, lyocell, or modal, and waste cotton fibers can be incorporated into composite materials to reinforce various products such as automotive parts, construction materials, and consumer goods [150]. Cellulose extracted from waste cotton can also be converted into CNCs, which are nanoscale particles with unique mechanical and optical properties. CNCs find applications in various fields including coatings, films, and biomedical materials [151]. Cellulose-based materials derived from waste cotton can be utilized in the development of flexible and eco-friendly electronic devices, such as wearable sensors and displays. Cotton waste can be processed to extract sugars, which can then be converted into biofuels through fermentation or other biochemical processes [152]. Waste cotton fibers can be used as an eco-friendly insulation material for construction applications. Waste cotton fibers can be incorporated into paper and packaging products, reducing the demand for virgin wood pulp. In addition to the mentioned applications, The diverse range of products that can be obtained from waste cotton showcases the potential of recycling and upcycling in minimizing waste and promoting sustainable practices across various industries. However, it is important to consider factors like technological feasibility, economic viability, and environmental impact when implementing these recycling processes [153, 154]. Moreover, the recycling of polyester/cotton mix textile waste and generating carbon materials with desirable electrochemical properties contribute to the development of sustainable approaches for textile waste management and the production of recycled carbon materials for various applications. The utilization of nanocellulose from textile waste reduces the demand for virgin resources and minimizes environmental impacts associated with textile disposal.

5.7 Mixed Solid Waste Recycling

The utilization of textile waste fibers as reinforcement in cement-based composites is a novel concept. These fibers can contribute to enhancing the mechanical properties of the composite material, such as ductility and crack resistance. This indicates the potential to create more durable and resilient construction materials. The proposed green chemistry strategy offers a potential solution for the recycling and regeneration of waste textiles. Separating and regenerating fibers at the molecular level opens up possibilities for reusing textile waste and reducing reliance on virgin resources. This approach aligns with the goals of sustainability and can contribute to minimizing the environmental impact of the fashion industry [155]. The use of geotextile ropes made from recycled natural and synthetic fibers demonstrates the application of recycled materials in environmental protection. These ropes effectively prevent slope erosion over a prolonged period [156]. By blending recycled fibers with virgin fibers, apparel manufacturers can achieve softer and more moisture-absorbent yarns suitable for knit top garments while also contributing to the recycling and sustainability goals of the textile industry.

Composite panels created from 100% wool waste fibers and bound with solutions like chitosan or gum Arabic exhibit favorable acoustic and non-acoustic properties. These panels have the potential to serve as effective building materials, enhancing thermal and acoustic comfort in construction [157]. The use of hemp textile waste to create stable lithium (Li) composite anodes offers a cost-effective and scalable solution aligned with sustainability goals [158]. Microalgae, particularly diatoms, offer a promising solution for dye remediation in textile wastewater. Nanoporous diatom frustules doped with nanocomposites enhance adsorption properties, improving the efficiency of wastewater treatment. Moreover, the integration of microalgae with photosynthetic microbial fuel cells demonstrates the potential for efficient, cost-effective, and scalable phytoremediation. This system not only removes dyes but also promotes microalgae growth and electricity generation [159]. The development of renewable polydopamine-functionalized cellulose aerogels (CAs) presents a green and feasible approach to generating clean water. These aerogels could potentially be used for water purification, showcasing the application of sustainable materials for environmental benefit method to recycle up to 50% of textile waste by incorporating marine bio-based calcium alginate fibers. The result is a fully bio-based composite textile achieved through a carding process. This approach demonstrates a sustainable way to repurpose waste and create new materials. The recycled marine bio-based composite textiles possess inherent flame retardancy, making them suitable for applications in fireproof structural filling and insulation materials for household textiles and construction. This not only repurposes textile waste but also contributes to greener and safer material solutions [160, 161].

5.8 Challenges and Prospects

Textile waste management has a number of difficulties and opportunities as the sector seeks to become more viable. Many people are uninformed of the environmental implications of textile waste and may overlook proper disposal or recycling. Inadequate collection infrastructure impedes effective textile waste collection, particularly in places with less developed waste management systems. Fast fashion encourages quick consumption and disposal of garments, resulting in increasing textile waste output. Textile waste is frequently contaminated with various materials, such as zippers, buttons, and accessories, complicating recycling. Developing efficient and cost-effective textile recycling processes, particularly for mixed-fiber garments, remains a challenge. The worldwide textile supply chain is complicated, making it challenging to detect and minimize waste at each level. It might be difficult to make textile recycling economically viable while maintaining quality standards. Inadequate legislation and incentives for appropriate textile waste disposal can stymie progress. To reduce waste and conserve resources, the industry is shifting toward a circular economy model that emphasizes reuse, repair, remanufacturing, and recycling. AI, machine learning, robotics, and material science advancements provide potential solutions for effective textile waste sorting and recycling. Increased use of sustainable materials such as organic cotton, recycled fibers, and biodegradable fabrics can help to lessen waste's environmental impact. Blockchain technology can improve transparency and traceability, ensuring that textiles are sourced and maintained sustainably. Awareness campaigns and education about the environmental impact of textile waste can encourage responsible usage and disposal. Continued collaboration among industry stakeholders, can drive collaborative efforts toward improved waste management practices. To reduce waste, many firms are implementing eco-friendly production practices, such as closed-loop procedures. Regulations and policies that support responsible waste management can encourage more sustainable practices. Brands are investigating rental, resale, and subscription models to extend the lifecycle of apparel and reduce waste. Textile recyclability can be improved by designing clothes with disassembly and recycling in mind. Community-based collection and recycling programs can address waste management issues at the ground level. While there are significant challenges in textile waste management, the industry's shift toward sustainability, along with technological advancements and increased stakeholder engagement, holds promise for a future with less textile waste and a more environmentally responsible approach to fashion and textiles. Challenges in nanoparticle-mediated wastewater treatment include optimization of the nanoparticle composition, size, and concentration to ensure efficient catalysis. Additionally, issues related to nanoparticle recovery and potential environmental impacts need to be addressed. The economic feasibility of this process will depend on factors such as nanoparticle cost, energy consumption, and the market demand for recycled products. Factors like catalyst stability, regeneration, and potential fouling need to be addressed to ensure the long-term effectiveness of the process. This approach holds promise for large-scale wastewater treatment in industries like textiles, where conventional methods might be less environmentally friendly. Sunlight irradiation illustrates the potential of harnessing solar energy and photocatalysis to address pollution challenges. However, successful implementation requires a thorough understanding of catalyst design, reaction kinetics, and process optimization. Despite their benefits, biosynthesized nanoparticles can have challenges related to reproducibility, control over properties, and scalability. The synthesis process can be sensitive to environmental factors, and achieving consistent results might require optimization. Factors like nanoparticle stability, optimization of functionalization, and scalability need to be considered for practical application. Additionally, potential impacts on other components of the wastewater and the overall process cost-effectiveness should be evaluated. This approach has potential applications in various industries, including textiles, where dye contamination is a concern. The successful implementation of this method requires collaboration between researchers, industry stakeholders, and regulatory bodies to ensure safe and effective treatment practices.

Many countries have regulations in place to limit the discharge of pollutants into water bodies. Textile industries are often required to treat their liquid waste to meet certain standards before releasing it into the environment. Textile manufacturers are adopting sustainable practices, such as using eco-friendly dyes, reducing water usage, and implementing closed-loop systems where water and chemicals are reused to minimize the generation of liquid textile waste. Efforts to address the challenges posed by liquid textile waste involve a combination of regulatory measures, technological innovations, and industry best practices to reduce its environmental impact and promote sustainable textile manufacturing.

5.9 Conclusion

The consumption barrier of 100 million tons highlights the enormous number of textiles used and destroyed each year. The fact that more than two-thirds of textiles end up in landfills reflects the fashion industry's major waste problem. Textile landfills not only waste valuable resources but also contribute to environmental damage and pollution. Another troubling aspect of the fashion industry's impact is its poor recycling rate of roughly 15%. This low rate is a missed chance to decrease waste, conserve resources, and reduce the industry's environmental impact. To produce desired colors, textures, and qualities, textile manufacturers utilize a vast variety of dyes, pigments, and chemicals. These compounds are potentially hazardous, carcinogenic, and persistent in the environment. There is an urgent need to transform the textile industry into a circular model. Implementing strategies for reuse, recycling, and repurposing of textile waste is crucial. Scientific breakthroughs, innovations, and disruptive strategies can contribute to establishing a circular textile economy. Pyrolysis is described as a thermochemical process capable of converting the complex and heterogeneous character of textile waste into value-added products. This process is under research due to its maximum recycling potential. There is a growing demand to recover materials and energy from textile waste, thereby reducing the carbon and water footprints associated with these products. This approach also aims to minimize environmental impact and work toward a zero discharge of hazardous chemicals (ZDHC) issues related to the collection, sorting, and processing of textile waste, as well as developing technologies for efficient and high-quality recycling.

References

1. Mamun, A.A., Bormon, K.K., Rasu, M.N.S., Talukder, A., Freeman, C., Burch, R., Chander, H., An Assessment of Energy and Groundwater Consumption of Textile Dyeing Mills in Bangladesh and Minimization of Environmental Impacts via Long-Term Key Performance Indicators (KPI) Baseline. *Textiles*, 2, 4, 511–523, 2022.
2. Shirvanimoghaddam, K., Motamed, B., Ramakrishna, S., Naebe, M., Death by waste: Fashion and textile circular economy case. *Sci. Total Environ.*, 718, 137317, 2020.
3. Peters, G.M., Sandin, G., Spak, B.r., Environmental prospects for mixed textile recycling in Sweden. *ACS Sustainable Chem. Eng.*, 7, 13, 11682–11690, 2019.
4. Niinimäki, K., Peters, G., Dahlbo, H., Perry, P., Rissanen, T., Gwilt, A., The environmental price of fast fashion. *Nat. Rev. Earth Environ.*, 1, 4, 189–200, 2020.
5. Pandit, P., Nadathur, G.T., Jose, S., Upcycled and low-cost sustainable business for value-added textiles and fashion, in: *Circular Economy in Textiles and Apparel*, pp. 95–122, Elsevier, 2019.
6. Khandaker, S., Bashar, M.M., Islam, A., Hossain, M.T., Teo, S.H., Awual, M.R., Sustainable energy generation from textile biowaste and its challenges: A comprehensive review. *Renew. Sustain. Energy Rev.*, 157, 112051, 2022.
7. Kazancoglu, I., Kazancoglu, Y., Yarimoglu, E., Kahraman, A., A conceptual framework for barriers of circular supply chains for sustainability in the textile industry. *Sustainable Dev.*, 28, 5, 1477–1492, 2020.
8. Kamble, Z. and Behera, B.K., Upcycling textile wastes: challenges and innovations. *Text. Prog.*, 53, 2, 65–122, 2021.
9. Zhang, J., Qian, X., Feng, J., Review of carbon footprint assessment in textile industry. *Ecofeminism Clim. Change*, 1, 1, 51–56, 2020.
10. Haseeb, M., Haouas, I., Nasih, M., Mihardjo, L.W., Jermsittiparsert, K., Asymmetric impact of textile and clothing manufacturing on carbon-dioxide emissions: Evidence from top Asian economies. *Energy*, 196, 117094, 2020.
11. Nunes, L.J.R., Godina, R., Matias, JCO., Technological innovation in biomass energy for the sustainable growth of textile industry. *Sustainability*, 11, 2, 528, 2019.
12. Stanescu, M.D., State of the art of post-consumer textile waste upcycling to reach the zero waste milestone. *Environ. Sci. Pollut. Res.*, 28, 12, 14253–14270, 2021.
13. Juanga-Labayen, J.P., Labayen, I.V., Yuan, Q., A review on textile recycling practices and challenges. *Textiles*, 2, 1, 174–188, 2022.

14. Dhir, Y.J., Hazards of fashion and textile waste: Approaches for effective waste management, in: *Waste management in the fashion and textile industries*, pp. 31–58, Elsevier, 2021.
15. To, M.H., Uisan, K., Ok, Y.S., Pleissner, D., Lin, C.S.K., Recent trends in green and sustainable chemistry: rethinking textile waste in a circular economy. *Curr. Opin. Green Sustain. Chem.*, 20, 1–10, 2019.
16. Liu, H., Fan, W., Miao, Y., Dou, H., Shi, Y., Wang, S., Zhang, X., Hou, L., Yu, X., Lam, S.S., Closed-loop recycling of colored regenerated cellulose fibers from the dyed cotton textile waste. *Cellulose*, 30, 4, 2597–2610, 2023.
17. Ma, Y., Zeng, B., Wang, X., Byrne, N., Circular textiles: closed loop fiber to fiber wet spun process for recycling cotton from denim. *ACS Sustain. Chem. Eng.*, 7, 14, 11937–11943, 2019.
18. Schmutz, M. and Som, C., Identifying the potential for circularity of industrial textile waste generated within Swiss companies. *Resour. Conserv. Recycl.*, 182, 106132, 2022.
19. Luque, A. and Herrero-García, N., How corporate social (ir) responsibility in the textile sector is defined, and its impact on ethical sustainability: An analysis of 133 concepts. *Corp. Soc. Resp. Environ. Manage.*, 26, 6, 1285–1306, 2019.
20. Gupta, V., Arora, M., Minhas, J., Innovating opportunities for fashion brands by using textile waste for better fashion, in: *Recycling from Waste in Fashion and Textiles: A Sustainable and Circular Economic Approach*, pp. 101–121, 2020.
21. Baptista, A., Abreu, L., Brito, E., Application of lean tools: Case study in a textile company. *Proc. Eng.*, 3, 1, 93–102, 2021.
22. Ütebay, B., Çelik, P., Çay, A., Textile wastes: Status and perspectives, in: *Waste in textile and leather sectors*, pp. 39–52, 2020.
23. Jäämaa, L. and Kaipia, R., The first mile problem in the circular economy supply chains–Collecting recyclable textiles from consumers. *Waste Manage. (Oxford)*, 141, 173–182, 2022.
24. Rani, S. and Jamal, Z., Recycling of textiles waste for environmental protection. *Int. J. Home Sci.*, 4, 1, 164–168, 2018.
25. Donkadokula, N.Y., Kola, A.K., Naz, I., Saroj, D., A review on advanced physico-chemical and biological textile dye wastewater treatment techniques. *Rev. Environ. Sci. Bio/Technol.*, 19, 543–560, 2020.
26. Slama, H.B., Chenari Bouket, A., Pourhassan, Z., Alenezi, F.N., Silini, A., Cherif-Silini, H., Oszako, T., Luptakova, L., Golińska, P., Belbahri, L., Diversity of synthetic dyes from textile industries, discharge impacts and treatment methods. *Appl. Sci.*, 11, 14, 6255, 2021.
27. Leal Filho, W., Ellams, D., Han, S., Tyler, D., Boiten, V.J., Paço, A., Moora, H., Balogun, A.-L., A review of the socio-economic advantages of textile recycling. *J. Cleaner Prod.*, 218, 10–20, 2019.

28. Bhatt, D., Silverman, J., Dickson, M.A., Consumer interest in upcycling techniques and purchasing upcycled clothing as an approach to reducing textile waste. *Int. J. Fash. Des. Technol. Educ.*, 12, 1, 118–128, 2019.
29. Marques, A.D., Moreira, B., Cunha, J., Moreira, S., From waste to fashion–a fashion upcycling contest. *Procedia CIRP*, 84, 1063–1068, 2019.
30. Yalcin-Enis, I., Kucukali-Ozturk, M., Sezgin, H., Risks and management of textile waste, in: *Nanoscience and biotechnology for environmental applications*, pp. 29–53, 2019.
31. Carrico, M., Sourcing Fabrics for New Designs from Post-Consumer Textiles. *International Textile and Apparel Association Annual Conference Proceedings*, Iowa State University Digital Press, 2018.
32. Velusamy, S., Roy, A., Sundaram, S., Kumar Mallick, T., A review on heavy metal ions and containing dyes removal through graphene oxide-based adsorption strategies for textile wastewater treatment. *Chem. Rec.*, 21, 7, 1570–1610, 2021.
33. Al-Mamun, M., Kader, S., Islam, M., Khan, M., Photocatalytic activity improvement and application of UV-TiO2 photocatalysis in textile wastewater treatment: A review. *J. Environ. Chem. Eng.*, 7, 5, 103248, 2019.
34. Pervez, M.N., Mondal, M.I.H., Cai, Y., Zhao, Y., Naddeo, V., Textile waste management and environmental concerns, in: *Fundamentals of Natural Fibres and Textiles*, pp. 719–739, Elsevier, 2021.
35. Boschmeier, E., Ipsmiller, W., Bartl, A., Market assessment to improve fibre recycling within the EU textile sector. *Waste Manage. Res.*, 42, 2, 2023, 0734242X231178222.
36. Bukhari, M.A., Carrasco-Gallego, R., Ponce-Cueto, E., Developing a national programme for textiles and clothing recovery. *Waste Manag. Res.*, 36, 4, 321–331, 2018.
37. Burton, K., Reducing textile waste in the apparel industry: Examining EPR as an option. *Clothing Cultures*, 5, 1, 33–45, 2018.
38. Degenstein, L.M., McQueen, R.H., Krogman, N.T., McNeill, L.S., Integrating Product Stewardship into the Clothing and Textile Industry: Perspectives of New Zealand Stakeholders. *Sustainability*, 15, 5, 4250, 2023.
39. Peters, S. and Samarasinghe, K., Waste management practices: innovation, waste to energy and e-EPR, in: *An Introduction to Circular Economy*, pp. 287–301, 2021.
40. Tyler, D., Hall, N., Street, C., Eco-system services and the circular economy for textiles. *The 91st Textile Institute World Conference*, Leeds University, UK, 2018.
41. dos Santos, P.S. and Campos, L.M., Practices for garment industry's post-consumer textile waste management in the circular economy context: an analysis on literature. *Braz. J. Oper. Prod. Manage.*, 18, 1, 1–17, 2021.
42. Rotimi, E.O.O., Topple, C., Hopkins, J., Towards a conceptual framework of sustainable practices of post-consumer textile waste at garment end of lifecycle: A systematic literature review approach. *Sustainability*, 13, 5, 2965, 2021.

43. Martikkala, A., Mayanti, B., Helo, P., Lobov, A., Ituarte, I.F., Smart textile waste collection system–Dynamic route optimization with IoT. *J. Environ. Manage.*, 335, 117548, 2023.
44. Anjum, M., Shahab, S., Umar, M.S., Smart waste management paradigm in perspective of IoT and forecasting models. *Int. J. Environ. Waste Manage.*, 29, 1, 34–79, 2022.
45. Tjhin, V., Riantini, R., Windihastuty, W., Utilization of Digital Technology in Managing Fabric Waste into Garments. *Proceedings of the 4th International Conference of Economics, Business, and Entrepreneurship, ICEBE 2021*, Lampung, Indonesia, 7 October 2021, 2022.
46. MobinHossain, M., Mukta, N., Akter, N., Tazin, T., Soroni, F., Khan, M.M., Microcontroller and Mobile App based Garments Environment Monitoring System for Workers. *2022 6th International Conference on Computing Methodologies and Communication (ICCMC)*, IEEE, pp. 616–622, 2022.
47. Tasnim, N.H., Afrin, S., Biswas, B., Anye, A.A., Khan, R., Automatic classification of textile visual pollutants using deep learning networks. *Alexandria Eng. J.*, 62, 391–402, 2023.
48. Chang, D., Second: Redesigning the Textile Recycling Experience Using a Circular Economy, 2020.
49. Clary, A.T., Alternative Clothing Disposal Methods: Analysis of the Millennial Participation in Take-Back Programs, North Carolina State University, 2020.
50. Moorhouse, D. and Moorhouse, D., Sustainable design: circular economy in fashion and textiles. *Des. J.*, 20, sup1, S1948–S1959, 2017.
51. Mallick, P., Pigosso, D., Salling, K., McAloone, T., Designing Configurator for Take-Back for a Circular Economy–A Conceptual Framework. *Proc. Des. Soc.*, 2, 1041–1050, 2022.
52. James, A., How Consumer Action can Help Build a Sustainable Future for Fashion. *J. Litter Environ. Qual.*, 4, 1, 46–50, 2020.
53. Patel, M., Sahu, A., Rajak, R., Solid waste management in textile industry, in: *Handbook of Solid Waste Management: Sustainability through Circular Economy*, pp. 1225–1256, Springer, 2022.
54. Zamani, B., Towards understanding sustainable textile waste management: Environmental impacts and social indicators, Chalmers Tekniska Hogskola (Sweden), 2014.
55. Rani, H. and Saha, G., Organizations and standards related to textile and fashion waste management and sustainability, in: *Waste Management in the Fashion and Textile Industries*, pp. 173–196, Elsevier, 2021.
56. Christensen, T.B., Towards a circular economy in cities: Exploring local modes of governance in the transition towards a circular economy in construction and textile recycling. *J. Cleaner Prod.*, 305, 127058, 2021.
57. Cheela, V.S., Ranjan, V.P., Goel, S., John, M., Dubey, B., Pathways to sustainable waste management in Indian Smart Cities. *J. Urban Manage.*, 10, 4, 419–429, 2021.

58. Ahmed, A.A.A. and Asadullah, A., Artificial intelligence and machine learning in waste management and recycling. *Eng. Int.*, 8, 1, 43–52, 2020.
59. Erkinay Ozdemir, M., Ali, Z., Subeshan, B., Asmatulu, E., Applying machine learning approach in recycling. *J. Mater. Cycles Waste Manage.*, 23, 855–871, 2021.
60. Du, W., Zheng, J., Li, W., Liu, Z., Wang, H., Han, X., Efficient recognition and automatic sorting technology of waste textiles based on online near infrared spectroscopy and convolutional neural network. *Resour. Conserv. Recycl.*, 180, 106157, 2022.
61. Zhang, Q., Yang, Q., Zhang, X., Bao, Q., Su, J., Liu, X., Waste image classification based on transfer learning and convolutional neural network. *Waste Manage. (Oxford)*, 135, 150–157, 2021.
62. Fang, B., Yu, J., Chen, Z., Osman, A.I., Farghali, M., Ihara, I., Hamza, E.H., Rooney, D.W., Yap, P.-S., Artificial intelligence for waste management in smart cities: a review. *Environ. Chem. Lett.*, 21, 1959–1959, 2023.
63. Yuldoshev, N., Tursunov, B., Qozoqov, S., Use of artificial intelligence methods in operational planning of textile production. *J. Process Manage. New Technol.*, 6, 2, 41–51, 2018.
64. Lievano-Martínez, F.A., Fernández-Ledesma, J.D., Burgos, D., Branch-Bedoya, J.W., Jimenez-Builes, J.A., Intelligent process automation: An application in manufacturing industry. *Sustainability*, 14, 14, 8804, 2022.
65. Rathore, B., Artificial Intelligence in Sustainable Fashion Marketing: Transforming the Supply Chain Landscape. *Eduzone: Int. Peer Reviewed/Refereed Multi. J.*, 8, 2, 25–38, 2019.
66. Zhang, F. and Demiris, Y., Learning grasping points for garment manipulation in robot-assisted dressing. *2020 IEEE International Conference on Robotics and Automation (ICRA)*, IEEE, pp. 9114–9120, 2020.
67. Ku, S., Choi, H., Kim, H.-Y., Park, Y.-L., Automated sewing system enabled by machine vision for smart garment manufacturing. *IEEE Rob. Autom. Lett.*, 8, 9, 5680–5687, 2023.
68. Corona, E., Alenya, G., Gabas, A., Torras, C., Active garment recognition and target grasping point detection using deep learning. *Pattern Recognit.*, 74, 629–641, 2018.
69. Nayak, R. and Padhye, R., Introduction to automation in garment manufacturing, in: *Automation in garment manufacturing*, pp. 1–27, Elsevier, 2018.
70. Calvini, R., Ulrici, A., Amigo, J.M., Growing applications of hyperspectral and multispectral imaging, in: *Data handling in science and technology*, vol. 32, pp. 605–629, 2019.
71. Li, J., Meng, X., Wang, W., Xin, B., A novel hyperspectral imaging and modeling method for the component identification of woven fabrics. *Text. Res. J.*, 89, 18, 3752–3767, 2019.
72. Brito, T., Queiroz, J., Piardi, L., Fernandes, L.A., Lima, J., Leitão, P., A machine learning approach for collaborative robot smart manufacturing inspection for quality control systems. *Procedia Manuf.*, 51, 11–18, 2020.

73. Avadanei, M., Olaru, S., Ionescu, I., Ciobanu, L., Alexa, L., Luca, A., Ursache, M., Olmos, M., Aslanidis, T., Belakova, D., ICT new tools for a sustainable textile and clothing industry. *Ind. Textila*, 71, 5, 504–512, 2020.
74. Maddikunta, P.K.R., Pham, Q.-V., Prabadevi, B., Deepa, N., Dev, K., Gadekallu, T.R., Ruby, R., Liyanage, M., Industry 5.0: A survey on enabling technologies and potential applications. *J. Ind. Inf. Integr.*, 26, 100257, 2022.
75. Hader, M., Tchoffa, D., El Mhamedi, A., Ghodous, P., Dolgui, A., Abouabdellah, A., Applying integrated Blockchain and Big Data technologies to improve supply chain traceability and information sharing in the textile sector. *J. Ind. Inf. Integr.*, 28, 100345, 2022.
76. ElMessiry, M. and ElMessiry, A., Blockchain framework for textile supply chain management. *Improving transparency, traceability, and quality, International conference on blockchain*, Springer, pp. 213–227, 2018.
77. Nazam, M., Hashim, M., Nută, F.M., Yao, L., Zia, M.A., Malik, M.Y., Usman, M., Dimen, L., Devising a mechanism for analyzing the barriers of blockchain adoption in the textile supply chain: A sustainable business perspective. *Sustainability*, 14, 23, 16159, 2022.
78. Pal, K., Blockchain Storage With Sharing of Internet of Things Data in Textile Production Supply Chains, in: *Blockchain Applications in Cryptocurrency for Technological Evolution*, pp. 33–59, IGI Global, 2023.
79. Faridi, M.S., Ali, S., Duan, G., Wang, G., Blockchain and IoT based textile manufacturing traceability system in industry 4.0. *International conference on security, privacy and anonymity in computation, communication and storage*, Springer, pp. 331–344, 2020.
80. Saxena, M., Saxena, T., Seth, N., Sustainability In Apparel and Textile Supply Chains With Blockchain Technology: A Bibliometric Analysis. *Acad. Mark. Stud. J.*, 26, 2, 1–9, 2022.
81. Wang, B., Luo, W., Zhang, A., Tian, Z., Li, Z., Blockchain-enabled circular supply chain management: A system architecture for fast fashion. *Comput. Ind.*, 123, 103324, 2020.
82. Lam, O.W.A. and Zhibin, L., Textile and apparel supply chain with distributed ledger technology (DLT). *2019 20th IEEE international conference on mobile data management (MDM)*, IEEE, pp. 447–451, 2019.
83. Gonçalves, A. and Silva, C., Looking for sustainability scoring in apparel: A review on environmental footprint, social impacts and transparency. *Energies*, 14, 11, 3032, 2021.
84. Turunen, L.L.M. and Halme, M., Communicating actionable sustainability information to consumers: The Shades of Green instrument for fashion. *J. Cleaner Prod.*, 297, 126605, 2021.
85. Lee, E.J., Bae, J., Kim, K.H., The effect of environmental cues on the purchase intention of sustainable products. *J. Bus. Res.*, 120, 425–433, 2020.
86. Balderjahn, I., Peyer, M., Seegebarth, B., Wiedmann, K.-P., Weber, A., The many faces of sustainability-conscious consumers: A category-independent typology. *J. Bus. Res.*, 91, 83–93, 2018.

87. Agrawal, T.K., Koehl, L., Campagne, C., A secured tag for implementation of traceability in textile and clothing supply chain. *Int. J. Adv. Manuf. Technol.*, 99, 2563–2577, 2018.
88. Barteld, M., Gebhardt, R., Siegel, F., Portsch, U., Value added creation with advanced mass-customized textile labels. *IOP Conf. Ser.: Mater. Sci. Eng.*, 827, 012010, IOP Publishing, 2020.
89. Gayialis, S.P., Kechagias, E.P., Papadopoulos, G.A., Masouras, D., A review and classification framework of traceability approaches for identifying product supply chain counterfeiting. *Sustainability*, 14, 11, 6666, 2022.
90. Ahmed, W.A. and MacCarthy, B.L., Blockchain-enabled supply chain traceability in the textile and apparel supply chain: A case study of the fiber producer. *Lenzing Sustainability*, 13, 19, 10496, 2021.
91. Rusinek, M.J., Zhang, H., Radziwill, N., Blockchain for a Traceable, Circular Textile Supply Chain: A Requirements Approach. *Software Qual. Prof.*, 21, 1, 4, 2018.
92. Boiten, V.J., Han, S.L.-C., Tyler, D., Circular economy stakeholder perspectives: Textile collection strategies to support material circularity, European Union's, Brussels, Belgium, 2017.
93. Ki, C.W., Chong, S.M., Ha-Brookshire, J.E., How fashion can achieve sustainable development through a circular economy and stakeholder engagement: A systematic literature review. *Corp. Soc. Resp. Environ. Manage.*, 27, 6, 2401–2424, 2020.
94. Pilot Region, A., Wegener, J., Uhlig, J., Bohne, M., Beinker, F., Pilot Region, B., Kiba-Janiak, M., Jagoda, A., Krzywonos, M., Kołakowski, T., Stakeholder map and regional readiness for a systemic territorial solution, 2023.
95. Chavez, R. and Sharma, M., Profitability and environmental friendliness of a closed-loop supply chain for PET components: A case study of the Mexican automobile market. *Resour. Conserv. Recycl.*, 135, 172–189, 2018.
96. Whicher, A., Harris, C., Beverley, K., Swiatek, P., Design for circular economy: Developing an action plan for Scotland. *J. Cleaner Prod.*, 172, 3237–3248, 2018.
97. Smith, N.C., Ansett, S., Erez, L., How GAP Engaged with Its Stakeholders, in: *Managing Sustainable Business: An Executive Education Case and Textbook*, pp. 213–226, Springer, 2018.
98. Nayak, R., Akbari, M., Far, S.M., Recent sustainable trends in Vietnam's fashion supply chain. *J. Cleaner Prod.*, 225, 291–303, 2019.
99. Fiksel, J., Sanjay, P., Raman, K., Steps toward a resilient circular economy in India. *Clean Technol. Environ. Policy*, 23, 203–218, 2021.
100. Brydges, T., Henninger, C.E., Amasawa, E., Hanlon, M., Jones, C., For waste's sake: Stakeholder mapping of circular economy approaches to address the growing issue of clothing textile waste. *Int. J. Sustainable Fashion Text.*, 1, 2, 175–199, 2022.

101. Javed, T., Yang, J., Gilal, W.G., Gilal, N.G., The sustainability claims' impact on the consumer's green perception and behavioral intention: A case study of H&M. *Adv. Manage. Appl. Econ.*, 10, 2, 1–22, 2020.
102. Kaner, G., Greenwashing: how difficult it is to Be transparent to the consumer—H&M case study, in: *Green Marketing in Emerging Markets: Strategic and Operational Perspectives*, pp. 203–226, 2021.
103. Cheng, I. and Charter, M., 28 A pathway towards future sustainable fashion, in: *Accelerating Sustainability in Fashion, Clothing and Textiles*, p. 237, 2023.
104. Moazzem, S., Crossin, E., Daver, F., Wang, L., Environmental impact of apparel supply chain and textile products. *Environ. Dev. Sustainability*, 24, 9757–9775, 2021.
105. Ganak, J., Chen, Y., Liang, D., Liu, H., Chi, T., Understanding US millennials' perceived values of denim apparel recycling: insights for brands and retailers. *Int. J. Sustainable Soc.*, 12, 4, 267–290, 2020.
106. Dezi, L., Hysa, X., Calabrese, M., Mercuri, F., Open Total Quality Management in the Circular Economy age: A social enterprise perspective through the case of Patagonia. *Total Qual. Manage. Bus. Excell.*, 33, 1–15, 2022.
107. Schabasser, C., Learning from fast & slow fashion supply chains. *Eurasian J. Higher Educ.*, 6, 1–12, 2022.
108. Semba, T., Sakai, Y., Ishikawa, M., Inaba, A., Greenhouse gas emission reductions by reusing and recycling used clothing in Japan. *Sustainability*, 12, 19, 8214, 2020.
109. Thasneema, K., Dipin, T., Thayyil, M.S., Sahu, P.K., Messali, M., Rosalin, T., Elyas, K., Saharuba, P., Anjitha, T., Hadda, T.B., Removal of toxic heavy metals, phenolic compounds and textile dyes from industrial waste water using phosphonium based ionic liquids. *J. Mol. Liq.*, 323, 114645, 2021.
110. Behera, M., Nayak, J., Banerjee, S., Chakrabortty, S., Tripathy, S.K., A review on the treatment of textile industry waste effluents towards the development of efficient mitigation strategy: An integrated system design approach. *J. Environ. Chem. Eng.*, 9, 4, 105277, 2021.
111. Damayanti, D., Wulandari, L.A., Bagaskoro, A., Rianjanu, A., Wu, H.-S., Possibility routes for textile recycling technology. *Polymers*, 13, 21, 3834, 2021.
112. Pattnaik, P., Dangayach, G., Bhardwaj, A.K., A review on the sustainability of textile industries wastewater with and without treatment methodologies. *Rev. Environ. Health*, 33, 2, 163–203, 2018.
113. Yousef, S., Tatariants, M., Tichonovas, M., Kliucininkas, L., Lukošiūtė, S.-I., Yan, L., Sustainable green technology for recovery of cotton fibers and polyester from textile waste. *J. Cleaner Prod.*, 254, 120078, 2020.
114. Abd Rashid, N.A. and Abustan, I., Application of Artificial Barrier as Mitigation of E. coli Which Pass through Riverbank Filtration, in: *New Insight into Brucella Infection and Foodborne Diseases*, IntechOpen, pp. 83–97, 2020.

115. Vinitha, V., Preeyanghaa, M., Anbarasu, M., Neppolian, B., Sivamurugan, V., Chemical Recycling of PET Polyester Textile Wastes Using Ag-Doped ZnO Nanoparticles: An Economical Solution for Circular Economy, 2023.
116. Badawy, A.A., Ghanem, A.F., Yassin, M.A., Youssef, A.M., Rehim, M.H.A., Utilization and characterization of cellulose nanocrystals decorated with silver and zinc oxide nanoparticles for removal of lead ion from wastewater. *Environ. Nanotechnol. Monit. Manage.*, 16, 100501, 2021.
117. Saleem, H. and Zaidi, S.J., Sustainable use of nanomaterials in textiles and their environmental impact. *Materials*, 13, 22, 5134, 2020.
118. Wong, S., Yac'cob, N.A.N., Ngadi, N., Hassan, O., Inuwa, I.M., From pollutant to solution of wastewater pollution: Synthesis of activated carbon from textile sludge for dye adsorption. *Chin. J. Chem. Eng.*, 26, 4, 870–878, 2018.
119. De León-Condés, C.A., Roa-Morales, G., Martínez-Barrera, G., Menchaca-Campos, C., Bilyeu, B., Balderas-Hernández, P., Ureña-Núñez, F., Toledo-Jaldin, H.P., Sulfonated and gamma-irradiated waste expanded polystyrene with iron oxide nanoparticles, for removal of indigo carmine dye in textile wastewater. *Heliyon*, 5, 7, E020271, 2019.
120. Ali, I., Peng, C., Naz, I., Removal of lead and cadmium ions by single and binary systems using phytogenic magnetic nanoparticles functionalized by 3-marcaptopropanic acid. *Chin. J. Chem. Eng.*, 27, 4, 949–964, 2019.
121. Ali, I., Peng, C., Ye, T., Naz, I., Sorption of cationic malachite green dye on phytogenic magnetic nanoparticles functionalized by 3-marcaptopropanic acid. *RSC Adv.*, 8, 16, 8878–8897, 2018.
122. Nizamuddin, S., Siddiqui, M., Mubarak, N., Baloch, H.A., Abdullah, E., Mazari, S.A., Griffin, G., Srinivasan, M., Tanksale, A., Iron oxide nanomaterials for the removal of heavy metals and dyes from wastewater, in: *Nanoscale materials in water purification*, pp. 447–472, 2019.
123. Adel, M., Ahmed, M.A., Elabiad, M.A., Mohamed, A.A., Removal of heavy metals and dyes from wastewater using graphene oxide-based nanomaterials: A critical review. *Environ. Nanotechnol. Monit. Manage.*, 18, 100719, 2022.
124. Abdellah, A.R., Abdelhamid, H.N., El-Adasy, A.-B.A., Atalla, A.A., Aly, K.I., One-pot synthesis of hierarchical porous covalent organic frameworks and two-dimensional nanomaterials for selective removal of anionic dyes. *J. Environ. Chem. Eng.*, 8, 5, 104054, 2020.
125. Mashkoor, F., Nasar, A., Inamuddin, Carbon nanotube-based adsorbents for the removal of dyes from waters: a review. *Environ. Chem. Lett.*, 18, 605–629, 2020.
126. de Araujo Scharnberg, A.R. and Ravanello, F., Wastewater Treatment Using Nanomaterials, in: *Environmental Applications of Nanomaterials*, pp. 17–31, 2022.
127. Domingues, F.S., Freitas, TKFS., de Almeida, C.A., de Souza, R.P., Ambrosio, E., Palácio, S.M., Garcia, J.C., Hydrogen peroxide-assisted photocatalytic

degradation of textile wastewater using titanium dioxide and zinc oxide. *Environ. Technol.*, 40, 10, 1223–1232, 2019.

128. Leelavathi, H., Abirami, N., Muralidharan, R., Kavitha, H.P., Tamizharasan, S., Sankeetha, S., Arulmozhi, R., Sunlight-assisted degradation of textile pollutants and phytotoxicity evaluation using mesoporous ZnO/gC 3 N 4 catalyst. *RSC Adv.*, 11, 43, 26800–26812, 2021.
129. Mohammad, I., Jeshurun, A., Ponnusamy, P., Reddy, B.M., Mesoporous graphitic carbon nitride/hydroxyapatite (g-C3N4/HAp) nanocomposites for highly efficient photocatalytic degradation of rhodamine B dye. *Mater. Today Commun.*, 33, 104788, 2022.
130. Hassani, A., Eghbali, P., Ekicibil, A., Metin, Ö., Monodisperse cobalt ferrite nanoparticles assembled on mesoporous graphitic carbon nitride (CoFe2O4/mpg-C3N4): a magnetically recoverable nanocomposite for the photocatalytic degradation of organic dyes. *J. Magn. Magn. Mater.*, 456, 400–412, 2018.
131. Wang, J., Wang, G., Wei, X., Liu, G., Li, J., ZnO nanoparticles implanted in TiO2 macrochannels as an effective direct Z-scheme heterojunction photocatalyst for degradation of RhB. *Appl. Surf. Sci.*, 456, 666–675, 2018.
132. Voisin, H., Falourd, X., Rivard, C., Capron, I., Versatile nanocellulose-anatase TiO2 hybrid nanoparticles in Pickering emulsions for the photocatalytic degradation of organic and aqueous dyes. *JCIS Open*, 3, 100014, 2021.
133. Yogalakshmi, K.N., Das, A., Rani, G., Jaswal, V., Randhawa, J.S., Nanobioremediation: a new age technology for the treatment of dyes in textile effluents, in: *Bioremediation of Industrial Waste for Environmental Safety: Volume I: Industrial Waste and Its Management*, pp. 313–347, 2020.
134. Debnath, P. and Mondal, N.K., Effective removal of congo red dye from aqueous solution using biosynthesized zinc oxide nanoparticles. *Environ. Nanotechnol. Monit. Manage.*, 14, 100320, 2020.
135. Ngoepe, N.M., Mathipa, M.M., Hintsho-Mbita, N.C., Biosynthesis of titanium dioxide nanoparticles for the photodegradation of dyes and removal of bacteria. *Optik*, 224, 165728, 2020.
136. Liu, H., Zhang, J., Lu, M., Liang, L., Zhang, H., Wei, J., Biosynthesis based membrane filtration coupled with iron nanoparticles reduction process in removal of dyes. *Chem. Eng. J.*, 387, 124202, 2020.
137. Ismail, G.A., Allam, N.G., El-Gemizy, W.M., Salem, M.A., The role of silver nanoparticles biosynthesized by Anabaena variabilis and Spirulina platensis cyanobacteria for malachite green removal from wastewater. *Environ. Technol.*, 42, 28, 4475–4489, 2021.
138. Hasan, I.M.A., Salman, H.M., Hafez, O.M., Ficus-mediated green synthesis of manganese oxide nanoparticles for adsorptive removal of malachite green from surface water. *Environ. Sci. Pollut. Res.*, 30, 10, 28144–28161, 2023.
139. Xiong, Y., Huang, L., Mahmud, S., Yang, F., Liu, H., Bio-synthesized palladium nanoparticles using alginate for catalytic degradation of azo-dyes. *Chin. J. Chem. Eng.*, 28, 5, 1334–1343, 2020.

140. Basnet, P., Chanu, T.I., Samanta, D., Chatterjee, S., A review on bio-synthesized zinc oxide nanoparticles using plant extracts as reductants and stabilizing agents. *J. Photochem. Photobiol., B*, 183, 201–221, 2018.
141. Menon, S., Agarwal, H., Shanmugam, V.K., Catalytical degradation of industrial dyes using biosynthesized selenium nanoparticles and evaluating its antimicrobial activities. *Sustainable Environ. Res.*, 31, 1, 1–12, 2021.
142. Noukelag, S.K., Razanamahandry, L.C., Ntwampe, S.K., Arendse, C.J., Maaza, M., Industrial dye removal using bio-synthesized Ag-doped ZnO nanoparticles. *Environ. Nanotechnol. Monit. Manage.*, 16, 100463, 2021.
143. Dhaka, A., Raj, S., Chand Mali, S., Trivedi, R., Balanites aegyptiaca leaf extract-mediated synthesis of silver nanoparticles and their catalytic dye degradation and antifungal efficacy. *Front. Bioeng. Biotechnol.*, 10, 977101, 2022.
144. Najafi, M. and Frey, M.W., Electrospun nanofibers for chemical separation. *Nanomaterials*, 10, 5, 982, 2020.
145. Sun, D., Meng, G., Sun, S., Guo, W., Hai, J., Su, J., Song, Y., Chen, F., Wang, B., Silk nanofibers-ZIF hybrid membrane with improved treatment efficiency and highly enhanced water permeability for excellent removal of multiple pollutants from water. *Environ. Sci.: Nano*, 8, 11, 3408–3420, 2021.
146. Aldalbahi, A., El-Naggar, M.E., El-Newehy, M.H., Rahaman, M., Hatshan, M.R., Khattab, T.A., Effects of technical textiles and synthetic nanofibers on environmental pollution. *Polymers*, 13, 1, 155, 2021.
147. He, H., Huang, M., Gao, Z., Zhou, Y., Zhao, Y., Chen, Y., Gu, Y., Chen, S., Yan, B., Mussel-inspired polydopamine-modified silk nanofibers as an eco-friendly and highly efficient adsorbent for cationic dyes. *New J. Chem.*, 47, 12, 5832–5842, 2023.
148. Haslinger, S., Hummel, M., Anghelescu-Hakala, A., Määttänen, M., Sixta, H., Upcycling of cotton polyester blended textile waste to new man-made cellulose fibers. *Waste Manage. (Oxford)*, 97, 88–96, 2019.
149. Pensupa, N., Leu, S.-Y., Hu, Y., Du, C., Liu, H., Jing, H., Wang, H., Lin, C.S.K., Recent trends in sustainable textile waste recycling methods: current situation and future prospects, in: *Chemistry and Chemical Technologies in Waste Valorization*, pp. 189–228, 2018.
150. Naik, V. and Kumar, M., A review on natural fiber composite material in automotive applications. *Eng. Sci.*, 18, 1–10, 2021.
151. Ganguly, K., Patel, D.K., Dutta, S.D., Shin, W.-C., Lim, K.-T., Stimuli-responsive self-assembly of cellulose nanocrystals (CNCs): Structures, functions, and biomedical applications. *Int. J. Biol. Macromol.*, 155, 456–469, 2020.
152. Matayeva, A. and Biller, P., Hydrothermal liquefaction of post-consumer mixed textile waste for recovery of bio-oil and terephthalic acid. *Resour. Conserv. Recycl.*, 185, 106502, 2022.

153. Johnson, S., Echeverria, D., Venditti, R., Jameel, H., Yao, Y., Supply chain of waste cotton recycling and reuse: A review. *AATCC J. Res.*, 7, 1_suppl, 19–31, 2020.
154. Lopatina, A., Anugwom, I., Blot, H., Conde, Á.S., Mänttäri, M., Kallioinen, M., Re-use of waste cotton textile as an ultrafiltration membrane. *J. Environ. Chem. Eng.*, 9, 4, 105705, 2021.
155. Sadrolodabaee, P., Claramunt, J., Ardanuy, M., de la Fuente, A., Mechanical and durability characterization of a new textile waste micro-fiber reinforced cement composite for building applications. *Case Stud. Constr. Mater.*, 14, e00492, 2021.
156. Nguyen, G., Grzybowska-Pietras, J., Broda, J., Application of innovative ropes from textile waste as an anti-erosion measure. *Materials*, 14, 5, 1179, 2021.
157. Rubino, C., Bonet Aracil, M., Gisbert-Payá, J., Liuzzi, S., Stefanizzi, P., Zamorano Cantó, M., Martellotta, F., Composite eco-friendly sound absorbing materials made of recycled textile waste and biopolymers. *Materials*, 12, 23, 4020, 2019.
158. Wang, Z., Zhang, P., Chen, S., Usman, K.A.S., Hegh, D., Kerr, R., Zhang, H., Qin, S., Zhang, C., Liu, D., Highly stable lithium anodes from recycled hemp textile. *Chem. Commun.*, 58, 12, 1946–1949, 2022.
159. Khan, M.J., Rai, A., Ahirwar, A., Sirotiya, V., Mourya, M., Mishra, S., Schoefs, B., Marchand, J., Bhatia, S.K., Varjani, S., Diatom microalgae as smart nanocontainers for biosensing wastewater pollutants: recent trends and innovations. *Bioengineered*, 12, 2, 9531–9549, 2021.
160. Zhang, T., Zhang, W., Xi, H., Li, Q., Shen, M., Ying, G., Zhang, J., Polydopamine functionalized cellulose-MXene composite aerogel with superior adsorption of methylene blue. *Cellulose*, 28, 4281–4293, 2021.
161. Weng, L. and Zhang, X., Fully bio-based fire-safety composite from cotton/viscose wastes and alginate fiber as furniture materials. *Waste Manage. (Oxford)*, 168, 137–145, 2023.

Part II

COMPREHENSIVE, SUSTAINABLE AND PRODUCTIVE RECYCLING OF WASTE USING NANOTECHNOLOGY

6

Role of Nanotechnology in Recycled Textiles

Muhammad Jahanzaib[1,2], Shambhavi Sharma[1,2] and Duckshin Park[1,2*]

[1]*Transportation Environmental Research Division, Korea Railroad Research Institute (KRRI), Chleodobangmulgwan-ro, Uiwang-si, Republic of Korea*
[2]*Transportation System Engineering, University of Science & Technology (UST), Yuseong-gu, Daejeon, Republic of Korea*

Abstract
The textile sector contributes significantly to the world economy by manufacturing products like garments, textiles, and other items that are essential to our everyday life. But the industry's explosive development has created significant environmental problems, notably regarding how to dispose of textile waste. In a world aiming for sustainability, the traditional "take, make, and dispose" method is no longer workable. In this setting, the incorporation of nanotechnology into textile recycling procedures has shown promise in changing the industry's waste management procedures and paving the way for more environmentally friendly textiles in the future. Discarded clothing, textiles, and other items that deplete resources and contribute to pollution makeup textile waste. The huge amount of garbage produced necessitates creative solutions that may both cut down on waste and recover valuable resources. The ability to manipulate matter at the nanoscale, or nanotechnology, provides tools and methods that have the potential to completely alter the textile recycling industry. The use of nanotechnology in textile recycling is complex. Increasing the efficiency of the mechanical and chemical recycling processes is one of the most important applications. Nanomaterials like nanoparticles and nano-catalysts can hasten the disintegration of complicated textile fibers into smaller, simpler parts. This encourages more effective recycling while lowering the amount of energy and chemicals needed for production. Additionally, smart fabrics with improved qualities can be created thanks to nanotechnology. Nanocoatings can be applied to textiles to increase their toughness, water resistance, and stain resistance, so extending their useful lives. This lowers the need for new resources overall and lessens the

**Corresponding author*: dspark@krri.re.kr

Prashansa Sharma and Shilpi Shree Sahay (eds.) Nanotechnology-Assisted Recycling of Textile Waste: Sustainable Tools for Textiles, (173–206) © 2025 Scrivener Publishing LLC

requirement for regular replacements. Additionally, nanotechnology can aid in the development of functional textiles with distinctive qualities, such as textiles that can self-clean or monitor vital signs. Additionally, advancements based on nanotechnology solve issues with dyeing and finishing procedures. The environmental effect of textile dyeing is diminished by the increased color fastness provided by nano-pigments and nano-dyes. Nano-coatings can take the place of traditional finishing treatments, which helps save water and energy. These developments adhere to the circular economy's guiding principles, which emphasize resource efficiency and reducing waste. However, the use of nanotechnology in textile recycling also calls into question the effects on the environment and human health. To guarantee that the advantages outweigh the hazards, the possible release of nanoparticles into the environment and their interactions with living things must be carefully assessed. Utilizing the promise of nanotechnology while protecting the environment and human health requires responsible innovation strategies and regulatory monitoring. As a sustainable instrument for the future of textiles, the fusion of nanotechnology and textile recycling has enormous potential. Nanotechnology has the potential to push the textile sector toward circular business models by boosting recycling procedures, enhancing textile characteristics, and resolving dyeing and finishing issues. As a result, there is less waste produced, less dependency on virgin resources, and the textile ecosystem is more sustainable. The use of nanotechnology in textile recycling stands out as a crucial facilitator of good change in the sector as the world works toward a greener and more sustainable future.

Keywords: Textile, nanotechnology, nanocoating, environment, recycling

List of Abbreviations

SiO2	Silica
NPs	Nanoparticles
DNA	Deoxyribonucleic acid
PDMS	Poly(dimethylsiloxane)
ATP	Adenosine triphosphate
PCM	Phase change materials
PANI	Polyaniline
PEDOT	Poly(3,4-ethylenedioxythiophene)
PLA	Poly lactic acid
PET	Polyethylene terephthalate

6.1 Introduction

The demand for energy, food, and resources rises drastically as the world's population expands at an exponential rate [1, 2]. In addition, this leads to an increase in environmental pollution and the depletion of nonrenewable resources such fossil fuels, water, and land [3, 4]. Environmental contamination is the cause of global warming. The textile business is a shining example of human ingenuity, inventiveness, and flexibility. The textile industry has advanced substantially throughout time, influencing cultures, economics, and communities as it went from the first types of cloth weaving to contemporary, technologically advanced production techniques. However, this development has not been without repercussions because the quick development of the sector has given rise to a critical environmental issue: textile waste. This sector is a substantial source of air and water pollution as well as solid waste production [5]. Pretreatment, dyeing, and finishing are only a few of the several procedures involved in textile production. Additionally, several hazardous chemicals are used, which produce a lot of toxic waste and greenhouse gases, and water is used extensively [6–8]. Because of this, the production of textiles—from their raw materials to their finished goods—is considered an unsustainable process [5, 9]. Making a non-toxic product is crucial, but pollution control is also important. The textile industry has a large environmental impact throughout its entire process, from growing raw materials to manufacturing fabrics, and disposing of waste after consumers discard their items. This industry is currently facing significant challenges related to sustainability and pollution, including concerns over water and energy consumption, natural resource scarcity, greenhouse gas emissions, and social responsibility. "This means that the industry needs to adopt more sustainable practices."

There are enormous amounts of materials left over after the process from cotton fields to fashion runways, from fiber extraction to garment fabrication and around 400 billion dollars' worth of clothes is wasted annually, according to recent industrial reports [10]. The management process for textile waste has historically been centered around the production, consumption, and disposal of textiles. Unfortunately, most discarded materials are natural fibers, which can take decades to decompose. In addition, polymer-based clothing can take up to 200 years to decompose in landfills [11]. Due to their detrimental effects on the environment, the old ways of managing textile waste, particularly landfilling and incineration, have become more and more unworkable. Because synthetic fibers can take hundreds of years to completely degrade, disposing of textiles in landfills leads to the polluting of soil and water. While the volume of physical garbage is reduced during incineration, hazardous contaminants and greenhouse gases are released.

Recycling has become a more environmentally friendly option with the ability to turn trash into useful materials. Traditional methods of recycling textiles, however, have their own problems. The quality of the fibers is frequently reduced during mechanical recycling, which entails shredding and reprocessing used textile waste, which restricts the possible uses of the recycled materials. Additionally, it performs worse for materials that have been mixed and include a variety of fibers. Contrarily, chemical recycling entails disassembling textiles into their chemical constituents for reuse. Even though there is little to no quality loss in the recovered fibers, this method frequently uses harsh chemicals and energy-intensive procedures.

Despite these difficulties, nanotechnology is proving to be a game-changing force in the recycling of textiles. Nanotechnology has made it easier to create materials with improved qualities including toughness, water resistance, and antibacterial activity in the textile industry. These developments have enabled the incorporation of textiles into a variety of applications, such as sportswear, medical textiles, and smart textiles, in addition to improving the performance of textiles. The ultimate promise of nanotechnology, though, lies in rethinking how we recycle textiles as well as in the development of innovative textiles.

In this perspective, the combination of nanotechnology and textile recycling seems as a viable avenue for reducing the environmental effect of the textile industry while setting the route for novel, environmentally friendly materials in the future.

6.2 Textile Waste Reuse or Recycle

Various aspects of recycling and reusing textile waste have been the focus of research teams from all over the world. Figure 6.1 provides information based on documents indexed by SCOPUS in the field of textile recycling. Many of these works concentrate on engineering, chemistry, material science, and environmental science. In recent years, the number of publications related to waste management and recycling, especially textile recycling, has increased significantly. In the early 1990s, only 20 publications were available, but this number has now grown to about 200 in 2019.

Recycling and reusing have become increasingly important in today's world as we strive to protect the environment and make the most of our resources. In the past, waste products were simply discarded in landfills, but now we are transitioning from a "linear" economy model to a "circular" one. This means that we are seeking to recover and restore resources and use products to their highest value.

In the context of textiles, reusing includes donating to charities or thrift stores, inheriting, renting, swapping, trading, and borrowing from others.

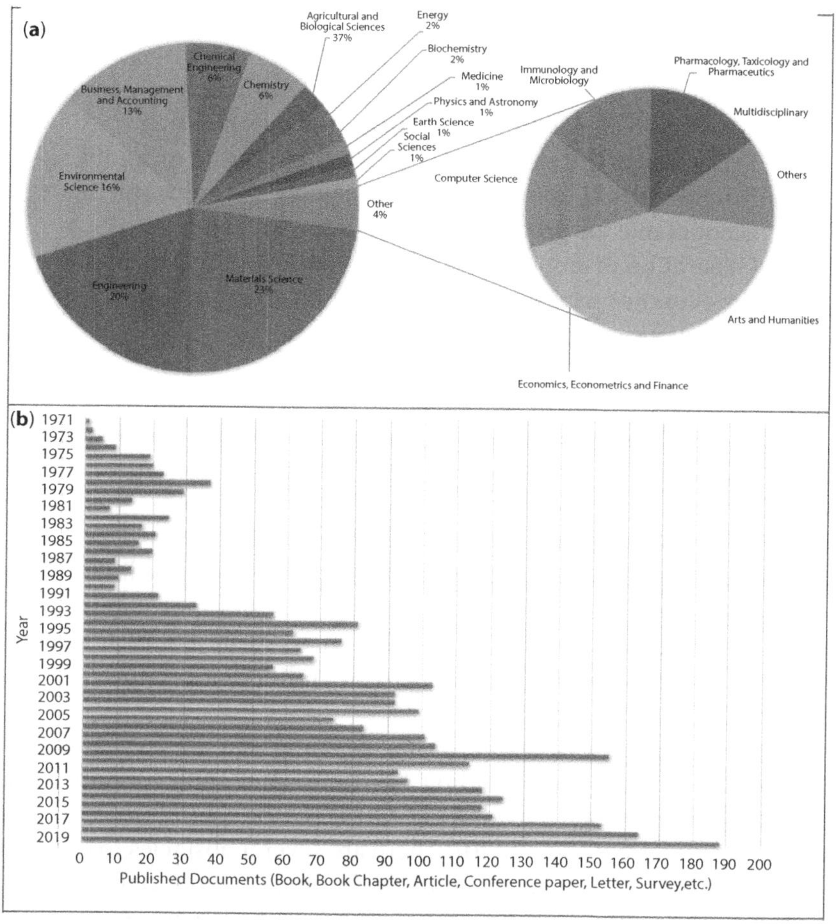

Figure 6.1 According to SCOPUS, publications on textile recycling were (a) compared based on subject matter and (b) across various years (December 2019).

According to Belk [12], there are four types of reuse: product-service systems, access-based consumption, collaborative consumption, and commercial sharing systems. The process of reusing pre- or post-consumer textile waste to create new textile or non-textile products is known as textile recycling [13]. Recycling includes reprocessing, reusing, or reproducing a product to conserve raw materials, water, energy, and other chemicals, waste reduction, and minimize environmental impacts, according to Muthu *et al.* [19].

Recycling and reusing textile and everyday clothing offer numerous benefits, including the following:

1) The ability to reuse valuable waste;
2) Conservation of resources;
3) Reduction of carbon footprint and greenhouse gas emissions associated with the production of new clothing;
4) Reduction of the amount of solid waste ending up in landfills;
5) Promotion of a more sustainable society where people feel accountable for protecting the environment; and
6) Generation of employment opportunities and stimulation of the economy as trash collection and recycling require a significant amount of manpower [14].

There are different ways to categorize the routes of textile recycling. One approach is based on the type of process used, which can be chemical, mechanical, or thermal. Another approach is based on the degree of material disassembly. Textile recycling can be downcycling, where the recycled product is of lower quality, or upcycling, where the recycled product is of higher quality. Downcycling is common in textile recycling because it reduces the length of fibers and the components of molecules. A third approach is based on the use of recycled material in a similar product, which is called closed-loop/cascading, or in a different product, which is called cascading [13].

Due to the high rate of textile waste, recycling and reusing can significantly reduce waste while providing environmental and economic benefits. Increasing the lifespan of clothes to lower carbon, water footprint and trash production by 5%–10% [20]. There are several obstacles to recycling and recovering textile waste, including technological, financial, legal, educational, infrastructural, and cost-effectiveness challenges [15, 18]. It is important to note that not all fibers are equally recyclable, and this is due to various factors [19] and some of these factors include the length of time required for material recycling and the chemicals used in the manufacturing process. There are over 8,000 chemicals used in the textile processing industry, including dyes, mordants, softeners, flame-retardants, and preserving agents. These chemicals can affect the recycling potential of different fibers [16]. As a result, recovering textile waste for fuel can lead to additional costs due to the presence of toxic chemicals and heavy metals [13].

To achieve successful textile recycling, fiber must be recycled at all points of its lifecycle, including production till post-consumer waste, due to the various obstacles hindering textile recycling. To address the growing problem of textile waste and its adverse effects on society, environment, and the economy, it is crucial to investigate methods for reducing textile waste and

Table 6.1 Various methods for lowering the production of textile waste and increasing opportunities for recycling in the manufacture, use, and post-consumption of textiles.

Step 1 Manufacturing	Rethink	Use renewable resources
		Made reusable and recyclable products
		Fully-bio gradable
		Increase product quality
	Reduce	Reducing the energy consumption in manufacturing process
		Decrease the use of natural resources and materials
Step 2 Consumption	Rethink	Buy high quality products
		Buy for long term
		Good maintenance of the product
		Multipurpose buying
		Support sustainable products and local suppliers
	Reuse	Swap
		Borrow
		Renting, trading (second hand shops, garage sell.)
		Donate to charities
	Repurpose	Use discarded product or its parts for new purpose (cleaning and making toys)
Step 3 Post-consumption	Recycle	Monomer, oligomer, and polymer recycling
		Fiber recycling
		Fabric recycling
	Repurpose	Use materials (recycled fabrics and fiber) for new purpose such as insulation, blanket, and rags
	Repair	Repair defective products and resell it with low price
	Recover	Energy recovery (heat or electricity) via incineration

recycling used textiles. You can find additional information on this topic in Table 6.1. Additionally, it should be recognized that effective recycling depends on community awareness, consumer demand for recycled goods, and the public willingness to participate in the recycling process [17].

6.3 Challenges in Textile Recycling

Recycling textiles has become an important answer to the growing environmental issues connected to the worldwide textile industry. Although recycling as a concept has enormous potential, the transition from theory to reality is fraught with difficulties that call for creative solutions. To overcome these challenges and pave the way for a more sustainable future, a multidimensional strategy is needed, considering everything from the complexity of textile materials to the nuances of recycling procedures.

6.3.1 Diversity of Textile Materials

The textile business uses a wide range of materials, from man-made fibers like polyester and nylon to natural ones like cotton and wool. When it comes to recycling, each material provides a different set of difficulties. Because the separation of these fibers can be complex and time-consuming, blended textiles that incorporate several fibers present unique challenges. To address this variability, sophisticated sorting processes and flexible recycling strategies that can handle a range of textile materials are needed.

6.3.2 Loss of Fiber Quality

One popular method of recycling textiles is mechanical recycling, which includes shredding and reprocessing used textiles to create recycled fibers. The mechanical stress and deterioration sustained during shredding, however, might cause a reduction in fiber quality because of this procedure. Recycled fibers end up being weaker and less effective, which restricts the uses they may be put to. A major issue in the sector is finding a balance between optimizing resource recovery and preserving fiber quality.

6.3.3 Complexity of Chemical Recycling

As an alternative to mechanical recycling, chemical recycling attempts to disassemble textiles into their chemical constituents for reuse. Compared to mechanical approaches, this technology can produce recycled fibers of greater quality, but it also has several drawbacks. The creation of efficient chemical recycling systems is challenging due to the complexity of textile compositions and the variety of chemical treatments given to fabrics. Additionally, careful management of the energy and chemical inputs needed for these processes is necessary to guarantee that the environmental advantages balance the disadvantages.

6.3.4 Scale and Efficiency

Moving from laboratory- to industrial-scale textile recycling methods is a daunting undertaking. Numerous recycling techniques that are promising in laboratory settings could not be easily transferable to larger scale applications because of things like energy usage, equipment needs, and cost-effectiveness. To develop effective recycling systems that can manage the enormous amounts of textile waste produced globally, it is imperative to bridge the gap between creative concepts and practical implementation.

6.3.5 Consumer Awareness and Behavior

The effectiveness of textile recycling projects is significantly influenced by consumer behavior. Even while a lot of customers show an interest in sustainable behaviors, not all of them are aware of their alternatives for recycling or the value of recycling textiles. To keep textiles out of landfills and incinerators, it is crucial to raise public knowledge of recycling programs and to encourage active involvement in them.

6.3.6 A Lack of Infrastructure and Collecting Networks

Strong collection networks and well-established infrastructure are necessary for efficient textile recycling. Unfortunately, many areas lack the infrastructure and methods required to effectively collect, classify, and handle textile waste. Improving recycling infrastructure requires creating easily accessible collection stations, working with municipalities, and engaging stakeholders.

6.3.7 Technological Advancements and Research

Continual research and innovation are necessary to advance textile recycling technology. To address current issues and create more sustainable and effective recycling processes, it is necessary to investigate new techniques, materials, and systems. To further research efforts in this area, funding and support are crucial.

To sum up, even while the idea of recycling textiles offers enormous potential for reducing the environmental effect of the textile industry, it is not without difficulties. The complex nature of textile materials, the difficulty of recycling procedures, and the demand for consumer involvement and education are all factors in the complexity of the recycling environment. Collaboration between researchers, producers, politicians, and consumers is necessary to overcome these obstacles. The textile industry can strive toward a future where recycling is not just a concept but a cornerstone of sustainable textile production and consumption by confronting these challenges head-on.

6.4 Nanotechnology in Textiles

The ability to manipulate materials at the nanoscale has opened new opportunities for a variety of sectors thanks to nanotechnology. One of its potentially game-changing uses is its incorporation into fabrics. The goal of using nanotechnology to textiles is to revolutionize the very nature of what materials are capable of by boosting its capabilities, performance, and sustainability.

Fabrics with characteristics that were previously thought to be the stuff of science fiction now exist thanks to the use of nanotechnology in textiles. Researchers and producers can create innovative properties that surpass the constraints of conventional textiles by modifying materials at the nanoscale. This has brought about the development of textiles that are more intelligent, robust, long-lasting, and able to fulfill a variety of functions.

The capacity of nanotechnology to improve textile qualities through surface modification is one of the major benefits of textiles. For example, nanocoating may be used to provide fabrics with amazing qualities like water repellency, stain resistance, and UV protection. These nanoparticle-based coatings change how the fabric interacts with the environment by adding a layer of protection to the fabric's surface. This helps to conserve water and energy by extending the useful life of textiles while also lowering the need for regular cleaning. With the use of nanotechnology, the idea of smart textiles, often known as "e-textiles," has gained popularity. Textiles can transmit electricity and contain electrical functions thanks to nanomaterials like carbon nanotubes and conductive polymers. This results in fabrics that can perceive and react to environmental changes, opening up possibilities for applications in wearable technology, sports performance measurement, and health monitoring. Beyond increasing already-existing qualities, nanotechnology has a significant impact on textiles because it makes it possible to design whole new functionality. For instance, antibacterial nanoparticles incorporated into fabrics can help to enhance cleanliness in medical facilities. Like this, thermoregulating fabrics with nanoparticles can support regulatory body temperature and comfort, making them useful in harsh weather. Nanotechnology's incorporation into textiles is in keeping with the current innovation ethos of sustainability. Nanotechnology can help create "green textiles" with longer lives and fewer replacement needs by improving the characteristics of textiles. In consequence, this lessens the negative environmental effects of textile manufacturing and disposal.

Nanotechnology has also made it possible to create fabrics that can clean themselves. Textiles can keep their cleanliness and eliminate the need for washing by using photocatalytic nanoparticles like titanium dioxide that degrade organic pollutants when exposed to light. By reducing the use of water and detergent, this invention not only affects personal convenience but also has substantial positive effects on the environment. The textile industry has seen a surge of innovation thanks to nanotechnology, which has completely changed the characteristics and functionality of materials. Here are a few notable applications of nanotechnology in clothing.

6.4.1 Self-Cleaning Fabrics

Nanotechnology has made it possible to create fabrics that can fend against stains, liquids, and dirt. When exposed to natural or artificial light, textiles can degrade organic pollutants by adding Nanocoating made of nanoparticles like titanium dioxide. This characteristic makes sure that the cloth stays fresh and clean without needing to be washed frequently, saving water and energy. In areas where cleanliness is important, self-cleaning textiles are used in outdoor clothes, furnishings, and even medical textiles.

6.4.2 Water-Resistant Fabrics and Stain-Resistant

Nano-sized hydrocarbon whiskers can make textiles water-repellent, giving the fabric a fuzzy appearance like peach fuzz [55]. Individual whiskers are spaced for surface tension to keep water on the surface, like the Lotus effect [56, 57]. As they absorb gases, they remain breathable. Water repellence can also be produced by adding gel-forming chemicals to the fabric or coating it with a nanoparticulate film to create three-dimensional surface structures on the fabric [58]. For instance, fluorocarbon derivative audio frequency plasma can be used to coat cotton fibers in nanoparticles [59]. Adding surface roughness to fabric generates super hydrophobicity without compromising softness or abrasion resistance. Silica nanoparticles combined with water-repellent substances can give textiles hydrophobicity. SiO_2 NPs with a water-repellent chemical created contact angles greater than 130° in cotton fabrics. Nanocoatings may be designed to provide textiles that are both water- and stain-resistant. By forming a barrier at the nanoscale, these coatings stop liquids from accessing the fabric's surface. This invention guarantees that spills and stains may be quickly cleaned up, retaining the beauty of the fabric, and increasing its lifespan in addition to keeping textiles dry. Outdoor clothing, work clothing, and home textiles all highly value fabrics that are water-repellent and stain-resistant. Textiles that repel oil have also been created. To give textiles super oleophobic qualities, polyester fabric could be treated with plasma fluorination and coated with silicone nanofilaments [60]. The fabric samples that were created resisted alkanes and had an oil repellency rating of 8. Textiles could have both hydrophobic and oleophobic qualities at once. After impregnating cotton fibers with SiO_2 particles, poly(dimethylsiloxane) (PDMS) was used to hydrophobize them, resulting in a surface roughness with two different sizes and a static water contact angle of 155° for droplets [61]. The fibers were treated with perfluoroalkyl chain SiO_2 particles to create oleophobicity. This was demonstrated by a static contact angle of 140° and a roll-off angle of 24° for oil droplets.

6.4.3 Textiles With Built-In Antibacterial and Antimicrobial Qualities

Nanotechnology has prepared the way for textiles with these properties. These textiles can prevent the development of bacteria and other microbes by incorporating nanoparticles like zinc, copper, or silver into the fabric. This is especially useful in hospital settings where antimicrobial fabrics may fight illnesses and encourage cleanliness. Hospital gowns, bed sheets, and wound dressings all employ these materials. The addition of Ag, TiO2, and ZnO NPs can confer antibacterial and antifungal properties to textiles [62]. Ag NPs have a large surface area that makes it easier for them to interact with bacteria and fungi. The antiseptic mechanism of Ag NPs is based on their reaction with proteins of these organisms that impairs their cellular function and stop cell division. In addition, they reduce respiration, which decreases the activity of the electron transfer system's basal metabolism and substrate trafficking into the cell membrane. When silver nanoparticles (Ag NPs) come into contact with moisture or bacteria, they stick to the cell wall and membrane. These nanoparticles are inert in their metallic state, but they ionize in the presence of moisture. This causes reactive silver ions (Ag+) to permeate into the cytoplasm through the cell membrane and wall. The interaction between Ag+ ions and proteins containing sulfur on the cell membrane alters the structure of the cell wall.

Structural changes in cells caused by variations in osmotic pressure can lead to the release of cellular components into extracellular fluid. Furthermore, the attachment of Ag+ ions to phosphate-containing proteins can cause compression of DNA, resulting in a reaction with proteins containing thiol groups that can ultimately lead to cell death. Enzymes are blocked and ATP production is halted by their action [63]. Silver nanoparticles (Ag NPs) could hinder the growth and reproduction of bacteria and fungi, which cause unpleasant odors and itching. For example, Ag NPs can be applied to socks to prevent the growth of bacteria and fungi. The effectiveness of the antibacterial properties of silver depends on the concentration, surface area, and the release rate of Ag+ ions.

Antimicrobial properties can be added to fabrics by using TiO_2 NPs. TiO_2 works as a photocatalyst by allowing its electrons to move from the valence band to the conduction band when it is illuminated with light that has an energy greater than its band gap (3.2 eV). On the surface of the photocatalyst, electrons and oxygen combine to form O_2, whereas positive electric holes and water produce hydroxyl radicals. This results in the formation of pairs of electrons and electric holes [64]. The surface volatile chemicals of the photocatalyst are oxidized into CO_2 and water, which breaks down organic material, including odor compounds, bacteria, and viruses. To give textiles antibacterial properties, the catalytic activity of TiO_2 NPs has been utilized. By using TiO_2/SiO_2 nanocomposites or Au-doped TiO_2 nanocomposites, cotton fabrics can be made with self-cleaning properties, thereby increasing the photocatalytic activity [65, 66]. To develop antibacterial characteristics, ZnO also exhibits behavior comparable to TiO_2. Reverse micelle cores made of poly(acrylic acid) and polystyrene (PS) were used to create ZnO NPs (21–25 nm).

6.4.4 Thermoregulating Textiles

Nanotechnology has made it easier to create textiles with the ability to control temperature, providing both cooling and heating capabilities. Harness the power of your environment with nanoparticles containing phase change materials (PCMs) that can absorb and release heat. These materials are appropriate for athletics, outdoor clothes, and beds because they maintain a comfortable temperature.

6.4.5 Smart Textiles and Wearable Technology

The development of smart textiles and wearable technology has been significantly influenced by nanotechnology. It is possible to make smart textiles using both natural (like cotton, silk, wool, and jute) and synthetic (like nylon, polyester, and elastane) fibers [67]. They offer the essential comfort, usefulness, and structure, which can be further improved by adding nanomaterials, electronic parts, or sensors. Smart textiles are revolutionizing the way that we interact with clothes. Conductive polymers like polyaniline (PANI) and poly(3,4-ethylenedioxythiophene) (PEDOT) are the backbone of this innovation. By incorporating these materials, we can create garments that are not only comfortable and stylish but also intelligent and interactive. Conventional components for smart textiles are frequently made with energy-intensive methods from fossil fuels. Additionally, they might include hazardous compounds, increasing environmental pollution. Additionally, they are less biocompatible than green nanomaterials, which limits their usage in some biomedical applications. They may also trigger allergic reactions or have negative effects on the skin. Green nanoparticles present a viable approach to creating sustainable and eco-friendly smart textiles in this context. Green nanomaterials, such as silver made from plant extracts, chitosan nanoparticles, and cellulose nanocrystals, have recently been studied for their possible use in smart textiles. Textiles can transmit electricity thanks to nanomaterials like carbon nanotubes and conductive polymers, making them into interactive surfaces. Incorporating sensors, LEDs, and other electrical elements enables these fabrics to perceive and react to a variety of inputs. Sports performance analysis, medical diagnosis, and even fashion all use smart fabrics.

6.4.6 UV-Protective Textiles

Nanotechnology has enabled the development of textiles with improved UV defense. Nanocoatings using nanoparticles like titanium dioxide or zinc oxide can efficiently filter the sun's damaging UV rays. These materials are appropriate for outdoor clothing, swimwear, and accessories because they add an extra layer of protection against skin aging caused by UV radiation.

Ozone depletion causes increased UV radiation on earth, damaging both clothes and skin. This exposure is linked to the development of harmful diseases, including skin cancer. With the increasing awareness of the harmful effects of ultraviolet radiation, it is now evident that clothing and textiles should have adequate protection against it [68]. When it comes to blocking ultraviolet radiation, certain metal oxide nanomaterials are proving to be more effective than organic UV absorbers. The likes of magnetite nanoparticles, titanium dioxide nanoparticles, zinc oxide nanoparticles, and nano-ceria are now frequently chosen for their superior qualities, including their safety, chemical stability under UV radiation, and resistance to high temperatures. The effectiveness of these nano-inorganic-ultraviolet additives is influenced by various characteristics, such as phase composition, particle size, crystallinity, surface characteristics, and crystal structure. Understanding how these factors impact the ability of nanoscale ultraviolet additives to block UV light can help to optimize their performance [69]. Particle attributes may seem like small details, but they can make a big difference in how particles behave and interact with their surroundings. From surface features to crystal structure, each attribute plays a crucial role in determining a particle's properties. Understanding these factors is key to unlocking the full potential of particle-based materials and technologies. Noorian *et al.* [70] conducted a recent study where they prepared zinc oxide nanoparticles on modified cotton fabric *in situ* to create multifunctional fabrics. After undergoing 100 abrasion cycles and 20 washing cycles, the zinc oxide-4-aminobenzoic acid ligand oxidized cotton fabrics exhibited superior ultraviolet protection and significant antibacterial effectiveness. Therefore, these fabrics may be useful in the production of innovative protective textiles. Dhineshbabu and colleagues (2019) developed a cotton fabric capable of protecting against UV rays while resisting fire damage. They achieved this by using MnO2-FeTiO3 nanocomposites that were covered with polyurethane. The cotton fabrics treated with a layer of MnO2-FeTiO3 coating demonstrated a remarkable and enduring ability to block UV radiation, in addition to exhibiting enhanced resistance to fire hazards, as determined by the limited oxygen index test. Additionally, throughout 10 water-laundering cycles, the coated cotton fabric maintained its characteristics, making it a smart, sustainable, and strong fabric.

Textiles that can collect energy from the environment have become possible because to nanotechnology. Mechanical vibrations may be converted into electrical energy by using nanomaterials like piezoelectric nanoparticles. These materials can be used to create wearable gadgets that harness movement to produce electricity, advancing the development of self-powered electronics. The enormous promise of nanotechnology in textiles does, however, come with many difficulties. Nanomaterials are a powerful tool for scientific advancement, but their power comes with great responsibility. Safety must always be our top priority when working with these materials, as their effects on the environment and human health could be significant. To ensure the proper development and application of nanotechnology in textiles, regulatory frameworks and norms must be developed.

The application of nanotechnology to textiles is proof of the countless opportunities that result from the fusion of science and creativity. By modifying textiles at the nanoscale, we can alter not only their essential characteristics but also a wide range of functions that have the potential to transform whole industries and ways of life. By combining nanotechnology with textiles, we can produce materials that are not only consumer goods but actively contribute to a sustainable, intelligent, and dynamic future. The realm of what textiles can accomplish will keep growing as researchers continue to investigate the possibilities of nanotechnology in textiles, spawning a new era of creativity that pushes the bounds of convention. Finally, the influence of nanotechnology on textiles goes beyond aesthetics, changing textiles into useful materials with properties that were previously only found in science fiction. These developments are improving the performance of textiles while also promoting convenience, hygienic practices, and sustainability in a variety of applications. Future fabrics will be shaped by even more amazing innovations as researchers continue to investigate the potential applications of nanotechnology in textiles.

6.5 The Circular Economy and Nanotechnology: The Case of Recycled Textile Waste

The concept of making things out of materials and using resources to use them and then discarding them is redesigned in a circular economy to prioritize reuse and recycling as much as possible [21]. The circular economy, which emphasizes extended product lives and material reuse, reduces waste production and helps consumers realize the complete value of their purchases [21]. Anna Bismar of Green Strategy has put forth the tenets of a circular e-fashion industry [22] (Figure 6.2) [21].

The comprehensive guidelines provide a holistic approach to the textile industry, encompassing every stage of the product lifecycle from the initial design phase to the final recycling process. By adhering to these guidelines, manufacturers can create products that are not only environmentally friendly but also sustainable. The guidelines put emphasis on the importance of decreasing the use of harmful chemicals, which can have a

Figure 6.2 Principles for a circular textile industry.

negative impact on the environment. Additionally, the guidelines encourage the minimization of waste to promote a cleaner and healthier planet for all. A sustainable civilization can benefit from the production of items that have an extended lifespan and are designed for multipurpose functionality. Furthermore, using natural materials that are biodegradable and environmentally friendly can contribute to a more sustainable future. To establish a circular economy, it is important to take into account the use of local human and material resources, as well as the avoidance of harmful chemicals during manufacturing and the utilization of sustainable resources. These are the fundamental principles of the circular economy from a producer's perspective, which also includes the idea of repurposing and reusing resources.

The circular economy and nanotechnology work together in a mutually beneficial way. The textile industry's environmental effect is reduced, resources are conserved, and waste is reduced by improved recycling procedures fueled by nanotechnology. The textile industry has the potential to shift its current linear, resource-intensive approach to a more efficient, sustainable, and environmentally friendly model. This transformation could bring about a significant positive impact on the industry, as well as the environment. Innovative uses of nanotechnology are redefining the potential for improving the effectiveness, sustainability, and quality of recycled textiles, in line with the principles of a circular economy. A variety of tools are available using nanotechnology to address difficulties in textile recycling. Nanomaterials can help separate fibers more successfully, get rid of colors, and break down materials into their individual parts during the recycling process. While nanofillers improve the characteristics of recycled fibers, nano-catalysts speed up enzymatic breakdown. Through techniques including recycling, remanufacturing, and reusing, the circular economy paradigm promotes the extension of product lifecycles. By making recycling processes more effective and resource-conserving, nanotechnology plays a crucial role. It raises the standard of recycled textiles, enabling them to be used in new applications without losing their performance.

6.5.1 Nanotechnology in Textiles Can Harm the Environment and Health

Throughout the day, textiles serve a variety of purposes such as decoration, protection and aesthetics, and come into direct and sustained contact with our skin. Thorough investigation is necessary to determine the impact of hazardous and carcinogenic materials found in textiles. One of the most recent subsets of potentially harmful compounds is the category of engineered nanomaterials. The textile industry has recognized the significance of engineered nanomaterials due to their ability to alter the chemical and physical characteristics of textile materials and fibers. This alteration provides benefits such as improved water and stain resistance, increased dye absorption capacity, and the ability to change wettability based on surface roughness and surface energy. The mentioned characteristics make the textiles highly advantageous and suitable for a wide range of applications, such as sportswear, protective apparel, and rainwear [23–26].

Despite several evaluations of the literature and academic papers, there is still insufficient research regarding safety concerns associated with nanotextiles. Two instances of recent research on nanotextiles are the work of Rovira *et al.* (2019) [27] and a chapter by Montazer *et al.* (2018) [28] in a book titled Nanofinishing of Textile Materials. According to the findings of Lorenz *et al.* (2012) and Kohler *et al.* (2014), as well as Geranio *et al.* (2009), and Lorenz *et al.* (2012), the use of nanotextiles can potentially harm human health, safety, the environment, and sustainability. According to where nanoparticles are integrated into textiles, they may be exposed to outside forces (Rather *et al.*, 2020 [29]). Artificial sweat was used to test the release of silver nanoparticles from antibacterial fabrics in numerous studies such as by Kulthong *et al.* (2010), Wagener *et al.* (2016), von Goetz *et al.* (2013), Kim *et al.* (2017), Stefaniak *et al.* (2014), Balakumaran *et al.* (2016), Spielman *et al.* (2018), and Milosevic *et al.* (2014). Wagener *et al.* (2016) [30] investigated how the release of silver nanoparticles into synthetic sweat is affected by textile functionalization. Six lab-prepared textiles and four commercial textiles have both undergone migration testing. Two of these lab-made textiles are composites, meaning that silver nanoparticles have been incorporated into the textile fiber, whereas the third lab-made textile has silver particles coated on the surface of a particular fiber. The test results showed that composites released less total silver than surface-coated textiles.

Lower levels of the particle fraction present in artificial sweat were observed in most fabrics, indicating that a large portion of the released silver is in a dissolved state. Furthermore, it is worth noting that the quantity of silver particles discharged from nanotextiles is not higher than that of traditional silver textiles. Moreover, the results confirmed that the type of functionalization utilized plays a vital role in migration.

Despite the published publications, it is still difficult to find comprehensive information on the dangers of nanoparticles. The underlying cause of this phenomenon can be attributed to a multitude of factors. These factors encompass a range of parameters such as the type of nanomaterials used, specific manufacturing processes employed, the size and shape of the nanoparticles, their degree of porosity, crystallinity, and aggregation, and other related parameters that play a crucial role. A comprehensive understanding of the product life cycle is necessary to evaluate the risks associated with nanotextiles and nano-finishing processes (refer to Figure 6.3). When creating nanocomposites and nanoparticles on textile substrates, it is important to consider the fundamental characteristics of the nanomaterials, the processes used to fabricate these materials, and the various application techniques employed. This also includes inserting nanoparticles into fiber formation processing, ensuring the durability of nano finishing on the surface, and considering the storage, transportation, and final usage of the treated textiles. It is important to consider the conditions that the final product will be exposed to as well as the pros and cons associated with using these techniques.

The way in which the nanotextiles are utilized will have consequences on the environment, such as impacting the soil, water, and posing a risk to human health if there is direct skin contact. There is also the possibility that nanoparticles may be accidentally released during the time the treated

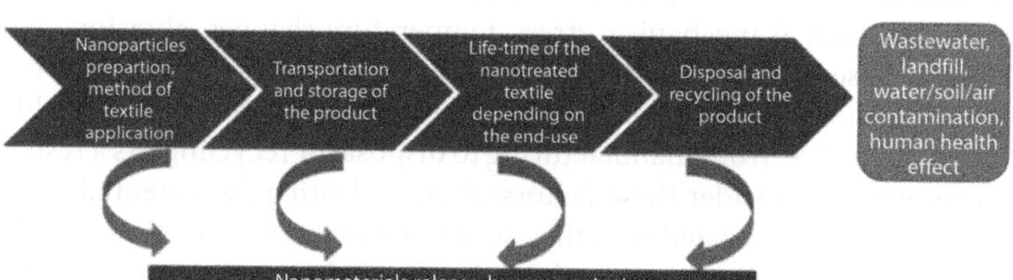

Figure 6.3 Nanomaterial lifecycle [28].

samples are in existence or during the fabrication process, putting workers at risk of exposure to harmful nanomaterials. Product design and life cycle play a significant role in controlling environmental and health exposure. For example, when geotextiles accidentally release engineered nanoparticles, they might end up in the soil, while those released from T-shirts could touch people before entering wastewater. To advocate for regulation based on the behavior and outcomes of nanotextiles, it is crucial to conduct a comprehensive evaluation of the product. This assessment should consider factors such as particle stability, surface morphology, shape, porosity, size, and chemical composition, as they all play a significant role in determining the properties and potential risks associated with nanotextiles.

6.6 Environmental Hazards Associated with Nanomaterials in the Textile Industry

Cucurachi *et al.* and Patnaik *et al.* (2019) [31, 32] expect that wastewater will contain a somewhat larger percentage of engineered nanoparticles found in textiles. For instance, concentrations of silver nanoparticles may get as high as 20%. However, the textile design that is employed, how the nanomaterials are placed into the textile fiber, has a significant impact on the amount of silver nanoparticles and other artificial nanomaterials that are mistakenly released into the environment. Abrasion-induced direct release of nanomaterials into the air appears to be of low importance (about 5%). While there may be other larger flows of nanomaterials, such as those from textiles that end up in garbage incineration, they are unlikely to cause much environmental impact. Therefore, textiles could potentially be a significant source of nanomaterials that are released into aquatic systems and soils due to the utilization of biosolids. The assessment of environmental risks associated with engineered nanomaterials is heavily influenced by the product life cycles and the volume of production. In the case of textile materials, the binding of nanomaterials to fibers is affected by various factors such as mechanical stress, temperature changes, abrasion, and detergents used during laundry or cleaning. These factors can significantly impact the durability and stability of the nanomaterials throughout the life cycle of the textile, from manufacturing to disposal or recycling. As a result, it is essential to consider these factors when evaluating the potential environmental risks associated with the use of nanomaterials in textiles. Due to their capability of being biodegradable, having higher environmental compatibility, and being non-toxic, the use of nano-biocomposites can serve as a viable alternative to most synthetic antibacterial agents. Employing a life cycle approach to study the probable release of silver nanoparticles from functionalized fabrics can help identify the importance of distinct phases in silver discharge over time. In a scientific study conducted by Mitrano and colleagues in 2016 [33], three distinct nanofabrics made in the laboratory were subjected to different washing cycles under varying conditions. The researchers found that the washing pattern and the incorporation of nanoparticles had a significant impact on the amount of metal released from the fabrics. The results indicated that the amount of nanoparticles released from the cloth surface was higher after washing than before and that the active landfill environment did not mobilize the particles as readily. This research highlights the importance of considering the entire life cycle of textiles, as increased nanoparticle release has been observed not only during disposal but also during washing.

Although nanotextiles have proven to be successful in breaking down pollutants from contaminated wastewater through photocatalysis and absorbing pollutants, as well as the potential use of nanofibrous membranes as filters for air and water, there have been plenty of concerns raised about the environmental risks associated with nanomaterials. Nanomaterials possess a unique property of having a significantly higher surface area compared to conventional materials. This characteristic makes them highly responsive to other materials throughout their life cycle. The interactions that nanomaterials have with other components can potentially lead to reduced toxicities, altering their original properties, which must be evaluated in terms of both potential dangers and opportunities. Separating the nanomaterials from wastewater effluent can be achieved through filtration or sedimentation methods in the event of nanoparticle aggregation or agglomeration. However, it is important to note that these nanomaterials may contain additional hazardous substances which could increase their toxicity.

Understanding the effects of sediment on the ecosystem is still incomplete. While it has not been proven that nanomaterial release can be completely avoided, it has been documented that burning sewage sludge has negative effects [34]. Some nanoparticles can dissolve in the environment without causing any harm; however, other metal-containing solutions are quite detrimental to the environment. Magnetic nanoparticles have been successfully separated from wastewater effluents by using magnetic separation techniques. This approach involves creating magnetic nanomaterials and applying them to textile substrates [35]. Notably, not all manufactured nanomaterials have an equally detrimental effect on human health and the environment. Carbon nanotubes, in particular, are known to exemplify this phenomenon.

Depending on their characteristics, engineered nanomaterials have a number of advantages, such as requiring less laundering of clothing due to their antibacterial capabilities or hastening the healing of wounds [36]. Piontek *et al.* carried out a study on the product-service systems and textile value chain, which involved analyzing two life-cycle assessment studies by the team [37]. The authors of this review paper have effectively presented their research findings concerning the dependence on life-cycle assessment studies, insufficiency of easily accessible information on chemicals, advantages of technological developments in this field, importance of the usage phase of clothing, and put forward additional areas to explore which include user behavior and methodological progress. By merging the outcomes of two studies, it is possible to develop a methodology for conducting life-cycle assessment investigations on a product-service system. The methodology can include an approach to appraise the ecological benefits of a product-service system by considering previous research and areas where knowledge is lacking.

In accordance with Som *et al.* (2011) [34], the crucial parameters for determining the environmental fate and effects of anthropogenic nanomaterials include the following: (1) explicit alerts regarding potentially harmful effects at present exposure concentrations; (2) the ability to dissolve in water, resulting in the disappearance of engineered nanomaterials and the production of dissolved metal ions; (3) the propensity to settle or aggregate under normal conditions; (4) destiny in wastewater facilities; and (5) stability during incineration.

In the realm of nanotechnology, the functioning and impact of manufactured nanomaterials in both technological and ecological settings have become a significant concern. The effects of these particles, particularly the negative ones, are further exacerbated by the dissolution of certain types of nanoparticles like alumina, zinc oxide, and nano-silver in water. Fortunately, the negative effects of nano-silicon dioxide are relatively lessened because the dissolved silica is considered safe at environmental levels. In contrast, all other types of nanomaterials tend to form dense agglomerations in natural waterways, which makes them relatively immobile and detached from the water system.

According to the few available study that used unreasonably high quantities of nanoparticles, the bulk of designed nanomaterials appear to be separated during wastewater treatment. Silica and carbon nanotubes, where a reduced separation rate was reported dependent on functionalization, may be an appropriate exclusion. Carbon nanotubes and silica are likely exclusions because they show a reduced separation rate dependent on functionalization. Metal-oxide or metal particles may endure unaltered after garbage incineration, whereas carbon-containing nanomaterials may be annihilated. It can be concluded that, among different nanoparticles, silver and zinc oxide are the most damaging to the environment, while the impact of titanium dioxide nanoparticles requires further study. When considering environmental hazards, the utilization of artificial nanomaterials such as silicon dioxide, carbon black, aluminum dioxide, and montmorillonite is currently perceived as either low-risk or risk-free [38].

6.7 Nanomaterials From Textile Industry and Health Effects

Multiple routes, including ingestion, inhalation, and skin absorption, could lead to exposure to manmade nanoparticles found in textiles. The workers in the textile industries are the ones who are most exposed to nanomaterials because they are exposed to more designed nanoparticles over a longer period. The respiratory system and the skin are the two most common entry points for manmade nanoparticles. After entering the body, nanomaterial builds up in the spleen, bone, kidney, and liver. According to several studies [39–41], nanomaterials can cause DNA damage and adverse toxic effects in living things. Because of its links to cancer, neurological disorders, and ageing, research on DNA damage has attracted increased attention. The evaluation of DNA damage caused by different nanomaterials in terms of types, detection methods, and DNA damage products will be conducted in the research conducted by [42]. According to the research carried out by [43], individuals who possess genetic polymorphisms and mutations in 8-oxoguanine DNA glycosylase 1 may be more susceptible to DNA damage caused by silver nanoparticles. According to Adabi *et al.* (2017) [43], biodistribution, biodegradation, biocompatibility, inflammation, as well as disruption of cell and organ function are the elements that control the potential toxicity of designed nanoparticles. The characteristics of the engineered nanoparticles such as their composition, reactivity, size, and shape are impacted by the constituent elements. Textiles are divided into categories based on whether they are used in products for newborns, items that come into direct touch with skin, articles that do not, articles that do, and articles that are just for decoration. There are varying limits for toxic and allergenic metals and compounds based on the amount of fabric contact with a consumer's skin and the degree of heavy metal toxicity.

These restrictions only apply to the portion of the chemicals that might be extracted rather than the whole number of compounds present in the fabric [44]. As the impact of nanoparticles on both the environment and human health remains uncertain, a similar method must be developed to determine if they pose a risk to fabrics. Currently, individuals are subjected to a sequence of nanoparticle discharges originating from textiles, textile production facilities, and textile cleaning services (as shown in Figure 6.4).

It is still uncertain how much harm nanotextiles can cause to human health and the environment because there is no conclusive evidence available due to the varying stability, porosity, form, size, dosage, and purpose of nanomaterials. Comparing the available reports is currently not feasible due to the lack of consistency. Further investigation is required to assess the potential hazards related to the implementation of nano finishing in the textile sector. Because there are no established regulations for the oversight of nanotechnology-based treatments, the supervisory institutions may become more concerned in the years to come. When nanoscale materials are used in textile garments, especially when there is direct human touch, ecolabeling is required. As research in nanotechnology advances, there is a growing interest in developing eco-friendly fabrication techniques that utilize green chemistry. Scientists are exploring the creation of nanomaterials with reduced toxicity levels and investigating *in situ* preparation methods

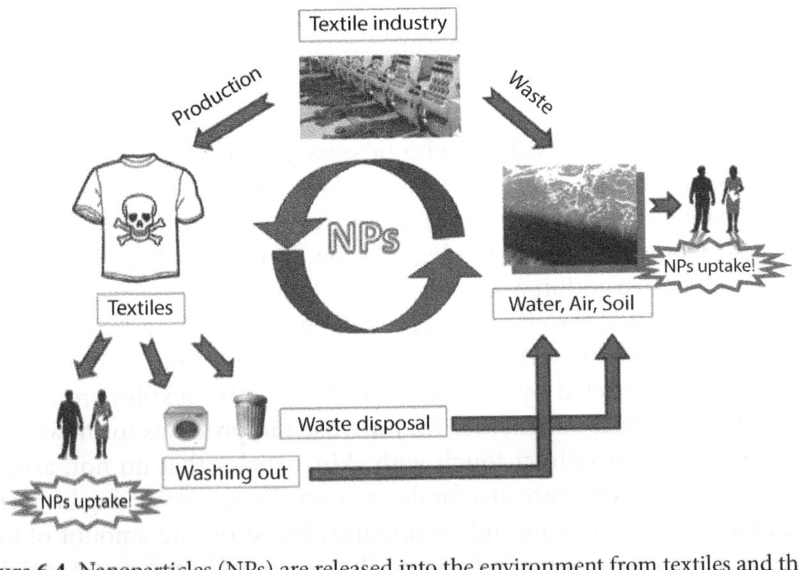

Figure 6.4 Nanoparticles (NPs) are released into the environment from textiles and the textile industry and are then absorbed by humans.

that minimize waste generation. Additionally, there is a need to enhance the durability of nanomaterials attached to textile substrates. Going forward, future studies will focus on recommending sustainable and environmentally conscious fabrication procedures to ensure a greener and cleaner future.

6.8 Industrial Applications of Recycled Textiles

The reuse and recycling of textiles are becoming increasingly popular in the textile industry due to the Ellen MacArthur Foundation's [45] recognition of their value in promoting a more sustainable circular economy (as shown in Figure 6.5). Below are several examples of recycled textile usage in different products and sectors:

- The building industry and construction sector can benefit greatly from the use of recycled textiles, specifically woven fabric waste and its residue. These materials can be repurposed for a variety of applications such as thermal and acoustic insulation, ventilation, conductivity, and even for making lightweight materials like concrete and bricks. The

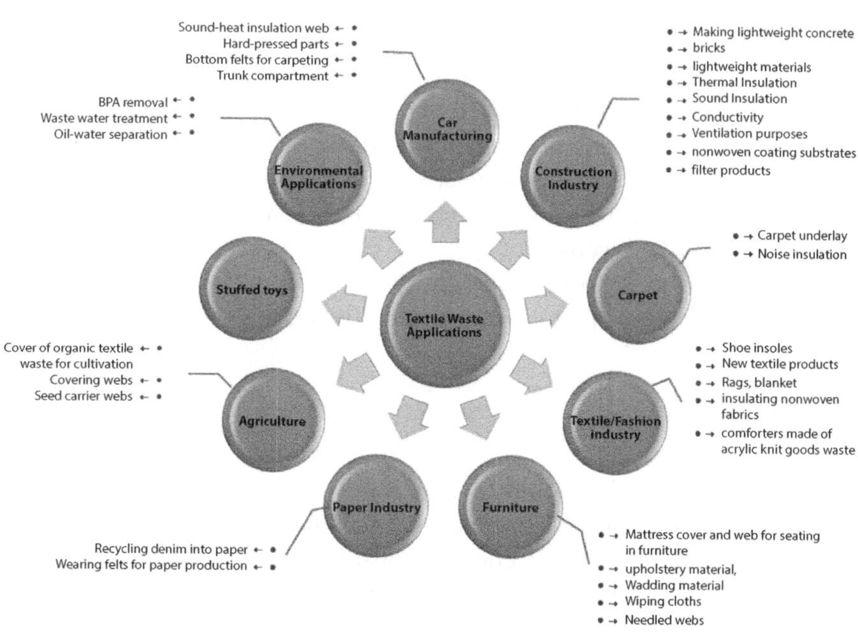

Figure 6.5 Uses of recycled textile waste across various sectors.

use of recycled textiles not only provides an eco-friendly solution but also offers an opportunity for cost-effective and sustainable construction practices [46].

- The utilization of textile scraps for ecological purposes: The utilization of discarded cotton as a cost-effective and eco-friendly agent for the treatment of water, remediation of pollution, and elimination of Bisphenol A.
- The paper industry has found creative ways to incorporate recycled denim into their production methods, such as using it to make paper [48] and creating felts for paper production from textile waste [47]. Similarly, the textile and fashion industry has also been making efforts to repurpose textile waste by creating new clothing items and accessories using sustainable practices within the major fashion brands.
- The carpet business uses textile waste as carpet underlay and noise insulation.
- In the automotive sector, recycled textile waste is used to create a web that insulates sound and heat, as well as hard-pressed components for floor coverings and carpeting's bottom felts [47].
- In the agricultural sector, textile waste is used as a layer to cover the cultivation surface in order to absorb more water from the humidity of the air and encourage microbial life [49]. In addition, a web for wrapping and transporting seeds can be made from recycled material [47].
- Textile waste can be repurposed in the furniture industry to create a variety of products, such as mattress covers, furniture seating webs, upholstery material, wadding material, and needled webs.
- It is possible to create stuffed toys using reclaimed textile waste, such as plain clothing and pile textile.

6.9 Conclusion

The production of fabrics and textiles has a considerable impact on the environment, from the creation of materials to the disposal of waste in landfills. As sustainability, energy consumption, pollution, resource availability, and greenhouse gas emissions become increasingly important issues, the textile industry faces significant challenges. To reduce the amount of solid waste that is dumped into landfills, conserve energy, and lower the environmental impact of textile waste generated worldwide, textile reuse and recycling can be a sustainable solution. Additionally, the manufacturing of new materials can be reduced by implementing textile recycling and reuse practices. The SDG goals and principles that form the foundation of the Fashion Climate Charter have received the backing of world fashion leaders [50]. Recognizing the crucial contribution of the fashion industry in both emitting greenhouse gases and embracing sustainable development, the statement is inclusive of other manufacturers who wish to participate in reducing emissions. The Charter is grounded on the Paris Agreement and identifies key areas that the signatories will tackle. These areas include reducing carbon emissions during the production phase, opting for sustainable and eco-friendly materials, using low-carbon transportation, enhancing communication and awareness among consumers, and collaborating with policymakers and the financial sector to explore practical solutions within the circular economy framework [50]. By 2030, they aim to reduce their overall greenhouse gas emissions by 30%. To achieve this, they have outlined specific measures such as eliminating coal-fired boilers or any other coal-fired sources of heat and power generation from their own operations and suppliers, commencing in 2025 [50]. It is important to focus on specific areas to reduce textile waste in the fashion industry. These areas include the initiation of key streams and the targeting of streams that require attention:

- Less harmful materials are being used in the fashion industry to reduce environmental impacts:
  - ❖ Bamboo is quick-growing, hypoallergenic, and absorbent. Bamboo is processed with less chemicals and fertilizers than other fabrics, making it a more environmentally friendly option.
  - ❖ Silk is a delicate fabric that does not use harsh dyes in its traditional manufacture.
  - ❖ Hemp is a versatile plant that is quick to grow and simple to cultivate [51].
- The process of on-site wastewater treatment can be improved through technological modifications [52–54]. Natural dyes can be used to replace chemical dyes as an input material in a process.
- Creation of a high-performance membrane nanoreactor for the treatment of textile effluent aiming for energy efficiency in production.

- The use of additive manufacturing and Industry 4.0 to lower the pace at which waste is produced during production.
- Using automated dosing devices and controllers in various industrial phases can optimize the use of chemicals.
- Process alternate methods innovative rinse pulsation.
- Water reuse and recycling involves reusing and recycling cooling water, recycling condensate water, and a system for recovering bleach bath.
- Decreasing the temperature during energy-intensive processes such as dyeing, printing, and finishing.
- Recycling PET bottles to create eco-friendly fabrics is an innovative and sustainable solution to the problem of waste. By choosing recycled PET-based textiles, we can contribute to a cleaner environment without compromising on quality or style.
- Safe biochemical processes that impact the characteristics of fibers and textiles.
- Treatment of fibers and textile materials using enzyme biocatalysts.
- Innovative materials for textile surface modification can include biopolymers and biomasses.
- Enzyme system growth and expansion of surface-treatment applications for functionalizing textile substrates.
- Smart textiles' multifunctionality allows their use in a larger range of applications.
- The circular plastic strategy presents an innovative solution to tackle plastic waste in the fashion industry, by focusing on repurposing materials like PLA and PET. This approach not only addresses the issue of plastic waste but also promotes sustainability and environmental responsibility in the fashion sector.

References

1. Kalkanis, K., Alexakis, D.E., Kyriakis, E., Kiskira, K., Lorenzo-Llanes, J., Themelis, N.J., Psomopoulos, C.S., Transforming Waste to Wealth, Achieving Circular Economy. *Circ. Econ. Sustain.*, 2, 1541–1559, 2022.
2. Hatzilyberis, K., Tsakanika, L.A., Lymperopoulou, T., Georgiou, P., Kiskira, K., Tsopelas, F., Ochsenkühn, K.M., Ochsenkühn-Petropoulou, M., Design of an advanced hydrometallurgy process for the intensified and optimized

industrial recovery of scandium from bauxite residue. *Chem. Eng. Process. Process Intensif.*, 155, 108015, 2020.
3. Psomopoulos, C.S., Kiskira, K., Kalkanis, K., Leligou, H.C., Themelis, N.J., The role of energy recovery from wastes in the decarbonization efforts of the EU power sector. *IET Renew. Power Gener.*, 16, 48–64, 2022.
4. Kiskira, K., Papirio, S., Pechaud, Y., Matassa, S., van Hullebusch, E.D., Esposito, G., Evaluation of Fe (II)-driven autotrophic denitrification in packed-bed reactors at different nitrate loading rates. *Process Saf. Environ. Prot.*, 142, 317–324, 2020.
5. Vercalsteren, A., Nicolau, M., Lafond, E., Textiles and the Environment in a Circular Economy. Eionet Report 2019- ETC/WMGE 2019/6, Available online: https://www.eionet.europa.eu/etcs/etc-wmge/products/etc-wmge-reports/textiles-and-theenvironment-in-a-circular-economy (accessed on 10 January 2023).
6. Sivaramakrishnan, C.N., Pollution in textile industry. *Colourage*, 16, 66–68, 2009.
7. Huang, B., Zhao, J., Geng, Y., Tian, Y., Jiang, P., Energy-related GHG emissions of the textile industry in China. *Resour. Conserv. Recycl.*, 119, 69–77, 2017.
8. WWF, Cotton Facts, Available online: https://www.worldwildlife.org/industries/cotton (accessed on 15 December 2022).
9. ten Wolde, A. and Korneeva, P., Circular Fashion Advocacy, A Strategy towards a Circular Fashion Industry in Europe. Report 2019, Ecopreneur.eu, Available online: https://ecopreneur.eu/wp-content/uploads/2019/03/EcoP-Circular-Fashion-Advocacy- Report-28-3-19.pdf (accessed on 15 January 2023).
10. Drew, D. and Reichart, E., These are the economic, social and environmental impacts of fast fashion, 2019.
11. ABC News, L., Fast fashion: Rivers turning blue and 500,000 tonnes in landfill, 2017.
12. Belk, R., You are what you can access: sharing and collaborative consumption online. *J. Bus. Res.*, 67, 8, 1595–1600, 2014.
13. Sandin, G. and Peters, G.M., Environmental impact of textile reuse and recycling e A review. *J. Clean. Prod.*, 184, 353–365, 2018, Elsevier.
14. Leblanc, R., Recycling and new job creation. The role of recycling businesses in opening new employment opportunities, 2018.
15. Gulich, B., Designing textile products that are easy to recycle, in: *Recycling in textiles*, Woodhead Textiles Series, Wang, Y. (Ed.), pp. 25–37, 2006.
16. Khan, M.M.R. and Islam, M.M., Materials and manufacturing environmental sustainability evaluation of apparel product: knitted T-shirt case study. *Text. Clothing Sustainability*, 1, 1, 8, 2015.
17. Ahmad, S.S. *et al.*, The application of Recycled Textile and Innovative Spatial Design Strategies for a Recycling Centre Exhibition Space. *Procedia Social Behav. Sci.*, 234, 525–535, 2016.

18. Echeverria, C.A. *et al.*, Cascading use of textile waste for the advancement of fibre reinforced composites for building applications. *J. Clean. Prod.*, 208, 1524e1536, 2019, Elsevier.
19. Muthu, S.S. *et al.*, Recyclability Potential Index (RPI): The concept and quantification of RPI for textile fibres. *Ecol. Indic.*, 18, 58–62, 2012.
20. Leblanc, R., Textile recycling facts and figures, 2018.
21. http://theconversation.com/explainer-what-is-a-circular-economy-29666.
22. https://circularfashion.com/key-principles/.
23. Harifi, T. and Montazer, M., Application of nanotechnology in sports clothing and flooring for enhanced sport activities, performance, efficiency and comfort: A review. *J. Ind. Text.*, 46, 1147–1169, 2017.
24. Shalaby, M.N. and Saad, M.M., Advanced Material Engineering and Nanotechnology for Improving Sports Performance and Equipment. *Int. J. Psychosoc. Rehabil.*, 24, 1–9, 2020.
25. Yu, L., A Study on Decomposition of Nanoparticle Finished Textiles. *Int. J. Nanosci. Nanoeng.*, 4, 80, 2018.
26. Kausar, A., Textile Nanocomposite of Polymer/Carbon Nanotube. *Am. J. Nanosci. Nanotechnol. Res.*, 6, 28–35, 2018.
27. Rovira, J. and Domingo, J.L., Human health risks due to exposure to inorganic and organic chemicals from textiles: A review. *Environ. Res.*, 168, 62–69, 2019.
28. Montazer, M. and Harifi, T., *Nanofinishing of Textile Materials*, Woodhead Publishing, Cambridge, UK, 2018.
29. Rather, L.J., Zhou, Q., Ganie, S.A., Li, Q., Environmental Profile of Nano-finished Textile Materials: Implications on Public Health, Risk Assessment, and Public Perception, in: *Advances in Functional Finishing of Textiles*, pp. 57–83, Springer, Singapore, 2020.
30. Wagener, S., Dommershausen, N., Jungnickel, H., Laux, P., Mitrano, D., Nowack, B., Schneider, G., Luch, A., Textile functionalization and its effects on the release of silver nanoparticles into artificial sweat. *Environ. Sci. Technol.*, 50, 5927–5934, 2016.
31. Cucurachi, S. and Rocha, C.F.B., Life-cycle assessment of engineered nanomaterials, in: *Nanotechnology in Eco-efficient Construction*, pp. 815–846, Woodhead Publishing, Cambridge, UK, 2019.
32. Patnaik, A. and Patnaik, S. (Eds.), *FibresFi to Smart Textiles: Advances in Manufacturing, Technologies, and Applications*, CRC Press, Boca Raton, FL, USA, 2019.
33. Mitrano, D.M., Limpiteeprakan, P., Babel, S., Nowack, B., Durability of nano-enhanced textiles through the life cycle: Releases from landfilling after washing. *Environ. Sci. Nano*, 3, 375–387, 2016.
34. Som, C., Wick, P., Krug, H., Nowack, B., Environmental and health effects of nanomaterials in nanotextiles and facade coatings. *Environ. Int.*, 37, 1131–1142, 2011.

35. Harifi, T. and Montazer, M., In situ synthesis of iron oxide nanoparticles on polyester fabric utilizing color, magnetic, antibacterial and sono-Fenton catalytic properties. *J. Mater. Chem. B*, 2, 272–282, 2014.
36. Temizel-Sekeryan, S. and Hicks, A.L., Global environmental impacts of silver nanoparticle production methods supported by life cycle assessment. *Resour. Conserv. Recycl.*, 156, 104676, 2020.
37. Piontek, F.M. and Müller, M., Literature reviews: Life cycle assessment in the context of product-service systems and the textile industry. *Procedia CIRP*, 69, 758–763, 2018.
38. Black Carbon, *Screening Assessment for the Challenge*, Chemical Abstracts Service Registry Number 1333-86-4, Environment Canada, Ottawa, ON, Canada, 2013.
39. Exbrayat, J.M., Moudilou, E.N., Lapied, E., Harmful effects of nanoparticles on animals. *J. Nanotechnol.*, 2015, 2015.
40. Roberto, M.M. and Christofoletti, C.A., How to Assess Nanomaterial Toxicity? An Environmental and Human Health Approach, in: *Nanomaterials-Toxicity, Human Health and Environment*, IntechOpen, London, UK, 2019.
41. Grumezescu, A.M. (Ed.), *Nanoscale Fabrication, Optimization, Scale-Up and Biological Aspects of Pharmaceutical Nanotechnology*, William Andrew, Norwich, NY, USA, 2017.
42. Nallanthighal, S., Chan, C., Murray, T.M., Mosier, A.P., Cady, N.C., Reliene, R., Differential effects of silver nanoparticles on DNA damage and DNA repair gene expression in Ogg1-deficient and wild type mice. *Nanotoxicology*, 11, 996–1011, 2017.
43. Adabi, M., Naghibzadeh, M., Adabi, M., Zarrinfard, M.A., Esnaashari, S.S., Seifalian, A.M., Faridi-Majidi, R., Aiyelabegan, H.T., Ghanbari, H., Biocompatibility and nanostructuredmaterials: Applications in nanomedicine. *Artif. Cells Nanomed. Biotechnol.*, 45, 833–842, 2017.
44. Yin, Y., Yu, S., Yang, X., Liu, J., Jiang, G., Source and pathway of silver nanoparticles to the environment, in: *Silver Nanoparticles in the Environment*, pp. 43–72, Springer, Berlin/Heidelberg, Germany, 2015.
45. Foundation, E.M., Circular Economy System Diagram, 2017.
46. Briga-Sá, A. *et al.*, Textile waste as an alternative thermal insulation building material solution. *Constr. Build. Mater.*, 38, 155–160, 2013, Elsevier.
47. Bhatia, D., Sharma, A., Malhotra, U., Recycled fibers: An overview. *Int. J. Fiber Text. Res.*, 4, 4, 77–82, 2014.
48. Travers, P., War on waste: Recycling denim into paper and fostering social inclusion in the process, 2017.
49. Eriksson, B.G., Organic textile waste as a resource for sustainable agriculture in arid and semi-arid areas. *Ambio*, 46, 2, 155–161, 2017, Springer.
50. https://unfccc.int/news/milestone-fashion-industry-charter-for-climate-actionlaunched.
51. https://info.esg.adec-innovations.com/blog/alternative-materials-for-sustainableclothing.

52. Jafari, H. *et al.*, Enhanced photocatalytic activities of TiO2–SiO2 nanohybrids immobilized on cement-based materials for dye degradation. *Res. Chem. Intermed.*, 42, 4, 2963–2978, 2016.
53. Tara, N. *et al.*, On-site performance of floating treatment wetland macrocosms augmented with dye-degrading bacteria for the remediation of textile industry wastewater. *J. Clean. Prod.*, 217, 541–548, 2019.
54. Hu, E., Shang, S., Chiu, K.-L., Removal of Reactive Dyes in Textile Effluents by Catalytic Ozonation Pursuing on-Site Effluent Recycling. *Molecules*, 24, 15, 2755, 2019.
55. Russell, E., Nanotechnologies and the shrinking world of textiles. *Text. Horiz.*, 9, 7–9, 2002.
56. Marmur, A., The lotus effect: superhydrophobicity and metastability. *Langmuir*, 20, 3517–3519, 2004.
57. Gao, L. and McCarthy, T.J., The "lotus effect" explained: two reasons why two length scales of topography are important. *Langmuir*, 22, 2966–2967, 2006.
58. El-Khatib, E., Antimicrobial and Self-cleaning Textiles using Nanotechnology. *Res. J. Text. Apparel*, 16, 156–174, 2012.
59. Zhang, J., France, P., Radomyselskiy, A., Datta, S., Zhao, J., van Ooij, W., Hydrophobic cotton fabric coated by a thin nanoparticulate plasma film. *J. Appl. Polym. Sci.*, 88, 1473–1481, 2003.
60. Artus, G.R., Zimmermann, J., Reifler, F.A., Brewer, S.A., Seeger, S., A superoleophobic textile repellent towards impacting drops of alkanes. *Appl. Surf. Sci.*, 258, 3835–3840, 2012.
61. Hoefnagels, H., Wu, D., De With, G., Ming, W., Biomimetic superhydrophobic and highly oleophobic cotton textiles. *Langmuir*, 23, 13158–13163, 2007.
62. Feng, Q.L., Wu, J., Chen, G.Q., Cui, F.Z., Kim, T.N., Kim, J.O., A mechanistic study of the antibacterial effect of silver ions on Escherichia coli and Staphylococcus aureus. *J. Biomed. Mater. Res.*, 52, 662–668, 2000.
63. Yamanaka, M., Hara, K., Kudo, J., Bactericidal actions of a silver ion solution on Escherichia coli, studied by energy-filtering transmission electron microscopy and proteomic analysis. *Appl. Environ. Microbiol.*, 71, 7589–7593, 2005.
64. Daoud, W.A. and Xin, J.H., Low temperature sol-gel processed photocatalytic titania coating. *J. Sol-Gel Sci. Technol.*, 29, 25–29, 2004.
65. Qi, K., Chen, X., Liu, Y., Xin, J.H., Mak, C.L., Daoud, W.A., Facile preparation of anatase/SiO2 spherical nanocomposites and their application in self-cleaning textiles. *J. Mater. Chem.*, 17, 3504–3508, 2007.
66. Uddin, M.J., Cesano, F., Scarano, D., Bonino, F., Agostini, G., Spoto, G., Bordiga, S., Zecchina, A., Cotton textile fibres coated by Au/TiO2 films: Synthesis, characterization and self cleaning properties. *J. Photochem. Photobiol., A*, 199, 64–72, 2008.
67. Dhir, Y.J. and Dhir, Y.J., Natural Fibres: The Sustainable Alternatives for Textile and Non-Textile Applications, in: *Natural Fibre*, InTech, London, UK, 2022.

68. Dhineshbab u, N.R. and Bose, S., UV resistant and fire retardant properties in fabrics coated with polymerbased nanocomposites derived from sustainable and natural resources for protective clothing application. *Compos. Part B Eng.*, 172, 555–563, 2019.
69. Dhineshbabu, N.R. and Bose, S., Smart textiles coated with eco-friendly UV-blocking nanoparticles derived from natural resources. *ACS Omega*, 3, 7454–7465, 2018.
70. Noorian, S.A., Hemmatinejad, N., Navarro, J.A., Ligand modified cellulose fabrics as support of zinc oxide nanoparticles for UV protection and antimicrobial activities. *Int. J. Biol. Macromol.*, 154, 1215–1226, 2020.

7

Contribution of Nanotechnology in Shaping Sustainable Textile Design and Future Fashion Trends

Shilpi Shree Sahay, Prashansa Sharma* and Amisha Singh

Department of Home Science (Clothing and Textile), MMV, Banaras Hindu University, Varanasi, U.P., India

Abstract

The genesis of cutting-edge explorations and discoveries in the field of nanotechnology has opened up new possibilities for the fabrication of multi-functional features in textiles. Nanotechnology, which operates at a scale millions of times smaller than a millimeter, offers the ability to manipulate molecules and introduce new functionalities to textiles, thereby revolutionizing the textile industry and paving the way for sustainable future fashion. The textile industry has long needed revolutionary materials that can enhance the performance and efficiency of fabrics [31]. Nanotechnology has emerged as a promising solution by exerting a profound impact on the inherent properties of textile materials [26]. This has captured the interest of scientists and researchers, making nanotechnology an incredibly fascinating field of study. One of the key advantages of nanotechnology in textiles is its ability to significantly increase the surface area–to–volume ratio of materials. This, in turn, offers lucrative economic and commercial potential for nano-enabled textiles. Researchers can alter, control, engineer, and synthesize future-generation textiles with diverse functionalities to cater to a wide range of end uses. Over the past decade, the textile sector has experienced a renaissance due to the remarkable properties and distinguishing characteristics of nanomaterials. Various types of nanomaterials have been developed, each exhibiting unique features and improved performance [26]. These nanomaterials, such as nano-whiskers or nanocrystals, nanofibers, metal nanoparticles, and graphene, have been identified as highly useful for textile applications. By incorporating nanomaterials into textiles, designers, and manufacturers can achieve multifunctional properties such as UV protection,

**Corresponding author*: prashansa.sharma@bhu.ac.in

Prashansa Sharma and Shilpi Shree Sahay (eds.) Nanotechnology-Assisted Recycling of Textile Waste: Sustainable Tools for Textiles, (207–234) © 2025 Scrivener Publishing LLC

self-cleaning, flame retardancy, water and spill repellency, wrinkle resistance, anti-odor and antimicrobial properties, electrical conductivity, thermoregulation, anti-static properties, breathability, durability, softness, and improved tear strength [12, 14, 31]. This fusion of technology and fashion has expanded the scope of textile applications in various fields, including industry, medicine, agriculture, military, sports, and personal clothing [14].

Multidisciplinary research in nanotechnology has been instrumental in promoting the application of advanced finishes in textiles. Researchers have been able to incorporate and impart a range of desirable features to textile products, thereby shaping a sustainable and eco-friendly textile design and fashion trend. Advanced nanotechnology offers high-performance and environmentally friendly solutions to conventional fibers and production methods, making it an indispensable tool for maintaining the credibility of the textile industry. This chapter sheds light on the significant impact of nanotechnology in shaping the future of textiles, with a focus on innovative and sustainable textile designs and fashion trends.

Keywords: Nanotechnology, textile, sustainable, innovation, design, fashion, industry, technologies

List of Abbreviations

UV	Ultraviolet
CNTs	Carbon nanotubes
NPs	Nanoparticles
3D	Three-dimensional
PLA	Polylactic acid
PHA	Polyhydroxyalkanoates
PMNC	Polymer matrix nanocomposites
MMT	Montmorillonites
ePTFE	Expanded polytetrafluoroethylene
OLED	Organic light-emitting diode
LED	Light-emitting diode
SMAs	Shape memory alloys
SMMs	Shape memory materials
TENG	Triboelectric nanogenerators
MOFs	Metal-organic frameworks
PEDOT	Poly(styrenesulfonate)

7.1 Introduction

The textile and fashion industries, with their intricate tapestry of creativity and commerce, have long been celebrated as conduits of self-expression and economic vitality. These industries, rooted in craftmanship and creativity, have clothed us, defined our identities, and served as a medium for self-expression for centuries [31]. The textile and fashion sectors have traditionally held importance in the global economy, offering consumers a wide array of clothing choices with the ever-evolving spectrum of styles, textures, and designs. They have simultaneously fuelled economic growth and sparked cultural revolutions. However, besides this rapid expansion and success lies a complex and deeply entrenched challenge-the sustainability challenge [9].

For decades, these textile and fashion industries have been recognized as major contributors to environmental degradation and resource depletion that have raised alarm bells across the globe. Excessive water and energy consumption, chemical pollution stemming from dyeing and finishing processes, rampant waste generation, and exploitative labor practices in certain regions have tarnished the allure of fashion. The era of "fast fashion," [29] characterized by disposable garments meant to be worn just a handful of times before being discarded, has become a hallmark of modern consumerism. Waste, a byproduct of the fast fashion revolution, piles up in landfills as garments fall out of favour almost as quickly as they come into vogue. The culture of "buy and discard" has become synonymous with fast fashion, reinforcing a cycle of overconsumption and environmental degradation. Amidst this paradigm shift, nanotechnology has emerged as a game-changing ally, offering cutting-edge techniques and materials to revolutionize textile design and manufacturing processes. From enhancing fabric durability and functionality to enabling novel applications such as smart textiles and wearable electronics, nanotechnology holds immense potential to shape the future of fashion in a sustainable and socially responsible manner [17].

7.1.1 The Concept of Sustainability in Textiles

Sustainability in the context of textiles refers to adopting strategies and methods that seek to reduce the environmental, social, and economic consequences of textile production, consumption, and disposal throughout the entire supply chain. Sustainable textile production focuses on reducing the environmental footprint associated with every stage of the textile life-cycle, from obtaining raw materials to manufacturing, distribution, usage, and eventual disposal or recycling.

As we confront the existential challenges of climate change, resource scarcity, and social inequality, the textile industry has a unique opportunity—and a moral obligation—to pioneer a path toward a more sustainable and compassionate world [22].

It addresses the pressing need to mitigate the environmental and societal repercussions linked to textile manufacturing. The adoption of sustainable practices fosters innovation and creativity, driving the development of new materials, technologies, and business models that redefine the boundaries of fashion and design. Moreover, embracing sustainability in fashion trends encourages innovation in materials and processes, leading to the development of eco-friendly alternatives to traditional textiles.

In line with above objectives, Figure 7.1 shows the sustainaibilty principles such as design for longevity and durability, modular and versatile design concept, technological innovation, circular fashion [29], closed-loop system, and conscious fashion choices are paramount. Design for durability ensures that products are built to last longer, reducing the need for frequent replacements and minimizing waste. Modular and versatile design concepts allow for adaptability and extended use, further enhancing sustainability.

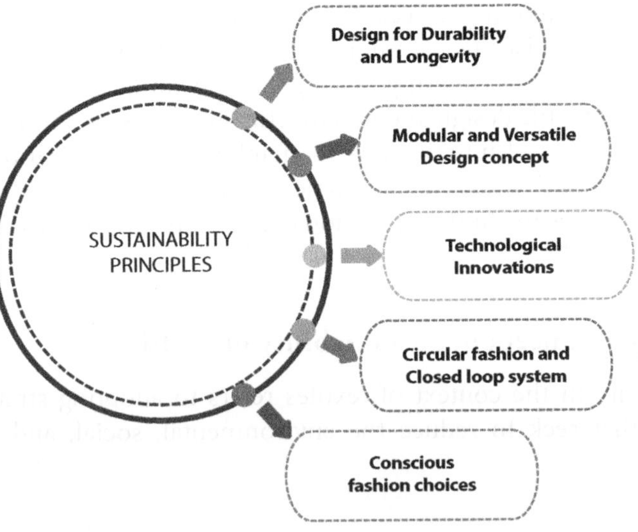

Figure 7.1 Sustainability principles.

Technological innovation drives the development of eco-friendly materials and manufacturing processes, pushing the boundaries of sustainability in fashion. Circular fashion and closed-loop systems promote the reuse and recycling of materials, closing the loop on resource consumption and waste generation [22]. Finally, conscious fashion choices empower consumers to make informed decisions that align with sustainability principles, driving demand for more environmentally friendly products and practices. Together, these principles form the foundation of a sustainable textile industry that not only minimizes its environmental impact but also contributes positively to societal well-being and economic prosperity.

7.2 Sustainable Design Principles

Sustainable design principles are fundamental guidelines that drive the textile sector toward more environmentally friendly, socially responsible, and economically viable practices. Sustainable design begins with the careful selection of materials and sourcing practices that prioritize environmental and social responsibility. It prioritizes the use of renewable, non-toxic, and locally sourced materials. It seeks to minimize waste generation by designing products with minimal packaging, optimizing material use, and incorporating circular economy principles. Designers opt for materials with lower environmental footprints, such as organic cotton, tencel, bamboo, and recycled fibers, which minimize water usage, chemical inputs, and greenhouse gas emissions [40]. Recycled polyester retains the same properties as virgin polyester, making it suitable for a wide range of applications in the textile industry. Sustainable design emphasizes modularity and flexibility to accommodate changing needs and preferences. Products are designed with interchangeable components or modular systems that allow for easy repair, upgrading, and customization. This approach extends the lifespan of products and reduces waste.

7.2.1 Design for Durability and Longevity

Designing for durability starts with selecting high-quality materials that are known for their strength, resilience, and longevity. Creating products with a focus on resilience and lasting performance is a key principle of sustainable fashion that focuses on creating garments and products that are built to last, withstand wear and tear, and remain relevant over time. The way garments are constructed plays a crucial role in their durability and longevity. Prioritizing strong and secure stitching techniques, such as reinforced seams, bar tacks, and flat-felled seams, to prevent unraveling and fraying over time, ultimately reduces the need for frequent replacements and minimizing waste in the fashion industry. Reinforced seams are stitches that provide extra strength and stability to critical areas of a garment, that experience high tension or stress. This technique provides a clean finish to the garment while ensuring durability and preventing fraying. Bar tacks are dense, reinforced stitches used to reinforce stress points and areas of high tension in a garment. They are typically placed at pocket corners, belt loops, and other areas subject to frequent wear and tear. Bound seams involve enclosing raw edges within fabric binding or bias tape to prevent fraying and add durability to the garment. This finishing technique is commonly used in high-quality garments to reinforce seams and enhance their longevity.

7.2.2 Modular and Versatile Design Concepts

These concepts allow garments and products to be transformed, reconfigured, or personalized to suit different occasions, preferences, and body types. Modular design involves creating garments or products with interchangeable or detachable components that can be mixed, matched, or combined in various ways to create different looks or functionalities. Clothing collections designed with modular components, such as interchangeable sleeves, collars, or hemlines, allow consumers to create multiple outfits from a limited number of pieces. Accessories with modular features, such as interchangeable panels or convertible styles, offer versatility and customization options. By embracing classic silhouettes, designers ensure that garments remain relevant season after season, offering enduring elegance and sophistication. Versatile design focuses on creating garments or products that can be worn or used in multiple ways, offering functionality and adaptability for different contexts or occasions. Convertible dresses can be styled in multiple ways, providing options for various events or settings. Reversible garments with different designs or colors on each side allow for easy transition between different looks. Shoes and sandals with interchangeable straps, insoles, or heel heights allow for customization and versatility. Modular and versatile design concepts often overlap with customization and personalization, allowing consumers to tailor products to their specific preferences and needs.

7.2.3 Biomimicry in Fashion Design

In the quest for sustainable innovation, designers and scientists are turning to nature as a source of inspiration. Nature provides a wealth of inspiration.

By studying the intricate patterns, structures, and functions found in the natural world, researchers are unlocking new possibilities for creating textiles that are not only eco-friendly but also functional and resilient. Material development in textiles holds great promise when approached through biomimicry techniques. Designers are exploring nature-inspired patterns, textures, and colors to create visually striking and environmentally conscious textiles. Inspired by the self-cleaning properties of lotus leaves, researchers have developed textile coatings and surface textures that repel water and resist dirt and stains [35].

7.2.4 Circular Fashion and Closed-Loop Systems

Circular fashion and closed-loop systems represent a fundamental shift in the fashion industry toward more sustainable and resource-efficient practices. It is based on the idea of creating clothing, accessories, and textiles that are designed to last longer, can be easily repaired, and are recyclable or biodegradable at the end of their lifecycle. Circular fashion emphasizes the importance of designing garments and products that are durable, timeless, and of high quality [38]. Circular fashion aims to minimize waste and maximize the use of resources throughout the production process. This includes using sustainable materials, reducing water and energy consumption, and optimizing manufacturing processes to minimize environmental impact.

Closed-loop systems in fashion refer to systems where materials and resources are kept in circulation for as long as possible, minimizing waste and maximizing value. These systems aim to create a closed-loop lifecycle for fashion products, where materials are reused, recycled, or regenerated to create new products, thus eliminating the concept of waste. Closed-loop systems focus on recovering materials from end-of-life products through recycling, upcycling, or repurposing. This may involve collecting and sorting used clothing and textiles, dismantling and processing them into raw materials, and reintroducing these materials into the production process.

7.3 Promoting Conscious Fashion Choices

Fair Trade certification ensures that products are produced according to strict social, environmental, and economic standards, promoting fair wages, safe and secure working environments, and community development for workers or individuals in the supply chain. Designers can opt for Fair Trade–certified materials to support ethical sourcing practices and promote social equity within the textile sector [38].

7.4 Sustainable Practices in Textile Production

Sustainable practices in textile production have become increasingly crucial in today's environmentally conscious world. This shift is evident across the entire supply chain, where efforts to reduce water consumption, minimize chemical usage, and adopt renewable energy sources are transforming traditional practices towards more sustainable alternatives. Sustainable initiatives are reshaping the way textiles are produced [28]. From the cultivation of raw materials to the manufacturing process, various initiatives are being undertaken to promote sustainability.

7.4.1 Innovative Fibers

Nanofibers, such as chitin nanofibers, silk nanofibers, alginate nanofibers, and polylactic acid (PLA) nanofibers, are revolutionizing textile applications with their biodegradability and unique properties. Chitin nanofibers, derived from crustaceans and insects, offer antimicrobial activity and are suitable for textiles, wound dressings, and environmental remediation. Silk nanofibers exhibit exceptional strength and flexibility, making them ideal for biomedical textiles and drug delivery systems [32]. Alginate nanofibers, sourced from seaweed or algae, provide biocompatibility and moisture absorption for wound dressings and tissue engineering. PLA nanofibers, derived from plant-based sources like corn starch, offer mechanical strength and thermal stability for textiles, medical devices, and packaging materials.

The sea-island microfiber is innovative fiber, also known as super-conjugated or matrix fibril fiber, which consists of one polymer dispersed as "islands" within another polymer matrix, resembling islands in a sea. Widely utilized in luxury fashion, building materials, filters, and automotive interiors, sea-island microfiber fabrics offer excellent breathability, softness, and drapability. Nanofiber production techniques often require fewer chemicals and solvents compared to traditional textile manufacturing processes [41].

Nanofiber production methods, such as electrospinning [16, 32], can be energy-efficient and require less energy compared to conventional textile manufacturing processes [13]. Yu and colleagues developed a nanofiber core-spun yarn using electrospinning, which exhibits remarkably effective antibacterial properties [33].

Plastic optical fibers enable data transmission and light emission without generating heat. The emerging trend in smart textile design revolves around the creation of multifunctional nanocomposite fibers. These fibers hold potential for various applications, including fiber optics and batteries integrated within textiles. By utilizing diverse materials such as biofunctional polymers, optical plastics, conductive polymers, metal alloys, and electrochemical substances, these fiber nanocomposites can be tailored during production. Electroluminescent wires, fluorescent fibers, optical fibers, and photonic band gap fibers offer advanced functionalities in smart textiles [21]. Beyond their aesthetic appeal in the fashion industry, these nano-engineered fibers can perform diverse functions, such as temperature and humidity sensing, pressure monitoring, optical displays, and data communication [33].

7.4.2 Recycled Polyester

Recycled polyester is produced from used plastic bottles or other discarded plastic materials, diverting these materials away from landfills and reducing the demand for virgin polyester. By repurposing post-consumer plastic materials, recycled polyester helps reduce environmental degradation and conserve natural resources. Additionally, recycled polyester retains the same quality and performance as virgin polyester, making it a viable alternative for a wide range of textile applications, including clothing, accessories, and home furnishings.

7.4.3 Vegan Leather Alternatives

Vegan leather is also referred to as faux leather or synthetic leather. With a growing demand for sustainable and ethical fashion, vegan leather has become increasingly popular among consumers seeking stylish and eco-conscious alternatives to conventional leather goods [38]. Innovations in biotechnology have led to the development of lab-grown or bio-based materials that mimic the look and feel of traditional leather without the need for animal products. With increasing demand for sustainable and ethical fashion choices, vegan leather has become a popular option for consumers seeking stylish, eco-friendly alternatives to conventional leather products.

The artificial leather made from sea-island microfiber is extremely similar to natural leather in terms of internal microstructure and appearance quality, which is superior to natural leather in terms of permeability, dimensional stability, mildew resistance, and various fastnesses. Companies like Zhejiang Furun Printing and Dyeing Co., Ltd., have developed specialized sea-island microfiber products for various high-end clothing items, including shoes, jackets, and leather bags [41].

7.5 Role of Technological Advancements in Achieving Sustainability

Innovations in materials science, manufacturing processes, and supply chain management enable the development and implementation of more sustainable practices. One significant area of technological advancement is the development of alternative materials derived from renewable sources or recycled materials. Technological innovations in textile manufacturing enable the implementation of more sustainable production methods. Advanced machinery and equipment equipped with energy-efficient features help reduce energy consumption. Digital printing technologies offer precise and resource-efficient methods of applying color to textiles, minimizing waste and chemical usage compared to traditional printing methods. Moreover, breakthroughs in nanotechnology are enabling the creation of functional textiles with enhanced properties. Another exciting development is the emergence of 3D printing technology in textile production [39]. 3D printing allows for the precise deposition of materials, enabling the creation of complex textile structures with minimal waste. Additionally, advancements in robotics are transforming textile manufacturing by automating repetitive tasks and increasing production efficiency. The Internet of Things enables greater transparency and traceability in the textile supply chain. Thus, technological advancements across various domains are driving sustainable practices in the textile industry, from the acquisition of raw material to the production process, distribution, and disposal at the end of its life cycle.

7.5.1 Waste to Value-Added Production Process

The processes involve the conversion of waste materials into new textile fibers and products. These processes aim to repurpose waste materials that would otherwise end up in landfills or incinerators [40]. Innovative technologies such as mechanical recycling, chemical recycling, and upcycling enable the transformation of waste materials into valuable resources. By repurposing waste materials into valuable resources, these technologies promote sustainability and resource efficiency in the textile industry. These processes help address the growing issue of waste management while reducing the reliance on virgin resources. Waste-to-textile processes contribute to a circular economy by closing the loop on material resources and reducing the environmental impact of textile production. Additionally, this offer opportunities for innovation and collaboration across various sectors, driving the development of new technologies and materials that further advance sustainability goals in the textile industry.

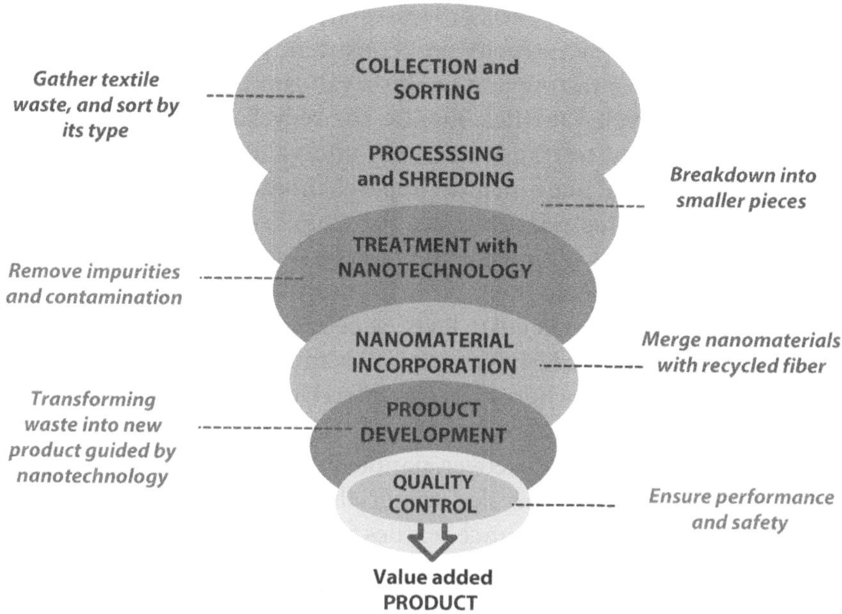

Figure 7.2 Process of textile waste recycling with nanotechnology.

Figure 7.2 shows that textile waste recycling with nanotechnology begins with collection and sorting. In this initial phase, discarded textiles are gathered from diverse sources including manufacturing scraps, post-consumer garments, and outdated textiles. This step ensures that discarded textiles from various sources are efficiently gathered and categorized based on material type, color, and condition. Following collection and sorting, the textile waste undergoes processing and shredding. This phase involves mechanical processing techniques aimed at breaking down the collected materials into smaller, more manageable pieces. Shredding machines efficiently transform large textile items into smaller fragments, facilitating subsequent treatment processes. Treatment with nanotechnology marks a pivotal stage in the textile waste recycling process. Here, nanotechnology-enabled chemical treatments are employed to transform the shredded textile waste at the molecular level. These nanotechnology-enhanced treatments ensure that the recycled materials exhibit improved properties and functionalities, contributing to the production of high-quality recycled textiles. The next step involves nanomaterial incorporation and quality control measures. Nanoparticles (NPs) or nanofibers are strategically incorporated to reinforce the recycled fibers, imparting enhanced strength, durability, and other desirable characteristics. Quality control processes are implemented throughout this phase to ensure the integrity and consistency of the recycled materials. Comprehensive testing and analysis validate the quality and functionality of the recycled textiles. Finally, the recycled materials undergo product development to transform them into value-added textile products. These products may include apparel, home textiles, automotive components, and industrial materials. Figure 7.3 shows the multifaceted role of nanotechnology in transforming textile waste into the value-added product. Initially, nanotechnology facilitates the enhanced breakdown of textile waste at the molecular level, accelerating decomposition processes and facilitating material separation. Nanomaterials are also instrumental in dye removal, where NPs can effectively adsorb or catalyze the degradation of dye molecules, resulting in cleaner and more sustainable recycling outcomes [25]. Moreover, nanotechnology enables the production of nanofibers from recycled materials, offering opportunities for the creation of high-performance textiles with enhanced strength, durability, and functionality. In the transformation phase, nanotechnology facilitates recycling or upcycling processes by enabling the integration of nanomaterials into recycled fibers, enhancing their properties and extending their lifespan. Additionally, nanotechnology-driven sustainable finishes, such as NP coatings or nano-encapsulation of active agents, impart desired functionalities to recycled textiles, further enhancing their value and sustainability.

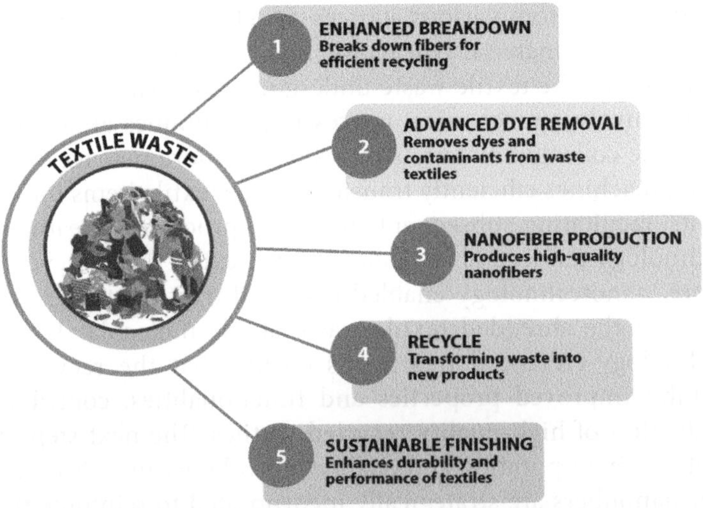

Figure 7.3 Role of nanotechnology in transforming textile waste.

7.5.2 Role of Automation and Robotics in Recycling

Automation and robotics play a significant role in the textile sector for solid waste management by improving efficiency, reducing labor costs, and minimizing environmental impact. Automation and robotics are employed in waste sorting and separation processes to streamline operations and increase accuracy. Automated sorting systems use sensors, cameras, and robotic arms to identify and separate different types of waste materials, such as fabric scraps, threads, and packaging materials, based on their composition, size, and color. Robotic arms equipped with suction or gripping mechanisms can pick and place waste materials onto conveyor belts or sorting bins, reducing the need for manual labor and increasing efficiency. Automation and robotics are utilized in recycling and upcycling processes to transform textile waste into new raw materials or products. Automated recycling systems shred, granulate, and melt down waste materials, such as fabric scraps and discarded garments, to produce recycled fibers, yarns, and fabrics. Robotic arms and automated machinery handle various tasks, such as feeding materials into recycling machines, monitoring processing parameters, and quality control inspections. These systems ensure consistent processing and high-quality output while also reducing labor costs and operational errors.

Advanced robotics systems can perform a wide range of repetitive tasks such as cutting, stitching, and assembly capabilities can create innovative designs and prototypes using recycled materials, such as bags, accessories, and home furnishings. Robotics also enable the integration of smart technologies for real-time monitoring, quality control, and predictive maintenance, further enhancing product quality, operational efficiency, and sustainability in textile production facilities [21]. Automated incineration or pyrolysis systems use robotics to handle waste feeding, combustion, and ash disposal, ensuring safe and efficient energy generation. By minimizing material waste, energy usage, and emissions, manufacturers can lower their carbon footprint and promote environmentally responsible practices.

7.6 Technological Innovations in Material Production

Advancements in material production technology have revolutionized various industries. These innovations encompass a wide array of techniques and processes aimed at enhancing material properties, improving production efficiency, and minimizing environmental impact.

7.6.1 The ChroMorphus Technology

Developed by researchers at the University of Central Florida, this pioneering fabric has the remarkable ability to change colors seamlessly. This groundbreaking technology integrates micro wires intricately woven into the fabric. ChroMorphus enables the application of electric current to induce heat, activating special pigments within the fabric and facilitating dynamic color changes. Remarkably, this technology extends beyond mere color alteration, allowing for the manipulation of patterns on the fabric's surface. In the context of evolving textile technologies, ChroMorphus offers a transformative solution to the conventional approach of selecting clothing based on seasonal or situational color preferences. This not only streamlines wardrobe management but also aligns with the growing emphasis on sustainability in the fashion industry. With its potential to redefine fashion consumption patterns, this transformative technology has the potential to revolutionize the fashion industry, offering consumers a sustainable and innovative solution [6].

7.6.2 3D and 4D Printing Technology

3D printing, referred to as additive manufacturing, is revolutionizing textile production by enabling the production of intricate designs and complex structures with precision and efficiency [21]. This technology facilitates the fabrication of customized or personalized garments, footwear, and accessories with minimal waste and faster production cycles. One of its key advantages in fashion lies in its capacity to produce complex shapes and structures that are challenging or impossible to achieve with traditional techniques. Designers can experiment with novel geometries, textures, and patterns, pushing the boundaries of creativity and innovation in fashion design [34]. Furthermore, 3D printing allows for customization at scale, enabling personalized products tailored to individual preferences and measurements. This level of customization not only enhances the consumer experience but also promotes inclusivity and diversity in fashion by accommodating a wide range of body shapes and sizes. By eliminating the need for traditional cutting and sewing processes, 3D printing reduces material waste and energy consumption, contributing to sustainability in textile manufacturing. In contrast to traditional manufacturing techniques, 3D printing offers various advantages, including heightened material utilization, financial profit, superior design adaptability, and improved efficiency in large-scale production. Additionally, the appeal of 3D printing lies in its ability to reduce raw material usage, automate processes, minimize post-processing needs, and enhance energy conservation. 3D printing allows for the creation of porous membranes, complicated catalytic structures, microreactors, bioreactors, spacers, microfluidic devices, separators, and filters all of which help to preserve the environment [34]. These polymer-based structures, developed by 3D printing, have accelerated progress in removing impurities from wastewater. Incorporating nanoscale fillers, such as NPs, nanofibers, nanotubes, or other nanostructures, into nanocomposites can lead to enhancements in mechanical, electrical, thermal, or optical properties when compared to traditional composites [3, 34].

Advancements in 3D printing technology coupled with integration of sophisticated functional shape memory materials (SMMs) have given rise to a novel concept known as "4D textiles." These 4D printed textiles exhibit the remarkable ability to reshape themselves and alter their properties in response to external stimuli. One notable example of this pioneering technology is demonstrated through a shirt designed by Susan Clowes' shirt design. By blending shape memory alloys (SMAs) fibers with nylon fibers, the shirt can undergo significant transformations. As temperatures rise, the SMAs retain its memory shape. Another pioneering effort comes from Vili, who utilized nickel-titanium SMAs to craft yarns and fabrics capable of undergoing shape changes through heating and cooling cycles. These textiles not only possess functional adaptability but also offer aesthetic appeal. Building upon this concept, Berzowska and Coelho have developed dresses incorporating nitinol SMAs wires. Upon cooling, the material reverts to its original state, showcasing the reversible nature of these transformations. The artistic potential of 4D textiles is exemplified in the floral dress "Kukkia." Here, silk petals embedded with nitinol wires create a visually captivating effect. Through controlled heating and cooling cycles, the petals contract and expand, evoking the illusion of a blooming flower on the fabric [6].

7.6.3 Dyeing and Finishing Technologies

Dyeing and finishing technologies are pivotal processes within the textile industry, seamlessly transforming raw fabrics into vibrant, textured, and functional materials [16]. Through a meticulous blend of chemistry, engineering, and creativity, these technologies imbue fabrics with desired colors, patterns, and performance attributes. Meanwhile, finishing technologies add the final touches to fabrics, enhancing their aesthetic appeal, durability, and functionality [16]. Advancements in nanotechnology have opened doors to a new era of functional finishes, enabling textiles to repel water, resist stains, and regulate temperature with unparalleled precision. Advanced dyeing techniques such as cold pad batch, air dyeing, and foam dyeing reduce water consumption by up to 90% compared to conventional methods. Innovative finishing processes, including plasma treatment and enzymatic finishing, offer alternatives to chemical treatments, reducing environmental impact and improving product sustainability. Recent studies focus on optimizing extraction methods, enhancing color fastness, and exploring new sources of natural dyes to expand the color palette while also maintaining color vibrancy and fastness. Waterless dyeing technologies, such as air dyeing and supercritical fluid dyeing, offer sustainable alternatives to traditional water-based dyeing processes in the textile industry. Air dyeing utilizes atomized dye fluid sprayed onto fabric using high-pressure airflow, eliminating the need for water baths and reducing energy consumption. This method significantly decreases water usage, minimizes environmental pollution, and lowers energy requirements, contributing to global responsibility goals. Similarly, supercritical fluid dyeing employs supercritical carbon dioxide, eliminating the need for water and chemicals, reducing energy consumption by 50%, and eliminating water contamination and treatment processes. Both techniques offer advantages such as uniform dyeing, high rubbing fastness, and the ability to operate in areas with water scarcity, making them promising solutions for sustainable textile dyeing practices. Recently, advancements in printing methods have facilitated the creation of electrical circuits through the utilization of a novel approach known as conductive ink. This innovative technique allows for the application of conductive inks as a flexible printing material, achieved by incorporating elastic polymers (elastomers) combined with conductive metallic additives [43].

7.6.4 Smart and Wearable Technologies

Smart textiles are also known as smart fabrics or e-textiles. These textiles can sense, react to, and adapt to changes in the environment [11, 36] or user inputs. These textiles modify their properties or generate responses such as changes in color, light intensity, fluorescence, or physical attributes like mechanical, electrical, thermal, chemical, and moisture transport properties [5]. These textiles incorporate a range of sensors, optical devices, nano-generators, and energy storage components. This includes LED displays, wearable sensors, and communication devices. Various sensors, including heat, touch, pressure, optical, and chemical sensors, can be seamlessly integrated into textiles to serve a myriad of applications [18]. Materials like conductive polymers, metallic yarns, and shape memory materials contribute to the development of advanced textiles. Techniques like photolithography and inkjet printing [36] enable the weaving of sensors directly into fabrics [28]. Moreover, miniature chips embedded within smart textiles function as analog-to-digital converters, multimeters, or amplifiers, further enhancing their capabilities. The integration of metal-organic frameworks (MOFs) with quantum nanorods within cotton fabrics enables efficient colorimetric sensing of toxic gases, adding yet another dimension to the versatility of smart textiles. These advancements promise breakthrough applications in healthcare, security, and fashion. Wearable garments equipped with specialized sensors can monitor the wearer's physiology and posture, offering early detection and correction of potential health issues [36]. Furthermore, innovations such as pressure and posture sensors in shoes or vibration nanomotors within garments aim to improve blood circulation and promote weight loss [5, 33]. Wearable capacitors offer the potential to supply power to both internal and external devices. One significant application lies in body cooling, particularly beneficial in hot climates. Smart textiles are emerging with applications in different industries, including fashion, sports and fitness, healthcare, military, and automotive [18].

7.7 Nanotechnology in Fabric Production

Nanotechnology, a multidisciplinary field that deals with materials and structures at the nanoscale, has emerged as a promising avenue. It is the science of engineering and manipulating matter at the nanoscale that has emerged as a groundbreaking field with vast potential across various industries. In recent years, it has made significant contributions to the textile and fashion industry revolutionizing the way fabrics are designed, produced, and utilized. Figure 7.4 represents the role nanotechnology contributing to sustainability in the textile sector by enhancing the durability, functionality, and longevity of textiles; this reduces the necessity for frequent replacement and maintenance. Traditional textile manufacturing processes often involve harmful chemicals, excessive water usage, and high energy consumption. It offers sustainable alternatives, such as the use of nano-sized enzymes for eco-friendly fabric treatments, and energy-efficient processes, such as enzymatic treatment for fabric finishing or dyeing [32].

Nanomaterials represent a remarkable intersection of science, engineering, and innovation at the nanoscale, where properties and behaviors diverge from those at larger scales. Figure 7.5 shows different types of nanomaterials. Researchers are constantly discovering new ways to apply nanoscale materials and structures to improve the properties and functionalities of textiles. Driven by the pursuit of developing textiles with novel functionalities, cutting-edge nanomaterials offer a wide array of potential applications. Functional nanomaterials have been seamlessly integrated into

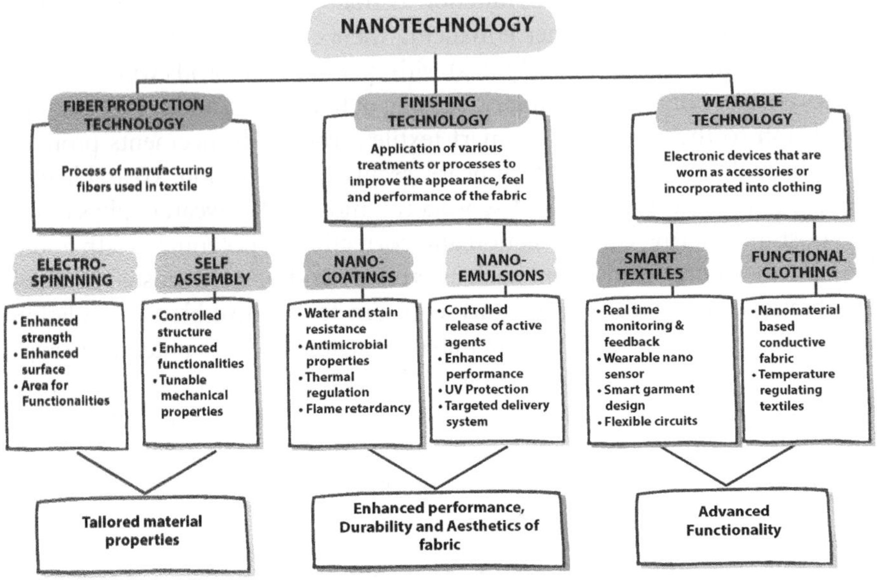

Figure 7.4 Role of nanotechnology in textile sector.

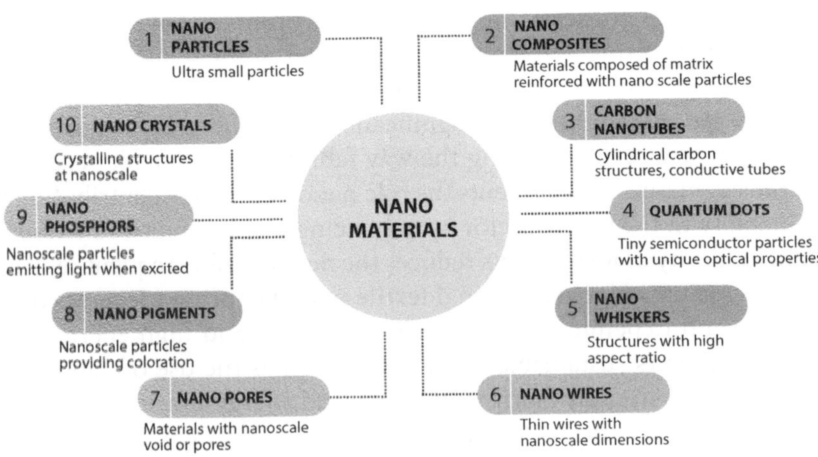

Figure 7.5 Types of nanomaterials.

fabrics, leading to enhanced properties and elevating their performance with attributes like stain resistance, waterproofing, UV protection, odor resistance, flame retardancy, anti-microbial, etc [16, 19, 28]. It has given rise to the development of innovative nano-fabrics with exceptional properties and aesthetic appeal [1]. Incorporating nanostructured materials and coatings allows textiles to achieve increased durability by improving resistance to wear, tear, and abrasion. These advancements ensure that textile products withstand repeated use and laundering, minimizing waste generation by reducing the need for frequent replacements [19].

7.8 Innovations in Nanotechnology Transforming Textile Sector

Recently, metal NPs have garnered attention for their applications in textiles, medicine, agriculture, environmental remediation, and physiochemical processes [2]. Nanoparticles can be incorporated into fabric fibers or applied as surface coatings. Metal oxide NPs, in particular, find extensive use in textile functionalization, serving as agents for antibacterial, antistatic, ultraviolet protection, and self-cleaning properties [6]. The integration of NPs into fabrics can be accomplished through diverse techniques, which can be broadly classified into *in situ* and *ex situ* methods. *In situ* methods encompass chemical reduction, solution immersion, surface impregnation, ultrasonication, magnetron sputtering, sol-gel synthesis, pulse laser ablation, and sono-chemical synthesis [23]. On the other hand, *ex situ* methods involve electrospinning, melt spinning, plasma coating, sol-gel processes, spray coating, and the pad-dry-cure method, among others [4, 23]. Nanoparticles can enhance the efficiency of water and energy usage during textile production, leading to reduced waste and lower carbon footprint. For instance, NPs of titanium dioxide can be woven into fabric fibers, giving rise to self-cleaning textiles that repel dirt and stains, thereby reducing the frequency of washing and conserving water resources.

Furthermore, nanotechnology enables the creation of functional finishes that extend the lifespan of textile products while minimizing the need for harsh chemical treatments and coatings, thus promoting environmental sustainability and reducing waste generation. Houshyar, Nayak *et al.* (2019) utilized dip coating to apply detonation nanodiamonds onto cotton fabrics, resulting in enhancements in tensile strength, abrasion resistance, and thermal stability. A similar coating method was employed for wool substrates, leading to improved durability and notable enhancements in surface wetting properties as the concentration of nanodiamonds in the coating suspension increased.

Nanotechnology has paved the way for the development of smart textiles with embedded sensors, actuators [36], and electronics that monitor, regulate, and respond to changes in the environment or user behavior.

These smart textiles can enhance comfort, safety, and performance while also promoting sustainability by optimizing energy consumption and resource utilization [11]. Additionally, nanotechnology-based sensors and coatings can detect and neutralize harmful substances, pathogens, and pollutants. Textiles can now possess capabilities such as monitoring vital signs, detecting changes in the environment, and even powering electronic devices [27]. We see that the impact of nanotechnology on textiles is profound and multifaceted.

7.8.1 Sensing Textiles

These textiles incorporate sensors or conductive materials to detect various stimuli such as temperature, pressure, moisture, or bio-signals like heart rate and muscle movement [10, 16]. Triboelectric nanogenerators (TENGs) integrated into wearable biomonitoring textiles offer a promising solution for real-time physiological monitoring [37]. These textiles can utilize readily available materials and scalable manufacturing techniques, providing a comfortable and non-invasive method for monitoring health parameters. Lin *et al.* introduced a TENG-based sensor array with a sandwich structure, utilizing wavy PET as an intermediate layer to assess sleep behavior and quality in real-time [37].

7.8.2 Actuating Textiles

These textiles include components that can respond to stimuli, such as shape memory materials that change shape in response to temperature or electrical stimuli. Shape-memory textiles can adapt to mechanical deformations, returning to their original form after being manipulated [17].

7.8.3 Energy-Harvesting Textiles

These textiles are equipped with technologies that enable them to harvest energy from their environment, such as solar cells [15] or piezoelectric materials that generate electricity from mechanical vibrations. Ni-Co layered double hydroxide nanosheets were cultivated on conductive textile substrates and suggested for use as electrodes in energy storage devices. Lu and colleagues created a cotton textile coated with nickel through electrodeposition of Ni-Al layered double hydroxide. This Ni-coated cotton textile displayed excellent conductivity and was introduced as a potential wearable energy storage device [44].

7.8.4 Thermoregulating Textiles

Thermochromic materials, known for their temperature-sensitive properties that enable reversible color changes. Thermochromic materials serve not only aesthetic and design purposes but also functional roles such as warning or data indicators. Recently, there has been a growing trend of incorporating thermochromic materials into textile products like t-shirts and hats to create both visually appealing and functional designs. Abou Elmaaty *et al.* (2020) have developed protective children's wear featuring thermochromic properties, achieved by printing thermochromic colorants onto cotton fabrics. Tozum and colleagues synthesized thermochromic microcapsules composed of poly (methyl methacrylate-co-ethylene glycol dimethacrylate-co-glycidyl methacrylate) using the emulsion polymerization technique [45]. The resulting fabrics treated with microcapsules demonstrated outstanding reversible thermochromic characteristics along with efficient energy storage capabilities [10, 44], whereas photochromic and thermochromic materials enable textiles to respond to changes in temperature or light intensity [28, 33].

7.8.5 Luminescent Textiles

These textiles incorporate materials that emit light, such as LED fibers or luminescent dyes, for applications ranging from fashion to safety. Phosphorescent films enable fabrics to glow in the dark. Retro-reflective inks are widely used for security clothing, and electroluminescent wires or optical fibers can be woven into textiles for illumination purposes [28]. Blosi and colleagues achieved the creation of versatile luminescent textiles through the novel application of newly developed iridium-supported silica (Ir@SiO2) compounds in a coating form [26].

7.8.6 Conductive Textiles

These textiles contain conductive materials such as metal fibers or conductive polymers [16], allowing them to transmit electricity and data, enabling applications like wearable electronics [15] and communication devices. Numerous conductive matrices have been devised for coating cotton to enhance its electrical conductivity. The cotton yarns are covered with thin layers of PEDOT [poly(styrenesulfonate)] containing an array of Au NPs. These layers make the cotton yarns conductive and able to carry solar energy throughout the entire garment.

Shim and colleagues pioneered a polyelectrolyte-based coating combined with multiwalled carbon nanotubes (CNTs) for rendering textiles conductive. Additionally, graphene has been integrated into textile fibers to impart conductive properties [32]. For instance, fabric was prepared by intertwining two sets of graphene micro ribbons [33].

7.8.7 Medical Textiles

These textiles are specifically designed for medical or healthcare applications, such as wound dressings with antibacterial properties or garments with embedded sensors for monitoring patient health [10, 16]. Examples such as cotton fabrics impregnated with silver NPs exhibit powerful antimicrobial properties [17]. Additionally, TiO_2 and ZnO NPs are utilized to confer antibacterial and fungicidal properties to textiles. It helps inhibit the growth of harmful microorganisms, making them ideal for healthcare professionals and individuals in high-risk environments [8].

7.8.8 Protective Textiles

These textiles are engineered to provide protection against various hazards, including fire-resistant fabrics, ballistic textiles, and textiles with UV-blocking properties. Yao *et al.* (2019) introduced inherently flame-retardant nylon fibers enhanced with nanoclay additives as a cost-effective component in a laminate thermal protective clothing system. Wu *et al.* (2018a; 2018b) investigated the efficacy of MMT nanoclay as an ingredient in flame retardant nylon 6 nanocomposite fibers. Their study demonstrated that, when adequately dispersed within the polymer matrix through high-shear mixing, nanoclay additives effectively mitigate the peak heat release and heat release capacity of the nanocomposite system. The researchers have utilized both pristine and surface-modified CNTs [18] on cotton fibers to emulate the surface texture observed in lotus leaves, known as the Lotus effect, resulting in the creation of superhydrophobic surfaces [31, 35]. Metal oxide NPs like ZnO, SiO_2, TiO_2, ZrO_2, and Al_2O_3 have been widely incorporated into textiles using the sol-gel method to create water-resistant fabrics [23].

7.9 Nanotextile Market Trend

In the textile industry, NanoTex (US) and Schoeller Technologies AG (Switzerland) are pioneers in commercializing nanotechnology finishes for fabrics [7, 30, 35]. Advancements in textiles have led to the development of innovative products. Philips has introduced a line of clothing called Bubelle, which changes color based on the wearer's mood or disposition. The Black-Eyed Peas have developed fashion garments using organic light-emitting diode (OLED) materials, allowing for dynamic lighting patterns and color effects. Gore-Tex utilizes expanded polytetrafluoroethylene (ePTFE) for waterproof and breathable applications across medical devices, fabrics, and electronics. NanoSphere, developed by Schoeller Textiles AG, mimics the self-cleaning effect of lotus leaves through hydrophobic surface properties [20, 35]. Aquapel, from Nanotex LLC, replicates natural water repellency seen in plant surfaces and animal coats. SmartSilver, produced by Nano Horizons, offers antimicrobial properties using silver NP additives in yarn. Nanosan, by SNS-Nano Fiber Technology and Schill & Seilacher, offers spun polymers for filters and adsorbent fabrics. Sarvint Technologies' Wearable Motherboard utilizes engineered fibers to monitor various body signals. Optical fibers are also employed in multifunctional garments for sports and fashion, exemplified by Cambridge Consultants' Xel flex fabric, which incorporates optical fiber sensors to track body movements [33]. This technology has applications in fitness, sports coaching, and physiotherapy. Moreover, there are demonstrations of gloves and garments equipped with fiber sensors capable of recognizing hand and body postures. On-garment fiber sensors can detect physical or chemical stimuli like strain, pressure, temperature, humidity, and metabolites [42].

7.10 Emerging Technologies and Future Trend

Emerging technologies are reshaping the landscape of the textile industry [39]. Nanotechnology stands at the forefront of this revolution, facilitating the development of smart textiles with remarkable capabilities [30]. Additive manufacturing technologies, including 3D printing, are another driving force behind the textile industry's transformation. These technologies allow for the production of textiles and garments with intricate geometries and customized designs [34]. In parallel, advancements in biotechnology are leading to the rise of biodegradable and bio-based textiles. These bio-based textiles not only minimize environmental impact but also offer desirable properties such as breathability, durability, and comfort, making them increasingly popular among eco-conscious consumers [24]. Augmented reality [36] and virtual reality [37] technologies are on the brink of transforming the way consumers engage with fashion and textiles. Additionally, integrating holograms into garments and wearable devices is a promising area. Nanotextiles can incorporate molecular dyes and analyte-sensitive compounds, with microfluidics integrated into thread-based channels for point-of-care diagnostics [42]. Robotics and automation are playing an increasingly vital role in textile manufacturing, automating tasks. As these technologies continue to evolve and converge, they hold the potential to reshape every aspect of the textile supply chain, from design and production to distribution and consumption, ultimately creating a more resilient, efficient, and inclusive textile ecosystem [9].

Innovations in wearable technology are transforming the way we interact with our clothing and accessories. The future of fashion trends is marked by a heightened focus on individualization and inclusivity. Personalization technologies and customization options will become increasingly prevalent. As fashion continues to evolve, futuristic designs will push the boundaries of creativity, embracing bold experimentation and innovative materials. From smart fabrics to wearable gadgets, these advancements are reshaping not only our fashion choices but also our daily routines and lifestyles. With gesture-controlled accessories, the possibilities are endless. In the future, textiles are expected to seamlessly integrate various functional components, leading to advancements in production processes that combine electronics, biomaterials, and optics with traditional weaving techniques.

These examples represent just a glimpse of the exciting possibilities that wearable technology brings, promising a future where our clothing and accessories are not only stylish but also intelligent and adaptive, enhancing our lives in unexpected ways.

7.11 Conclusion

The integration of nanotechnology into textile design holds immense promise for addressing the sustainability challenges in the fashion industry. The exploration of nanotechnology in textile design reveals promising avenues for innovation and sustainability in the fashion industry. By leveraging nanomaterials, functional finishes, and smart textiles, the textiles can achieve enhanced properties such as strength, conductivity, and flexibility, leading to the creation of fabrics with futuristic textures and patterns. The implications of nanotechnology in shaping sustainable textile design are profound, offering opportunities to reduce environmental impact, enhance product performance, and shape future fashion trends that prioritize sustainability [32].

Looking ahead, future directions should focus on optimizing nanomaterials for scalability, cost-effectiveness, and recyclability while also addressing concerns related to health and safety [7]. Recommendations include further research into biodegradable nanomaterials, collaboration between academia and industry, and the implementation of regulatory frameworks to ensure the responsible use of nanotechnology in textile design.

References

1. Akbar, M., Masood, B., Price, G., Shibiru, A.L., Qurashi, A., Applications of nanotechnology in smart textile industry: A critical review. *J. Adv. Res.*, *38*, 55–75, 2022, https://doi.org/10.1016/j.jare.2022.01.008.
2. Andra, S., Balu, S.K., Jeevanandam, J., Muthalagu, M., Emerging nanomaterials for antibacterial textile fabrication. *Naunyn-Schmiedeberg's Arch. Pharmacol.*, *394*, 7, 1355–1382, 2021, https://doi.org/10.1007/s00210-021-02064-8.
3. Andrew, J.J. and Dhakal, H.N., Composites Part C: Open Access Sustainable biobased composites for advanced applications: recent trends and future opportunities – A critical review. *Composites, Part C: Open Access*, 7, 100220, 2022, https://doi.org/10.1016/j.jcomc.2021.100220.
4. Ayatullah, A.K.M., Asif, H., Hasan, M.Z., Application of Nanotechnology in Modern Textiles: A Review. *Int. J. Curr. Eng. Technol.*, *8*, 2, 227–231, 2018, http://inpressco.com/category/ijcet.
5. Basel, Y., Textronics: a review of their technological aspects and applications. *J. Text. Inst.*, 0, 0, 1–17, 2023. https://doi.org/10.1080/00405000.2023.2236320
6. Biswas, M.C., Chakraborty, S., Bhattacharjee, A., Mohammed, Z., 4D Printing of Shape Memory Materials for Textiles: Mechanism, Mathematical Modeling, and Challenges. *Adv. Funct. Mater.*, *31*, 19, 1–25, 2021, https://doi.org/10.1002/adfm.202100257.
7. Black, S., The role of nanotechnology in sustainable textiles, in: *Sustainable Textiles: Life Cycle and Environmental Impact*, pp. 302–328, 2009, https://doi.org/10.1533/9781845696948.2.302.
8. Broadhead, R., Craeye, L., Callewaert, C., The future of functional clothing for an improved skin and textile microbiome relationship. *Microorganisms*, *9*, 6, 1192ps, 2021, https://doi.org/10.3390/microorganisms9061192.
9. Chourasiya, R., Pandey, S., Malviya, R.K., Sustainable manufacturing adoption in textile industries: A systematic state-of-art literature review and future research outline. Sustainable Development, 31, 2, 612–638, 2023. https://doi.org/10.1002/sd.2434
10. Das, S.K., Ulhuq, I., Ahamed, S., Das, S., An Overview on Multifunctional Smart Textiles and Future Perspective for Mankind Development with Nano-technology An Overview on Multifunctional Smart Textiles and

Future Perspective for Mankind Development with Nano-technology. *Int. J. Res. Cult. Soc., 4*, 1, 12–16, 2020, https://doi.org/10.2017/IJRCS.2456.6683/202001004.

11. Dong, X., Cao, Y., Wang, C., Wu, B., Zheng, M., Xue, Y., Li, W., Han, B., Zheng, M., Wang, Z., Zhuo, M., MXene-Decorated Smart Textiles with the Desired Mid-Infrared Emissivity for Passive Personal Thermal Management, *ACS Appl. Mater. Interfaces, 15,* 9, 12032–12040 2023, https://doi.org/10.1021/acsami.2c21696.
12. Hossain, M.A., Simultaneous Thermoregulated and Sensorial effect of Smart Textiles with Artificial Composite Phase Change Materials (CPCM) incorporated with Carbon Nano Conductive Materials for special workers and extreme weather conditions. School of Fashions and Textiles, RMIT University, Victoria 3056, 2023, https://doi.org/10.13140/RG.2.2.32289.79204.
13. Garcia-Orue, I., Gainza, G., Garcia-Garcia, P., Gutierrez, F.B., Aguirre, J.J., Hernandez, R.M., Delgado, A., Igartua, M., Composite nanofibrous membranes of PLGA/Aloe vera containing lipid nanoparticles for wound dressing applications. *Int. J. Pharm.*, 556, 320–329, 2019, https://doi.org/10.1016/j.ijpharm.2018.12.010.
14. Syduzzaman, M., Hassan, A., Anik, H.R., Akter, M., Islam, M.R., Nanotechnology for High-Performance Textiles: A Promising Frontier for Innovation. *Chem. Nano. Mat.*, *9*, 9, e202300205, 2023. https://doi.org/10.1002/cnma.202300205
15. Hatamvand, M., Kamrani, E., Lira-Cantú, M., Madsen, M., Patil, B.R., Vivo, P., Mehmood, M.S., Numan, A., Ahmed, I., Zhan, Y., Recent advances in fiber-shaped and planar-shaped textile solar cells. *Nano Energy*, *71*, 104609, February, 2020, https://doi.org/10.1016/j.nanoen.2020.104609.
16. Jatoi, A.S., Ahmed Khan, F.S., Mazari, S.A., Mubarak, N.M., Abro, R., Ahmed, J., Ahmed, M., Baloch, H., Sabzoi, N., Current applications of smart nanotextiles and future trends, in: *Nanosensors and Nanodevices for Smart Multifunctional Textiles. Inc.*, 2020, https://doi.org/10.1016/B978-0-12-820777-2.00019-4.
17. Joshi, M. and Bhattacharyya, A., Nanotechnology - A new route to high-performance functional textiles. *Text. Prog.*, *43*, 3, 155–233, 2011, https://doi.org/10.1080/00405167.2011.570027.
18. Kaounides, L., Yu, H., Harper, T., Nanotechnology innovation and applications in textiles industry: Current markets and future growth trends. *Mater. Technol.*, *22*, 4, 209–237, 2007, https://doi.org/10.1179/175355507X250014.
19. Köhler, A.R. and Som, C., Risk preventative innovation strategies for emerging technologies the cases of nano-textiles and smart textiles. *Technovation*, 34, 8, 420–430. https://doi.org/10.1016/j.technovation.2013.07.002
20. Mahmud, R. and Nabi, F., Application of Nanotechnology in the field of Textile. *IOSR J. Polym. Text. Eng.*, *04*, 01, 01–06, 2017, https://doi.org/10.9790/019x-0401010106.

21. Malik, S., Muhammad, K., Waheed, Y., Nanotechnology: A Revolution in Modern Industry. *Molecules*, 28, 2, 661 2023, https://doi.org/10.3390/molecules28020661.
22. Mech, A., Gottardo, S., Amenta, V., Amodio, A., Belz, S., Drbohlavov, J., Farcal, L., Jantunen, P., Ma, A., Rasmussen, K., Riego, J., Rauscher, H., Safe-and sustainable-by-design : The case of Smart Nanomaterials. A perspective based on a European workshop. *Regulatory Toxicology and Pharmacology, 128*, 105093, 2022, https://doi.org/10.1016/j.yrtph.2021.105093.
23. Meda, U.S., Soundarya, V.G., Madhu, H., Bhat, N., Nano-engineered textiles: Development and applications. *Mater. Sci. Eng.: B, 296*, June, 116636, 2023, https://doi.org/10.1016/j.mseb.2023.116636.
24. Nitschke, M. and Marangon, C.A., Microbial surfactants in nanotechnology: recent trends and applications. *Crit. Rev. Biotechnol., 42*, 2, 294–310, 2022, https://doi.org/10.1080/07388551.2021.1933890.
25. Palani, G., Arputhalatha, A., Kannan, K., Lakkaboyana, S.K., Hanafiah, M.M., Kumar, V., Marella, R.K., Current trends in the application of nanomaterials for the removal of pollutants from industrial wastewater treatment—a review. *Molecules, 26*, 9, 2799, 2021, https://doi.org/10.3390/molecules26092799.
26. Paramsothy, M., Nanotechnology in Clothing and Fabrics. *Nanomaterials*. 12, 1, 2022, 67. https://doi.org/10.3390/nano12010067.
27. Iliescu, B.F., Mancasi, V.N., Ilie, I.D., Mancasi, I., Costachescu, B., Rotariu, D.I., Design Principle and Proofing of a New Smart Textile Material That Acts as a Sensor for Immobility in Severe Bed-Confined Patients. *Sensors, 23*, 5, 2573, 2022. https://doi.org/10.3390/s23052573
28. Gajendra Kumar Tyagi and Amit Madhu. Nanotechnology in Textiles: Unleashing Innovation and Value Addition. *Fashion Technol.*, 9, 1. TTEFT. 000702. 2023. DOI: 10.31031/TTEFT.2023.09.000702
29. Rahman, O., Hu, D., Fung, B.C.M., A Systematic Literature Review of Fashion, Sustainability, and Consumption Using a Mixed Methods Approach, *Sustainability, 15*, 16, 12213, 2023. https://doi.org/10.3390/su151612213.
30. Riaz, S., Ashraf, M., Hussain, T., Hussain, M.T., Rehman, A., Javid, A., Iqbal, K., Basit, A., Aziz, H., Functional finishing and coloration of textiles with nanomaterials. *Color. Technol., 134*, 5, 327–346, 2018, https://doi.org/10.1111/cote.12344.
31. Saleem, H. and Zaidi, S.J., Sustainable use of nanomaterials in textiles and their environmental impact. *Materials, 13*, 22, 1–28, 2020, https://doi.org/10.3390/ma13225134.
32. Sbai, S.J., Boukhriss, A., Majid, S., Gmouh, S., The recent advances in nanotechnologies for textile functionalization, in: *Advances in Functional and Protective Textiles. LTD*, 2020, https://doi.org/10.1016/b978-0-12-820257-9.00020-5.
33. Shah, M.A., Pirzada, B.M., Price, G., Shibiru, A.L., Qurashi, A., Applications of nanotechnology in smart textile industry: A critical review. *J. Adv. Res., 38*, 55–75, 2022, https://doi.org/10.1016/j.jare.2022.01.008.

34. Siddiqui, S., Surananai, S., Sainath, K., Zubair, M., Raja, R., Kuppusamy, P., Kempaiah, Y., Emerging trends in development and application of 3D printed nanocomposite polymers for sustainable environmental solutions. *Eur. Polym. J.*, *196*, April, 112298, 2023, https://doi.org/10.1016/j.eurpolymj.2023.112298.
35. Sawhney, A.P.S., Condon, B., Singh, K.V., Pang, S.S., Li, G., Hui, D., Modern Applications of Nanotechnology in Textiles. *Text. Res. J.*, 78, 8, 731–739, 2008. doi:https://doi.org/10.1177/0040517508091066
36. Singha, K., Kumar, J., Pandit, P., Recent advancements in wearable & smart textiles: An overview. *Mater. Today: Proc.*, *16*, 1518–1523, 2019, https://doi.org/10.1016/j.matpr.2019.05.334.
37. Tian, X., Dong, S., Yang, M., Ng, H., Liu, Y., Hu, H., Textile-Based Triboelectric Nanogenerators for Smart Wearable Systems : Comfort, Integration, and Application. 2201294, 2023, https://doi.org/10.1002/admt.202201294.
38. Todeschini, B.V., Cortimiglia, M.N., Callegaro-de-Menezes, D., Ghezzi, A., Innovative and sustainable business models in the fashion industry: Entrepreneurial drivers, opportunities, and challenges. *Bus. Horiz.*, *60*, 6, 759–770, 2017, https://doi.org/10.1016/j.bushor.2017.07.003.
39. Waheed, M. and Khalid, A., Impact of Emerging Technologies for Sustainable Fashion, Textile and Design, in: *Proceedings of the 2nd International Conference on Intelligent Human Systems Integration (IHSI 2019): Integrating People and Intelligent Systems,* San Diego, California, USA, February 7-10, 2019, 10.1007/978-3-030-11051-2_104.
40. Zhang, L., Leung, M.Y., Boriskina, S. *et al.*, Advancing life cycle sustainability of textiles through technological innovations. *Nat. Sustain.*, 6, 243–253, 2023. https://doi.org/10.1038/s41893-022-01004-5
41. Fu, H., Zhang, T., Zhang, S., Qian, X., Current advances on sea-island microfiber nonwoven materials preparation technology and its applications: a review. *J. Text. Inst.*, *0*, 0, 1–11, 2023, https://doi.org/10.1080/00405000.2023.2278373.
42. Yetisen, A.K., Qu, H., Manbachi, A., Butt, H., Dokmeci, M.R., Nanotechnology in Textiles, *ACS Nano*, *10*, 3, 3042–3068, 2016, https://doi.org/10.1021/acsnano.5b08176.
43. Shaheen, T.I., Nanotechnology for modern textiles: highlights on smart applications. *J. Text. I.*, 113, 10, 2274–2284, 2022, https://doi.org/10.1080/00405000.2021.1962625.
44. Krifa, M., and Prichard, C., Nanotechnology in textile and apparel research – an overview of technologies and processes. *J. Text. I.*, 111, 12, 1778–1793, 2020. https://doi.org/10.1080/00405000.2020.1721696
45. Chen, X.X., Yu, J.C., Chen, K. *et al.*, Facile and Large-scale Fabrication of Biodegradable Thermochromic Fibers Based on Poly(lactic acid). *Chin. J. Polym. Sci.*, 40, 1242–1251, 2022. https://doi.org/10.1007/s10118-022-2781-6

8

Understanding and Characterization of Functional Properties of Novel Recycled Nano-Textile

Chet Ram Meena

Department of Textile Design, NIFT Jodhpur, (Ministry of Textiles, GOI), Rajasthan, India

Abstract

Global population, industrial growth, and urbanization have generated vast amounts of textile waste. As textile products are the basic need of human beings, they should test them for their toxicology and acceptance levels. Innovative fabrics have negative consequences for human health and the environment as well. This is an international conflict that requires action to find a solution and put it into practice. The textile industry's present needs urge it to intensify its cooperation with other emerging sectors and emphasis more on nanotechnology research and development. Nanotechnology usually involves the novelty, synthesis, characterization, and application of desirable properties on the base materials. Nanotechnology also offers novel applications on textiles such as biological detection, decomposition of toxic gases, and self-decontamination and military protective gear. Recent applications of nanotechnology for textiles include biological detection, the breakdown of hazardous gasses, self-decontamination, and military protective gear. Nanotechnology expands the horizon of the textile industry with better productivity, efficiency, and profitability.

Nano-textiles appealed to scientists, academics, and businesses due to their vast economic potential and unique properties. Novel nano-textiles introduced functional properties like antimicrobial, chemical, and biological self-cleaning, ultraviolet protection, electrically conductive, optical, hydrophobic, flame-retardant, coloration, and other novel properties to textiles without negotiating the inherent properties of materials. Nanomaterials (NMs) are commonly integrated into novel textiles, including nanoparticles, nanofibers, nanocomposites, carbon nanotubes,

Email: chetram.meena@nift.ac.in

Prashansa Sharma and Shilpi Shree Sahay (eds.) Nanotechnology-Assisted Recycling of Textile Waste: Sustainable Tools for Textiles, (235–264) © 2025 Scrivener Publishing LLC

nanocapsules, nanosols, and nanoemulsions. End applications for NMs are going to include athletic apparel, military gear, medical supplies, and fire safety. Novel nano-textiles are distinguished by a variety of techniques, including spectroscopic, microscopic, X-ray, and imaging techniques that are used to observe the surface, micro, and nano size of materials. These techniques include scanning electron microscopy, transmission electron microscopy, atomic force microscopy, and scanning tunneling microscopy. This chapter focuses on understanding NM's functionality and characterization expected to increase awareness of the functional properties of novel recycled nano-textiles with environment-friendly practices.

Keywords: Nanotechnology, nano-textile, nanomaterials, characterization, environment-friendly

List of Abbreviations

AFM	Atomic force microscopy
ATO	Antimony-doped tin oxide
CNTs	Carbon nanotubes
NMs	Nanomaterials
NPs	Nanoparticles
NRs	Nanorods
NT	Nanotechnology
PANI	Polyaniline
PET	Polyester
PPy	Polypyrrole
ROS	Reactive oxygen species
SEM	Scanning electron microscopy
STM	Scanning tunneling microscopy
TEM	Transmission electron microscopy
TiO_2	Titanium dioxide
UV	Ultraviolet
XRD	X-ray diffraction

8.1 Introduction

Nanotechnology (NT) is a current top priority for research in many nations because of its immense potential and significant economic impact. One of the most cutting-edge and inventive technologies that has transformed the textile and apparel sectors is NT, which increases functionality. As a result, it has found use in a number of textile industries. A new spectrum of superior materials has emerged from the fields of nanoscience and NT. It is anticipated that this new, interdisciplinary field will have significant effects on all areas of science and technology. While maintaining the essential bulk, tactile, and mechanical aspects of textiles, the incorporation of NT into textile substrates has produced remarkably amazing practical and aesthetic qualities. A revolutionary and futuristic idea, NT has the power to drastically alter several important industrial areas. In the textile sector, it is primarily used in four key areas: nanofinishing, nanocoatings, nanofibers, and nanocomposites. With the ability to significantly improve the functional and performance features of different textile categories, including casual wear, leisure wear, vanity clothes, technical clothing, personal care items, and smart textiles, NT holds great promise for the textile industry. NT has evolved into an icon for the textile and fashion industries, with applications in a wide range of textile sectors, such as smart wearable textiles, casual wear, antimicrobial clothing, athletics, medical textiles, protective textiles, and extreme cold weather apparel, among others. Nanotechnology's uses in the contemporary textile industry not only show how relevant it is now, but they also provide motivation for more study and development in this dynamic area [1].

Urbanization, industrial growth, and global overpopulation have all resulted in enormous garbage production. The significant impact of NT on the textile industry is related to the radical transformation of material properties brought about by scaling down to the nanoscale level. However, due to ambiguities and discrepancies in their size, structure, and chemical makeup, certain nanoparticles (NPs) may have detrimental effects on the environment and human health. This is viewed as a big global issue that has to be resolved quickly [2].

8.2 Current and Future Trends in Novel Nano-Textiles

When introducing different functional qualities to textile materials, NT overcomes the drawbacks of traditional techniques. Modern fabrics that exhibit high levels of functioning in terms of hydrophobicity, anti-wrinkle, UV resistance, antibacterial, antistatic, stain resistance qualities, waterproofness, and electrical conductivity have been produced in addition to "conventional" textiles. The discovery of additional textile nanomaterials (NMs) and better understanding of their activities have piqued the curiosity of young scientists and researchers. Nanomaterials have been used in textiles in a variety of ways, although more study and advancements are still needed. It is in charge of expanding the use of functional textiles in items like athletic apparel, military gear, and clothing for escape from fire, among other things. The main goal is to launch new, highly effective applications without sacrificing the fabric's comfort, flexibility, or light weight [3].

The most promising future usage of NT in textiles would include a summary of the advancements achieved in rapid, easy, novel, and inventive characterization methods for nanocomposites, NMs, and nanocoatings and utilized in textile applications. Table 8.1 shows the list of desirable characteristics in novel nano-textiles The wearer's comfort and maintenance are significantly improved by these functional qualities, which are of the utmost importance.

Table 8.1 List of desirable characteristics in novel nano-textiles [2].

Sr. no.	Characteristics in smart nano-textiles
1	Flexibility
2	Biocompatibility
3	Wash ability
4	Mechanical toughness
5	Conductivity
6	A comfortable wearing experience
7	Resistance to washing and aging
8	Danger detection and mitigation
9	Biometric data collection
10	Increasing sports performance with wearable technology that incorporate sensors
11	Medical care for illnesses, conditions, and injuries
12	Military personnel communication channels

The following are the primary implications of applying nanotechnologies to textiles:

1. To make existing materials' qualities and capabilities better;
2. To create innovative, intelligent, and smart fabrics;
3. Increasing the usage of fibers in applications such as technical textiles, biomedical textiles, and healthcare;
4. To create fresh opportunities for fibers as sensors [4].

8.3 Nanotechnology (NT)

Since the 1980s, NT has been a concept. Nanotechnology is the use of structures having at least one nanometer-sized dimension to create new or significantly enhanced materials, devices, or systems, according to the National NT Initiative. Comprehending, controlling, and manipulating matter at sizes ranging from 1 nm to 100 nm is the primary objective of NT. Due to their enormous economic potential, NT is not just for academics and researchers but also for businesses. Table 8.2 shows the global contribution of nano products in textile industry. There is great potential for commercial uses of NT in the textile industry. Table 8.3 shows the types of nanomaterials with contribution in textiles. The United States now has the most Nano goods in the textile business, with distinct products [5].

Table 8.2 Global contribution of nano products in textile industry [6].

Sr. no.	Nano products	Contribution in terms of %
1	Coating	46%
2	Technical textiles	24%
3	Clothing textiles	18%
4	Healthcare textile	6%
5	Sport textile	3%
6	Home textile	3%
7	Leather and shoes	1%

Table 8.3 The types of nanomaterials with contribution in textiles [7].

Sr. no.	Nanomaterials	Contribution in textiles
1	Nanoparticles (NPs)	39%
2	Nanofibers	21%
3	Nanocomposites	12%
4	Carbon nanotubes (CNTs)	7%
5	Nanosols	3%
6	Nanocapsules	3%
7	Nanoemulsion	2%
8	Others	13%

8.4 Nanomaterials (NMs)

Nanomaterials (NMs) are among the most alluring substances of the 21st century and have several uses in a variety of industries and sectors, including biology, medicine, industry, and energy. Much work has been put into examining the noteworthy characteristics of NMs and their prospective uses ever since the word "nanotechnology" was coined. Materials with one nanoscale dimension (100 nm) or more are classified as NMs.

There are two basic causes for why nanoscale properties of NMs differ from bulk properties. The first is called the surface effect; the surface area–to–volume ratio of NPs is larger than that of bulk materials. They thus have more accessible reactive sites per unit mass, which raises the reactivity of the material and improves its physical, mechanical, and chemical characteristics. Quantum effect, which intensifies with decreasing size in NMs, is the second justification. Electronic energy levels that align differently produce dynamic changes with unique optical, electrical, and magnetic capabilities [3].

Nanomaterials have been actively used to offer textiles multifunctional properties like physical, chemical, self-cleaning, ultraviolet resistance, antimicrobial, flame-retardant, wrinkle resistance, and coloration without compromising the intrinsic qualities of the textile materials. Table 8.4 shows the list of nanomaterials and functionalities in textiles.

Nanomaterial-based smart devices are increasingly being incorporated into textiles to perform a range of functions, such as energy storage, sensing, drug release, optics, and biological applications including drug

Table 8.4 List of nanomaterials and functionalities in textiles [5].

Sr. no.	Type of nanomaterials	Functionality on textiles
Metal Nanoparticles		
1	Iron	The ability to conduct
2	Copper	Antimicrobial resistance Dyeing/printing Conductivity of electricity UV protection
3	Silver bromide (AgBr)	Antimicrobial resistance
4	Silver	Electrical conductivity Antimicrobial resistance Dyeing/printing Conductivity of electricity UV protection Self-sanitizing
5	Gold	Dyeing/printing Conductivity of electricity
Carbon Nanotubes		
6	Single/multi-walled nanotubes	Antimicrobial resistance Conductivity of heat and electricity The ability to resist static Surface roughness and cleaning ability
7	Clay	Conductivity of electricity
Metal Oxide		
8	Zinc oxide (ZnO) & Titanium dioxide (TiO_2)	UV resistance Antimicrobial resistance Surface roughness and cleaning ability The ability to resist static Resistance to wrinkles
9	Silica (SiO_2)	Resistance to wrinkles UV Protection Resistance to wrinkles Surface roughness and cleaning ability
10	Dioxocerium (CeO_2)	Resistance to wrinkles

(Continued)

Table 8.4 List of nanomaterials and functionalities in textiles [5]. (*Continued*)

Sr. no.	Type of nanomaterials	Functionality on textiles
11	Dioxocerium (CeO_2)	Surface roughness/self-cleaning
12	Magnesium oxide (MgO)	Fire resistance
13	Copper oxide	Antimicrobial resistance
14	Iron oxide (FeO)	The ability to resist static
15	Alumina (Al_2O_3)	Cleaning ability Water resistance

delivery, immunizations imaging modalities, and biosensors. In addition to being widely used in the fashion industry, these advancements are being explored for widespread use in the military, healthcare, and on-body power harvesting applications [8]. Regarding the overall energy dependence of clothing, NMs are used in current textiles. The different processes and procedures for functionalizing NMs and incorporating them into textiles, with a focus on affordability, comfort, wear ability, energy conversion effectiveness, and environmental sustainability.

8.4.1 Novel Applications of Nanomaterials

Using NMs in textiles usually requires two steps. Nanomaterials are created in the first phase and then formed as a stable dispersion. The produced dispersion is applied to textiles in the second phase by spraying or coating them with a traditional impregnation method. NPs are typically applied as a dispersion in a carrier medium made up of the necessary ingredients in addition to a suitable surfactant. People are showing an increasing amount of interest in using improved lifestyle items as a result of technical advancement and increased media awareness. As a result, several businesses are using NPs to create intelligent and useful products for consumers. Figure 8.1 shows the systematic representation of nanomaterials on textile materials for necessary functionality. Electronic fabrics that are wearable, protective, sanitary, sensory, smart, and intelligent are in fashion today.

In the processing of textile materials, the NM may have a major role, which have a wide range of potential applications in daily life. Table 8.5 shows the applications of nanomaterials in the various fields. Therefore, both at the professional and commercial levels, research and growth are needed to market the items with specific aims about their intended attributes.

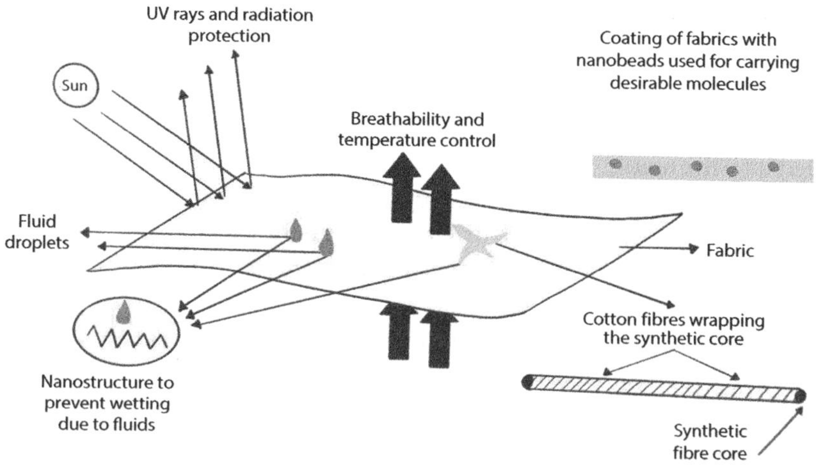

Figure 8.1 Systematic representation of nanomaterials on textile materials for necessary functionality [9].

Table 8.5 Applications of nanomaterials in the various fields [5].

Sr. no.	Nanomaterials	Applications
1	Nanoparticles/films	Textiles and clothing Environment and energy Technology of information Medical and pharmaceutical Nutrients and food Biotechnology and agriculture Aviation and transportation
2	Nanocomposite materials/carbon nanotubes	
3	System supramolecular	
4	Nanowires/nonorods/ nanowhiskers	
5	Surfaces with nanolithographic patterns	

NMs in the textile sector have a promising future because they can produce outcomes that are superior to those of traditional techniques. For example, silicone softener nanoemulsions are much softer than conventional microemulsions, and little amounts of these nano additives may provide goods exceptional functionality at very low cost. Even if it is challenging to reduce the price, success in this area would be a significant step toward the future commercialization of these products. The practical use of NMs is hampered by their high synthesis costs.

8.5 Synthesis of Nanomaterials (NMs)

NMs are produced via many methods. As artificial NPs are added to the coating or nanoscale fibers, applied as a coating on the surface, or integrated into the textile or fibers, that is the primary way that they differ from one another. Nanotechnology is used to create textiles that have improved mechanical, electrical, optical, or biological qualities. However, nanocomposite coatings can use a wide range of nanofillers, whiskers, and nanofibers with structural alteration.

Nanoparticles are frequently applied to textile materials through coating. Typically, chemicals, a carrier medium, and NPs are used in coating compositions that can alter the surface of textile fibers. Novel techniques for coating, such as sol-gel and layer by layer, may greatly increase the weather resistance, durability, and multifunctionality of textiles. Nanotextile products include digital fashion, chromatic textiles, stimulating textiles, thermal and cooling textiles, communicating textiles, textile sensors and actuators, athletics and recreation, the armed forces, emergency services consumers, and intelligent applications in buildings.

"Top-down" and "bottom-up" are the two fundamental methods for producing NPs. "Top-down" refers to the method of mechanically crushing the raw material using a milling process. In the "bottom-up" approach, chemical processes are used to create the structures. Table 8.6 shows the list of desired properties for material selection for nanoparticle synthesis best. The best method is selected based on the chemical composition and the desired characteristics of the NPs [6].

Table 8.6 List of desired properties for material selection for nanoparticle synthesis [2].

Material Selection for Nanoparticle Synthesis	Desired properties
	The ideal nanoparticle size
	Surface properties
	Biodegradability level
	Compatibility with biological systems
	Levels of toxicity
	Essential qualities, such as stability and solubility

8.5.1 Overview of the Production Process for Recycled Nano-Textiles

Recycled nano-textiles signify a promising avenue in sustainable and innovative textile production. This innovative approach involves recycling discarded textiles while harnessing the potential of NT to fill fabrics with enhanced properties. The production process for recycled nano-textiles involves several key steps, from collecting and processing discarded materials to integrating NMs for improved performance.

The production process of recycled nano-textiles, straddling from material collection to final product creation such as the following:

1. **Collection and Sorting of Raw Materials:** The meticulous gathering of raw materials, most of which are discarded textiles. These textiles may come from a variety of places, such as post-consumer collections, textile recycling facilities, or even garbage from the textile industry. The collected textiles are meticulously sorted upon arrival, according to fiber type, color, and quality. This guarantees compatibility and effectiveness in later production phases.
2. **Cleaning and Preparation:** The fabrics go through a thorough cleaning procedure after sorting to get rid of pollutants, stains, and impurities. Cleaning is necessary to remove any extraneous materials from the recycled fibers that can affect the final product's quality.
3. **Mechanical and Chemical Processing:** Mechanical processing techniques like shredding and cutting are used to convert the gathered textiles into their raw fiber form. In some cases, chemical techniques may be used to further decompose the fibers into their constituent parts, providing more material choice and recycling alternatives.
4. **Integration of Nanomaterials:** The creation of nano-textiles depends heavily on NT. The matrix of recycled textiles contains NMs like NPs or nanofibers. Numerous beneficial qualities, including as improved strength, durability, moisture-wicking abilities, and even antibacterial properties, are introduced by these NPs. Depending on the desired properties, integration techniques could include electrospinning, dip-coating, or other specialized procedures.
5. **Final Product Manufacturing:** The recycled nano-textile can undergo additional processing and be put together to

create finished goods. For instance, it might be used to create apparel, accessories, or footwear in the fashion sector. It might be used to make filtration materials, safety equipment, or automobile components in industrial settings.

6. **Quality Control:** To make sure the finished product meets the necessary criteria, quality control procedures are crucial throughout the production process. Testing might entail analyzing the recycled nano-textile's mechanical qualities, chemical make-up, and functional traits.
7. **Sustainability Considerations:** Sustainability is important at every phase of the industrial process. There is very little waste, less energy is required, and ecologically beneficial manufacturing techniques are used. Even the use of recycled materials themselves helps reduce waste and conserve resources.

Recyclability concepts and cutting-edge NT capabilities are combined in the manufacturing of recycled nano-textiles. The process starts with the collecting and meticulous treatment of used textiles, and, then, NT is added to improve the textile qualities. By reusing resources, using NT to increase lifespan, and meeting consumer needs for practical, high-performance textiles, recycled nano-textile products might help lessen the environmental effect of the textile industry. Sustainability considerations are included throughout the process, ultimately resulting in the development of novel, environmentally friendly fabrics with enhanced functionality and diminished environmental effect. Recycled nano-textiles are an appealing solution that not only repurposes materials but also improves their functioning for a variety of textile industry applications as the demand for sustainable textiles rises.

8.6 Characterization of Nanomaterials

According to the definition of NT, this field of research focuses on systems and substances whose structures and constituents, because of their nanoscale size, reveal novel and significantly enhanced physical, chemical, and biological phenomena and processes. Utilizing a variety of tactics is essential to ensuring NMs' characteristics. Textile NMs are examined through spectroscopic, X-ray, and other traditional methods. Materials that are micro and nano in size have been seen using imaging techniques such as scanning tunneling microscopy (STM), scanning electron microscopy (SEM), transmission electron microscopy with high resolution (TEM/HRTEM), and atomic force microscopy (AFM). The development of AFM and STM in the early 1980s has opened up new avenues for the manipulation, measurement, and characterization of NMs and their bulk structures. Each method observes a surface with a high degree of magnification, but their underlying concepts vary. Many different spectroscopic methods, including Raman, infrared spectroscopy, ultraviolet-visible, and ultraviolet-visible spectroscopy, have been used to examine the optical properties of micro- and nano-sized textile materials. Energy dispersive X-ray analysis (EDX), wide-angle X-ray diffraction, XPS, and particle size analyzers are a few of the X-ray methods that may be used to evaluate textile NMs. Some more popular techniques are also utilized to investigate the physical properties of textile NPs.

8.6.1 Characterization of Functional Properties of Recycled Nano-Textiles

Recycled nano-textiles offer a wide range of functional qualities that make them highly adaptable and appealing, and they represent a remarkable fusion of sustainability with cutting-edge technology. For determining whether certain functional features are appropriate for different applications, it is essential to characterize such properties. The following Table 8.7 lists the main functional characteristics of recycled nano-textiles.

To establish whether recycled nano-textiles are suitable for a certain application, their functional qualities must be characterized. In addition to improved strength and durability, these fabrics also provide antibacterial, moisture-management, and other beneficial qualities. Careful testing and analysis allow manufacturers and customers to decide if these fabrics are appropriate for their intended application, promoting high-performance and sustainable textile solutions across a range of industries.

For the assessment of textile NMs applying an alternative approach, the textiles underwent further physical testing to determine their tensile properties, bending length, crease recovery angle, air permeability, and other structural and chemical changes [11].

Table 8.7 Characterization of functional properties of recycled nano-textiles [11].

Sr. no.	Functional properties	Description
1.	Enhanced Strength and Durability	The enhanced strength and durability of recycled nano-textiles is one of their most notable characteristics. The textile matrix is strengthened with nanomaterials like nanoparticles or nanofibers, creating fabrics that can endure more mechanical stress. Tensile testing is one method of characterization that assesses a fabric's capacity to tolerate stretching and pulling without rupturing. Higher tensile strength and enhanced tear resistance are frequently the results of the inclusion of nanoparticles.
2.	Moisture Management	Excellent moisture-wicking qualities are frequently seen in recycled nano-textiles. Fabrics can effectively transport moisture away from the skin when hydrophobic or hydrophilic surfaces made of nanomaterials are used. Characterization entails testing the fabric's capacity to keep the wearer dry while also monitoring the rate of moisture absorption and desorption.
3.	Antimicrobial Properties	Antimicrobial substances can be incorporated into recycled nano-textiles thanks to nanotechnology. These substances prevent germs and fungi from growing, making the fabrics ideal for uses such as medical apparel, athletic wear, and personal protective equipment. To determine the fabric's resistance to microbial colonization, characterization techniques include microbial challenge tests.

(Continued)

Table 8.7 Characterization of functional properties of recycled nano-textiles [11]. (*Continued*)

Sr. no.	Functional properties	Description
4.	UV Protection	UV protection can be added to nano-textiles by adding nanoparticles like zinc oxide or titanium dioxide. Measuring the fabric's UV-blocking effectiveness during characterization is important for applications like outdoor garments and sun blinds.
5.	Thermal Regulation	Phase-change materials (PCMs) or thermoregulating nanoparticles can be included into nano-textiles to provide heat regulation capabilities. The ability of the fabric to absorb, store, and release heat is evaluated as part of the characterization process to ensure comfort in a range of temperature settings.
6.	Self-Cleaning and Stain Resistance	Textiles can become stain- and self-cleaning thanks to certain nanoparticles. Testing the fabric's resistance to liquid absorption and the adhesion of dirt and stains is part of the characterization process. Measurements of the fabric's hydrophobicity using water droplet contact angles are possible.
7.	Flame Retardancy	Nanotechnology can give nano-textiles flame-resistant characteristics. Tests for flame resistance can be used to measure a textile's capacity to thwart ignition and limit the spread of a flame.
8.	Gas and Chemical Sensing	Nano-textiles can detect particular gases or chemicals with the help of certain nanomaterials. Characterization entails determining the fabric's responsiveness, selectivity, and sensitivity to the desired compounds, which makes them useful for applications in environmental monitoring and safety.

(*Continued*)

Table 8.7 Characterization of functional properties of recycled nano-textiles [11]. (*Continued*)

Sr. no.	Functional properties	Description
9.	Sustainability Metrics	Characterization of recycled nano-textiles must take into account both their practical qualities and sustainability. Life cycle assessments and environmental impact evaluations examine how environmentally friendly textiles are by accounting for factors including greenhouse gas emissions, water, and energy usage.
10.	Conductivity and Smart Textile Capabilities	Recycled nano-textiles can be made into smart textiles that can sense, heat up, or provide electronic connectivity by integrating conductive nanomaterials. Tests for electrical conductivity and assessments of the textile's compatibility with electronic components are all included in characterization.
11.	Wear Comfort and Skin Sensitivity	Evaluation of characterization also includes assessing skin sensitivity and clothing comfort. In order to make sure that the fabric is comfortable for prolonged usage, this involves testing the fabric's softness, breathability, and hypoallergenic qualities.

8.6.2 Determination of Nanomaterial Toxicity

Based on the present global market scenario, NT is being produced at a very rapid pace. Nanowastes and nanobyproducts will inevitably circulate in the environment and our surrounds as NMs become more and more integrated into our daily lives through consumer products. These include the waste materials created when products are processed using NT. The trash produced by numerous industries, such as NMs or nanostructured byproducts. When compared to their bulk equivalent, NPs that are discharged into the atmosphere are thought of as various sorts of waste. With the global adoption of NT, recycling of NMs can be a crucial step toward achieving sustainability in N2T. The potential use of recycled NMs in industrial applications is discussed, along with scalable technology for recovering NPs from composite systems. To adhere to the norms and obtain the authorization of recycled NMs, pertinent waste management legislation is also mentioned.

Examining the toxicity of NMs can be done *in vitro* (using cell cultures) or *in vivo* (using live species like rats, fish, tiny mussels, and other animals). It is possible to measure the innate reaction of a living organism to

Table 8.8 Characterization of nanoparticles by various techniques [10].

Sr. no.	Characterization techniques	
Physical Parameters		
1	XRD and the melting point device	Lattice parameters and temperatures
2	TEM and AFM	Mechanical factors
3	Spectroscopy using optics	Quantum size effects and surface plasm on resonance
Chemical Parameters		
4	Spectroscopy using optics	The chemical nature and bonding of the nanomaterials
5	Spectroscopy of electrons	Chemical composition analysis
6	Ionic spectrometry	Characterization of thin films and elemental analysis
Structural Parameters		
7	XRD	The size of crystallites and the crystal structures of nanoparticles
8	Small angle X-ray scattering	The surface area or particle size per unit volume
9	SEM with EDX	Topographical information of nanomaterials and their chemical composition
10	TEM with EDX	Particle size and chemical composition
11	Scanning probe microscopy (SPM)	Real-space three-dimensional visuals

a substance using distinctive standardized toxicological tests. Additionally, there are no accepted criteria for determining the toxicity of NMs, which makes it difficult to review data and comprehend its toxicity. The majority of study done so far has been updates of practices used for other materials [12, 13]. Table 8.8 shows the characterization of nanoparticles by various techniques.

Recycling in NMs has several advantages in terms of the economy and the environment. Due to the fact that NMs can be recycled from both old and discarded items (waste from integrated products) and new, pure products (from production). Figure 8.2 shows the pictorial representation of recycled nanomaterials. Segments discuss effective methods for recycling nanowaste, rules governing nanowaste (such as requirements for disposal and recycling), and advantageous uses for these reused NPs.

List of various methods used for the separation of NPs [14]:

1. Using pH/thermal responsive materials
2. Centrifugation
3. Nanostructured colloidal
4. Solvent evaporation

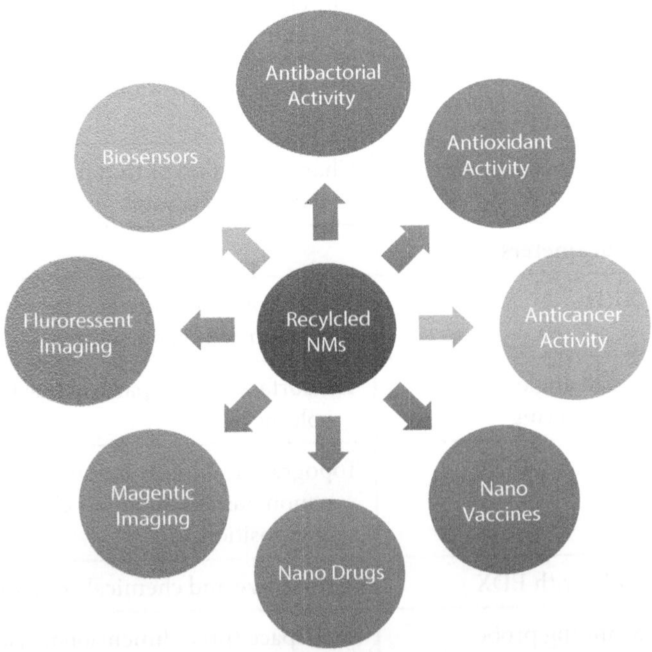

Figure 8.2 Pictorial representation of recycled nanomaterials (Self-developed).

5. Magnetic separation
6. Molecular anti solvents

Therefore, by giving particular examples from various domains and taking environmental concerns into account, recent research breakthroughs connected to recycling NMs for industrial applications are explored.

8.7 Applications of Nanotechnology

The textile industry is still in the early stages of using NT. The following categories essentially describe the textile uses for NT.

8.7.1 Water Repellence Properties in Textiles

As a result of the fabric's whiskers' narrower spacing than a regular drop of water but yet greater size than water molecules, water stays above the fabric's surface and on top of the whiskers. Liquid may still travel through the cloth, however, if pressure is applied. Nano-whiskers, which are hydrocarbons 1/1000 the size of cotton fibers, provide textiles water resistance [2, 6].

8.7.2 Antimicrobial Properties in Textiles

The antimicrobial textile coatings are quite fascinating and good for human health. Typically, triclosan, chitosan, nano-sized Ag, TiO_2, ZnO, and triclosan are used to make textiles antibacterial [15].

8.7.3 Antistatic Properties in Textiles

Due to their hydrophobicity, synthetic fibers display higher static charges because they do not absorb a lot of water, like nylon and polyester. To produce antistatic properties in synthetic fibers, researchers have used a variety of NMs, including nano antimony-doped tin oxide (ATO), ZnO whiskers, TiO_2 NPs, Sb-doped SnO_2 NPs, and others. A material like this effectively aids in dispersing the static charge that has built up on the synthetic fabric. Conversely, silane nanosol improves anti-static properties because the silane gel particles on fiber bind water molecules and draw moisture and water from the air via amino and hydroxyl groups [3, 16].

8.7.4 Water and Oil Resistance Properties

The genuine source of inspiration for innovative, useful materials is nature. It has frequently motivated academics to mimic biological occurrences. For instance, the preening oil coating on the feathers of the ducks allows them to survive in the water.

The scientists were able to mimic this natural phenomenon by covering cotton and polyester fabrics with chitosan. Precipitation was used to make the chitosan coating solution, which was then treated with a silicone compound to provide a reduced surface energy. Similarly, scientists placed both unmodified and surface-modified carbon nanotubes (CNTs) on cotton fibers to produce very hydrophobic surfaces. They achieved this by imitating the lotus leaf's surface roughness [17].

8.7.5 Self-Cleaning Aspects

Self-cleaning textile surfaces made possible by NT offer considerable potential for the development of novel materials, as well as for the invention of new products and applications for already-existing ones. Insects and plant leaves are sources of self-cleaning concepts. Figure 8.3 shows the scheme illustration of the synthesis of super hydrophobic polyester fabric. Super-hydrophobic surfaces have attracted special uses such as anti-contamination, anti-sticking, and self-cleaning coating [6].

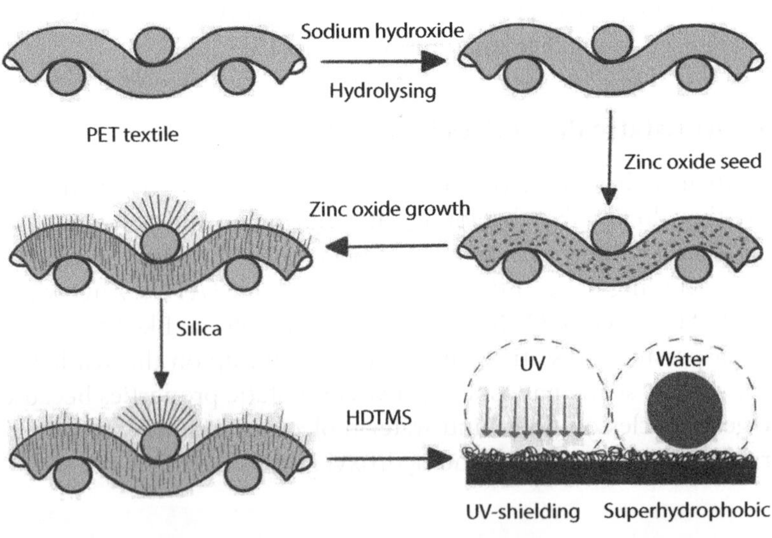

Figure 8.3 Scheme illustration of the synthesis of super hydrophobic polyester fabric [18].

8.7.5.1 *Physical Self-Cleaning*

Biomimicry is the copying and adaption of natural phenomena in order to produce artificial materials and goods. The distinctive nanoscale features of butterfly wings, desert beetles, mosquito eyes, etc., are what give these materials their exceptional hydrophobicity and self-cleaning surfaces. It is accomplished by depositing nanostructures made of gold, silica, ZnO, TiO_2, and other materials on textile surfaces. The silica NPs are among these compounds that are frequently utilized to give textiles physical self-cleaning and extremely hydrophobic properties [19].

8.7.5.2 *Self-Cleaning Chemical*

The photocatalytic fading of coloring agents brought on by light is referred to as "chemical self-cleaning." Reactive oxygen species (ROS) are produced by photocatalysts such as ZnO and TiO_2 when they are placed as NMs on fabric surfaces and exposed to light. Light, especially UV light, causes an electron in a photocatalyst's valence band to jump to the conduction band, leaving a positive hole in its stead. Oxidation and reduction take place at the photocatalyst's surface [5].

8.7.6 Ultraviolet Protection Properties in Textiles

Human skin may be harmed by prolonged UV light exposure. It might result in inflammatory disorders, skin cancer, and aging. Reducing skin exposure to UV radiation can lessen the devastating effects these diseases have on human health. The range of wavelengths for ultraviolet radiation is 150 nm to 400 nm. UV-protective textiles function in two ways: either by scattering or reflecting UV radiation or by absorbing UV light and turning it into heat. Numerous metal and metal oxide NPs have been coated or adsorbed on textile materials in order to provide a shielding effect against UV light. Nanomaterials like ZnO and TiO_2 that are sensitive to UV light may either absorb or scatter UV energy. Because nano-sized materials have higher surface area, mass, and volume than conventional materials, they are more effective at blocking UV radiation. The efficiency of UV protection is measured by the UPF, which is dependent on the specifications of the textile materials [2, 20].

8.7.7 Flame-Resistant Properties in Textiles

To protect textiles with a flame-retardant substance, halogenated flame retardants are combined with nano antimony pentoxide. Flame-retardant properties may also be added by applying polymer nanocomposites based on nanometric fillers, such as layered silicates, metal oxides, layered double hydroxides, and nanotubes, among others.

8.7.8 Wrinkle Resistance Properties in Textiles

A unique wrinkle-free treatment based on NT has been presented by Nano-Tex as an alternative to harsh conventional procedures. It is said to improve performance while maintaining the strength and integrity of fabrics. The tendency of cellulosic materials is for them to crease and shrink quickly. The majority of chemicals and processing techniques weaken the tensile and rip strength of a fabric. The challenges of using typical resin application techniques, which lower the tenacity, abrasion resistance, moisture vapor transfer, dye ability, and breathability of resin treated fabrics, are partially addressed by the introduction of NPs in textiles to avoid wrinkles. Zirconia (ZrO_2), nano TiO_2, silver, and silica have all been used as crosslinking agents in wrinkle-free finishes as catalysts. TiO_2 NPs are widely employed because of their photocatalytic activity, which makes it easier for crosslinking agents to attach to substrates. A wrinkle-preventing finish consisting of TiO_2 NPs spread in self-crosslinking DMDHEU was applied to a textile substrate using the pad-dry-cure method [21].

8.7.9 Active and Leisure Properties in Smart Sportswear

Investigations on NT are being conducted to enhance the functionality of various sports apparel and accessories for skydiving, parachuting, snowboarding, skiing, cardiovascular activity, sports, climbing, hiking, and undersea activities, among other activities [22].

8.7.10 Anti-Odor Properties in Textiles

Tourmaline is typically used as an anti-odor finish. When tourmaline comes into contact with oxygen (O_2), CO_2, and water (H_2O) molecules, electrolytic dissociation occurs, producing negative ions. The creation of negative magnetic fields decreases the textile substrate's odor.

8.7.11 Electrically Conductive Properties in Textiles

Nanotechnology has revolutionized the textile industry by introducing sensors and actuators based on materials' conductive qualities. Wearable textiles can be designed to monitor physiological characteristics like heartbeat and pulse rate. Various types of sensors can be embedded onto textiles for various applications. Margaret Frey developed a textile that can function as a sensor after learning about nanofibers' increased surface area and porosity. Nano-modified color-changing chameleon fabrics are fastened with sensor tools to adapt to their environment and help military soldiers in battle.

Conducting polymers have many uses in the textile sector. These materials' tunable resistivity of the textile surface produces an electric response when it is subjected to an external stimulus. Desired qualities may be altered by adding various NPs to the matrix of these polymers. Three widely utilized conducting polymers with enhanced mechanical, optical, and conducting capabilities include PT, PANI, and PPy with nanostructures. Cotton yarn's conductivity may be altered by coating it with polyelectrolyte base CNTs, which makes it appropriate for electronic textiles. Furthermore, materials for medical usage and protective clothing may be made using nanoscale cells, which can also be used to contain insecticides and pharmaceuticals [23, 24].

8.7.12 Energy Storage Properties in Textiles

In textile technology, supercapacitors are employed for energy storage applications. The best way to incorporate supercapacitor electrodes within fabric without compromising its wearability and elasticity is being investigated [25].

8.7.13 Photonics Properties in Textiles

The fashion industry has been greatly intrigued by the new innovative textiles and photonic technology. Various optical materials including optic fibers, optical films, and NPs have been used to produce a range of elegant and creative textile textiles. By adjusting the intensity of color and light pattern, the photonic substance in the cloth will change how the dress looks. When examined from different angles, optical films made from periodic dielectric multilayers can produce highly reflecting and vibrant patterns on the fabric when forcefully deposited on the textile fibers. Intriguing 3D visual effects are provided by holographic film coatings on textile substrates [3].

8.7.14 Textile Coloration Properties in Textiles

Particle sizes in the nanoscale range cause some metals, such as copper, gold, silver, platinum, and palladium, to show color. Depending on the particle size, colors may change. This phenomenon is called surface plasmon resonance, or SPR. When light hits NPs (NPs) of the above metals and their size is less than the wavelength of light, the oscillating electromagnetic field interacts with the free metal electrons and causes coherent oscillation of the metal electrons. It is responsible for charge separation with respect to the dipole-forming ionic lattice. The oscillation of the dipole is parallel to the direction of the electric field. This oscillation's amplitude peaks at a certain frequency because to SPR. The potential uses of the structural colors in a variety of industries, including paint, textiles, cosmetics, and automobiles, make them intriguing. It takes the form of NP-based pigments and disperse dyeing [5].

Green silver NPs are revolutionarily dyed into polyester fabric using chitosan, a nontoxic and environmentally beneficial reductant. Furthermore, it endows the fabric with antibacterial properties, which lead to an 80% reduction in germs even after 10 cycles of washing. In general, the suggested approach of coating with green NPs might contribute to a reduction in the cost of manufacturing sustainable textiles [26].

8.8 Novel Nano-Textile

The outstanding performance characteristics of nano-enhanced materials make nano-textiles indisputably advantageous for the garment, accessory, and technical textile industries. The exploration of the practical and aesthetic potential of NT-enabled textiles is extensive, including high-performance apparel, smart textiles, sensors for clothing, athletics, digital clothing, medicinal, and military applications. Nanotechnology has advanced significantly as a result of the latest innovations in fabric coatings, utilizing a mixture of organic and inorganic chemicals along with NPs. NPs incorporated in natural or synthetic fibers expand the variety of textiles available, from everyday clothing to technological and smart textiles, without changing the tactile qualities of the fabric. Increased comfort, sanitary qualities, durability, and reduced production costs are a few notable characteristics of nanofinished fabrics. In terms of performance and functionality, nano-textiles outperform more traditional materials like Teflon and Gore-Tex [27, 28]. Major players in the novel nano-textile industry include Texcote Technology, Nano-Tex, and Schoeller Textiles AG, which are based in Sweden, the USA, and Switzerland, respectively [10].

8.9 Environmental and Health Concerns Connected with Nanomaterials

The unparalleled qualities of NPs have drawn the attention of businesses as well as scientists and researchers due to their enormous economic potential and opportunities for the textile industry, which might increase the profitability and size of this enormous market. Among the many risks associated with NT are its impact on humans and the atmosphere. NPs and electrical devices are frequently found in smart clothes, which must have a negative impact on the environment. When substances are discharged into the environment, whether on purpose or not, they primarily constitute a serious safety risk because they are either naturally occurring or artificially created by humans. Since they do not come into touch with people or the environment, NPs incorporated in the finished product are less of a problem. Table 8.9 shows the list of novel nano-textiles.

The size, location of contact, and length of contact determine how easily NPs may penetrate the skin, lungs, or intestines. After that, the circulation transports them to additional tissues and vital organs. Particles of NPs raise the risk of cardiovascular disorders and produce tissue-damaging respiratory system infections. Recycling systems may need to be adjusted to keep up with these technical improvements if pure NPs are generated or if materials with NP finishes are extensively utilized [6].

Concerns about negative effects on people and the environment are raised by the widespread usage of NPs and NMs in the creation of smart cloth. Fabric characteristics including drape, hand-feel, elasticity, and stiffness may be negatively impacted by nanocoating. Depending of the concentration, solid content, viscosity, and coating thickness of the nanosols, the fabric may become stiffer. Following sol-gel coating, tear strength could also decline. They are produced using a variety of hazardous chemicals, and after washing the textiles, NPs from the finishing products might leach into the water sources. An estimated 5%–20% of a textile's total weight is lost during the laundering process as a result of abrasion, the effects of

Table 8.9 List of novel nano-textiles [2, 29].

Sr. no.	Novel nano-textiles	Description
1	Nanotex	It was developed by a US-based corporation and exhibits excellent performance qualities for textiles, such as resilience to abrasion and creasing, tearing strength, and softness and suppleness.
2	Nano-engineered crease resistance agents	To provide cotton textiles resistance to creases.
3	Nano-sized antimicrobial particles	To use in medical arena for designing wound dressings.
4	*De novo* fibers	On cotton and synthetic textile substrates, it provides outstanding comfort qualities in addition to strength, antibacterial characteristics, wrinkle resistance, and soil resistance.
5	Nano-based soft shells fabrics	This product offers dynamic climate control and superior comfort due to its soft interior layer and durable exterior, making it suitable for extreme cold clothing, sportswear, mountaineering, and ski sportswear.
6	Nano-sphere finished fabrics	Soft shell technology and nano-sphere finish may be used to create textiles that are resistant to rain and snowfall.
7	Cellulose based nano-composites	A variety of flame-retardant end products incorporate nanoparticles as a nanofiller material because of their improved thermal stability. The use of nano-filtration membrane technology improves both the environment and the economy by facilitating dye recovery and water saving.
8	Wrap nano sheet	The fibers are wrapped to create the wrap nano sheet, which may improve a fabric's strength, color fastness, wrinkle resistance, and stain resistance.
9	Nano-fresh	To capture body odor.
10	Nano-care	The product provides stain repellent, crease resistance, and water repellency on cotton.

washing detergents, etc. During the washing process, NMs from nanotextiles are also released into the environment via discharged water. The risk of silver NPs, which are harmful to soil organisms and aquatic life, entering wastewater after the first wash was estimated to be at 35% [3].

8.10 Overpowering the Threats Connected with Nanomaterials

Despite the revolutionary advances, this technology has brought about in the textile industry, researchers and scientists still confront challenges in adopting sustainable techniques for the incorporation of NPs in fabrics and in developing prototypes. When using traditional pad-dry-cure techniques for textile nanofinishing, a number of drawbacks cannot be ignored, including high energy consumption from renewable and non-renewable sources, creation of hazardous liquid waste, and multiple batch operations. Further research is needed to determine cost-effective, energy-efficient, and biocompatible manufacturing processes as well as how to utilise NPs at room temperature. To predict the effects of NMs on biology and the environment, reliable models should be created to predict their release, transport, transformation, and uptake in relation to their physical and chemical properties. Reliable models that forecast a NM's release, transportation, transformation, and absorption in connection to its physical and chemical characteristics should be developed in order to forecast its impacts on biology and the environment. Risk management techniques like adequate ventilation, filtration, and discharge control must be used in workplaces. Workers' exposure hours may be shortened by changing shifts, and they must be instructed to wear goggles, face masks, and gloves. Instructions for product disposal and recycling should be made available to consumers. It is essential to create sophisticated sensors that respond in real time to safety concerns about the environment and people's health. All stakeholders involved, including manufacturers, decision-makers, and consumers, should be informed technically about the potential hazards, benefits, and drawbacks of NMs [5].

Conventional techniques for synthesizing NMs by chemical or physical means are risky, labor-intensive, and neither biocompatible nor sustainable. The plausible method to get over the constraints of traditional procedures is to produce NPs in the gas phase and apply them to textile fibers. Aerosol NPs are introduced throughout the process to increase the finished nano-textiles' washability and provide the way for the development of textiles with better functionalities. Sustainability may be enhanced via the creation of nature-inspired fibers and fabric coatings for nano-based smart textiles. Plasmonic NPs are incorporated into textile substrates to enhance their aesthetic appeal by modifying their optical appearance [30].

8.11 Conclusion

Nanotechnology has found diverse applications across numerous sectors, owing to its innovative nature, practicality, and cost-effectiveness, particularly within the textile industry. Nanomaterials (NMs) have introduced fresh possibilities in various domains, with textiles being a prominent example. NMs provide textiles with a number of beneficial qualities, yet certain NMs may be harmful to human health and the environment. The primary concerns associated with the widespread adoption of these materials revolve around human and environmental safety, in addition to production expenses. To address these concerns, it is essential to conduct safety assessments of NMs on an individual basis. Thus, it is necessary to conduct in-depth life-cycle evaluations and a careful analysis of the possible toxicity of NPs that might leach from textiles.

As a result, the issue of NM risks comes into focus. Many NMs are already present in various commercial products, with some being clearly labeled as such, whereas others remain unlabeled. It is foreseeable that more such products will emerge in the future. Therefore, it is crucial for the textile industry to possess a sound understanding of NMs in order to make informed and responsible choices. This chapter aims to heighten awareness regarding the potential impact of textile waste containing NMs and underscores the importance of establishing robust protocols for the sustainable recycling of novel textile waste. This initiative ultimately serves to protect the environment and the well-being of humanity.

References

1. Saleem, H. and Zaidi, S.J., Sustainable use of nanomaterials in textiles and their environmental impact. *Materials*, 13, 5134, 1–28, 2020.
2. Jhanji, Y., Nanomaterials for Textile Applications - A Critical Review. *J. Text. Assoc.*, 83/1, 414–421, 2022.
3. Shah, M.A., Pirzada, B.M., Price, G., Applications of nanotechnology in smart textile industry: A critical review. *J. Adv. Res.*, 38, 55–75, 2022.

4. Holme, I., Nanotechnologies for textiles, clothing, and footwear. *Text. Mag.*, 1, 711, 2005.
5. Riaz, Functional finishing and coloration of textiles. *Color. Technol.*, 1–20, 2018.
6. Temesgen, A.G., Tursucular, .F., Eren, R., Ulcay, Y., Novel application of nanotechnology in modification of textile fabrics properties and apparel. *Int. J. Adv. Multidiscip. Res.*, 5, 12, 49–58, 2018.
7. Norberg, K., *Inst. Fibre J.*, 20, 52, 2005.
8. Elkodous, M.A., Recent advances in waste-recycled nanomaterials for biomedical applications: Waste-to-wealth. *Nanotechnology*, 10, 1662–1739, 2021.
9. Sawhney, A.P.S., Condon, B., Singh, K.V., Pang, S.S., Li, G., Hui, D., Modern applications of nanotechnology in textiles. *Text. Res. J.*, 78, 731, 2008.
10. Vigneshwaran, N., Modification of textile surfaces using nanoparticles, Surface Modification of Textiles, Woodhead Publishing Series in Textiles, pp. 164–184, 2009
11. Verma, A., Arif, R., Jadoun, S., Synthesis, characterization, and application of modified textile nanomaterials, in: *Frontiers of Textile Materials: Polymers, Nanomaterials, Enzymes, and Advanced Modification Techniques*, pp. 167–188, Wiley, 2020.
12. Ju-Nam, Y. and Lead, J.R., Manufactured nanoparticles: An overview of their chemistry, interactions and potential environmental implications. *Sci. Total Environ.*, 400, 396–414, 2008.
13. Drasler, B., Sayre, P., Steinhäuser, K.G., Petri-Fink, A., Rothen-Rutishauser, B., *In vitro* approaches to assess the hazard of nanomaterials. *Nano Impact*, 8, 99–116, 2017.
14. Rai, M. and Nguyen, T.A., *Nanomaterials Recycling*, Elsevier, 1, 2021
15. Geetika, J., Vikas, S., Avinash, C., Nanotechnology in The driver's seat of sportswear industry: A review of current trends and future applications. *Latest Trends Text. Fashion Designing*, 2, 4, 206, 2018.
16. Shishoo, R., Recent developments in materials for use in protective clothing. *Inst. J. Clothing Sci. Technol.*, 14, 3/4, 201–15, 2002.
17. Liu, Y., Chen, X., Xin, J.H., Hydrophobic duck feathers and their simulation on textile substrates for water repellent treatment. *Bioinspir. Biomim.*, 3, 4, 2008. DOI 10.1088/1748-3182/3/4/046007
18. Xue, C.-H., Yin, W., Zhang, P., Zhang, J., Ji, P.-T., Jia, S.-T., Fabrication of adhesive superhydrophobic Ni-Cu-P alloy coatings with high mechanical strength by one step electrodeposition. *Colloids Surf., A Physicochem. Eng. Asp.*, 7, 427, 2013.
19. Xue, C.-H., Jia, S.-T., Zhang, J., Tian, L.-Q., Superhydrophobic surfaces on cotton textiles by complex coating of silica nanoparticles and hydrophobization. *Thin Solid Films*, 517, 4593, 2009.

20. Yang, H., Zhu, S., Pan, N., Studying the mechanisms of titanium dioxide as ultraviolet-blocking additive for films and fabrics by an improved scheme. *J. Appl. Polym. Sci.*, 92, 5, 3201–3210, 2004.
21. Mondal, K., Recent advances in soft e-textiles. *Inventions*, 3, 2, 23, 2008.
22. Thilagavathi, G., Raja, A., Kannaian, T., Nanotechnology and protective clothing for defence personnel. *Def. Sci. J.*, 58, 4, 451, 2008.
23. Ullah, N., Mansha, M., Khan, I., Qurashi, A., Nanomaterial-based optical chemical sensors for the detection of heavy metals in water: Recent advances and challenges. *TrAC, Trends Anal. Chem.*, 100, 155–66, 2018.
24. Patra, J. and Gonda, S., *J. Eng. Technol. Res.*, 5, 5, 104, 2018.
25. Jost, K., Perez, C.R., McDonough, J.K., Presser, V., Heon, M., Dion, G. *et al.*, Carbon coated textiles for flexible energy storage. *Energy Environ. Sci.*, 4, 12, 5060–5067, 2011.
26. Faridul Hasan, K.M., Pervez, N., Talukder, E., Sultana, Z., Mahmud, S., Meraz, M., Bansal, V., Genyang, C., A novel coloration of polyester fabric through green silver nanoparticles (G-AgNPs@PET). *Nanomaterials*, 9, 569, 2019.
27. Rai, M. and Yadav, A., Silver nanoparticles as a new generation of antimicrobials. *Biotechnol. Adv.*, 27, 1, 76, 2008.
28. Sawhney, A. and Condon, B., Modern applications of nanotechnology in textiles. *Text. Res. J.*, 78, 8, 731, 2008.
29. Yongho, C., Sang-Duk, L., Shin Hwan, Y., High efficiency, transparent, reusable, and active PM2.5 filters by hierarchical ag nanowire percolation network. *Int. J. Appl. Eng. Res.*, 12, 12, 6704, 2017.
30. Jurgen, A., Military uses of nanotechnology: Perspectives and concerns. *Secur. Dialogue*, 35, 1, 61, 2004.

9

Future Perspective of Nanotechnology in Relation to Textile Applications Using Textile Waste

Ruchi Kholiya[1*], Shefali Massey[2] and Mansi Hans[3]

[1]Department of Fashion Technology, National Institute of Fashion Technology, Jodhpur, Rajasthan, India
[2]Department of Textiles & Apparel Designing, College of Community Sciences, G.B.P.U.A.&T., Pantnagar, Uttarakhand, India
[3]Center of Home Science, MMPG College, Haridwar, Uttarakhand, India

Abstract

Over the last few years, there has been tremendous growth in customers' awareness of upgrading their traditional clothes and apparel necessary to be smarter than ever to improve their lives. Nanotechnology helps make strong and multipurpose fabrics that can be utilized to manufacture smart clothes. Nanotechnology is now being used in important fields like medicine, engineering, and electronics. Recently, it is also finding its way into fashion and textiles as well. The production system in the fashion industry exerts pressure on valuable natural resources as the production model is linear in operation, i.e., "*take, make, dispose*." This affects societies worldwide by polluting the environment, damaging natural habitats, exhausting natural energy, etc. Therefore, new advanced technologies such as nanotechnology could be vital in controlling the adverse impact of the fashion industry on our planet. The high-end application of nanotechnology particularly in making protective and medical textiles is being researched. Scientists are developing even more advanced finishes as well as materials for this purpose. So, textile industries need to utilize the opportunities of nanotechnology for growth and make the textile market more profitable at the global level. This chapter gives an overview of the top-tired applications of nanomaterials in textiles. It aims to raise awareness about the potential impact of textile waste containing nanomaterials.

**Corresponding author*: ruchi.kholiya@nift.ac.in

Prashansa Sharma and Shilpi Shree Sahay (eds.) Nanotechnology-Assisted Recycling of Textile Waste: Sustainable Tools for Textiles, (265–284) © 2025 Scrivener Publishing LLC

Keywords: Nanotechnology, textile waste, protective textiles, smart clothes, nanomaterials

List of Abbreviations

USD	United States Dollar
NNI	National Nanotechnological Initiative
PCM	Phase-change materials
USA	United States of America
UV	Ultraviolet
TiO_2	Titanium dioxide

9.1 Introduction

The economic value generated by the textiles industry was estimated at approximately USD 1.6 trillion in 2022, and the technical textile market share itself is estimated to be USD 218.88 billion (Sector Report 2023). The trends in the textile industry world show that traditional textile businesses are at risk due to the demand for technical textiles with advanced production methods, new materials, and innovative products. Also, conventional textile products are easily available in the market with competitive prices that result in a significant reduction in profit percentage [1]. To survive in a global scenario, it is important to develop new textile products or materials with sustainable production routes.

Nanotechnology is among one of the promising technologies known for its novel approach and high-tech applications [2]. Nowadays, as per consumer demand, multifunctionality in fabrics with predominant features of comfort are need of the hour. Nanotechnology will fulfil customized customer demand for fabrics or smart clothing because of the advanced nanomaterials and nano finishes. Moreover, it is now crucial for textile industries to review the technologies and chemicals that they use in manufacturing to meet both environmental and consumer needs. Nano-textiles, in general, are considered emerging innovative textiles that deal with nanofibers at atomic and molecular levels. At its core, the technology alters the properties of the fibers, ultimately leading to a transformation in the fabric's overall characteristics [3]. This new technology can greatly improve the properties and worth of textile processing and products, leading to significant technological, economic, and environmental advantages [4]. The conversion of regular textiles into multifunctional textile structures without altering their essential properties like breathability, strength, and resiliency is the major innovativeness of nanotechnology at par [5]. Although this technology is still in its infancy stage, it is gaining popularity and generating interest in scientific and technological fields like optics, material science, textiles, and many more. Nanomaterials with unlimited benefits expanded the potential of nanotechnology with a wide range of novel and innovative textile products [6].

9.2 Nanotechnology and Its Emergence

9.2.1 Background of Nanotechnology

Mr. R. Feynman, a professor at the Californian Institute of Technology, first mentioned the term "nanotechnology." He purposely introduced and briefed about nanotechnological processes in his lecture during the session held by the American Physical Society in 1959. He was the first person who commenced the possibility of creating nanosized particles using atoms as the building blocks. His lecture is now considered the origin of the nanotechnological paradigm. Later in 1974, the term nanotechnology again captured the attention of the scientific fraternity during the International Conference on Industrial Production held in Tokyo. It was then described as the technology used to process the materials with nanometer accuracy. The concept of a nanotechnological strategy was introduced by E. Drexler in his 1986 book, "Engines of Creation: The Coming Era of Nanotechnology."

From the second half of the 1980s to the early 1990s, many important discoveries and inventions were made, impacting nanotechnology's further development. Since then, there has been a significant increase in nanotechnological research and developments. The number of publications on research related to nanotechnology has sharply risen giving more ideas for wider applications of this field. The substantial advantage of nanotechnology spurred several organizations and countries to engage in production processes and applications associated with nanomaterials.

The United States of America (USA) was the first country to initiate the first nanotechnological program funded by the National Scientific Fund in year 1991 along with the National Nanotechnological Initiative (NNI) in 2001. The core idea behind these programs was to create a strategic connection between various federal departments in the USA working for nanotechnology development. This also served as the foundation for the economic and national security of the country in the initial part of the 21st century [7].

Although the definition of nanotechnology broadened up over the passage of years with intense research, related to the field and application of nanomaterials. The ability of nanotechnology to detect, measure, and manufacture materials at the nanometer scale exhibits objects with unique features [8–10]. Nanotechnology encompasses the investigation and advancement of materials, devices, and systems that possess distinctive emergent physical, chemical, and biological characteristics as a result of their tiny scale [11]. The basic principle of nanotechnology lies in the fact that as soon as the dimensions of materials are reduced to the nanometer scale, the properties of materials drastically change as per the purpose needed. The process not only controls the structure but also produces miniature structures from 1-100 nm in size, bound to perform the functions at the basic atomic level. This significant size reduction, improved structure, or material properties with promising future of nanotechnology in wider fields of textile technology, materials, and products [12].

Nanomaterials' vast economic potential has attracted scientists, researchers, and businesses to produce multifunctional textiles and improve their performance, hence generating worldwide interest [13]. A new technological industrial revolution is to be witnessed in the near future owing to the diversity in nanotechnological production routes and materials.

9.2.2 Advantages of Nanotechnology

Nanotechnology is slowly and steadily attracting global massive attention due to its potential and application in a wide range of end uses (Figure 9.1) [14]. The utilization of nanotechnology and processes involving nanomaterials is experiencing a remarkable expansion across various domains within the realms of science and technology, some of which are listed below:

- The broader application of nanotechnology in the form of nanocomposites and nanofibers.
- Smart polymeric nanocoatings used in high-end textile products to provide improved functionality and performance (Figure 9.2).
- Functionalizing nanoparticles during various textile processing enhances the product's performance and provides unachieved functionality [5].
- Nanotechnology provides high durability as nano-particles have a large surface area–to–volume ratio, which offers better affinity for fabrics and increased durability.

- Permanent finish with excellent functionality after wash and wear without affecting the breathability or feel of the fabric [14].
- Production of smart next-generation multifunctional textiles with properties such as UV (ultraviolet) protection, anti-odor, antibacterial, water/spill repellency, wrinkle resistance, etc.
- Energy saving, economy, eco-friendliness, and control release of the nano substance are a few other advantages to offer [3].

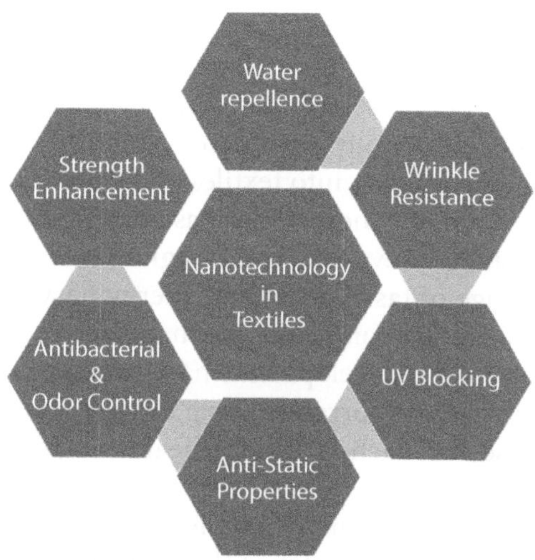

Figure 9.1 Application of nanotechnology in textiles.

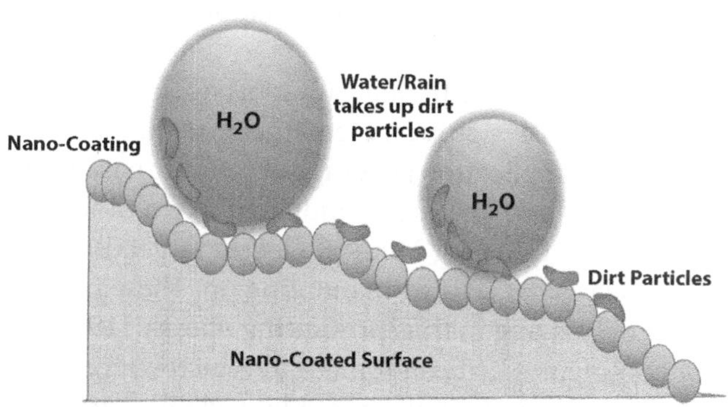

Figure 9.2 Smart polymeric nanocoatings.

The field of nanotechnology has emerged as a significant force in the textile industry, offering a multitude of advantageous opportunities. Another notable benefit of nanotechnology is its ability to address the issue of textile waste and pollution. According to a recent survey, an estimated annual production of around 92 million tons of textile waste has been reported. This statistic is of significant magnitude. It constitutes around 5% of the annual landfill volume. The application of nanotechnology has the potential to address this concern through its capacity to mitigate environmental degradation and promote resource conservation [15].

9.2.3 Future of Nanotechnology

In recent years, the field of nanotechnology has emerged as a significant contributor to the development of intelligent textiles. Nanomaterials have been utilized to incorporate several desirable features, including antibacterial, UV-resistant, electrically conductive, optical, hydrophobic, and flame-retardant characteristics, into textiles and garments in a sustainable manner. The integration of nanomaterial-based smart devices with textiles has become increasingly prevalent. This integration allows for the performance of a wide range of tasks, including energy harvesting and storage, sensing capabilities, medication release mechanisms, and optical functionalities. These technological developments have been extensively utilized in the fashion industry and are currently being further developed for broader implementation in sectors such as defence, healthcare, and on-body energy harnessing [16].

It offers an innovative means of textile processing that could transform the entire apparel industry [3]. Approximately 30 countries worldwide are actively engaged in the innovation and advancement of nanotechnology-driven textile materials. Several futuristic applications of nanotechnology can be attained in developing advanced customer-oriented textile products [17].

The application of nanotechnology in sportswear and spacesuits has become one of the most significant emergent segments [18]. Another potential research and application are usage of cationic nanoparticles to target fat cells for weight loss. Research conducted in Columbia Engineering and Columbia University Irving Medical Centre targets specific locations of fat and prevents large fat cells from building up. These materials modify the fat instead of destroying it, thus preventing obesity [19].

The field of cooling is a critical domain in which energy sources based on nanotechnology can potentially provide a viable answer. The development of densely packed fabric batteries or solar cells is necessary to provide efficient cooling capabilities while maintaining optimal comfort levels. Phase-change materials (PCMs) have the potential to be employed in combination with various power sources to facilitate the cooling of the human body in heated environments or the electrical components integrated within textiles.

Future research will also focus on developing green manufacturing techniques and chemistries for the synthesis of nanomaterials that can withstand washing. It is necessary to develop novel nano additives and stabilizers for formulations and finishing treatments because agglomeration of deposited nanoparticles is a formidable obstacle. Maintaining environmental equilibrium requires the immobilization of nanoparticles on fabric without impairing their chemical, optical, or electrical properties. To attach nanoparticles to cotton or synthetic fibers, for example, it would be beneficial to develop new covalent binding mechanisms [20].

Nanomaterials derived from silver, titanium, iron, and other commonly utilized elements have been found to exhibit cytotoxic properties in certain instances, potentially posing a threat to the fragile ecological equilibrium if the uncontrolled release of these nanomaterials persists. The release of nano-textile products with biocidal capabilities has been observed to destroy significant soil bacteria. Moreover, the recycling of these textile goods is a significant challenge, as they are predominantly disposed of in landfills, worsening soil pollution. Hence, the imperative to develop analytical methods capable of accurately quantifying the emission of nanoparticles from textile goods is increasingly relevant. These hazards and problems underscore the need for authorities to implement more stringent regulations and conduct toxicity analyses across all stages of usage, to prevent nano-textiles from posing significant risks to humans, animals, and the environment [21].

In addition, scientists are developing bio-reactive plastic compounds that can protect the wearer from chemical and biological attacks. Researchers are developing apparel layers that could alert an individual to any unseen attack. Changing the color of military uniforms can also conceal their wearers from their adversaries [22]. The application of nanotechnology is being employed to fabricate garments embedded with sensor technology. Garments have been designed to monitor various physiological parameters such as body temperature and critical clinical measures. Self-ironed nanomaterials are acquired utilizing nanomaterials that react to temperature. In conclusion, the future of nanotechnology holds boundless potential to revolutionize numerous fields, from medicine to electronics, and textiles to environmental conservation. We can anticipate breakthroughs by unlocking the secrets of the nanoscale world, which will reshape our understanding of materials, enhance our capacity to manipulate matter at the atomic level, and ultimately lead to innovations beyond our current imagination.

9.3 What is Textile Waste?

Over the past few decades, there has been a noticeable surge in the global population. Interestingly, despite this trend, society has experienced a notable enhancement in overall living standards. These two developments have amplified the overall consumption of textiles, which has led to an increased demand for textile products with remarkable properties. Owing to the large consumption and production of apparel and textiles, it has been estimated that there is the generation of 110 million tons of textile waste annually [23].

Textile waste encompasses the waste generated within the fashion and textile industry during various manufacturing stages, such as spinning, weaving, dyeing, finishing, and garment manufacturing. It also extends to garbage generated by consumers. Textile waste encompasses both pre-consumer and post-consumer forms. Pre-consumer textile waste, also known as production waste, is a byproduct that arises during the initial stage of the supply chain. This encompasses a comprehensive range of waste materials that are produced as a result of various textile production processes, as well as the utilization of chemicals in textile processing, and the application of sizing or starch during the weaving process, among other factors. On average, around 15% of the fabric utilized in the creation of garments is subjected to cutting, rejected, and ultimately rendered as waste [23, 24]. Pre-consumer textile waste is commonly regarded as "clean waste" as they are produced by the original manufacturers and its little exposure to end users [22]. On the other hand, post-consumer textile waste refers to the accumulation of textile or apparel trash that arises subsequent to its utilization by customers. These items refer to household products or apparel that are no longer needed by the owner and are subsequently disposed of. This category encompasses garments that are considered out-of-fashion, are damaged, and have fitting concerns [24, 25]. In general, post-consumer textile waste exhibits good quality and can be recovered or reused as second-hand apparel. Consequently, these clothes are usually purchased by customers with limited financial resources. There exists the possibility of transforming unwanted textile items into shredded fibers, which can subsequently be utilized in the manufacturing process for further reuse [25, 26].

Industrial textile waste constitutes a distinct category within the broader domain of textile waste. The items generated by industrial applications encompass a range of components such as conveyor belts, filters, geotextiles, and wiping cloths. The waste constituents of those materials which are discarded when a specific component becomes worn out. Typically, the process involves the sequential stages of downcycling, upcycling, and recycling. Numerous companies are actively engaged in the management of waste materials, striving to incorporate them into an asset, thus leading to a circular economy [23].

9.3.1 Reasons Behind the Textile Waste Generation

There are five general reasons for the sharp increase in textile waste:

1. Change of Perception: Post modernization, everyone is influenced by the Western lifestyle and we have lost our rich tradition of using and reusing textiles to its maximum extent, which was deeply rooted in our society.

   We now have an abundant supply of materials. Today, a specific product for everything is available, so we need not recycle our old things and this is significantly contributing to landfills [24, 27]. The consumption of textile materials is not only substantial, but these items also tend to be excessively packaged, resulting in increased waste generation and the utilization of additional natural resources for packaging purposes [28].
2. Fast-Forward Fashion: A significant population is following the fast fashion culture, which means buying cheap fancy dresses repeatedly to make a collection. The production of these clothes involves harmful dyes and chemicals. Finally, these low-quality materials end up in a landfill which leads to the emission of carbon gases into the environment.
3. Increased Shopping Frequency: A tremendous increase in demand and supply of textile products is supplemented by rapidly changing fashion and lucrative shopping options, which attract customers to purchase new dresses frequently.
4. Lack or No Awareness of Sustainability: Sustainability is the primary reason several online clothing brands and designers especially women's wear are switching to upcycling garments. But still, a major population is unaware of the term and does not adopt upcycling to reduce textile waste.
5. Lack of Quality Materials and Eco-Friendly Steps: There is a lack of awareness of eco-friendly measures for discarding a fiber or dress. The case worsens when you handle low-quality material without knowledge of how to dispose of it in an environment-friendly manner at the same time.
6. Lack of Strict Government Norms: Developed nations are actively engaged in efforts to optimize waste management processes, ensure

appropriate trash disposal practices, enhance consumer consciousness, and enforce stringent regulatory standards. Regrettably, the enforcement of environmental legislation in less developed countries remains inadequate. Hence, the governmental initiatives and stringent regulations pertaining to the trade of pre-owned garments, trash management, and educational campaigns possess significant potential to contribute toward environmental conservation [24].

9.3.2 Magnitude of Textile Waste Generation

The textile industry places a significant strain on our environment, consuming substantial quantities of natural resources. Additionally, it employs toxic chemicals in the manufacturing process, contributing to the release of 10% of total carbon emissions globally, worsening this issue further [29]. Remarkably, the clothes manufacturing industry has been identified as a prominent consumer and polluter of fresh water resources. During the manufacturing process, around 70 to 150 liters of fresh water are utilized for dyeing a single kilogram of cloth [30, 31]. Furthermore, the prevailing linear model of "take, use and dispose of" in the textile and clothing industry results in a significant quantity of textile products being discarded. Specifically, it has been approximated that in Europe and America alone, approximately 10 million tons of textile products are disposed of. Notably, the anticipated quantity for China is twice as high as this figure [28].

On a global scale, the estimated quantity of textile consumption exceeds 100 million tons [32]. However, the recycling rate remains rather low, with only approximately 13% of the overall textile material being recycled in some manner after usage. Of the total, 13% is allocated to the production of new apparel, constituting a fraction that is less than 1%. The residual material is repurposed into alternative things of lesser value, such as insulating material, cleaning cloths, or mattress stuffing [33].

The current linear manufacturing structure hinders our ability to effectively utilize resources and address the significant issue of environmental pollution. Based on the ongoing worldwide patterns, it is estimated that the textile industry will account for approximately 26% of the global carbon budget by the year 2050. The figures are gigantic, and, if current trends persist within the textile and apparel industry, it is estimated that by the year 2050, the utilization of non-renewable raw materials will escalate to about 300 million tons. Additionally, the quantity of microplastic discharged into the oceans is anticipated to reach a level of 22 million level [23, 33].

Textile waste generation is prominent at almost every stage of textile manufacturing, processing, production, and consumption levels. Based on the findings of the World Resources Institute, the production of a single cotton shirt typically necessitates the utilization of around 2,700 liters of water. Furthermore, the disposal of clothing by customers in conventional garbage containers not only results in the squandering of financial and material resources but also requires a decomposition period exceeding 200 years within landfill environments. Throughout the process of decomposition, textile waste produces methane gas, a potent greenhouse gas, and releases highly harmful chemicals and colors into the groundwater and soil [34]. Apart from this, textile aftercare also has a significant adverse environmental impact on the textile life cycle on a home and commercial scale. The over-consumption and mass production of clothing resulted in more industrial pre-consumer textile waste [23, 35].

The use of energy and water consumption to clean up dirt, surface coatings, etc., utilize many resources and pollute the air and water in the form of fast fashion not only damages the environment but also creates various negative impacts socially [23, 36]. The effective management of textile waste poses a significant challenge. The environmentally preferred guiding principles for waste management encompass reduction, reuse, recycling, energy recovery, and waste disposal [37].

9.4 Potential of Nanotechnology in the Utilization of Textile Waste

From raw materials to finished products and from manufacturers to consumers, this industry leaves a substantial carbon footprint due to its extensive use of natural resources. According to the U.S. Environmental Protection Agency, textile wastes occupy nearly 5% of all landfill space; however, recycled postconsumer textile wastes account for less than 15% annually; consequently, an astounding 85% of the waste is disposed of in landfills [38]. A significantly lower percentage of clothes go into recycling as technologies for converting used clothes into virgin fibers are in their nascent stages. The future of sustainability requires alternative solutions to address these issues. Nanotechnology is an emerging discipline that addresses the issue of sustainability and explores new opportunities [15]. Due to the long-lasting effect and durability of nanomaterials, nanotechnology has significant commercial potential for the textile industry, which is limited by the impermanence of the effects of conventional techniques used to impart various properties to fabrics.

The textile industry needs huge water usage in the processing, manufacturing, and finishing of textiles and apparel. Most of the treatments, like wet finishing treatments, involve the use of chemicals. The textiles and apparel are dyed mainly with synthetic dyes, which are not biodegradable. The various types of waste generated through industrial processes such as bleaching, weaving, dyeing, and finishing include wastewater, chemicals, dyes, organic solvents, chlorinated solvents, non-degradable surfactants, and other substances. These waste materials are often disposed of by being discharged into water streams causing a significant threat to the well-being of marine organisms, as well as the livelihoods of countless individuals who rely on water resources for their daily sustenance (Figure 9.3).

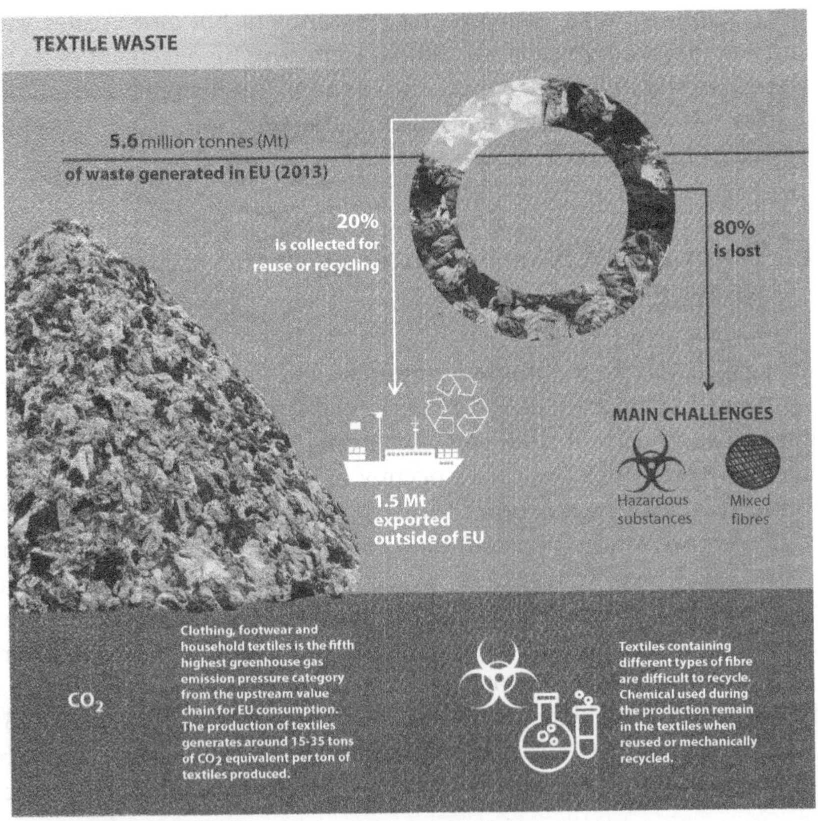

Figure 9.3 Harmful impact of textile waste.

Conventional water treatment technologies have limitations in effectively removing these contaminants, so the emergence of nanoscience and nanotechnology has promising prospects for the advancement of novel techniques in the domain of water treatment. The utilization of small-scale nanomaterials presents novel prospects for enhancing the efficacy of dye-contaminated wastewater treatment technologies [39]. Nanotechnologies possess notable benefits over traditional technologies due to their large surface area, quantum effects, electrochemical and magnetic properties, and various other size-dependent physical and chemical attributes [40]. The emergence of nanomaterials has led to the advent of a few promising materials for the removal of dyes from water, including carbonaceous nanomaterials, nanosized TiO_2, and graphitic carbon nitride [41].

9.4.1 Nanosized Adsorbents

Adsorption is a widely employed method for the purification of wastewater, specifically in industrial settings, to eliminate a diverse array of detrimental substances. This approach involves the binding of molecules in a liquid to the surface of a solid material. Adsorbent materials possess a significantly elevated interior surface area, hence enabling the process of adsorption. In the past, conventional adsorbents such as zeolites and activated carbon were employed for the purification of textile wastewater. However, these adsorbents were found to possess some limitations, primarily characterized by their relatively poor adsorption capacity and rate. The utilization of various nanomaterials, including carbon nanotubes and nanoscale zeolites, offers distinct benefits in the context of adsorption of compounds in textile wastewater. These advantages encompass a substantial specific surface area, an abundance of selective adsorption sites, a reduced intraparticle diffusion distance, an adjustable pore size, and an easy regeneration process that leads to enhanced adsorption of compounds in the textile wastewater.

9.4.2 Nanofiltration

The filtration process is widely applied to remove pollutants in textile wastewater treatment. Nano filters, with different kinds of membranes, focus on higher selectivity and permeability than conventional membranes, thus filtering the small size particulates, hardness ions, and small organic matter [42]. Incorporating functional nanomaterials into members increases the permeability and performance of the membrane. Adding metal oxide nanoparticles involving zeolite, alumina, silica, and titanium oxide to the membranes resulted in membrane surface hydrophilicity and water permeability [43].

9.5 Environmental, Health, and Safety Concerns of Nanotechnology

Environment degradation is undoubtedly the need of the hour as our natural resources (soil, water, and air) are being polluted by several means. One of the contributors to polluting the environment is the textile waste generated by industries. The dust, solid particulates, chemicals, dyes, insoluble organic compounds, etc., are discharged directly into water streams, air, and soil, thus causing significant environmental damage. Because these contaminants are difficult to remove from conventional methods, new technologies are constantly being explored. Environment remediation needs different materials to break and remove complex pollutants from the soil, air, and water.

Nanotechnology has gained wide application because of its unique size, surface properties, selective nature, potential to combine easily, etc. Presently, the textile industry is a significant user of nanotechnology, and there are many nanotextiles and nanotechnologies, including nanomaterials [44]. Various nanomaterials, nanotubes, nanocomposites, etc., are used for environmental remediation. However, it is of utmost importance that the materials employed to save the environment should not be harmful or other pollutants. Thus, nanotechnology processes, materials, and functions should be safe for the environment and human health. Therefore, biodegradable materials are critical and should be used in the particular technology so that no problem arises with disposal after treatment.

Various types of nanomaterials, including inorganic, polymeric, and carbonaceous materials, have been effectively employed in the process of environmental restoration. Despite the demonstrated efficacy of nanotextiles in photocatalytic processes and the absorption degradation of pollutants in contaminated wastewaters, as well as the potential use of nanofibrous membranes as filters for air and water, there have been numerous concerns raised regarding the ecological risks associated with nanomaterials. The selection of nanomaterials for restoration of the environment necessitates a thorough investigation of the material, including an assessment of its possible subsequent consequences, considerations regarding the applicability of the nanomaterial, and an evaluation of its properties. The issue of nanosafety has garnered increasing attention because of the association between exposure to manufactured nanomaterials and various health consequences, such as genotoxicity, lung inflammation, carcinogenicity, and circulatory impacts [45].

The high reactivity of nanomaterials with other substances throughout their life cycle can be attributed to their increased surface area. Nanomaterials are likely to be linked with both positive and negative implications, encompassing both potential risks and potential benefits. The interaction between nanoparticles and other components can potentially result in reduced toxic effects. Conversely, these nanomaterials may also include hazardous substances that contribute to increased toxicity. Certain types of nanomaterials have the ability to dissolve in their surrounding environment without causing any discernible effects. However, it should be noted that nanoparticles composed of metals can have significant detrimental impacts on the surrounding environment.

At the moment, there are very few studies that investigated the impacts of nanomaterials on human health and the environment. Hence, the investigation of nanomaterials and their impact on the environment, health, and safety is a subject of ongoing research.

9.6 Conclusion

Nanotechnology has a more comprehensive application in nearly all fields especially catering to the broader aspects of the clothing and textiles industry. With the advent of new fibers and fabrics and the need for multifunctionality, nanotechnology is a field that has created a niche in producing nanotextiles. Nanomaterials, nanowhiskers, nanoclays, and nanocomposites are all such advanced high-tech nanomaterials with a lot of potential in several fields, including sportswear, Meditech, special garments manufactured for defence personnel, etc. However, more research is still waiting to see the other side of the coin, i.e., the effects of nanotechnology on health, safety, environment, and related aspects. The current increase in our contact with nanoparticles originating from textile materials is a subject of ongoing research. However, it is concerning that there is a lack of sufficient awareness and/or safety regulations pertaining to the toxicological information associated with these nanoparticles. It is foreseeable that this condition may exhibit variability in the near future. Still, in the present scenario, nanomaterials have ample opportunities for making highly functional, durable textiles and apparel. With careful consideration of ethical and safety implications, coupled with collaborative efforts across scientific, industrial, and regulatory domains, we stand poised to harness the full power of nanotechnology, ushering in an era of unprecedented progress and transforming the way we live, work, and interact with the world around us.

References

1. Chinta, S.K., Landage, S.M., Jain, S., Water repellency of textiles through nanotechnology. *Int. J. Adv. Res. IT Eng.*, 2, 1, 36–57, 2013.
2. Grebler, S., Simko, M., Gazso, A., Fiedeler, U., Nentwich, M., The Nano Trust Dossiers, Institute of Technology Assessment of the Austrian Academy of Sciences, No. 015en, 1-5, 2010, http://epub.oeaw.ac.at/ita/nanotrust-dossiers/dossier015en.pdf.
3. Srinivasan, K., Rajanikumar, K., Bhardwaj, S.K., Kumari, L.B., Chavali, M., Nanotechnology Trends in Fashion and Textile Engineering. *Curr. Trends Fashion Technol. Text. Eng.*, 2, 3, 0056–0059, 2018, https://juniperpublishers.com/ctftte/pdf/CTFTTE.MS.ID.555590.pdf
4. Mahmud, R. and Nabi, F., Application of Nanotechnology in the field of Textile. *J. Polym. Text. Eng.*, 4, 1, 01–06, 2017, https://iosrjournals.org/iosr-jpte/papers/Vol4-Issue1/A04010106.pdf.
5. Asif, A.H. and Hasan, Z., Application of nanotechnology in modern textiles: A review. *Int. J. Curr. Eng. Technol.*, 8, 2, 227–231, 2018, https://pdfs.semanticscholar.org/92f2/b4913a981b89a7792878c8818196254 e56fc.pdf.
6. Musante, C. and White, J.C., Toxicity of silver and copper to Cucurbitapepo: Differential effects of nano and bulk-size particles. *Environ. Toxicol.*, 27, 9, 510–517, 2012.
7. Tolochko, N.K., Nanoscience and Nanotechnologies - History of Nanotechnology - Encyclopedia of Life Support Systems (EOLSS), 2017, http://www.eolss.net/sample-chapters/c05/E6-152-01.pdf.
8. Song, X.Q., Liu, A., Ji, C.T., Li, H.T., The Effect of Nano-Particle Concentration and Heating Time in the Anti-Crinkle Treatment of Silk. *J. Jilin Inst. Technol.*, 2001, 22, 24–27, 2001.
9. Russell, E., Nanotechnologies and the Shrinking World of Textiles. *Text. Horiz.*, 9, 10, 7–9, 2002.
10. Ütebay, B., Çelik, P., Çay, A., Textile Wastes: Status and Perspectives, in: *Waste in Textile and Leather Sectors*, Körlü, A. (Ed.), 2020, https://www.intechopen.com/chapters/71971.
11. Helwegen, W. and Escoffier, L. (Eds.), *Nanotechnology Commercialization for Managers and Scientists (1st ed.)*, Jenny Stanford Publishing, Singapore, 2012, https:// doi.org/10.1201/b11619.

12. Sayed, A., What Is Textile Waste? Definition, Meaning and Types, 2018, https://textileapex.blogspot.com/2018/12/what-is-textile-waste-definition.html.
13. Joshi, M. and Bhattacharyya, A., Nanotechnology – a new route to high-performance functional textiles. *Text. Prog.*, 43, 3, 155–233, 2011.
14. Kiron, M.I., Application of Nanotechnology in Textile Industry, 2013, https://www.fibre2fashion.com/industry-article/7135/application-of-nanotechnology-in-textile-industry.
15. Dokania, A., Nanotechnology & Its Potential in the Textile Industry, 2021, https://www.onlineclothingstudy.com/2021/06/nanotechnology-its-potential-in-textile.html.
16. Shah, M.A., Pirzada, B.M., Price, G., Shibiru, A.L., Qurashi, A., Applications of nanotechnology in smart textile industry: A critical review. *J. Adv. Res.*, 38, 55–75, 2022, https://doi.org/10.1016/j.jare.2022.01.008, https://www.sciencedirect.com/science/article/pii/S2090123222000194#:~:text=Background,properties%20into%20textiles%20and%20garments.
17. Kumar, V.S., Paul, S.S., Nozar, D.S., Li, G., Pang, S., Brian, C., Parachuru, R., Applications and future of nanotechnology in textiles. *Beltwide Cotton Conferences*, San Antonio, Texas, pp. 2497–2503, 2006, https://www.cotton.org/beltwide/proceedings/2005-2022/data/conferences/2006/pdfs/2497-2503.pdf.
18. Chandrasekar, S., Vijayakumar, S., Rajendran, R., Application of chitosan and herbal nanocomposites to develop antibacterial medical textile. *Biomed Aging Pathol.*, 4, 1, 59–64, 2014.
19. Krishanga, Nanotechnology can now assist you in losing weight. The Medicinal Medical Review, 2022, Accessed on 27 Sept. 2023 https://www.medindia.net/news/lifestyleandwellness/nano-technology-can-now-assist-you-in-losing-weight-209523-1.htm.
20. Yetisen, A., Hang, Q., Manbachi, A., Butt, H., Dokmeci, M.R., Hinestroza, J.P., Skorobogatiy, M., Khademhosseini, A., Seok, H.Y., Nanotechnology in Textiles. *ACS Nano*, 10, 3, 3042–68, 2016. DOI:10.1021/acsnano.5b08176,https://www.researchgate.net/publication/296055764_Nanotechnology_in_Textiles.
21. Chakrabarty, S. and Jasuja, K., Applications of Nanomaterials in the Textile Industry, in: *Nanoscale Engineering of Biomaterials: Properties and Applications*, pp. 567–587, 2022, DOI:10.1007/978-981-16-3667-7_20, https://www.researchgate.net/publication/358668534_Applications_of_Nanomaterials_in_the_Textile_Industry.
22. Haque, M., Nano Fabrics in the 21st century: a review. *Asian J. Nanosci. Mater.*, 2, 2, 131–148, 2019, http://www.ajnanomat.com/article_81050.html.
23. Goworek, H., Fisher, T., Cooper, T., Woodward, S., Hiller, A., The sustainable clothing market—An evaluation of potential strategies for UK retailers. *Int. J. Retail Distrib. Manage.*, 40, 935–955, 2012.

24. Aishwariya, S. and Jaisri, J., Harmful Effects of Textile Waste, 2020, https://www.fibre2fashion.com/industry-article/8696/harmful-effects-of-textile-waste#:~:text=compatibility%20with%20nature.-,Textile%20waste%20is%20produced%20in%20every%20phase%20of%20the%20textile,from%20combing%2C%20drawing%20and%20spinning.
25. Sakthivel, S., Ramachandran, T., Gowthami, R., Chandhanu, R., Padma Priya, J., Vadivel, P., Gokulraj, S., Source and Effective Utilization of Textile Waste in Tirupur, 2010, https://www.fibre2fashion.com/industry-article/5251/source-and-effective-utilization-of-textile-waste-in-tirupur.
26. Textile waste, 2019, https://www.eea.europa.eu/media/infographics/textile-waste/view.
27. Snehal, Textile Waste - an Environmental Crisis, 2021, ://www.onlineclothingstudy.com/2021/03/textile-waste-environmental-crisis.html.
28. Wang, Y. (Ed.), *Recycling in Textiles*, 1st ed., p. 248, Woodhead Publishing, Cambridge, 2006, ISBN: 9781855739529.
29. Forbes, Making climate change fashionable—The garment industry takes on global warming, 2015, http://www.forbes.com/sites/jamesconca/2015/12/03/makingclimate-change-fashionable-the-garment-industry-takes-onglobal-warming/#2a20a979778.
30. Kant, R., Textile dyeing industry an environmental hazard. *Nat. Sci.*, 4, 22–26, 2012.
31. Chakraborty, S., De, S., DasGupta, S., Basu, J.K., Adsorption study for the removal of a basic dye: Experimental and modeling. *Chemosphere*, 58, 1079–1086, 2005.
32. Harder, J., Throwaway Clothing: The Limitations of Textile Recycling, 2019, https://www.recovery-worldwide.com/download/1473851/recovery_4_2019_online.pdf.
33. Ellen MacArthur Foundation, A New Textiles Economy: Redesigning Fashion's Future, 2020, https://www.ellenmacarthurfoundation.org/assets/downloads/publications/A-New-Textiles-Economy_Full-Report.pdf.
34. Brown, R., The environmental crisis caused by textile waste, 2021, https://www.roadrunnerwm.com/blog/textile-waste-environmental-crisis.
35. Lau, Re using pre-consumer textile waste. *SpringerPlus*, 4, Suppl 2, O9 1–2, 2015.
36. Aqilah, N.H., From Fiber to Consumers: How Nanotechnology Helps In The Sustainability of Textile Value Chains?, 2021, Available online at https://www.nanotextileinnovation.com/from-fiber-to-consumers-how-nanotechnology-helps-in-the-sustainability-of-textile-value chains/#:~:text=Stain%20resistant%2C%20abrasion%20resistant%2C%20water,to%20eliminate%20dirt%20and%20smells.
37. Weber, S., Lynes, J., Young, S.B., Fashion interest as a driver for consumer textile waste management: reuse, recycle or disposal. *Int. J. Consum. Stud.*, 8, 17, 1–9, 2016, https://onlinelibrary.wiley.com/doi/full/10.1111/ijcs.12328.

38. Vadicherla, T. and Saravanan, D., Textiles and apparel development using recycled and reclaimed fibers, in: *Roadmap to Sustainable Textiles and Clothing Eco-Friendly Raw Materials, Technologies, and Processing Methods*, S.S. Muthu (Ed.), pp. 139–160, Springer Science-Business Media, Singapore, 2014.
39. Uddin, M.K. and Rehman, Z., Application of Nanomaterials in the Remediation of Textile Effluents from Aqueous Solutions, in: *Nanomaterials in the Wet Processing of Textiles*, pp. 135–161, 2018.
40. Cai, Z. *et al.*, Application of nanotechnologies for removing pharmaceutically active compounds from water: development and future trends. *Environ. Sci.: Nano*, 2018, 5, 27–47, 2018.
41. Cai, Z. *et al.*, An overview of nanomaterials applied for removing dyes from wastewater. *Environ. Sci. Pollut. Res. Int.*, 24, 19, 15882–15904, 2017.
42. Qu, X., Pedro, J.J., Alvarez, Li, Q., Applications of nanotechnology in water and wastewater treatment. *Water Res.*, 47, 12, 3931–3946, 2013.
43. Bea, and Tak, Interpretation of fouling characteristics of ultrafiltration membranes during the filtration of membrane bioreactor mixed liquor. *J. Membr. Sci.*, 264, 1, 151–160, 2005, DOI:10.1016/j.memsci.2005.04.037.
44. Saleem, H. and Zaidi, S.J., Sustainable Use of Nanomaterials in Textiles and Their Environmental Impact. *Materials*, 13, 5134, 2020, doi: 10.3390/ma13225134.
45. Johnston, L.J., Gonzalez-Rojano, N., Wilkinson, K.J., Xing, B., Key challenges for evaluation of the safety of engineered nanomaterials. *NanoImpact*, 18, 100219, 2020, https://doi.org/10.1016/j.impact.2020.100219.

Part III

ADVANCED APPLICATION OF RECYCLED AND NANO-ASSISTED NOVEL TEXTILE GENERATED THROUGH WASTE

10

Utilizing Textile Waste in the Production of Nanotechnology-Based Sports Textiles

Janmay Singh Hada

Department of Textile Design, National Institute of Fashion Technology, Jodhpur, (Ministry of Textiles, Govt. of India), Rajasthan, India

Abstract

Nanotechnology is continuously advancing, and its applications in sports textiles represent an exciting area of research and innovation. Nevertheless, it is essential to take into account the potential environmental impacts linked to the utilization of nanoparticles. This study examines eco-efficiency in sustainable waste management, concentrating on the 3R approach (reduce, reuse, and recycle) to alleviate ecological consequences. In particular, it investigates the utilization of textile waste for consumption and production, placing special emphasis on sports textiles. Nanotechnology treatments, including coatings on fibers, yarns, and finished products, offer various enhancements such as antibacterial properties, self-cleaning abilities, moisture absorption, and controlled release of chemical agents. Several companies are already repurposing textile waste to create recycled sports products, although concerns about carcinogenic and toxic substances necessitate the nanotechnology role to counter these issues. Technological innovations from companies like Adidas, Nike, and VF Corporation, employing Gore-Tex nanotechnology are discussed in relation to scalability, durability, cost reduction, end-user awareness, and health concerns. The study aims to explore the potential impact of nanocoatings on advanced applications of recycled and nano-sports textiles derived from waste materials. Eco-efficiency in sustainable waste management involves reducing ecological impact through the 3R approach: reduce (waste and by-products), reuse (items repeatedly), and recycle (materials and proper disposal). The growing awareness and favorable mindset demand greater technological intervention in the utilization of textile waste for consumption and production. Numerous companies such as Eco-spun, Lutrader Eco, Sequel Eco, Safe Leigh, K-sorb, Barnhart, and Stein are utilizing textile waste to create recycled textile products.

Email: janmay.hada@nift.ac.in

Prashansa Sharma and Shilpi Shree Sahay (eds.) Nanotechnology-Assisted Recycling of Textile Waste: Sustainable Tools for Textiles, (287–310)

Sports textiles encompass a segment where physiological comfort and product performance play crucial roles, spanning fiber production, yarn spinning, fabric weaving and knitting, and the design and production of finished products (garments, etc.). The application of nanotechnology involves treatments, predominantly coatings, applied to fibers, yarns, or finished products. The integration of functional nanomaterials into raw materials during fiber production, coating, or textile manufacturing, as well as electrospinning for nanometric fiber production, can yield non-woven fabrics with enhanced properties for diverse applications. These properties include antibacterial features, self-cleaning capabilities, moisture absorption, UV-blocking, wear resistance, tear resistance, thermal insulation, conductivity, and the controlled release of active chemical agents such as drugs, cosmetics, and fragrances. Sports textiles, a segment placing significant emphasis on performance, employ recycling approaches using PET bottles, ocean waste, self-disposal sites, or other sources to create sportswear. However, these applications carry the negative consequence of carcinogenic and toxic substances. Nanotechnology is harnessed to mitigate the adverse effects of these recycled materials. Innovative technologies like Gore-Tex nanotechnology, embraced by companies such as Adidas, Nike, and VF Corporation, develop solutions that require scalability, wash durability, cost reduction, increased user awareness, and concern for potential negative health effects. The convergence of clothing functionalities with wearer comfort is an emerging market trend, particularly crucial for achieving high-performance levels in terms of chemical resistance, flame retardancy, corrosion resistance, and UV light blocking.

The objective of this paper is to uncover the potential impact of nanocoatings on the advanced applications of recycled and nano-sports textiles derived from waste materials.

***Keywords*:** Nanomaterials, composites, application on sports textiles, textiles waste, health and safety

List of Abbreviations

CFRP	Carbon fiber-reinforced polymer composites
CNT	Carbon nanotubes
MSN	Micro- and mesoporous nanoparticle silica
MTBR	Made to remade
NF	Nano cellulose fibers
PE	Polyethylene
PET	Poly ethylene terephthalate
PP	Polypropylene
SnO_2	Tin oxides
STNF	Silver-titania nanoparticles
TENG	Triboelectric nano generator
TiO_2	Titanium dioxides

ZC Zirconium carbide
ZnO Zinc oxides

10.1 Introduction

Sports textiles using nanotechnology have revolutionized the sports industry by incorporating advanced materials and processes on a nanoscale to enhance the performance, comfort, and durability of sports clothing and gear. Figure 10.1 depicts the different stakeholders working in the sports textiles industry. It involves manipulating materials at scale (1 Nm = 1 billionth of a meter) to create unique properties and functionalities. To address the ecological impact caused by the fast-changing global fashion and the resulting exponential growth of fast fashion, a waste management approach focused on eco-efficiency is crucial. The core principles of this approach are embodied in the 3R's: reduce, reuse, and recycle.

10.1.1 Reduce

The first step is to minimize the generation of waste and by-products in the fashion industry. This involves implementing practices that promote sustainable production, responsible sourcing of materials, and conscious design to create durable and longer-lasting products.

10.1.2 Reuse

Emphasizing the reuse of items is essential to extend their lifespan and reduce waste. Encouraging consumers to embrace a culture of repairing,

Figure 10.1 Key stakeholders in the relocation of major sports textile categories [2].

refurbishing, and upcycling clothing can significantly reduce the environmental impact of fast fashion.

10.1.3 Recycle

Proper recycling practices are crucial for managing the waste generated in the fashion industry. By recycling materials and employing eco-friendly production techniques, we can lessen the burden on landfills and incinerators, thereby reducing their negative environmental effects. By adopting an eco-efficient waste management strategy that centers around the 3R's, we can effectively mitigate the harmful consequences of fast fashion on the environment. This approach will not only help us reduce the amount of waste ending up in landfills and incinerators but also contribute to a more sustainable and environmentally responsible fashion industry [1].

As highlighted during the Copenhagen Summit's "Pulse of the Fashion Industry" event, the fashion industry plays a role in generating 4% (92 MT) of total waste. This waste is categorized into pre-consumption and post-consumer sources. Within the waste stream, 35% is traced back to pre-consumption sources, encompassing 9% from fiber production and 91% from technological processes such as spinning, weaving, and textile processing. In the realm of post-consumer waste, a substantial 80% finds its way to landfills and incinerators, leaving only 20% available for reclamation through recycling initiatives [3].

Synthetic fibers dominate the market with a significant share of 62.2%, whereas natural fibers account for 37.7%. Notably, within the synthetic fiber category, polyester alone constitutes 51.5%. Figure 10.2 illustrates how the market share is distributed among various types of the fibers.

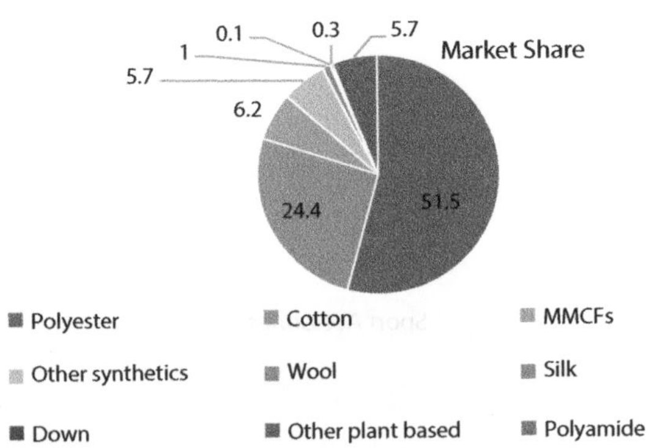

Figure 10.2 Market share among different fiber types.

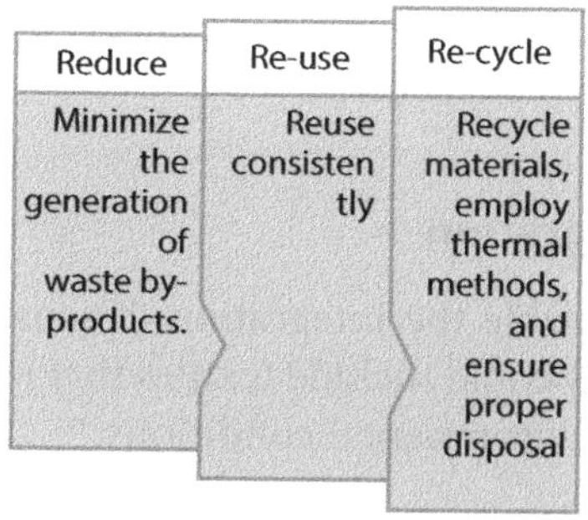

Figure 10.3 Promoting sustainable waste management through the 3R's eco-efficiency model.

The prevailing linear models in the textile and fashion industry, characterized by the *Take-Make-Use-Dispose* approach, are acknowledged for their role in escalating waste production. In a concerted effort to bolster ecological sustainability and alleviate environmental impacts, adopting the 3R model—reduce, reuse, and recycle—emerges as a pivotal and eco-efficient strategy for effective waste management as shown in the Figure 10.3.

10.2 Integration of Textile Waste and Nanotechnology

The integration of textile waste and nanotechnology can take several forms nanomaterials and nano-scale additives, such as nanoparticles or nanofibers, can be added to textiles during manufacturing. These nanomaterials can enhance the performance and properties of the resulting sports textiles.

10.2.1 Coating Techniques

Textile waste materials can be coated with nanomaterials to impart specific functionalities. For instance, a textile made from recycled fibers could be coated with nanoparticles to achieve water repellence or stain resistance.

10.2.2 Hybrid Structures

Textile waste and nanomaterials can be combined to create hybrid structures with improved properties. For example, a composite fabric might combine recycled cotton fibers with nanofibers to achieve both comfort and enhanced moisture management.

10.2.3 Sensor Integration

Nanotechnology can enable the integration of sensors and smart textiles into sports apparel. This could include textiles that monitor biometric data or adapt to changing environmental conditions.

10.3 Sport Textiles

Sports textiles represent a vital subset within technical textiles, prioritizing optimal performance. These textiles offer advanced attributes that enhance both functionality and comfort as shown in Figure 10.4. This specialized sector encompasses an extensive value chain, spanning from initial fiber production to yarn spinning, fabric weaving, and knitting, as well as the design and manufacture of final apparel. Notable applications include various garments such as t-shirts, jogging suits, and tracksuits, alongside sports footwear like athletic shoes. Additionally, this domain extends to composite materials, sailing and aviation products, parachute fabrics, artificial turf, and materials for ballooning, among others.

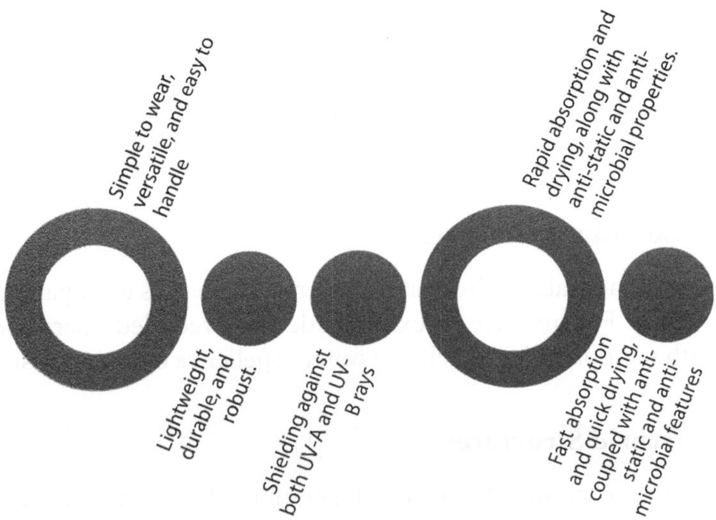

Figure 10.4 Key characteristics of high-performance sports textiles.

10.4 Nanotechnology and Nanomaterials in Sport Textiles

Nanofibers and nanoparticle finishing revolutionized the sportswear market. The nanotechnology in textiles is used for preventing the build-up of static electric charge, water impermeable, wrinkle resistance, strength enhancement, ultraviolet (UV) blocking, anti-bacterial and odor control, sensors, optical displays, etc. In the industry, technology is categorized into three groups: nanofibers, nanocomposite fibers, and finishing textile materials, measuring less than 100 nm, boasting a significantly larger surface area (1000 times greater than traditional fibers) and exceptional mechanical properties. They find application in the production of sports apparel and sports shoes, contributing to the manufacturing of waterproof and windproof products. It can be produced by electrostatic spinning, electrospinning, and interfacial polymerization [4].

Nanocomposite fibers exhibit superior mechanical and physical toughness with good abrasion properties by incorporating nanoparticles, thereby increasing surface area and adopting a high aspect ratio interlining with a polymer matrix. Properties like antibacterial features, conductivity, and anti-static capabilities can be imparted to the fibers through the application of nanoparticles as fillers. Enhanced mechanical properties are attainable by modifying the shape, volume fraction, and size of nano-fillers [5]. In the production of sports equipment such as tennis and badminton rackets, carbon fibers filled with carbon nanotubes (CNTs) are employed to enhance impact strength and rebounding performances.

Nano finishing: To improve the end uses of various physical, chemical, mechanical, and general wear by nanoparticles by method *ex situ* and *in situ*. In the *ex situ* nano-finishing procedure, the fabric undergoes treatment with nanoparticle precursors using methods such as exhaustion, spraying, padding, foaming, and printing. The process involves nanoparticle dispersion, application, and fixation steps. In the case of *in situ* nanostructures, nanoparticles are synthesized by the textile material in a single step. The mechanical properties are improved through the physical and chemical interaction between nanoparticles and nanofibers [6].

Sportswear integrated with nanotechnology exhibits multifunctional properties, protecting both cold and hot water, unpleasant odors, and bacteria. Additionally, these garments are characterized by flexibility, lightweight design, and high-impact strength. The comfort of sportswear is positively influenced by nanotechnology, contributing to enhanced wearer performance and efficiency.

The widespread application of nanocomposites and nanomaterials in the field of sports has significantly enhanced the strength, stiffness, and lifespan of sports materials and equipment when compared to traditional counterparts. This technological advancement brings numerous advantages to various aspects of our daily lives, particularly in sports-related domains such as human movement, sports halls, equipment, and facilities [7]. The utilization of nanotechnology in sports presents a multitude of benefits, contributing to the improvement of athletic equipment and materials. This advancement aims to enhance athletes' performance, promoting increased activity and comfort [8].

Organic and inorganic nanoparticles undergo processing through various physical and chemical techniques. The implementation of nanotechnology introduces treatments, primarily coatings, applied to fibers, yarns, or finished products by the method or techniques shown in Figure 10.5. Integrating functional nano-materials into raw materials during fiber production, coating, or textile manufacturing, and employing electrospinning for nanometric fiber production, can result in non-woven fabrics with enhancements for diverse applications. These improvements encompass antibacterial (biocidal or photoactive) properties, self-cleaning attributes, oleo repellence, absorption, UV blocking, wear and tear resistance, thermal insulation, conductivity, and controlled release of active chemical agents such as drugs, cosmetics, and fragrances.

Sports shoes have always been very tough to design because they need to be both soft (to absorb the impact whenever a ball is kicked, a jump is taken, or the impact of a sprint) and tough (so they do not soften and lose

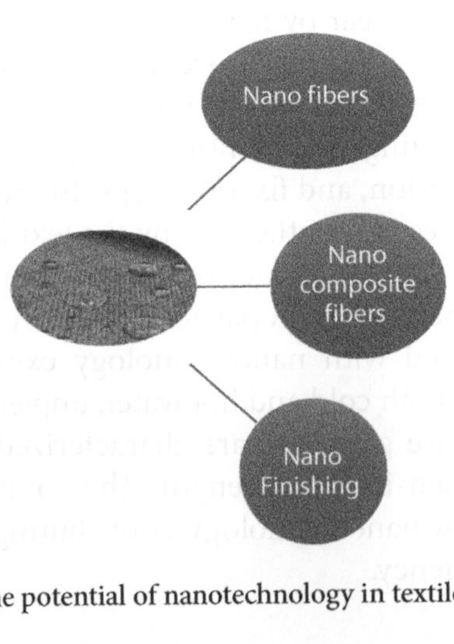

Figure 10.5 Exploring the potential of nanotechnology in textile.

their shape after being used for a while). However, acknowledges to nanotechnology this has changed [9].

Sporting equipment incorporating various nanoparticles demonstrates as shown in the Table 10.1 enhanced performance, greater flexibility, and durability while remaining lightweight. The presence of nanoparticles on textile substrates imparts distinctive multifunctional properties without compromising breathability or tactile sensation [10].

The recycle approach using PET bottles, ocean or self-waste disposal sites, or others is used by companies to make sportswear. The negative effects of these applications are carcinogenic and toxic substances. They use nanotechnology to dissipate the negative effects of these recycled materials.

Integrating clothing functions with wearer comfort is an emerging market trend, and, for active athletes, it is a crucial factor in achieving high performance. This includes chemical resistance, flame retardancy, anti-corrosive properties, and the ability to block UV light.

Table 10.1 Exploring textile material utilization for functionalization in textiles.

S. no.	Nanoparticles	Use
1.	Silver, titanium dioxide, zinc oxide, and TiO_2/Ag and ZnO/Ag nanocomposites	Self-cleaning and antibacterial
2.	Cu and Cu_2O, and CuO nanoparticles	Properties such as antibacterial capabilities, conductivity, and chemical sensing abilities.
3	Fe_2O_3 and TiO_2/Ag/FeO	Magneto, photo, and bio-active textiles
4	TiO_2, NPs, ZnO, Sb, and SnO_2	Antistatic properties in synthetic fibers
5	Nano clays	Active ingredient support, flame retardance, and abrasion resistance
6	Al_2O_3	Flame retardancy and abrasion resistance
7	SiO_2	Reinforcement improves dyeability, abrasion resistance, water repellence, and dirt repellence

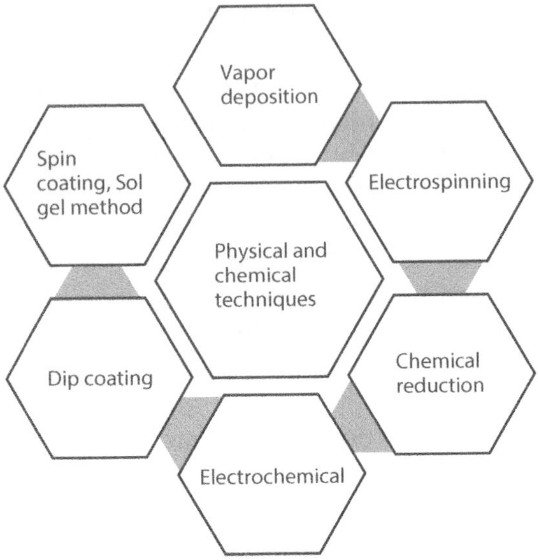

Figure 10.6 Deposition of nanoparticles: techniques from physical and chemical methods.

Innovative materials, including nano-cellulose fibers (NFs), CNTs, micro- and mesoporous nanoparticle silica (MSN), graphene, and nano-clays, are seamlessly integrated into sports equipment by different physical and chemical methods as shown in Figure 10.6, to elevate both performance and longevity. Beyond enhancing key properties like reduced weight, increased strength, and heightened durability, these nanoparticles play a pivotal role in boosting the physical resilience and lifespan of sporting gear. Notably, NFs and CNTs stand out, offering superior performance compared to conventional-sized materials commonly used in sports equipment, such as mountain bikes, tennis rackets, and bicycles.

10.4.1 Performance and Properties of Novel Sports Textiles

The realm of sports textiles is transforming with the integration of novel materials and advanced technologies. The primary objective is to evaluate the performance attributes of the newly developed sports textiles. This includes analyzing factors such as moisture management, breathability, thermal regulation, and comfort during physical activities. The properties of interest include anti-odor capabilities, UV protection, impact resistance, and durability.

Comparative Analysis: A comparative analysis will be conducted between traditional sports textiles and the novel variants. This analysis aims to highlight the advantages and potential drawbacks of innovative textiles in terms of performance and properties.

User Experience: Understanding the user experience is vital. It provides insights from athletes, sports enthusiasts, and potential users regarding their perceptions and satisfaction with the performance and properties of the new textiles.

Material Selection: A diverse range of novel sports textiles will be selected, considering various materials, weaving techniques, and treatments.

Laboratory Testing: Rigorous laboratory testing will be conducted to evaluate performance factors such as moisture wicking, breathability, and thermal regulation. Specialized equipment will be used to measure attributes like UV protection and impact resistance. By bridging the gap between scientific analysis and user experience, the study endeavors to contribute to the evolution of sports textiles for improved athlete comfort and performance.

Here are some ways in which nanotechnology has been applied to sport textiles:

10.4.1.1 Moisture Management and Breathability

Nanoscale coatings and treatments can be applied to fabrics to create hydrophobic or hydrophilic surfaces. This allows moisture to be quickly transported away from the body, keeping athletes dry and comfortable during intense physical activities. Nanofibers can also be integrated into textiles to enhance breathability.

10.4.1.2 Odor Resistance

Nanoparticles like silver or zinc oxide can be embedded into textiles to inhibit the growth of odor-causing bacteria. This helps in maintaining freshness and reducing the need for frequent washing.

10.4.1.3 UV Protection

Nanotechnology allows for the incorporation of UV-absorbing nanoparticles into fabrics, providing enhanced protection against harmful UV rays from the sun. This is particularly important for outdoor sports where athletes are exposed to sunlight for extended periods.

10.4.1.4 Temperature Regulation

Phase-changing materials and nanoparticles can be incorporated into textiles to create fabrics that respond to changes in temperature. These fabrics can help regulate body temperature by absorbing or releasing heat as needed.

10.4.1.5 Abrasion Resistance

Nanocoatings and reinforcements can enhance the durability of sportswear, making them more resistant to wear and tear, especially in high-contact sports.

10.4.1.6 Lightweight and Performance Enhancement

Nanofibers and nanocomposites can be used to create lightweight yet strong materials, contributing to improved performance in sports where speed and agility are crucial.

10.4.1.7 Stretch and Flexibility

Nanotechnology can be employed to create fabrics with increased elasticity and flexibility, providing athletes with greater freedom of movement.

10.4.1.8 Wearable Sensors

Nanoscale sensors and conductive materials can be integrated into textiles to create "smart" sportswear capable of monitoring vital signs, movement, and other data, which can be valuable for performance tracking and injury prevention.

10.4.1.9 Water and Stain Resistance

Nanocoatings can render fabrics water- and stain-resistant, helping to keep sportswear clean and dry in various conditions.

10.4.1.10 Biomechanics and Injury Prevention

Nanofiber-reinforced textiles can be designed to provide targeted support to specific muscle groups, reducing the risk of injuries and enhancing recovery.

Table 10.2 Overview of leading companies and their key innovations in cutting-edge sports textiles.

Company	Notable technologies or innovations
Nike	Dri-fit moisture-wicking fabrics, Nike Aero react
Adidas	Cimcool, Prime Knit
Under Armour	Heat Gear, Cold Gear, UA Rush
Columbia	Omi-Heat Reflective, Omni-Wick
Lululemon	Luon, everlux
Puma	Dry cell, PWRSHAPE
Rhone	Gold fusion (anti-odor), Sliver Tech (breathability)
Skins	Gradient compression technology
Coolibar	UPF 50+ fabrics for UV protection
G-Form	Reactive Protection technology
Sensoria Fitness	Textile Sensors for Real-Time Feedback
Arc'teryx	Gore-Tex Laminated technology

10.4.1.11 Aerodynamics

Nanostructured textiles can contribute to improved aerodynamics in sports such as cycling and swimming, where reducing air or water resistance is crucial. Table 10.2 provides the overview of different technologies adopted by various companies in this field.

10.5 Textile Waste

The industry suffers a large amount of waste due to the variable nature of fibers, fiber machine interaction, and lack of purity of raw materials. Forty-three percent of waste is recycled, 39% of the waste goes to landfill, and the rest is incinerated or disposed of by other means in the US only [11].

Textile waste can be broadly classified into two categories: pre-consumer and post-consumer. Pre-consumer waste encompasses by-products generated within the fiber and textile industries. On the other hand, post-consumer waste includes items discarded by owners due to wear, damage, or other reasons. To manage textile waste, four primary approaches are commonly employed: waste reduction, recycling, incineration, and landfill disposal. Notably, when it comes to synthetic fibers, only thermoplastics like polyester, polypropylene (PP), and polyethene (PE) can be effectively recycled, whereas materials like thermoset resins are not suitable for recycling. The key-factors essential for successful implementation for this are shown in Figure 10.7.

Effective waste management includes reclamation purposes and clear distinction between textile waste such as new production waste, new production waste with a non-textile content, discarded textiles from domestic sources, discarded textiles from municipal refuse sources, and industry and technical commodities [12].

The recycling technology includes solvent extraction, where solvent extraction has been used for carpet recycling, cryogenic fracture, powdering (solid state sheer extrusion), and pyrolysis kiln. In cryogenic fracture, the temperature of the polymer is reduced to below glass transition temperature with liquid nitrogen [13].

The textile sector has incorporated nanomaterials owing to their substantial surface area–to–volume ratio, elevated surface energy, and heightened durability, all achieved without compromising the textile's tactile qualities. These nanomaterials play a transformative role in modifying the

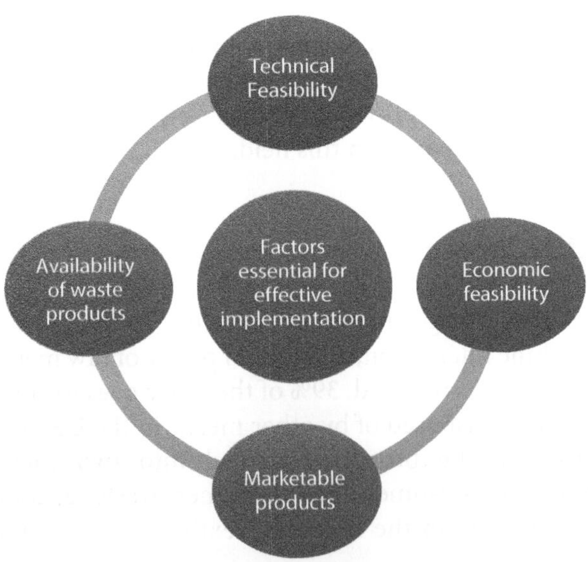

Figure 10.7 Critical factors for successful implementation of textile applications.

chemical and physical characteristics of textile materials and fibers. They enhance water and stain resistance, improve dye absorption capacity, and influence wettability based on surface roughness and energy. These advantageous properties find particular application in non-woven and woven textiles used in sportswear, protective clothing, and rainwear [14].

Textile waste serves as a valuable source of raw materials for diverse applications in composites, contributing to insulation materials for noise and temperature, as well as fillers. Carbon fiber-reinforced polymer composites (CFRPs) are widely employed in industrial settings due to their lightweight nature and high specific strength. Repurposing fibrous carpet waste into a value-added product is a notable practice. Additionally, textile-reinforced composite concrete emerges as a construction material utilizing textiles for reinforcement in applications such as precast constructions, repair, rehabilitation, and structural strengthening of existing structures. This material offers advantages such as thin size, robust load-bearing capacity, corrosion resistance, excellent ductility, absence of magnetic disturbances, and lightweight properties. Furthermore, textile waste finds application in the production of sturdy ropes designed for slope protection against sliding and erosion. Utilizing scraps of insulating materials made from low-quality wool and nonwoven scraps from a blend of recycled fibers, these ropes contribute to sustainable solutions [15].

10.6 Health and Safety Measures

Evaluating the adverse effects of nanoparticles on both the wearer's skin and the environment is crucial. Given that textile materials are in direct contact with the skin throughout the day, any release of nanoparticles poses potential health risks and environmental threats. Understanding the impact of nanoparticles on human and environmental safety is paramount [16].

Engineered nanoparticles, extensively utilized in the sports sector for enhancing water and stain resistance, as well as improving dye absorption capabilities, raise concerns regarding their impact on human health, safety, the environment, and overall sustainability [17]. There is a growing apprehension about the potential health effects associated with engineered nanomaterials, including pulmonary inflammation, circulatory effects, genotoxicity, carcinogenicity, and more [18]. It is imperative to address and comprehensively assess these concerns to ensure the responsible and sustainable use of nanotextiles in various applications.

10.7 Present Scenario

The present scenario regarding textile waste utilization:

10.7.1 Mechanical Recycling

Mechanical recycling involves breaking down textile waste into smaller fibers that can be spun into new yarns and fabrics. This method is well-established and commonly used for recycling post-consumer garments. However, it may have limitations in terms of maintaining fiber quality and performance, especially if the textiles have been through multiple cycles of wear and washing.

10.7.2 Chemical Recycling

Chemical recycling encompasses various processes that use chemical reactions to break down textile waste into its basic components. This allows for the recovery of high-quality fibers that can be used in new textiles. Chemical recycling methods include solvolysis, hydrolysis, and depolymerization. These methods have the potential to address the challenges associated with mechanical recycling, such as fiber degradation.

10.7.3 Upcycling

Upcycling involves transforming textile waste into higher-value products without completely breaking down the original fibers. This approach encourages creativity and innovation, as designers and manufacturers can repurpose textiles into new products with enhanced aesthetics or functionalities.

10.7.4 Collaboration and Innovation

Many fashion brands, textile manufacturers, and research institutions are collaborating to develop innovative solutions for textile waste utilization. Startups and research projects are exploring novel techniques, such as 3D knitting and digital fabrication, to transform textile waste into new fabrics and products.

10.7.5 Consumer Awareness

Increasing consumer awareness of the environmental impact of fast fashion and disposable textiles has led to a greater demand for sustainable and circular fashion practices. This shift in consumer mindset has pushed brands to adopt recycling and waste reduction initiatives, including the use of recycled materials in their products.

10.7.6 Regulatory and Industry Initiatives

Governments and industry associations in various countries have been implementing policies and initiatives to encourage sustainable practices in the textile industry. These initiatives may include regulations on waste disposal, incentives for recycling, and certifications for sustainable textile production.

10.7.7 Challenges and Considerations

While progress has been made, challenges remain in terms of scaling up textile waste utilization techniques, ensuring consistent quality, and addressing the environmental impact of recycling processes. Balancing sustainability, cost-effectiveness, and performance in recycled textiles is an ongoing focus of research and development.

Nano-biocomposites offer an eco-friendly substitute for non-natural antibacterial agents due to their biodegradability and non-toxicity. Exploring the release of Ag nanoparticles from treated textiles reveals discharge patterns. Companies like Eco-spun, Lutrader Eco, Sequel Eco, Safe Leigh, K-sorb, Bernhard, Asics, and Stein are recycling textile waste. Nanotechnologies can elevate economies in low-tech clothing manufacturing industries [19].

Due to their expansive surface area, these materials hold the potential to enhance both the properties and functionalities of materials. High-performance fibers and fabrics are used in uniforms, equipment and sports textiles. Strength, abrasion, and breathability are some parameters for athletes' prototypes. The composites, like roller blades, frames (cycle, bike etc.), golf clubs, tennis rackets, and surfing equipment, rely heavily on composites due to their high strength and durability [20].

Asics has introduced a sustainable sports shoe line that incorporates 5 tons of recycled textile waste, equivalent to repurposing 25,000 t-shirts, for the production of new shoes. Notably, these shoes showcase sock linings crafted through a resource-efficient process known as solution dyeing. This innovative technique results in a remarkable 45% reduction in CO_2 emissions and approximately 33% less water usage compared to traditional dyeing methods [21].

Adidas is embracing sustainability by incorporating 90% recycled polyester in its products, operating under the label "made to remade (MTBR)," with a focus on circularity in design. Among the MTBR offerings are the Stan Smith MTBR and Terrex Free Hiker MTBR, along with MTBR apparel products tailored for running and Adidas Stella McCartney. The brand plans to introduce more products following a similar sustainable approach in the future.

10.8 Composites

Composites are the go-to material when technology demands a mix of qualities. For continuous fiber composites, fiber choice greatly impacts

Table 10.3 Cutting-edge products and applications in the field of sports textiles.

S. no.	Products	Applications
1	Winter ACTi wear	Microporous membrane-treated high-filament nylon
2.	Camping and hiking	Ultrafine microfiber (0.001 denier), polyester, and polyamide
3	Baseball	Wood and graphite fiber composites
4	Tennis	Composite fiberglass, Kevlar, and graphene
5	Tennis ball and net	Ball: woven and needle tennis felts tennis net: polyethylene monofilament
6	Golf and hockey equipment	Textile structure composites
7	Skates	Glass-reinforced thermoplastic resin
8	Bikes	Epoxy resin composites
9	Marine products	Acrylic, polycarbonate, teflon, etc.
10	Diving suits	Laminated elastic suits, etc.
11	Hot air balloon	Rip-stop woven nylon fabrics

properties, especially in load-bearing roles for textile reinforcements. These reinforcements, woven into 2D and 3D structures, have different characteristics. 2D structures lack in-plane shear strength due to crimp, whereas 3D structures offer more strength. Composite production involves wetting the textile with resin and curing, enhancing mechanical properties shown in Table 10.3. These waste-based composites find use in aerospace, military, and sports gear. With varied fibers, yarns, and manufacturing methods, design flexibility is ample [22].

The disposal of thermoplastic waste could be redirected toward less critical land uses, such as the production of fiberfill, melt-blown, or spunbonded products. Recycling methods need to meet the criteria of being technically feasible, environmentally friendly, and economically viable.

10.9 Nanotechnology is Applied to Enable Self-Cleaning Properties in Textile Waste

By employing nanotechnology on textured solid surfaces, nanoparticles intermingle with the voids on the textile substrate, leading to the creation of oxygen and hydroxyl radicals, both of which are highly reactive and unstable. These radicals actively bind with impurities, dirt, and pollutants present on the textile surface, effectively eliminating them [23]. For self-cleaning purposes in textile waste, silver-titania nanoparticles (STNFs) are used. The antimicrobial properties of nano silver (Ag) find application in areas such as wound dressings and surface coatings like catheters. Additionally, the nano-TiO_2 coating facilitates the self-cleaning of stains like wine, coffee, and juices through the action of solar-stimulated light. Moreover, silver (Ag) nanoparticles exhibit an antibacterial effect on textile products, making them valuable for various applications [24].

10.10 Innovations

Employing groundbreaking technologies like Gore-Tex nanotechnology as well as industry giants such as Adidas, Nike, and VF Corporation are spearheading technological advancements that demand scalability, unwavering wash and wear durability, cost reduction, heightened end-user consciousness, and vigilant monitoring of potential adverse health impacts. These innovations are designed to deliver optimal heat and humidity regulation, excellent air and water vapour permeability, and supple, skin-like flexibility. Moreover, they prioritize a lightweight, delicate tactile experience while maintaining the highest standards of performance.

Adidas has introduced cutting-edge data-driven footwear tailored to specific runner profiles, boasting a seamless, lightweight design for a precise fit. Another noteworthy development comes from Stan Smith Milo, who has incorporated mushroom-based materials into their creations [25].

In a bold stride, Descente, a prominent Japanese sportswear brand, has engineered Solar A, a premium line of skin wear. This innovative fabric integrates particles like zirconium carbide (ZC) that harness solar energy and convert it into thermal power [26].

A remarkable leap in technology is represented by the triboelectric nanogenerator (TENG), a mechanism that adeptly transforms mechanical energy into electrical energy. Within the sports realm, TENG serves as a pioneering platform for self-powered sensors, reducing reliance on batteries and intricate sensor systems. Positioned as an alternative solution, TENG, positioned as an alternative solution, stands out as an ideal candidate for efficient energy harvesting, impact monitoring, and athlete performance evaluation within the intelligent sports landscape.

The potential of waste materials underwent a thorough examination utilizing various techniques, including structural analysis, morphological examination, and fourier transform infrared spectroscopy (FT-IR) spectroscopy. A single-electrode mode TENG was carefully designed, integrating positive triboelectric textile layers with corresponding negative counterparts. The inventive four-finger knitted TENG (F-TENG) configuration was introduced, demonstrating outstanding performance. This structure achieved an impressive peak-to-peak voltage of 4.2 Volts, along with a peak-to-peak current output of 2.7 nanoamperes, utilizing a straightforward releasing and stretching mechanism [27].

Practical applications of this technology span various sports activities, including ball dribbling, running, and stretching exercises. The implementation of the F-TENG system demonstrates versatility across different sports facilities, contributing to functionalities such as a self-powered punching power recognition system and an edge ball judgment mechanism.

10.11 Conclusion

The progress of nanotechnology has played an important role in enhancing the visual appeal, functionality, and user interfaces of clothing and textiles. This advancement is driven by a dual objective: not only to elevate product quality but also to address environmental concerns and promote resource preservation. This chapter focus into the potential impact of nanocoatings on the advanced utilization of recycled and Nano-enhanced sports textiles, derived from waste materials. The growing concern among consumers about environmental issues has prompted manufacturers to incorporate textile waste into their production methods, aiming to reduce the ecological footprint. This convergence of consumer behavior and the urgent matter of climate change set the stage for the emergence of environmentally neutral nano-enhanced textiles, designed for recyclability and alignment to reduce greenhouse emissions.

The trajectory of nanotechnology in textiles is poised to transcend mere stylistic shifts, ushering in a crucial paradigm shift in the domain of couture waste management. As nanotechnology continues to shape the evolution of textile products through innovative applications, manufacturers and industries face the responsibility of ensuring the responsible stewardship of these technologies. Striking a delicate balance, they must foster innovation without compromising the environment or human well-being.

By harnessing nanotechnology, this chapter showcases the potential to enhance the performance and functionality of sports textiles, meeting the evolving demands of athletes and consumers. The integration of nanomaterials not only improves textile properties like moisture-wicking, breathability, and durability but also opens up new avenues for creating smart and adaptive sports apparel. The findings underscore the importance of sustainable practices in the textile sector, emphasizing the value of repurposing waste materials to create high-value products. This approach aligns with the principles of a circular economy and resource efficiency, contributing to a more responsible and eco-friendly industry. In conclusion, the exploration of utilizing textile waste in conjunction with nanotechnology represents a significant step forward in sustainable innovation within the textile industry.

References

1. Ailenei, E.C., Ionesi, S.D., Dulgheriu, I., Loghin, M.C., New Waste-Based composite material for construction applications. *Materials*, 14, 1–14, 2021, MDPI.
2. Anand, K.J., Technological Developments in Sports Textiles. *Text. Value Chain*, 9, 8, 22–24, August 2021, ISSN No: 2278-8972.
3. Eder-Hansen, J. *et al.*, Pulse-of-the-Fashion-Industry, 2017, Available online: https://globalfashionagenda.com.

4. Huang, Z., Zhang, Y.Z., Kotaki, M. *et al.*, A Review on Polymer Nanofibers by electrospinning and their applications in nanocomposites. *Compos. Sci. Technol.*, 63, 2223–2253, 2003.
5. Tjong, S.C., Structural and Mechanical properties of polymer nanocomposites. *Mater. Sci. Eng. R*, 53, 73–197, 2006.
6. Harifi, T. and Montazer, M., Application of Nanotechnology in sports clothing and flooring for enhanced sports activities, performance, efficiency and comfort: A Review, Sage pub.co.uk. *J. Ind. Text.*, 46, 1147–1169, 2017, DOI: 10.1177/1528083715601512.
7. Zhang, Y., Zhang, L., Liu, X., The status and prospect for Nanotechnology in tumour diagnosis and treatment. *China Med. Eng.*, 30–36, 2004.
8. Costa, L.P., Engineered nanomaterials in the sports industry, in: *Handbook of nanomaterials for manufacturing applications*, pp. 309–320, Institute of Exact Sciences and Technology (ICET), Federal University of Amazonas—UFAM, Elsevier, Itacoatiara, Brazil, 2020.
9. Nano-Tech Sports Shoes, 2012, [Online], [Cited: 27 December 2023.] https://sites.psu.edu.
10. Nazari, A., Montazer, M., Rashidi, A. *et al.*, Nano TiO_2 photo-catalyst and sodium hypophosphite for cross-linking cotton with poly carboxylic acids under UV and high temperature. *J. Molecul. Catal. A*, 237, 101–108, 2005.
11. Waste, US Textile Mills Recycle, 1993.
12. Cadieux, R.D. and Godfrey, J.F., Waste Prevention and Reclamation. *Int. Fibre J.*, 4, 15–19, 1990.
13. Ravnitzky, M., Tackling the edge trim problem, in: *Industrial Fabric Products Review*, 1994.
14. Ghosh, S., Smith, T., Goswami, P., Nanofnishing of Textiles Sportswear, World Textile Information Network, Dublin, Ireland, 2020, [Online].
15. Velde, K., *Textile composites in sports products*, Woodhead Publishing series in Textiles, pp. 309–322, 2005.
16. Vishwakarma, V., Synthesis, Design, and Characterization of Controlled Nanoparticles for the construction industry, in: *Handbook on Nanomaterials for manufacturing applications*, Chapter 15, pp. 323–341, Elsevier, s.n., 2020.
17. Kohler, A.R., Risk preventative innovation strategies for emerging technologies the cases of nano-textiles and smart textiles. *Technovation*, 2014.
18. Leong, D.T., *Nanosafety Issues of Nanomaterials*, Elsevier, Amsterdam, The Netherlands, 2017.
19. Saleem, H. and Zaidi, S.J., Sustainable Use of Nanomaterials in Textiles and Their Environmental Impact. *Materials*, 13, 22, 5134, 2020, https://doi.org/10.3390/ma13225134.
20. Adanur, S., *Wellington Sears Handbook of Industrial Textiles*, Johnston Industries Group, Pennsylvania, USA, 1995.
21. Moore, D., https://www.circularonline.co.uk.asics-turns-5-tonnes-of-recycled-textile-waste-into-earth-day-running-shoe. [Online] 16 April 2021. [Cited: 27 December 2022.]

22. Watson, J.C. and Raghupati, N., Glass Fibers, in: *Engineered material Handbook*, ASM International, 1993.
23. Khatib, E.M., Anti Microbial and self-cleaning using Nanotechnology. *Res. J. Text. Apparel*, 2012.
24. Subhranshu, S.S., Jeyaram, P., Vinita, V., Sonochemical Coating of Ag-TiO_2 Nanoparticles on Textile Fabrics for Stain. *J. Miner. Mater. Charact. Eng.*, 9, 6, 519–525, 2010.
25. Stories, Adiddas Sustainability. 7 simple solutions to fight pollution, 2021, [Online] 15 November 2021. [Cited: 03 October 2022.] https://www.adidas.com/us/blog/sustainable.
26. Manisha, S. *et al.*, Waste textiles as the versatile triboelectric energy-harvesting platform for self-powered applications in sports and athletics. *Nano Energy*, 97, 107208, June 2022, ISSN 2211-2855, 97.
27. https://www.adidas-group.com. More Sustainable Materials and circular services, [Online] [Cited: 27 December 2022.]

11

Functional Textiles from Agro-Industrial Waste

Shubham Joshi[1]*, Neelu Kambo[2] and Saurabh Dubey[3]

[1]*Department of Textile Technology, M.L.V. Textile and Engineering College, Bhilwara, India*
[2]*Department of Basic Science, Uttar Pradesh Textile Technology Institute, Kanpur, India*
[3]*R. S. Formulation and Consultant, Pune, India*

Abstract

Every day, tons of residues are produced by agriculture-based industries. Most of the wastes come from mills producing rice, pulses, grains, sugarcane, and oilseed. Usually, agro-industrial waste is thrown away by burning, dumping, or unplanned land filling. This can cause serious problems related to climate change by increasing the amount of greenhouse gases. Several researchers studied on the efficient utilization of agro-industrial waste. Some of researchers reported the applications of agro-industrial wastes for enzymes, fertilizers, building materials, polymers, geology, composite materials, dyeing, textile finishing, etc. To reduce carbon footprints and decrease reliance on virgin goods, it is important to utilize agricultural residues on a large scale. The popularity of biocompatible textile materials has increased recently. There is ample scope in this field, as agricultural residues are available in large quantities at low prices, and research efforts are underway to find new sources and value-added products. Agricultural residues such as corn husk, rice bran, and corn husk and linseed cake have wide application in the treatment of dye waste materials. The extraction of natural dyes from various agro-processing residues, such as gramme peel, pomegranate peel, walnut shell, and green coconut shell, has been studied earlier. Agricultural residues are rich in bioactive compounds. Furthermore, it has been reported that these natural compounds help in improving the physical and functional properties. All these factors are responsible for the increase the popularity of agriculture waste in several application areas. This chapter provides a discussion on various natural materials, their extraction

**Corresponding author*: shubhambmjoshi@gmail.com

Prashansa Sharma and Shilpi Shree Sahay (eds.) Nanotechnology-Assisted Recycling of Textile Waste: Sustainable Tools for Textiles, (311–350) © 2025 Scrivener Publishing LLC

procedures, and effective utilization to achieve maximum functionality. The primary objective of this chapter is to study various significant agro-industrial waste products and explore their extensive potential for recycling applications.

***Keywords*:** Agro-industrial wastes, green technology, antimicrobial activity, value-added textiles

List of Abbreviations

CAGR	Compound annual growth rate
PHWE	Pressurized hot water extraction
MAE	Microwave-assisted extraction
UAE	Ultrasonic-assisted extraction
WSP	Water-soluble polysaccharides
CPI	Chemical process industries
SC-CO_2	Super critical carbon dioxide
NP	Nanoparticle
ISC	Institute for Sustainable Communities

11.1 Introduction

Currently, the disposal of organic waste originating from agro-industries is a prominent contributor to environmental contamination. Industrial trash and waste from agriculture and forestry are both examples of organic waste. Agricultural and forestry waste encompasses various organic materials such as cattle slurry, compost, agricultural leftovers, as well as trash generated by trimming activities and forestry preservation [1, 2]. Organic wastes are generated by many industrial entities and encompass residual materials derived from the agrifood sector, such as leftover of coffee beans, bagasse, degummed fruits and legumes, milk serum, wool sludge, cellulose, and other similar substances [3, 4]. Numerous nations have instigated restrictions on such organic wastes out of concern for the environment, as their accumulation poses a threat to the natural world. The mitigation of adverse consequences and the implementation of regulations pertaining to the utilization of organic waste as a fertilizer in agricultural practices are of paramount importance. The term "waste" pertains to waste and includes important definitions and principles governing waste management. It emphasizes the importance of waste evaluation and elimination without endangering water, air, land, or flora and fauna. Clean technology can decrease industrial pollution by reducing biological debris through recycling. Since the beginning of time, most biowaste products, such as used vegetables, fruits, peels, seeds, manure, and slurry, have been added to the soil for cultivation [5]. Indeed, the utilization of biowaste facilitated the restoration of nutrients and an increase in the volume of biological material within the Earth's lithosphere [6]. It is apparent that the historical availability of animal dung was comparatively lower than the current levels of discharge, suggesting that the environmental impact associated with the use of this kind of waste would have been relatively less severe. In the last five decades, the expansion of farming, animal husbandry, and manufacturing operations has led to a substantial rise in population and the production and accumulation of significant quantities of trash [7, 8]. The increase in livestock production has disrupted the mutually beneficial relationship between livestock and agricultural productivity, which is associated with the utilization of mineral fertilizers and pesticides in the manufacturing of animal feed [9]. Therefore, the incorporation of biological waste into the ground represents a significant issue linked to environmental disruption. In order to mitigate the existing environmental challenge, scientific organizations have implemented strategies such as reuse, green technologies, and waste management techniques to reuse agricultural and industrial wastes and effluents. Alternatively, these materials may be reintroduced into their natural ecosystems.

In today's fast-growing world, waste can cause a significant impact on the health of ecosystems and, ultimately, humanity. As a result, it is essential that all agro-industrial sectors find safe ways to use agro-materials, such as garbage recycling. For example, the agro-business produces a significant quantity of waste that can be used in industries such as energy production, composting, and textile manufacturing [10]. Energy consumption and economic pressures on companies have made it critical to use resources responsibly and fully. It is a viable option to use agro-industrial waste to satisfy the needs of the current generation without jeopardizing the needs of future generations. Consequently, there is an urgent need to focus on the complex problems of utilizing and recycling agro-industrial waste. This chapter focuses on some of the most common agro-industrial waste materials and their various recycling processes. Figure 11.1 depicts applications for recycled agro-industrial waste with additional value.

Agro-industrial waste is a cheap and sustainable way to create renewable approaches for high-value products. Microbial conversions provide a viable pathway for producing a variety of natural flavoring ingredients from waste products [11, 12]. Vanillin is a commonly used flavoring ingredient in the food business [13]. Promising outcomes have been discovered

Figure 11.1 Applications for recycled agro-industrial waste.

from previous studies conducted on the biotechnological synthesis of agro-products using ferulic acid derived from by-products and waste in the agro-food sector [14]. Various cereal by-products, including corn, rice, wheat, and sugar beet pulp, have been investigated as potential abundant sources of ferulic acid that can be extracted from waste materials [15]. The current advancements allow to produce vanillin using by-products from the agro-food sector in a manner that is innovative, environmentally sustainable, and economically feasible while also facilitating the utilization of waste materials. Ascorbic acid, commonly referred to as vitamin C, is a naturally occurring compound obtained from diverse plant tissues [16–18]. It has been employed for centuries as a traditional remedy and as a means of preserving food. Several antimicrobial food packaging solutions have been employed to safeguard cheese and other food items by impeding the growth of microorganisms that cause spoilage while preserving their beneficial attributes [19]. These methods encompass the utilization of extracts derived from the waste of citrus species. These high-value products are not limited to food industry but now used in variety of high-end products including biofuel, biocomposites, and bioactive and biodegradable products for various applications [20, 21].

11.2 Socio-Economic Perspective

In recent years, the use of agro-industrial wastes in the creation of new items has advanced remarkably. Modern waste removal techniques are used on materials from the food and manufacturing industries. Green technology is used to recover high-value compounds from agro-industrial wastes. Because traditional methods often use more organic solvents, high energy, and take a long time, the current trend is to slowly replace them with more environmentally friendly methods. These environmentally friendly extraction methods are more versatile, selective in terms of product recovery, and flexible. Furthermore, the use of agro-industrial waste for pharmaceutical sector using current technologies has been widely accepted. In this view, the synthesis of medicines using unconventional methods and materials can be a viable alternative to conventional methods, adding value and providing a sustainable deterrent to the massive amounts of biomass generated mostly by the food industry. The agri-food generate massive amounts of waste daily. These residues are a product of the harvesting, transporting, storing, industrial processing, marketing, and consumption stages of the production cycle, respectively [20]. Nevertheless, these leftovers such as peels and seeds can provide a variety of important elements, including vitamins, minerals, fibers, carbohydrates, and proteins, and, in most cases, in greater quantities than other sections of the plant. The majority of agro-industrial wastes consist of complex polysaccharides/proteins, carbohydrates, polyphenolic compounds, etc. Lignocellulosic raw materials are natural, renewable, biodegradable, and ecologically acceptable matrices that are abundant and inexpensive and have good and promising qualities for industrial applications in a variety of market segments. These matrices have advantages over synthetic inputs. A variety of extractable residue from forest sources and agro-industrial by-products are low in density and reduced energy consumption and have specific strength and modulus. Several researchers have proposed agro-industrial wastes as an alternate source of energy. Agriculture is one of the world's key sources of food production, and it has a considerable impact on income distribution and social dynamics. It has a greater impact on the global economy [22].

In last decade, the agricultural commodities that generated the greatest agro-industrial waste are legumes, oilseeds, sugar, and tubers. Cereals were also the agricultural commodity that grew the most. In contrast to future innovations and new investment opportunities, the fundamental pillars for the development of a sustainable society must expressly pass through the construction of cost-effective, environmentally responsible, and scalability-promoting technology. The objective of an environmentally responsible waste reuse industry is to protect the health of the planet and to contribute to the economy and society.

There are still regions where these abundant agricultural by-products are dumped as a waste. These valuable agricultural by-products are burned because they have no other economic application. The income earned by farmers is decreasing, whereas income is increasing in locations where small- and medium-sized handloom businesses operate and contribute considerably to the generation of revenue. If viable solutions for agro-residues are investigated, farmers will profit socially and environmentally. Agro-residues can therefore help farmers generate more revenue. Technology advancements for the best use of these agro-residues would benefit rural populations by raising the value of their harvests while also preserving the environment. Agro-residue utilization requires strong coordination among industrialists and research groups. Textile applications necessitate high-quality fibers, notwithstanding the relatively high-value addition and large market for agro-residue fiber utilization [23]. In the textile business, the adaptability of agro-residues is determined by the fineness and strength of the fibers generated. The extraction methods employed have a significant effect on the extraction of high-quality fibers. In order to add value to agro-residues such as bagasse and cornhusk, traditional extraction techniques must be upgraded with more robust procedures and machinery. Sometimes, it is believed that residues are garbage and therefore cheap. However, this is a foolish assumption to make. Even if residues are freely accessible in a monetized economy, everything that has a use will eventually acquire monetary value.

11.3 Agro-Industrial Waste

Agricultural residue, often referred to as agro-residue, encompasses any organic matter that is generated as a result of the cultivation and processing of agricultural crops. Agro-residues refer to non-wood lignocellulosic materials that are abundant in cellulose and lignin content [24]. The mentioned botanical components include the stem, straw, seed pod, and leaves. Agricultural leftovers provide the characteristics of sustainability and cost-effectiveness, making them a viable and economical source for obtaining natural cellulosic fibers. There are two distinct categories of agro-residues: field residue and process residue. Field residues refer to the remnants of organic components that remain in an agricultural field following the completion of the crop harvesting process. Some examples of natural materials that can be utilized include the leaves of pineapple plants, the pseudo stems and leaves of banana plants, cornhusks, cotton seed pods, kapok seed pods, and cotton stalks. Process residues refer to the residual materials that remain after the processing of a crop into a resource that may be effectively utilized [25]. Following the extraction of the core agricultural commodity, a substantial quantity of by-product is produced concurrently. Process wastes, such as bagasse, pea peel, wheat, and rice straw, serve as notable illustrations. Agriculture-based industries produce a huge amount of waste every year. Their proper disposal and utilization are a big concern among humans and environment. Mostly, the disposal of these untreated agro-industrial waste either by burning, dumping, or unplanned landfilling. These agro-industrial waste are rich in bioactive compounds. This waste can be used to produce different products as discussed in Figure 11.1.

Most of the agricultural residues like banana stems, pineapple leaves, orange peels, corn husks, fish scales, and lotus stems are increasingly being used in the textile industry [25]. Eco-friendly materials are long-lasting, and the usage of agricultural leftovers provides farmers with additional money. The conversion of agricultural waste into fibers, spun yarns, and woven fabrics has the potential to preserve the environment by reducing food waste and water usage [26]. Figure 11.2 illustrates the categorization of agro-industrial waste. The utilization of agro-industrial waste has drawn increased attention within the textile business.

According to Agricultural Textiles Global Market Report 2023, the global agro textiles market rose at a compound annual growth rate (CAGR) of 9.1% from \$10.88 billion in 2022 to \$11.87 billion in 2023 (Figure 11.3). Despite the Corona pandemic and war disasters, the agro-textile industry is expanding. The agro-textiles market is projected to expand at a 7.1% CAGR to \$15.62 billion by 2027 [27].

Agro-waste fibers and textiles have emerged as a viable and ecological substitute for synthetic fibers, which are known for their high energy consumption [28]. A variety of agricultural residue, including bamboo, corn husks, orange peels, pineapples, soya beans, eucalyptus, lotus stems, betel nut husks, nettle, hemp, aloe vera, rose petals, sugarcane, milk, and fish scales, are increasingly being utilized as feedstock in the textile industry.

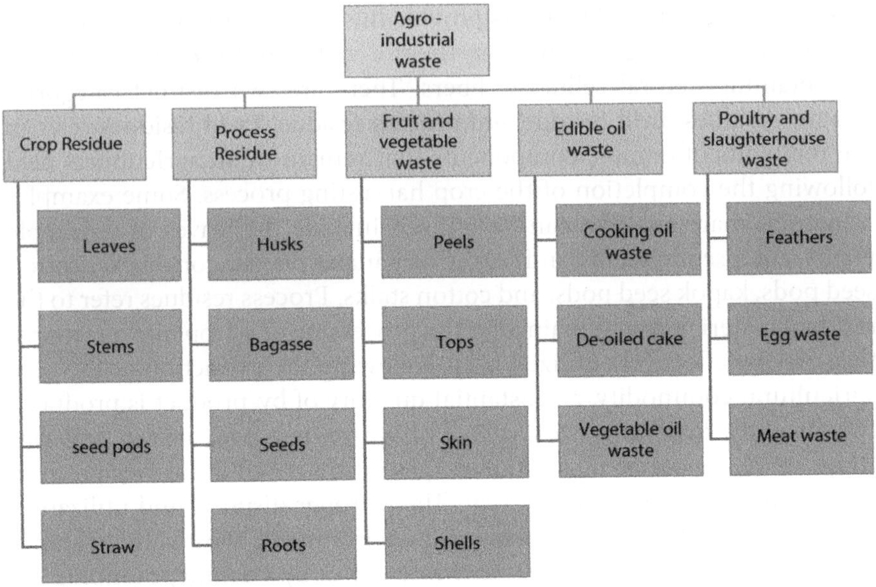

Figure 11.2 Classification of agro-industrial waste.

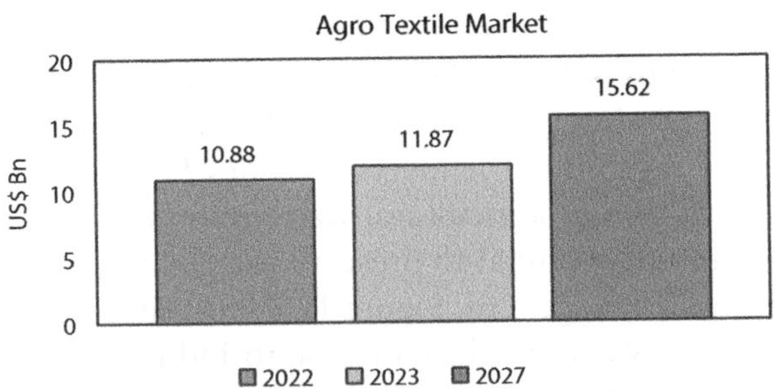

Figure 11.3 Agro-textile market growth.

Agriculturally derived textiles offer three primary benefits, namely, water conservation, waste management, and biodegradability. The predominant fibers utilized in the garment industry are derived from petroleum-based sources. Based on recent research conducted by the World Resources Institute (WRI) in India, the Institute for Sustainable Communities (ISC), and Wageningen, it has been found that natural fibers such as traditional cotton, which ranks as the second most widely used textile fiber, exhibit a substantial reliance on agrochemicals and necessitate a considerable volume of water. Additionally, the waste obtained from the poultry farm and abattoir can be effectively utilized in the production of cosmetics, food and additives, materials for biomedical purposes, medication delivery applications, and various other goods of both low and high value. Table 11.1 presents an overview of the diverse residues obtained from different sources, including information on their composition and the practical applications in which they can be effectively utilized.

Before learning about the improvements in agro-industrial waste-based products, readers should become acquainted with the traditional methods of utilizing agricultural waste. Agro-residues are commonly employed for

Table 11.1 Different types of agro-industrial waste, their composition, and applications.

Residue	Source	Composition	Applications
Leaves	Forest and agriculture lands	Antioxidants, vitamins, minerals, and fibers	Bioactive, fertilizers, biofuel, bioplastics, dyes, insect repellent, mosquito repellent, charcoal, and ink
Stems	Forest and agriculture lands	Minerals, vitamins, and fibers	Composites and dyes
Seed pods	Forest and agriculture lands	Soluble and insoluble fibers, proteins (tryptophan and lysine), complex carbohydrates, vitamin B, folate, minerals	Biochar and biomass material, in food industry, and dyes
Straw	Agriculture lands, grass lands	Rich in fiber, protein, minerals	Ethanol production, building material, composites
Husks	Agriculture lands and process industry	Protein and minerals	Insulation material, biofuel, fertilizer, ash, and healthcare products

(Continued)

Table 11.1 Different types of agro-industrial waste, their composition, and applications. (*Continued*)

Residue	Source	Composition	Applications
Bagasse	Agro-industry process waste	Cellulosic fiber	Biodegradable, compostable material for food service, fertilizer, biofuel, and paper pulp
Seeds	Fruits, vegetables, and food industry	Protein, healthy fats, and fiber; minerals (such as magnesium, potassium, calcium, plant iron, and zinc); and vitamins B1, B2, B3, and E	Chemicals; compounds including protein, carbohydrates, lipids, and small molecules; cosmetics; and healthcare
Roots	Agriculture fields	Fiber, antioxidants, carotenoids, and vitamins	Antibacterial compounds, dyes, pigments, and cosmetics
Peels	Fruits, vegetables and food industry	Carbohydrates, fiber, proteins, and phytochemicals, phenolic compounds, and antioxidant compounds	Antimicrobial and antibiotic compounds, dyes, pigments, and healthcare
Tops	Fruits, vegetables and food industry	Fiber, minerals, vitamins, and antioxidants	Biofuel, skin care, healthcare, and finishes
Skins	Food industry and home applications	Fiber, minerals, vitamins, and antioxidants	Biodegradable and antimicrobial

(*Continued*)

Table 11.1 Different types of agro-industrial waste, their composition, and applications. (*Continued*)

Residue	Source	Composition	Applications
Shells	Fruits, vegetables, and food industry	Polyphenols, epicatechin, and procyanidin B2	Agriculture, medicine, chemical production, construction, environmental protection, cosmetic industry, and food and feed industry
Cooking oil waste	Edible oil waste	Lipids and quercetin-3-O-glucoside	Composting, soaps, personal care, lubricants
De-oiled cake	Vegetable oil process industry	High phenolic and fiber content	Cattle/poultry feeds, waste water treatment, antifungal, antibacterial, and high-performance composites
Vegetable oil waste	Food industry	Free fatty acid	Biofuel, lubricants, skin care, soaps, and paint remover
Feathers	Poultry farm	Keratin protein	Nitrogen fertilizer, pillow stuffing, diapers, insulation, upholstery padding, plastics, and decorative purpose
Egg waste	Food industry and bakery	Minerals and calcium	Biodiesel, Bone repair, tissue generation, and nanoparticles
Meat waste	Slaughter house	Nutrient, proteins, and carbohydrate	Feed supplements

a variety of purposes, many of which are site-specific. Agricultural residues are used differently depending on how and where they are harvested. Apart from fuel, agro-residues can also be utilized as fodder, fertilizer, fiber, and feedstock [29]. Coconut coir dust is used as a fertilizer and soil conditioner, and straw is utilized as a mushroom growing substrate. The utilization of coconut husk as both a growing medium for orchids and as a material for packaging purposes has been observed [30]. Rice husks offer many kinds of applications. It serves as a source of energy. The produced ash is utilized by the steel industry as a carbonaceous material and serves as an insulating agent in feedstock. Rice straw serves as animal bedding through the utilization of its fibers, which are then incorporated into compost for fertilization purposes. Additionally, agricultural waste can be employed as a feedstock for the creation of biogas, a kind of fuel, whereas the resulting sludge can be used as fertilizer [31]. Nevertheless, the application of such agricultural leftovers in the textile industry is now constrained or relatively obscure. In order to facilitate the extraction and utilization of cellulose-rich fibers from agro-residues for textile production, it is imperative to develop and provide sophisticated technological solutions within this technologically advanced industry.

The agricultural industry generates millions of tons of agricultural waste annually, which is a major concern for environmental preservation. It is quite difficult to manage solid waste effectively and utilize these residues in a useful way. The textile industry is recognizing a comparable challenge. The textile sector is primarily reliant on cotton, accounting for 54% of global fiber use. Cotton is one of the most commercially and environmentally costly commercial commodities to grow. The cultivation of cotton is the world's second-largest agricultural pesticide user, relying on carcinogenic pesticides like cyanide, dicofol, naled, and propargite [32]. Cotton continues to represent 60% of global fiber output, notwithstanding the notable advancements made by man-made fibers in advancing onto cotton's market share in recent years. Cotton demand is increasing rapidly due to the population expansion. The current situation becomes more complex and confusing due to the increasing prices associated with the production of cotton and other cellulosic fibers derived from natural sources, such as jute and linen. It is essential to seek alternative sources of cellulose. The search for alternate cellulose resources is crucial. To address this critical issue, there is a need for environmentally benign and cost-effective alternatives to conventional cellulosic fibers. Numerous studies have been undertaken to investigate the progression of fibers derived from agricultural by-products, including maize stalks, wheat straw, banana leaves, sugarcane straw, hop stem, and similar materials [33–35]. Cellulosic fibers consist primarily of cellulose, pentosan, and lignin as their principal constituents. For a fiber to be suitable for use in textiles, it is necessary for it to possess a substantial amount of cellulose and a minimal amount of lignin. The presence of cellulose in agro-residues is observed in a composite arrangement, wherein cellulose is tightly woven within layers of hemicellulose and lignin. Lignin is classified as an aromatic polymer, whereas cellulose and hemicellulose are categorized as carbohydrate polymers. Nevertheless, the extraction of natural fibers containing pure cellulose necessitates the processing of agro-residues to exclude a significant amount of hemicellulose and lignin. Hemicellulose and lignin are two polymers that get in the way of using cellulose from agro-residues. In addition to the availability of extracted cellulose, the following factors influence the development and use of cellulosic fibers in the textile industry:

- ✓ Spinnability
- ✓ Availability
- ✓ Cost or economy of production
- ✓ Desirable properties

The molecular structure of the agro-residues varies based on the crop type, crop-specific environmental circumstances under which they are cultivated. Furthermore, the duration of fiber growth, biological classification of fibers, height of the stalk, environmental circumstances, rain, and various other elements also exert an influence on the chemical and physical characteristics of numerous natural fibers. Agro-residues exhibit a high concentration of cellulose, making them a viable candidate for utilization in the textile sector. Particularly, they show major potential as a source of natural or regenerated cellulosic fibers. Nevertheless, the full extent of its capabilities remains largely unexplored due to a lack of comprehensive research and limited technological progress [36].

11.4 Extraction and Formulation of Natural Compounds

There are various methods for obtaining organic compounds from agriculture. During the extraction process, the procedure of transferring a solute from one phase to another is employed with the objective of isolating it from unprocessed materials or impurities. Extraction can be employed as a technique to isolate a solute from a reactive solvent that is difficult to

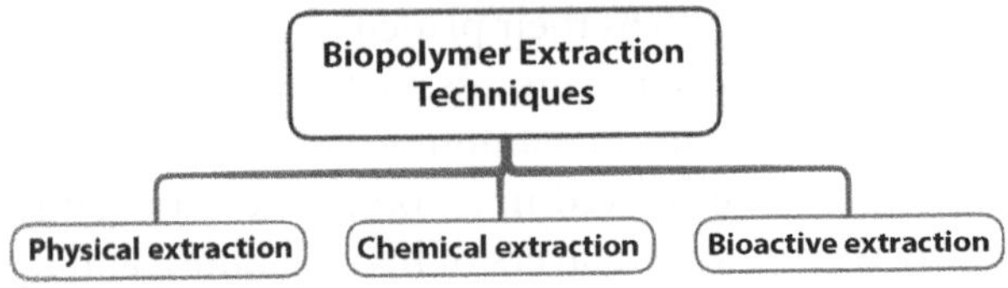

Figure 11.4 Classification of biopolymeric extraction.

remove through evaporation, particularly in situations when the solvent has a high boiling point. Figure 11.4 illustrates three distinct classifications of biopolymeric extraction [37].

11.4.1 Conventional Methods

11.4.1.1 Physical Method

Physical extraction procedures are often used to prevent the usage of chemicals. These techniques are used to increase efficiency and yield. The Soxhlet extraction approach is the most used conventional method for heating the solvent (ethanol) to boiling point (>78°C) (Figure 11.5). The use of the solvent reflux and syphon idea in Soxhlet extraction, which constantly removes the solid matter using pure solvent, achieves high efficiency and solvent conservation.

11.4.1.2 Pressurized Hot Water Extraction

Pressurized hot water extraction (PHWE) is an extraction technique that use liquid water as the extractant (extraction solvent) at temperatures higher than the ambient boiling point of water (100°C/273 K, 0.1 MPa) but lower than the critical point of water (374°C/647 K, 22.1 MPa) [38]. This method is used to recover more phenolic chemicals (Figure 11.6).

11.4.1.3 Microwave Extraction

Microwave-assisted extraction (MAE) is a method that use microwave energy to induce heating of solvents in direct contact with a sample, thereby facilitating the extraction of analytes from the sample matrix and their subsequent transfer into the solvent (Figure 11.7). The main benefit of MAE is that it can quickly heat the sample fluid mixture. Microwaves induce thermal energy within the extraction medium by interacting with its polar molecules, so elevating the temperature and augmenting the

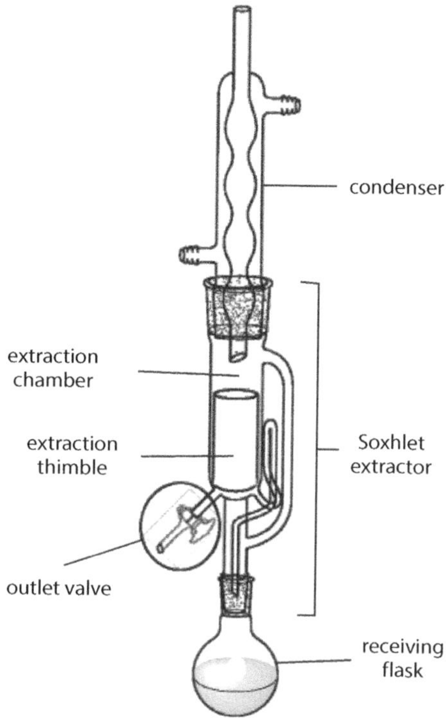

Figure 11.5 Soxhlet apparatus.

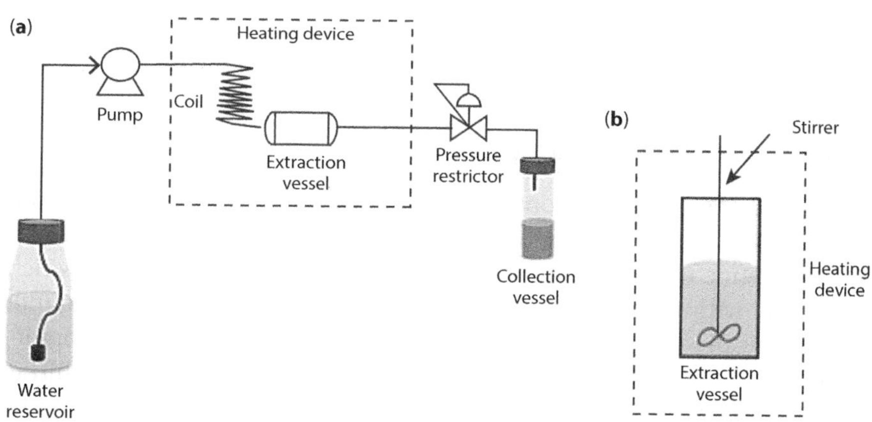

Figure 11.6 Pressurize hot water extraction technique.

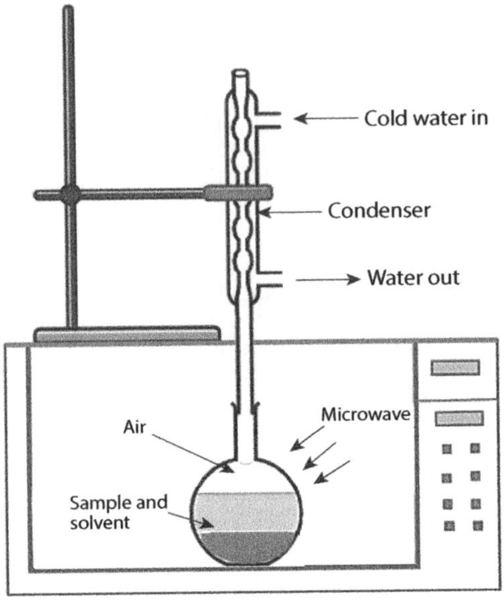

Figure 11.7 Solvent-free microwave extraction.

internal pressure of the solid material in the context of microwave-assisted extraction (MAE). This methodology has been employed for the purpose of extracting pigments from organic substances. Apart from its conventional use, microwaves have demonstrated utility in the extraction of water-soluble polysaccharides (WSPs) from several fruits.

The utilization of microwave energy in the oil extraction process has several benefits, such as enhanced heating efficiency, accelerated energy transfer, improved responsiveness to process heating management, reduced start-up time, heightened production output, and the elimination of certain process steps [39].

11.4.1.4 Ultrasound Extraction

Ultrasonic-assisted extraction (UAE) extracts compounds of interest from various plant matrices using ultrasonic energy and solvents (Figure 11.8). The frequency ranges of ultrasound encompass power ultrasound, which often falls within the range of 20–100 kHz, and diagnostic ultrasound, which typically falls within the range of 1–10 MHz. There are two significant factors that contribute to the increased efficacy of utilizing ultrasonic waves, including cell rupture, and there is an observed escalation in the velocity of mass transfer. Ultrasound is a fast and efficient extraction

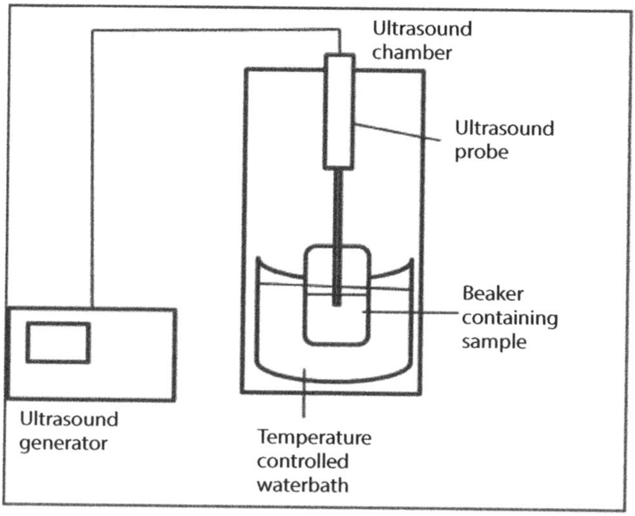

Figure 11.8 Ultrasound extraction technique.

method because it causes solvents to flow swiftly, which speeds up mass transfer and the extraction process. Many researchers have studied into methods of extracting colors from natural molecules [40]. The utilization of microwave-assisted extraction for the retrieval of WSPs from fruits also presents advantageous outcomes. The utilization of sonication as an extraction technique is also prevalent. A yield of 0.021 w/w can be attained when the substance is in powdered form. The schematic representation of the ultrasonic extraction technique is depicted in Figure 11.8. The advantages of ultrasound-assisted extraction—low energy consumption, shorter extraction times, less active component breakdown, and greater extraction yields—have made it more enticing than traditional extraction approaches [41].

11.4.2 Chemical Methods

Organic solvents are used in many processes, extractions, separations, and other operations in the chemical process industries (CPIs). Chemical procedures are used to extract the majority of bioactive substances due to their long history of dependability and relative ease of execution. Acetone-water combinations have proven to be good antioxidant extraction solvents. Non-polar organic solvents, such as hexane or dichloromethane, are commonly employed for the extraction of lipophilic substances. Polar solvents, including acetone, methanol, and ethanol, are commonly employed for the purpose of extracting lignans and other hydrophilic constituents. Polar solvents, such as water, are effective in the recovery of polar molecules, including lignan glycosides.

The organic solvents employed in the separation of bioactive compounds can be classified into two distinct categories, namely polar and non-polar solvents. Cyclohexane, hexane, toluene, benzene, ether, chloroform, and ethyl acetate are instances of solvents that exhibit non-polar characteristics. Polar solvents cover a range of substances, such as acetone, acetonitrile, butanol, propanol, ethanol, and methane [42]. The extraction efficiency is determined by the following:

- Solvent property
- Energy input
- Agitation to improve the chemical solubility
- Efficiency of mass transfer

11.4.2.1 Super Critical Carbon Dioxide Extraction

Super critical carbon dioxide (SC-CO_2) is one of the most efficient ways for obtaining high-yielding flavonoids. One of the environmentally friendly extraction methods is supercritical carbon dioxide extraction [43]. Figure 11.9 illustrates the supercritical extraction process. The chamber containing the plant material is first filled in this process of extraction, and, then, pressurized carbon dioxide is injected into the chamber. Despite its gaseous state, pressurized carbon dioxide displays liquid-like properties. Carbon dioxide acts

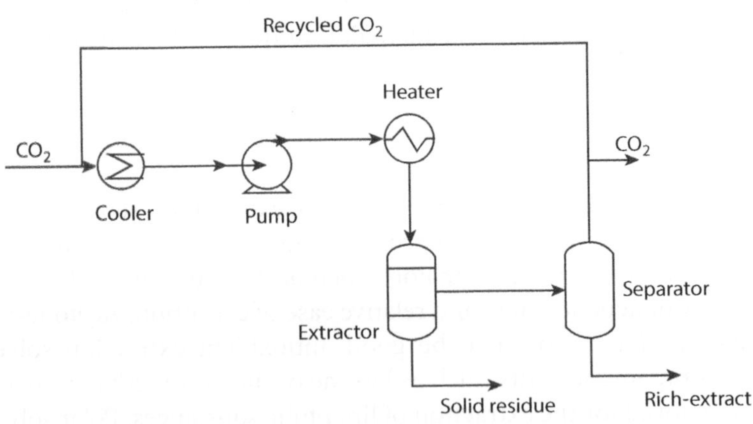

Figure 11.9 Supercritical extraction process.

as a lipophilic solvent in the supercritical state, making it possible to extract most nonpolar solutes from it. The carbon dioxide may be extracted from the extract quickly and easily. In the supercritical extraction procedure, both the fluid's temperature and pressure are higher than the critical point. In contrast to organic solvent extraction, supercritical extraction leaves no solvent residue. Supercritical CO_2 is a readily available, low-cost, non-toxic, and non-explosive solvent that allows gentle product treatment at moderate temperatures (100°C) as well as easy separation of solvent and extract. The only disadvantage of this approach is the high pressure (>3,800 psi) [43].

11.4.3 Enzyme-Assisted Techniques

The utilization of enzymes for the extraction of bioactive chemicals from plants presents a viable alternative to conventional solvent-based extraction methods. Enzymes serve as very efficient catalysts in the extraction, modification, and synthesis of complex bioactive molecules derived from natural sources. The utilization of enzymes in the extraction procedure is grounded on their intrinsic capacity to facilitate chemical reactions with notable precision and regioselectivity, alongside their capability to function in aqueous solutions at comparatively mild processing circumstances. Enzyme-assisted extraction techniques are popular because of their compatibility with the environment. Few enzyme applications represent a quantifiable feature of industrial enzymatic processing. As the food and pharmaceutical industries struggle to find cleaner ways to extract innovative chemicals, this approach holds the potential of greener chemistry. The use of enzymes. Laccase is utilized in the process of pulp and paper bleaching, protease and lipase are employed in leather production, lipase is utilized in the formulation of skincare products, and phospholipase is employed in the degumming process of soybean oil. Pectinases, cellulases, and hemicelluloses are enzymatic agents that find application in the digestion of juice and the clarification of beer. The utilization of enzymes prior to solvent extraction has been found to decrease the quantity of solvent needed for extraction or improve the yield of extractable chemicals, rendering this application highly appealing [44, 45].

11.5 Functional Properties of Agro-Industrial Waste-Based Formulations

The development of value-added products, such as the utilization of agricultural waste for pigment production and extraction of bioactive compounds for functional application is an effective way of waste management and pollution control, and it is currently a preferred technology.

11.5.1 Antimicrobial Properties

The use of novel antimicrobial agents, chemicals, natural sources, and cutting-edge technology is propelling the continual development of antimicrobial fabrics. Antimicrobial fabrics are in high demand due to cleanliness and health problems in a variety of industries, including medicine and athletic gear [46]. Natural fibers are more typically damaged by bacteria than synthetic fibers. Protein-derived fibers provide a source of food for both mould and moths. Microorganisms in cellulosic fibers require humidity, dust, and dirt to grow. Synthetic fibers are particularly water repellent and antibacterial. Microorganisms cause body odor and diseases to damage or stain textiles [47]. In both bio state and biocide applications, antimicrobial compounds are used. Biostatic substances slow the growth of an organism without killing it. As long as the bacterium is exposed to biostatic substances, biostatic fabric, or biostatic fibers, it will be unable to reproduce (Figure 11.10). In contrast, if the bacterium is not exposed to the biostatic substance, then it will be able to grow. A biocide agent is any substance that can harm living organisms. Biocides suited for textiles should have certain fundamental features, such as being effective against a broad spectrum of bacteria at low concentrations and being non-hazardous to humans and the environment [48].

Natural bioactive antimicrobial compounds are required to produce non-toxic and environmentally friendly fabrics [49]. Antimicrobial compounds derived from various plant sources, such as aloe vera, tea tree, eucalyptus oil (EO), neem, grapefruit seed, and tulsi leaf extracts, encompass a range of chemical constituents including phenolics and polyphenols (such as simple phenols, phenolic acids, quinones, flavonoids, flavones, flavonols, tannins, and coumarins), terpenoids, essential oils, alkaloids, and lectin. Tannins, alkaloids, and quinones are bioactive constituents that play a crucial role in the composition of this substance [49, 50]. It possesses antibacterial properties attributed to the presence of azadirachtin, salannin, and meliantriol, which effectively impede insect growth and function as antifeedants. The inherent characteristics of polyester, including its semi-crystalline and highly dense structure, lack of active polar groups, and low surface energy, pose challenges in achieving improved and long-lasting antibacterial properties. The bactericidal activity of neem oil, bark, and seed extract has been investigated on fabrics composed of 100% cotton and polyester/cotton blend. The antibacterial properties of neem oil

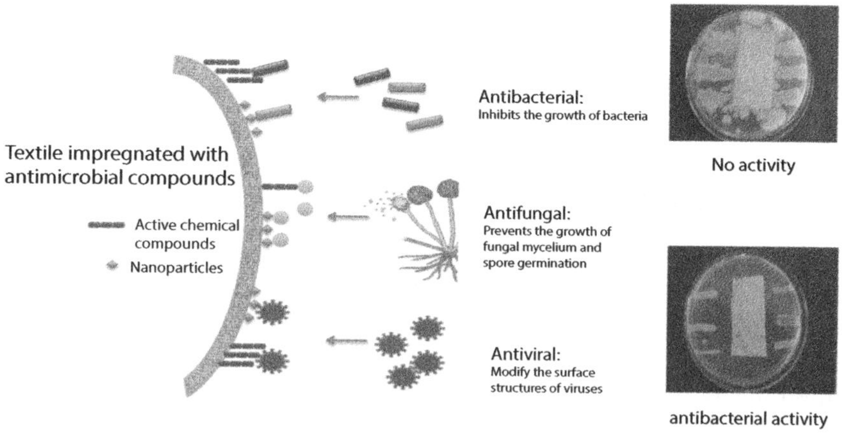

Figure 11.10 Biocidal and biostatic activity [50, 51].

combined with clove, tulsi, and karanga on cotton fabrics have been investigated by some researchers [51–53]. Surface modification can improve the durability of antimicrobial finish [54, 55]. High concentrations of antibacterial neem components in the polyester structure increased antimicrobial efficiency while maintaining crystallinity and tensile strength. Aloe vera is another bioactive component used in cosmetics and skin care. Several investigations have demonstrated that the antibacterial efficacy of aloe vera gel against both Gram-positive and Gram-negative bacteria is enhanced with increasing concentration. Aloe vera exhibits antimicrobial properties, effectively inhibiting microbiological growth on treated cotton garments. The study of naturally occurring bioactive compound helps for the purpose of treating and preventing various diseases and disorders in humans. These compounds exhibit interactions with many biological components, particularly DNA and proteins, in a manner that effectively produces the intended effects. Consequently, these interactions can be utilized in the development of natural medicinal treatments. In recent years, there has been increasing awareness of the potential of fruit and vegetable waste as a viable resource to produce bioactive chemicals, with a particular emphasis on phenolic compounds. The tomato rinds possess a significant concentration of bioactive phytochemicals, including carotenes, terpenes, sterols, tocopherols, and polyphenols. These compounds have been demonstrated to have significant antibacterial and antioxidant properties. Hence, the extracted value-added residues derived from these waste materials possess the potential to serve as natural bioactive agents for the advancement of functional food products or as additives to enhance the longevity of food items. Hence, researchers investigated on the chemical composition of exhausted coffee waste generated in a soluble coffee manufacturing facility. Their findings revealed that the exhausted coffee waste contains around 6% total polyphenols and 4% tannins [56].

11.5.2 Aroma Properties

There is a growing trend in utilizing plant-based methods to mitigate the emission of unpleasant odors generated by industrial sources. Indeed, the act of deodorizing has become a deeply entrenched social demand. As a result, textile substrates, specifically cotton, silk, and wool, have been treated with natural agents obtained from various plant extracts, including Gardenia, coffee sludge, coffee pod, and pomegranate. The resulting fabrics had excellent deodorizing qualities. In the instance of pomegranate-treated fabrics, it has been estimated to be around 99%. The deodorizing capacities of natural textiles can be significantly enhanced through the utilization of extracts derived from many plant species, such as the Amur cork tree (Dryopteris crassirhizoma), Chrysanthemum boreale, and Artemisia. Numerous biopolymers, such as Aloe vera, tea tree oil, and neem oil, have been the subject of investigation due to their capacity to absorb odors and their potential utilization in the context of wound dressings. Many compounds, including cyclodextrin, are being studied for their ability to complex smell molecules found in human sweat. The application of polycarboxylic acids, such as citric acid, in the cross-linking process of cotton fabric with smells derived from rosemary, jasmine, lavender, and vanillin has been found to enhance the longevity of fragrance retention in these materials. The application of microencapsulation technology has demonstrated promising outcomes in enhancing the efficacy and longevity of fragrances when applied to fabrics. Many companies have used this technique to create new fabrics that keep the original aroma of herbal goods after many wash cycles [57, 58].

11.5.3 Dyeing Properties

Dyes contain pigment molecules that are widely employed as colorants, particularly in the textile sector. They can be classified into two main categories: natural dyes and synthetic dyes. Today, most dyes employed are of synthetic origin, derived from aromatic chemicals. In industrial applications, synthetic dyes are predominantly utilized due to their inherent resistance to various environmental factors such as light, water, dilute acids, alkalis, and organic solvents commonly employed in dry cleaning processes. These characteristics contribute to the enhanced durability and increased functionality of dyes. A significant number of synthetic dyes have been identified as having hazardous properties, including carcinogenicity, posing a substantial risk to both public health and the environment. Consequently, there is an immediate demand for environmentally sustainable natural dyes that provide medicinal properties when applied to textile materials. Nevertheless, the variety of colors offered by these dyes is restricted, their resistance to fading is inadequate, and their efficacy is limited to cellulosic materials. The utilization of synthetic dyes in many industries is influenced by their characteristics such as color fastness, widespread accessibility, and affordability. Certain natural hues have been found to possess significant efficacy in the mitigation of germs and blood-borne infections. Natural dyes, as well as natural materials, are skin-friendly. These can boost the aura of the human body while also inhibiting bacterial growth caused by sweat and other factors. Natural dyes produce a lighter color and are less colorfast. However, these issues can be solved by utilizing substances known as mordents. Mordents are metal salts that create a bond between the fabric and the dye. Metal ions of mordents operate as electron acceptors for electron donors to establish coordination bonds with the dye molecule, rendering them insoluble in water. Since ancient times, various natural dyes have been utilized. Nature offers a wealth of plants that offers their color for dyeing. Natural dyes are becoming more popular due to their low environmental and health risks, as well as their biodegradability. Industrial by-products may provide potential new sources of natural dyes for textile coloring. Consequently, source mitigation has emerged as a new strategy for waste reduction. As a result of these legislation, several goods have been banned, and new policies on synthetic colorants have been introduced. Initially, research focused on developing wastewater treatment technology. These technologies enabled the dye manufacturing and textile dying sectors to reduce waste. Many issues have been raised regarding the use and ecotoxicity of azo dyes. Natural colors are typically derived from plant, mineral, or animal sources. The majority of the dyes employed have various effects such as diuretic, digestive stimulant, hepatoprotective, antipyretic and antiperiodic, eye infection, and antibacterial. Nowadays, consumers desire hygienic products at any cost. Textiles with antimicrobial property are essential to everyone, which is why research into antimicrobial properties is expanding day by day. Some natural colors with antibacterial properties have been tested and combined with some natural antimicrobial compounds to boost their efficiency. Every color is composed of a distinct component that possesses the ability to eradicate or impede the proliferation of hazardous germs. Fabrics composed of cotton, silk, and wool, which have been dyed using pomegranate extract and other

Table 11.2 Various natural extract and their applications.

Common name	Chemical composition	Botanical name	Part used	Dye color	Medicinal properties
Annatto 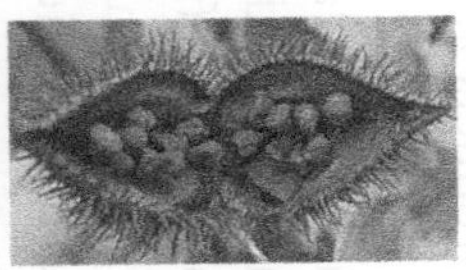	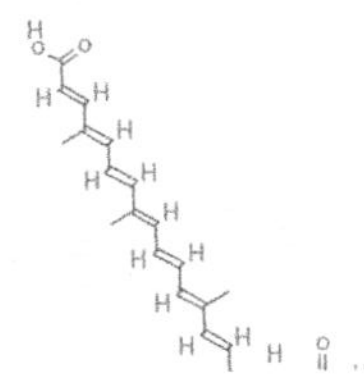 $C_{24}H_{28}O_4$	Bixa orellena	Leaves and seeds	Red orange	Antimicrobial, diuretic, digestive stimulant, hepatoprotective, antipyretic, and antiperiodic
Cutch (Khair) 	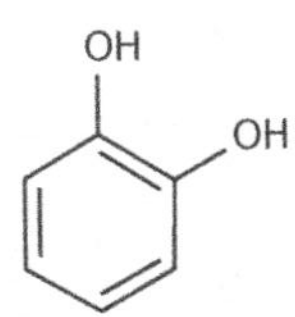 $C_6H_4(OH)_2$	Senegalia catechu	Heartwood extract	Reddish brown	Eyes blood shot, skin diseases, antidermatitis, antiasthmatic, and antibacterial

(*Continued*)

Table 11.2 Various natural extract and their applications. (*Continued*)

Common name	Chemical composition	Botanical name	Part used	Dye color	Medicinal properties
Celandines 	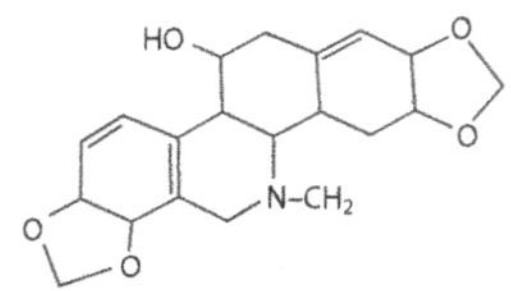	Chelidonium majus	Roots	Yellow orange	Analgesic, cholagogic, antimicrobial, oncostatic, and central nervous system sedative
Henna 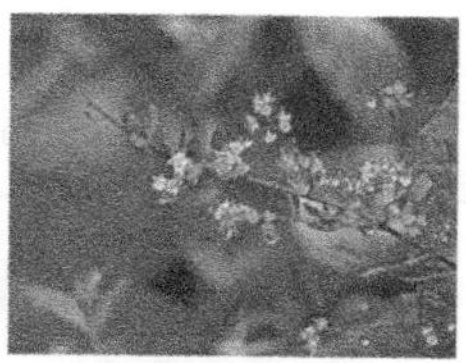	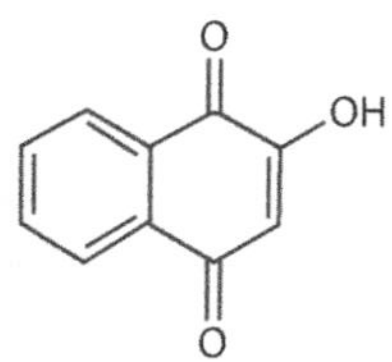 $C_{10}H_6O_3$	Lawsonia inermis	Leaves	Gold	Antidiarrheal, antidysenteric, astringent, Emmenagogue liver tonic, and antifungal.
Logwood 	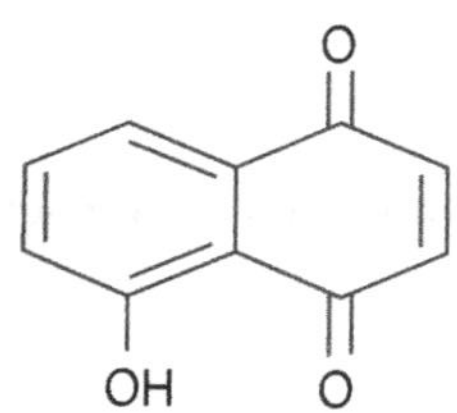	Haematoxylum Campechianum	Bark and leaves	Purple, reddish, yellow, and brownish	Antimicrobial

(*Continued*)

Table 11.2 Various natural extract and their applications. (*Continued*)

Common name	Chemical composition	Botanical name	Part used	Dye color	Medicinal properties
Madder 	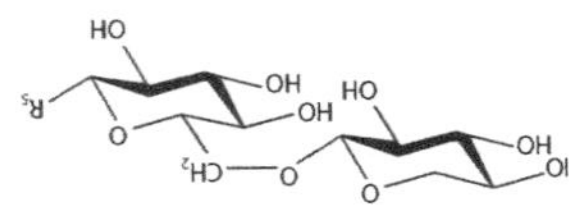	Rubia tinctora	Whole plant	Orange red	Treatment of kidney, bladder stones, and antibacterial
Pomegranate 	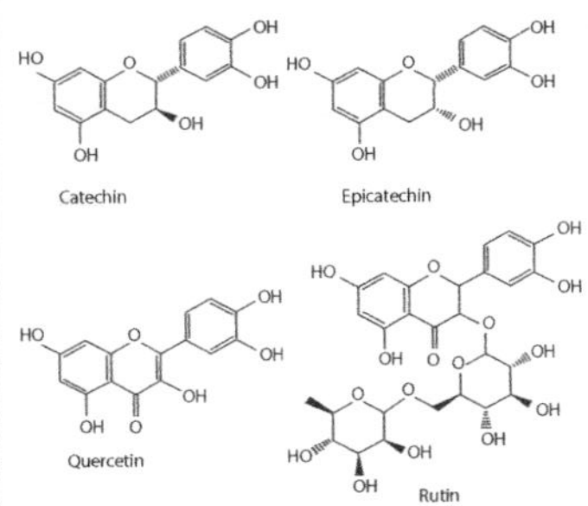	Punica granatum	Fruit	Red and purple color	Antibacterial, antiviral, astringent, cardiac, demulcent, Emmenagogue, refrigerant; stomachic, and vermifuge

(*Continued*)

Table 11.2 Various natural extract and their applications. (*Continued*)

Common name	**Chemical composition**	**Botanical name**	**Part used**	**Dye color**	**Medicinal properties**
Red Onion Peels 	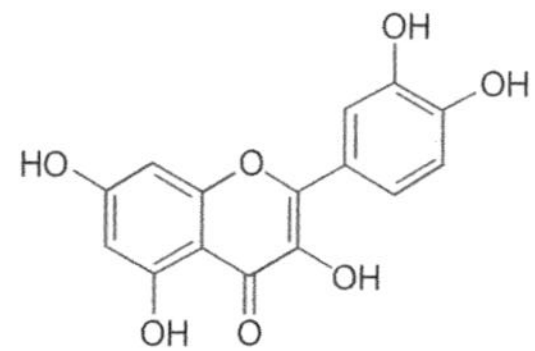	Allium Cepa	Skin	Yellow Orange	Reduce risk of cancer, colorectal, stomach, antifungal, and antibacterial
Tea 	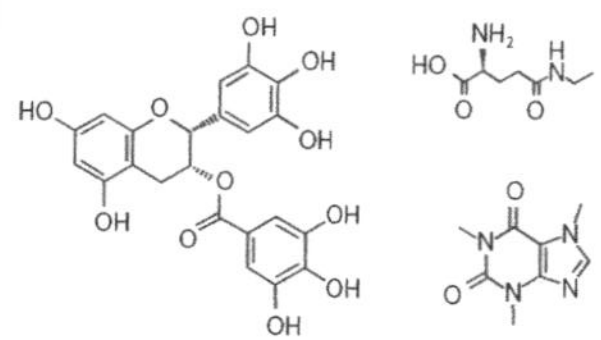	Camelia sinensis	Leaves	Beige	Antibacterial
Turmeric 	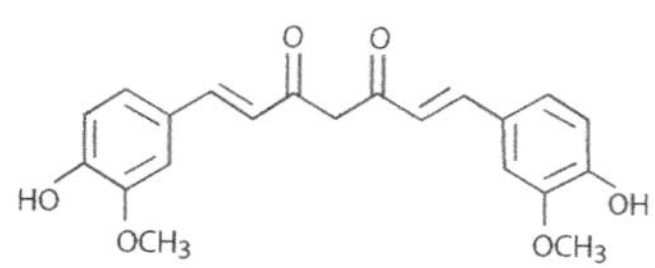 $C_{21}H_{20}O_6$	Curcumin Curcuma longa	Root	Yellow	Anti-oxidant, anti-inflammatory, antiplatelet, anti-cancer, anti-viral, anti-fungal, anti-bacterial effects

(*Continued*)

Table 11.2 Various natural extract and their applications. (*Continued*)

Common name	Chemical composition	Botanical name	Part used	Dye color	Medicinal properties
Tomato 	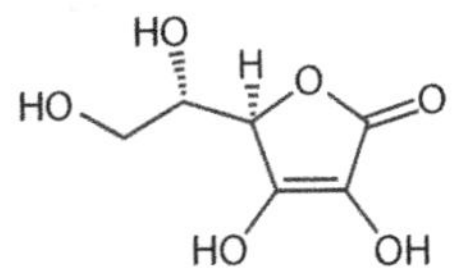 $C_{40}H_{56}$	Solanum lycopersicum.	Fruit	Red	Anti-bacterial, anti-fungal, anti-mutagenic prostate cancer, arteriosclerosis, and diabetes
Saffron 	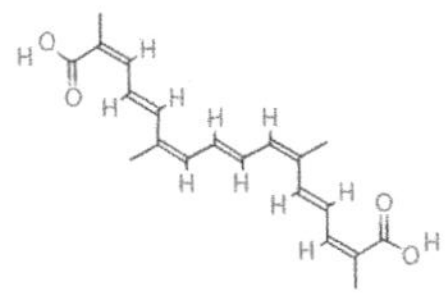 $C_{20}H_{24}O_4$	Crocus sativus L.	Flower	Yellow red	Used as sedative and Emmenagogue antimicrobial
Walnut 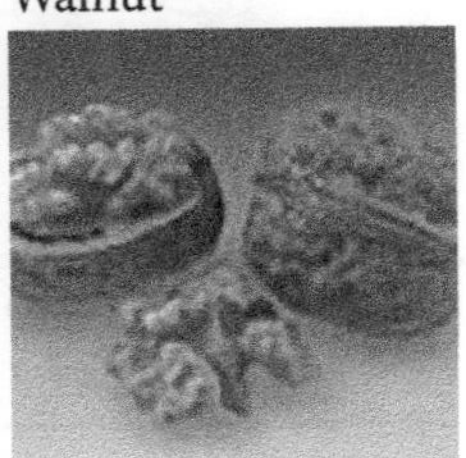	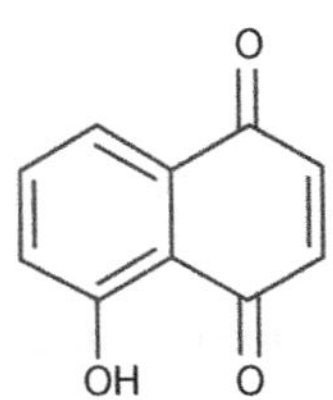	Juglans regia	Husk	Shell	Antimicrobial

natural colorants, have exhibited antibacterial properties against *S. aureus* and *K. pneumonia*. Pomegranate's antibacterial effect has been linked to the presence of phenolic and flavonoid chemicals. Similarly, other agriculture products are also used [59]. Table 11.2 discusses various natural products used for dyeing and other medicinal applications.

11.5.4 Flame Retardant Properties

Flame retardants are being recognized as innovative substances utilized to reduce property and personal harm caused by fire occurrences. There exists a wide array of textile items that incorporate these chemicals. The worldwide investigation into flame-retardant textiles is currently in progress. Numerous studies have been reported. Natural products, particularly cyclodextrin, are now frequently employed in flame retardant fabrics. In another study, the flame retardant is composed of phytic acid and chitosan, which are both made from agricultural and food waste and are high in carbon, phosphorus, nitrogen, and other elements. These naturally occurring polymers function as a physical barrier that restricts the transfer of heat, fuel, and oxygen between the flame and the textile substrate, therefore providing an environmentally friendly approach [60].

11.5.5 Mosquito Repellent Properties

The mosquito repellent (insecticide) textiles are a category of protective textiles designed to mitigate the harmful effects caused by insects. These textiles serve as a protective layer, shielding persons from mosquito and other insect bites, thereby offering defence against diseases such as malaria, dengue fever (DF), Nile fever, hemorrhagic fever induced by dengue (DHF), chicken gunia, and filariasis. Besides contributing to potential health risks, insect repellents have the capacity to mitigate the risk of damage to valuable textile items, including carpets, clothing, upholstered furniture, blankets, and precious antiques, by inhibiting insect attack. There is currently a growing consumer preference for insect repellents that possess attributes of safety, user-friendliness, and environmental sustainability. Natural products, particularly plant-based compounds, are gaining popularity for this purpose. Plant-based repellents are generally considered as safer than long-established synthetic substitutes since they consist of natural ingredients.

The use of pyrethrum has recently been suggested as a prospective substitute for the impregnation of mosquito nets and fabrics in regions where resistance to pyrethroids has emerged as a concern. Pyrethrum is derived from the desiccated blossoms of the Chrysanthemum Cineraria folium flower and exhibits insecticidal properties through the induction of persistent depolarization in the presynaptic sodium channels of insects. In recent years, several essential oils extracted from aromatic plants have been described as novel mosquito repellents. The significant limitation associated with plant-based repellents is the volatility of certain active compounds. This volatility entails that, although these compounds exhibit efficacy as repellents shortly after application, they rapidly evaporate, thereby exposing the user to potential vulnerability. In order to address this limitation, the textile industry has implemented microencapsulation technology. The use of microcapsules encapsulating citronella essential oil has been found to offer enhanced and prolonged efficacy in pest control, surpassing that of fabrics treated with an ethanol solution of the essential oil. This approach has demonstrated a pest protection rate of over 90% for a duration of 3 weeks. In addition, naturally occurring chemicals abundant in quinone, flavonoid, and indigoid, which are produced from diverse plant extracts including madder, walnut, chestnut, logwood, and indigoid, have exhibited encouraging outcomes in the suppression of insect proliferation that poses significant harm to wool-based textiles. Most of the natural dyes employed textiles are also responsible for insect repellent properties [61].

11.5.6 Surfactants

Surfactants are chemical substances with different solubility properties that lower the surface tension at the interfaces of solid-liquid or liquid-liquid systems. These molecules have functional groups that are both hydrophobic and hydrophilic. Biosurfactants have the capacity to reduce surface tension, stabilize emulsions, and enhance foaming. These materials are environmentally friendly and renewable in most cases. Biosurfactants have acquired popularity over the years due to their diversity, sustainability and eco-friendliness, mass-production capability, specificity, performance under extreme conditions, and imminent use in environmental protection. These compounds have the potential to be used in a variety of industries, including agriculture, medicine, humectants, food preservatives, and textiles. Biosurfactant synthesis is currently prohibitively expensive due to the usage of synthetic culture medium [62].

Consequently, there has been an increased focus on utilizing cost-efficient agro-industrial waste as substrates, including by-products derived from sugar and waste generated from cooking oil. With the aid of contemporary technology, these can be used as substrates for making biosurfactants on a massive scale. Glycolipids are carbohydrates linked to

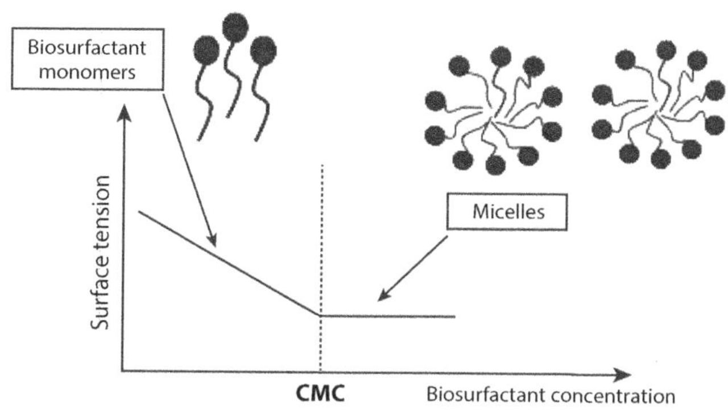

Figure 11.11 Relationship between biosurfactant concentration, surface tension, and formation of micelles.

elongated aliphatic acids, are the most prevalent biosurfactants derived from microbes. Besides glycolipids, there are other biosurfactants derived from microorganisms that possess higher levels of intricacy, such as lipopeptides, lipoproteins, and heteropolysaccharides. In addition, biosurfactants cover a range of compounds including neutral lipids, fatty acids, phospholipids, and polymeric biosurfactants. The biosurfactant activity is dependent upon the concentration of surface-active molecules until the critical micelle concentration (CMC) is reached. Biosurfactant molecules can form micelles, bilayers, and vesicles at concentrations above the CMC, as depicted in Figure 11.11. Biosurfactants possess the ability to diminish surface and interfacial tension, hence enhancing the solubility and bioavailability of hydrophobic organic compounds through the formation of micelles. The activity of biosurfactants is also affected by environmental conditions including as pH, salinity, and temperature.

11.5.7 Green Synthesis of Nanofibers and Nanoparticles from Agro-Industrial Waste

Plants are great biological sources for nanostructures because they are rich in bioactive phytochemicals such as alkaloids, glycosides, flavonoids, tannins, terpenoids, resins, carbohydrates, vitamins, phytoestrogens, and so on. Since ancient times, the bioactivity and uses of many plant species have been known. In recent decades, scientific research has proven this. Plants are bio-renewable, sustainable, and diverse resources and platforms for creating beneficial nanostructures and nanoparticles (NPs) due to their high yields, moderate reaction conditions, cleaner reaction profiles, simple work-up approaches, and use of harmless solvents such as water. Plants have several benefits as natural NP manufacturers. Plants can build a wide range of nanostructures of different sizes and shapes. Plant extracts contain bioactive phytochemicals. Phytochemicals safeguard and stabilize nanostructures developed by plants. Photosynthetic plants create sugars and lipids as main metabolites, followed by secondary metabolites. The quantity and kind of antioxidant components in the plant media, such as glycosides, flavonoids, vitamins, and phenolics as reducing agents, affect the ability of the plant extract to synthesize new molecules. Green nanotechnology aids in the elimination or reduction of polluting substances during the manufacturing of nanostructures. Biosynthesized NPs are used in medicine, healthcare, and as building blocks for future products. The most lucrative NPs are photosynthesized nanostructures, which have uses in medicine, food and feed, cosmetics, biomedical science, energy science, gene transfer, and environmental healthcare. Plant-based sources are advantageous because of their low cost, environmental friendliness, long shelf life, simplicity of access, and diversity. Metallic NPs are biologically synthesized using plant parts such as stems, fruits, roots, calluses, peel, seeds, flowers, and leaves. Metal concentrations and plant extract levels in the reaction media can be changed to alter biosynthesis reactions. In comparison to the numerous studies on metallic NPs, there have been few investigations on the extraction of nanostructures and NPs from plants. It is difficult to biosynthesize nanostructures with defined properties. There are several synthetic routes for creating nanostructures of various sizes, shapes, morphologies, and functionalities. These techniques have limitations and usually have adverse impacts because to the presence of hazardous substances on their surfaces, notably in the pharmaceutical and medical areas. Researchers demanded the development of environmentally friendly methods. Biological activities do not require the use of hazardous ingredients, expensive machinery, or harsh synthesis conditions [62].

Green nanotechnology aids in the elimination or reduction of polluting substances during the manufacturing of nanostructures. Plants are excellent sources of nanomachines because they respond to physical, chemical, and biological stimuli. The micro- and nanoscale complexity of natural plant structures mimics that of modern artificial materials and systems. Plants can create biocompatible nanostructures that can be used in the pharmaceutical and healthcare fields in a safe, cost-effective, and environmentally friendly manner. This includes the synthesis of green water-dispersible fluorescent carbon dots from Saccharum officinarum juice, physical methods for producing lignin-based NPs, and the use of fluorescent biosensors to image bacteria such as *Escherichia coli* and yeast such as *Saccharomyces cerevisiae*. A quick and efficient way of calcining the rice husks to produce spherical biogenic silica NPs (10–30 nm). Amorphous silica, according to a photoluminescence study, is an ideal source of silicon NPs for solar cells and medicine. Tea NPs are produced (100–300 nm; spherical) at pH 7 using an infusion-dialysis method. *In vitro*, spherical tea NPs could be used as cancer therapy carriers. The cellulose nano paper has been used as a filter and a flexible transparent substrate for Ag nanowires and CNTs, allowing for the construction of next-generation paper electronics. These methods minimize the pollution and toxicity by utilizing safe and natural chemical components. Plant extracts have a high concentration of bioactive chemicals such as alkaloids, flavonoids, glycosides, tannins, vitamins, and porphyrins. Biofuels made from sugarcane and jatropha may be a feasible substitute for fossil fuels due to their ability to purify air, water, and soil [63].

Phytochemicals exhibit a broad distribution within the plant, including various plant parts such as leaves, stems, flowers, roots, glands, and fruit peel, which can serve as viable sources for the synthesis of nanomaterials. The production of green nanostructures involves the utilization of many antioxidants, such as phenolic chemicals, nitrogen compounds, vitamins, reducing sugars, terpenoids, and other metabolites. The concentration of antioxidant phytochemicals is crucial for NP synthesis as well as the speed and efficiency of bio-reduction. The hydroxyl and carboxyl groups in phenolic antioxidants may act as a reducing or stabilizing agent during NP production. Nanomaterials might degrade when exposed to proteins, terpenoids, or carbohydrates. Plant components, which act as stabilizing and capping agents, can protect NP surfaces from deformation and disintegration. Numerous studies have shown that phytochemicals play an important role in phytochemical reduction as well as the size and stability of nanostructures. The reduced toxicity and higher bioactivity of nanostructures made from plants compared to those formed through nongreen synthesis favor their use in medical and healthcare applications. Bioactive plant components have antibacterial, antioxidant, antifungal, and anticancer properties on nanostructured surfaces. The reducing agents in plant-mediated NP production are water-soluble metabolites and coenzymes. The drawbacks of biosynthesizing NPs include longer reaction durations, time-consuming purification procedures, agglomeration and aggregation, and bigger NP sizes. The reaction conditions (temperature, pressure, extracting agent volume, raw material amount, pH, stirring rate, etc.) should be altered prior to starting green synthesis in order to attain any advantages of the synthesis depicted in Figure 11.12 [64].

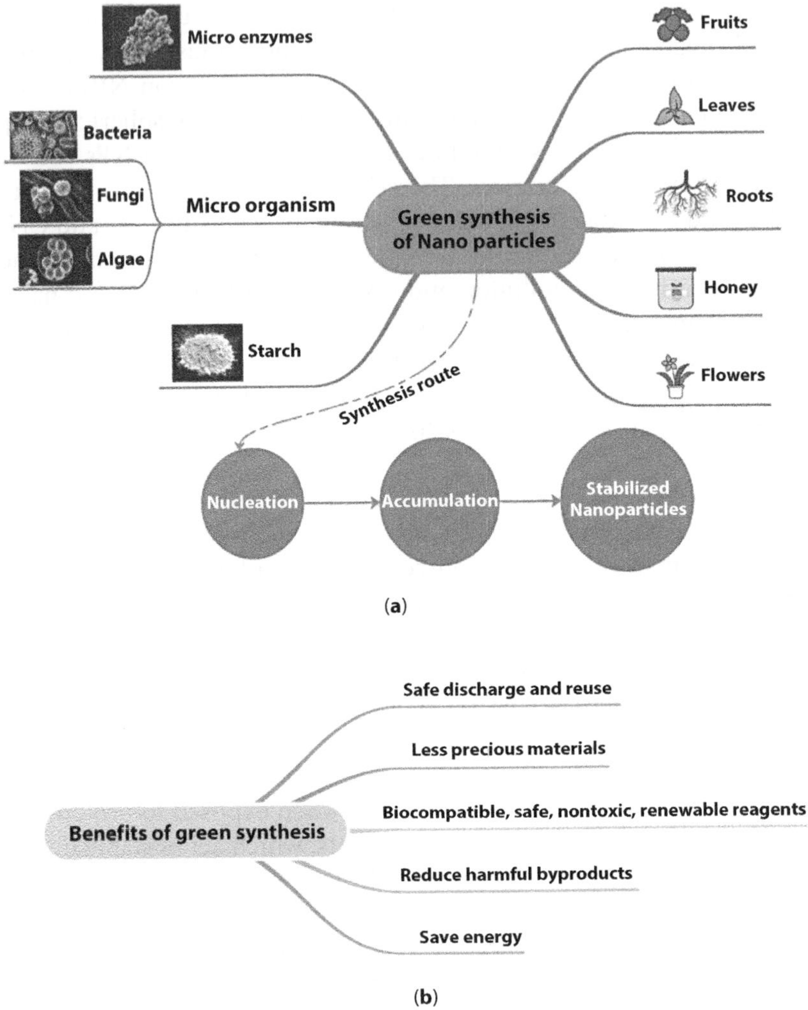

Figure 11.12 Green synthesis route, applications, and advantage.

11.6 Future and Challenges

An emphasis is being set upon the investigation and the development of materials obtained from agro-industrial waste. Finding new materials, improving the properties of existing materials and products, and developing unique solutions all necessitate basic and applied research. The development of new products from agro-industrial waste is the focus of continuing high-technology research at multiple levels. Applied research has proven to be beneficial in addressing crucial concerns within an industry. The outcome is applicable in several application areas in addition to traditional and technical textiles. The procedure is an efficient way to cope with the high cost of materials utilized in a wide range of technological applications, such as medical textiles and biodegradable polymeric systems. The following factors may have an impact on agro-industrial waste in the near future:

- Government policies
- Industrial regulations
- Developments in the management of industrial waste
- Global requirements
- Environmental concern (ISO 9000)

Rapid development in the efficient utilization of agro-industrial waste will raise the demand for innovative alternatives to chemically generated products. This field's advancement will expand employment prospects in numerous industries, including chemical, fertilizer, textile, cosmetic, medical, and pharmaceutical. Numerous studies have demonstrated the efficacy of recycling in the utilization of agro-industrial waste. Several significant application areas include the manufacture of biofuels, enzymes, citric acid, pigments, food and preservatives, and agricultural composting, among others.

11.7 Conclusion

Indeed, the utilization of waste materials not only addresses the challenges associated with waste disposal but also offers a viable solution to pollution-related issues. The global attitude toward agro-industrial waste is undergoing rapid transformation considering the imperative for environmental sustainability and conservation. Today, there has been a widespread utilization of agro-industrial residue in numerous ways, particularly in the creation of diverse value-added goods. The primary uses of recycled trash encompass a range of applications, such as the manufacture of biofuels, enzymes, organic acids, pigments, food flavorings, preservatives, bioactive compounds, biodegradable polyhydroxyalkanoates (PHA), and agricultural composting, among others. Hence, the commercialization of these value-added products necessitates additional regulatory permission and capital inputs. The utilization of agro-industrial leftovers to produce valuable chemicals has the potential to offer new avenues for researchers and mitigate existing environmental risks.

Acknowledgment

The authors would like to express their gratitude to their respective institutes for providing all facilities.

References

1. Omran, B.A. and Baek, K.-H., Valorization of agro-industrial biowaste to green nanomaterials for wastewater treatment: Approaching green chemistry and circular economy principles. *J. Environ. Manage.*, 311, 114806, 2022.
2. Manna, M.C., Rahman, M.M., Naidu, R., Sahu, A., Bhattacharjya, S., Wanjari, R.H., Patra, A.K., Chaudhari, S.K., Majumdar, K., Khanna, S.S., Bio-waste management in subtropical soils of India: future challenges and opportunities in agriculture. *Adv. Agron.*, 152, 87–148, 2018.
3. Karić, N., Maia, A.S., Teodorović, A., Atanasova, N., Langergraber, G., Crini, G., Ribeiro, A.R.L., Olić, M., Bio-waste valorisation: Agricultural wastes as biosorbents for removal of (in) organic pollutants in wastewater treatment. *Chem. Eng. J. Adv.*, 9, 100239, 2022.
4. Kumar, A., Kumari, S., Dindhoria, K., Manyapu, V., Kumar, R., Efficient Utilization and Bioprocessing of Agro-Industrial Waste, in: *Sustainable Agriculture Reviews 56: Bioconversion of Food and Agricultural Waste into Value-Added Materials*, pp. 1–37, Springer, Cham, 2022.
5. Zabaleta, I. and Rodic, L., Recovery of essential nutrients from municipal solid waste–Impact of waste management infrastructure and governance aspects. *Waste Manage. (Oxford)*, 44, 178–187, 2015.
6. Koneswaran, G. and Nierenberg, D., Global farm animal production and global warming: impacting and mitigating climate change. *Environ. Health Perspect.*, 116, 578–582, 2008.
7. Deelstra, T. and Girardet, H., others, Urban agriculture and sustainable cities, in: *Growing Cities, Growing Food. Urban Agriculture on the Policy Agenda*, N. Bakker, M. Dubbeling, S. Gündel, U. Sabel-Koshella, H. de Zeeuw (Eds.), pp. 43–66, Zentralstelle Für Ernährung Und Landwirtschaft (ZEL), Feldafing, Germany, 2000.
8. Hu, Y., Cheng, H., Tao, S., Environmental and human health challenges of industrial livestock and poultry farming in China and their mitigation. *Environ. Int.*, 107, 111–130, 2017.
9. Yu, M.-H. and Tsunoda, H., *Environmental toxicology: biological and health effects of pollutants*, CRC Press, New York, Washington D.C., 2004.
10. Dhamodharan, K., Konduru, T., Kannan, M., Malyan, S.K., Techno-economic feasibility and hurdles on agricultural waste management, in: *Emerging Trends to Approaching Zero Waste*, pp. 243–264, Elsevier, Netherlands, 2022.

11. Gupta, C., Prakash, D., Gupta, S., A biotechnological approach to microbial based perfumes and flavours. *J. Microbiol. Exp.*, 2, 34–42, 2015.
12. der Schaft, P., Approaches to production of natural flavours, in: *Flavour Development, Analysis and Perception in Food and Beverages*, pp. 235–248, Elsevier, Netherlands, 2015.
13. Zamzuri, N.A. and Abd-Aziz, S., Biovanillin from agro wastes as an alternative food flavour. *J. Sci. Food Agric.*, 93, 429–438, 2013.
14. Bonnin, E., Saulnier, L., Brunel, M., Marot, C., Lesage-Meessen, L., Asther, M., Thibault, J.-F., Release of ferulic acid from agroindustrial by-products by the cell wall-degrading enzymes produced by Aspergillus niger I-1472. *Enzyme Microb. Technol.*, 31, 1000–1005, 2002.
15. Radenkovs, V., Juhnevica-Radenkova, K., Jakovlevs, D., Zikmanis, P., Galina, D., Valdovska, A., The release of non-extractable ferulic acid from cereal by-products by enzyme-assisted hydrolysis for possible utilization in green synthesis of silver nanoparticles. *Nanomaterials*, 12, 3053, 2022.
16. Chemat, F., Abert-Vian, M., Fabiano-Tixier, A.S., Strube, J., Uhlenbrock, L., Gunjevic, V., Cravotto, G., Green extraction of natural products. Origins, current status, and future challenges. *TrAC, Trends Anal. Chem.*, 118, 248–263, 2019.
17. Sharma, A., Sharma, A., Singh, J., Sharma, P., Tomar, G.S., Singh, S., Nain, L., A biorefinery approach for the production of ferulic acid from agroresidues through ferulic acid esterase of lactic acid bacteria. *3 Biotech*, 10, 1–10, 2020.
18. Oddou, J., Stentelaire, C., Lesage-Meessen, L., Asther, M., Colonna Ceccaldi, B., Improvement of ferulic acid bioconversion into vanillin by use of high-density cultures of Pycnoporus cinnabarinus. *Appl. Microbiol. Biotechnol.*, 53, 1–6, 1999.
19. Coelho, M.C., Pereira, R.N., Rodrigues, A.S., Teixeira, J.A., Pintado, M.E., The use of emergent technologies to extract added value compounds from grape by-products. *Trends Food Sci. Technol.*, 106, 182–197, 2020.
20. Casalini, S. and Giacinti Baschetti, M., The use of essential oils in chitosan or cellulose-based materials for the production of active food packaging solutions: a review. *J. Sci. Food Agric.*, 103, 1021–1041, 2023.
21. Nargotra, P., Sharma, V., Tsai, M.-L., Hsieh, S.-L., Dong, C.-D., Wang, H.-M.D., Kuo, C.-H., Recent Advancements in the Valorization of Agro-Industrial Food Waste for the Production of Nanocellulose. *Appl. Sci.*, 13, 6159, 2023.
22. de Jong, E., Higson, A., Walsh, P., Wellisch, M., others, Bio-based chemicals value added products from biorefineries. *IEA Bioenergy, Task42 Biorefinery*, vol. 34, pp. 1–33, 2012.
23. Jagtap, A.S. and Meena, S.N., Seaweed farming: a perspective of sustainable agriculture and socio-economic development, in: *Natural Resources Conservation and Advances for Sustainability*, pp. 493–501, 2022.
24. Adhia, V., Mishra, A., Banerjee, D., Appadurai, A.N., Preethan, P., Khan, Y., de Wagenaar, D., Harmsen, P., Elbersen, B., van Eupen, M., others, Spinning

future threads: the potential of agricultural residues as textile fibre feedstock, 2021.

25. Bhuyan, N., Narzari, R., Gogoi, L., Bordoloi, N., Hiloidhari, M., Palsaniya, D.R., Deb, U., Gogoi, N., Kataki, R., Valorization of agricultural wastes for multidimensional use, in: *Current Developments in Biotechnology and Bioengineering*, pp. 41–78, Elsevier, Netherlands, 2020.
26. Shahid, M. and Mohammad, F., others, Perspectives for natural product based agents derived from industrial plants in textile applications–a review. *J. Clean Prod.*, 57, 2–18, 2013.
27. Agricultural Textiles Market Size, Share, Growth Rate And Global Forecast 2032.
28. Ribul, M., Lanot, A., Pisapia, C.T., Purnell, P., McQueen-Mason, S.J., Baurley, S., Mechanical, chemical, biological: Moving towards closed-loop bio-based recycling in a circular economy of sustainable textiles. *J. Clean Prod.*, 326, 129325, 2021.
29. Shaker, K., Nawab, Y., Jabbar, M., Bio-composites: eco-friendly substitute of glass fiber composites, in: *Handbook of Nanomaterials and Nanocomposites for Energy and Environmental Applications*, pp. 1–25, 2020.
30. Okafor, C.C., Madu, C.N., Ajaero, C.C., Ibekwe, J.C., Nzekwe, C.A., Okafor, C.C., Madu, C.N., Ajaero, C.C., Ibekwe, J.C., Nzekwe, C.A., Sustainable management of textile and clothing. *Clean Technol. Recycl.*, 1, 70–87, 2021.
31. Prabhu, S.R. and Thomas, G.V., Biological conversion of coir pith into a value-added organic resource and its application in Agri-Horticulture: Current status, prospects and perspective. *J. Plantation Crops*, 30, 1–17, 2002.
32. Dar, R.A., Parmar, M., Dar, E.A., Sani, R.K., Phutela, U.G., Biomethanation of agricultural residues: Potential, limitations and possible solutions. *Renew. Sustain. Energy Rev.*, 135, 110217, 2021.
33. Mohan, T.P. and Kanny, K., Green and sustainable advanced biopolymeric and biocomposite materials, in: *Green and Sustainable Advanced Materials, Volume 1: Processing and Characterization*, p. 67, 2018.
34. Sfiligoj Smole, M., Hribernik, S., Stana Kleinschek, K., Kreže, T., Plant fibres for textile and technical applications, in: *Advances in Agrophysical Research*, pp. 369–398, 2013.
35. Pongtratic, M., Greening the supply chain: A case analysis of Patagonia, IR/PS CSR, Case, pp. 7–22, 2007.
36. Holmatov, B., Schyns, J.F., Krol, M.S., Gerbens-Leenes, P.W., Hoekstra, A.Y., Can crop residues provide fuel for future transport? Limited global residue bioethanol potentials and large associated land, water and carbon footprints. *Renew. Sustain. Energy Rev.*, 149, 111417, 2021.
37. Li, P., Sakuragi, K., Makino, H., Extraction techniques in sustainable biofuel production: A concise review. *Fuel Process. Technol.*, 193, 295–303, 2019.
38. Wiboonsirikul, J. and Adachi, S., Extraction of functional substances from agricultural products or by-products by subcritical water treatment. *Food Sci. Technol. Res.*, 14, 319, 2008.

39. Macías-Garbett, R., Sosa-Hernández, J.E., Iqbal, H.M.N., Contreras-Esquivel, J.C., Chen, W.N., Melchor-Martínez, E.M., Parra-Saldívar, R., Combined pulsed electric field and microwave-assisted extraction as a green method for the recovery of antioxidant compounds with electroactive potential from coffee agro-waste. *Plants*, 11, 2362, 2022.
40. Selahvarzi, A., Ramezan, Y., Sanjabi, M.R., Namdar, B., Akbarmivehie, M., Mirsaeedghazi, H., Azarikia, F., Optimization of ultrasonic-assisted extraction of phenolic compounds from pomegranate and orange peels and their antioxidant activity in a functional drink. *Food Biosci.*, 49, 101918, 2022.
41. Buvaneshwaran, M., Radhakrishnan, M., Natarajan, V., Influence of ultrasound-assisted extraction techniques on the valorization of agro-based industrial organic waste–A review. *J. Food Process Eng.*, 46, e14012, 2023.
42. Marcus, Y., *Solvent mixtures: properties and selective solvation*, CRC Press, New York, Washington D.C., 2002.
43. Albarelli, J.Q., Rabelo, R.B., Santos, D.T., Beppu, M.M., Meireles, M.A.A., Effects of supercritical carbon dioxide on waste banana peels for heavy metal removal. *J. Supercrit. Fluids*, 58, 343–351, 2011.
44. Nadar, S.S., Rao, P., Rathod, V.K., Enzyme assisted extraction of biomolecules as an approach to novel extraction technology: A review. *Food Res. Int.*, 108, 309–330, 2018.
45. Bubalo, M.C., Vidović, S., Redovniković, I.R., Jokić, S., New perspective in extraction of plant biologically active compounds by green solvents. *Food Bioprod. Process.*, 109, 52–73, 2018.
46. Ran, B., Ran, L., Wang, Z., Liao, J., Li, D., Chen, K., Cai, W., Hou, J., Peng, X., Photocatalytic Antimicrobials: Principles, Design Strategies, and Applications. *Chem. Rev.*, 123, 22, 12371–12430, 2023.
47. Midha, V.K., Joshi, S., Dakuri, A., others, Surgical gown fabrics in infection control and comfort measures at hospitals. *Indian J. Fibre Text. Res. (IJFTR)*, 47, 96–104, 2022.
48. Joshi, S., Midha, V., Rajendran, S., Grapeseed Oil (Vitis vinifera L.) Treatment on Polyester-Based Fabrics to develop Antibacterial and Physiologically Comfortable Health Care and Hygiene Textiles. *J. Inst. Eng. India Ser. E*, 102, 339–351, 2021.
49. Joshi, S., Kambo, N., Dubey, S., Shukla, P., Pandey, R., Effect of onion (Allium cepa L.) peel extract-based nanoemulsion on anti-microbial and UPF properties of cotton and cotton blended fabrics. *J. Nat. Fibers*, 19, 8345–8354, 2022.
50. Joshi, S., Midha, V., Rajendran, S., Investigation of Durable Bio-polymeric Antimicrobial Finishes to Chemically Modified Textile Fabrics Using Solvent Induction System. *Tekstilec*, 64, 55–69, 2021.
51. Enawgaw, H., A compressive review on different surface finishing of cotton fabrics. *Res. J. Text. Apparel*, 06, 68, 2023.

52. Joshi, M., Ali, S.W., Purwar, R., Rajendran, S., Ecofriendly antimicrobial finishing of textiles using bioactive agents based on natural products. *Indian J. Fibre Text. Res. (IJFTR)*, 34, 295–304, 2009.
53. Joshi, S., Midha, V., Rajendran, S., Multifunctional waterproof breathable coating on polyester-based woven protective clothing for healthcare application. *Prog. Org. Coat.*, 178, 107482, 2023.
54. Joshi, S., Midha, V., Rajendran, S., Influence of trichloroacetic acid-methylene chloride on structure of polyester fabrics. *Indian J. Fibre Text. Res. (IJFTR)*, 46, 341–348, 2021.
55. Regar, M.L., Amjad, A.I., Joshi, S., Effect of Solvent Treatment on Siro and Ring Spun TFO Polyester Yarn. *Tekstilec*, 64, 1, 47–54, 2021.
56. Pujol, D., Liu, C., Gominho, J., Olivella, M.À., Fiol, N., Villaescusa, I., Pereira, H., The chemical composition of exhausted coffee waste. *Ind. Crops Prod.*, 50, 423–429, 2013.
57. Verma, M., Gahlot, N., Singh, S.S.J., Rose, N.M., UV protection and antibacterial treatment of cellulosic fibre (cotton) using chitosan and onion skin dye. *Carbohydr. Polym.*, 257, 117612, 2021.
58. Jarrige, F. and Le Roux, T., *The contamination of the earth: A history of pollutions in the industrial age*, MIT Press, England, 2020.
59. Chengaiah, B., Rao, K.M., Kumar, K.M., Alagusundaram, M., Chetty, C.M., others, Medicinal importance of natural dyes-a review. *Int. J. Pharmtech. Res.*, 2, 144–154, 2010.
60. Betts, K.S., New thinking on flame retardants, *Environ health perspect*, 116, 5, 210–213, 2008.
61. Sajib, M., Banna, B., Mia, R., Ahmed, B., Chak, R., Alam, S.S., Rasel, M.A., Islam, M.T., Mosquito repellent finishes on textile fabrics (woven & knit) by using different medicinal natural plants. *J. Text. Eng. Fashion Technol.*, 6, 164–167, 2020.
62. Gayathiri, E., Prakash, P., Karmegam, N., Varjani, S., Awasthi, M.K., Ravindran, B., Biosurfactants: potential and eco-friendly material for sustainable agriculture and environmental safety—a review. *Agronomy*, 12, 662, 2022.
63. Dubey, K.V., Charde, P.N., Meshram, S.U., Shendre, L.P., Dubey, V.S., Juwarkar, A.A., Surface-active potential of biosurfactants produced in curd whey by Pseudomonas aeruginosa strain-PP2 and Kocuria turfanesis strain-J at extreme environmental conditions. *Bioresour. Technol.*, 126, 368–374, 2012.
64. Yadav, M., Dwibedi, V., Sharma, S., George, N., Biogenic silica nanoparticles from agro-waste: Properties, mechanism of extraction and applications in environmental sustainability. *J. Environ. Chem. Eng.*, 10, 6, 108550, 2022.

12

Potential Application of Recycled Waste in Technical Textiles

Blesson Tom Mathew, Advitiya Kumar and Archana Samanta*

Department of Textile and Fibre Engineering, Indian Institute of Technology Delhi, Hauz Khas, New Delhi, India

Abstract
This chapter examines the complex issue of recycled textile waste, including its history, current state, and projected trajectory. The use of recycled waste materials in textiles is promoted as a powerful method for both environmental protection and economic sustainability. Concerns about pollutants, inconsistent quality, and consumer impression are addressed, and advanced recycling technology, the adoption of the circular economy, sustainability certifications, and waste minimization programs are some of the upcoming developments. This chapter highlights the importance of recycled textiles in reducing environmental impact while promoting cost savings and creating new jobs. All these factors work together to make recycled textile waste an essential component of the textile industry's push toward a greener, more environmentally friendly future.

***Keywords*:** Recycled textile waste, technical textile, smart textile

List of Abbreviations

CAGR	Compound annual growth rate
MSW	Municipal solid waste
PET	Polyethylene terephthalate
rPET	Recycled polyester
SF	Silica fume
PFF	Poultry feather fiber

**Corresponding author*: archana@iitd.ac.in

Prashansa Sharma and Shilpi Shree Sahay (eds.) Nanotechnology-Assisted Recycling of Textile Waste: Sustainable Tools for Textiles, (351–378)

JBO	Jute batching oil (JBO)
ASTM	American Society for Testing and Materials
ISO	International Organization for Standardization
TKC	Tamarind kernel cellulose

12.1 Introduction

The textile industry is currently facing two significant challenges: first, it must meet the increasing demands of its customers; second, it must address the environmental effects of its waste management and manufacturing processes [1]. According to a *World Bank* report, global municipal solid waste (MSW) production is expected to reach 2.59 billion tons by 2030 and 3.4 billion tons by 2050 [2]. In the face of escalating environmental concerns and a global push toward sustainable practices, industries worldwide seek innovative solutions to address the challenges posed by waste management and resource depletion. The textile industry stands at the intersection of consumer demand, industrial production, and environmental impact. As the demand for textiles, especially technical textiles, continues to rise across diverse sectors such as automotive, aerospace, healthcare, and sports, the need for sustainable and eco-friendly materials has never been more pressing [3]. In this dynamic industry, adopting sustainable practices and utilizing state of the art technology have emerged as crucial factors in defining a sustainable future.

This chapter examines the use of recycled materials in high-technology textile manufacturing, a crucial aspect of environmentally friendly textile manufacturing. This study, which focuses on recycling waste materials, considers the financial advantages of employing recycled resources as well as the broader environmental effects of using traditional textile production methods. This chapter investigates the transformative potential of recycling in the textile industry by examining various types of waste materials, recycling procedures, and how they are incorporated into technical textiles. This chapter presents an in-depth review of the benefits, difficulties, and future potential related to applying recycled waste to technical textiles against the backdrop of increased environmental consciousness. This chapter also encourages academics, professionals in the textile sector, and policymakers to create a more ecologically friendly future by implementing sustainable practices. Through creative waste material recycling, technical textiles have the potential to revolutionize textile manufacturing and consumption for a more sustainable future.

12.2 Overview of Technical Textiles

Technical textiles are a category of textile products that are developed to perform functional responsibilities other than those found in apparel and household items. Technical textiles are specialized fabrics developed for specific use and characterized by their practicality, performance, and longevity. Figure 12.1 illustrates the wide range of subcategories designed to fulfil many industries' special needs. These include Agro-textiles, Meditech, Geotextiles, Protective textiles, Automotive textiles, Aerospace textiles, Sportech, Buildtech, and Pack-Tech. As reported by the Business Research Company, the worldwide technical textiles market is projected to experience a compound annual growth rate (CAGR) of 5.6%, rising from \$201.71 billion in 2022 to reach \$213 billion in 2023 [4]. This section explores the intricate matter of recycled textile waste, encompassing its present condition and projected trajectory. Using recycled waste materials in textiles is marketed as an effective way of both environmental and economic sustainability [5]. The implementation of the circular economy, waste minimization initiatives, enhanced recycling technologies, and sustainability certifications are some of the next advancements that address concerns about contaminants, inconsistent quality, and customer perception. The study emphasizes the value of recycled textiles in lowering environmental pollution, enhancing global economy, and generating employment. These

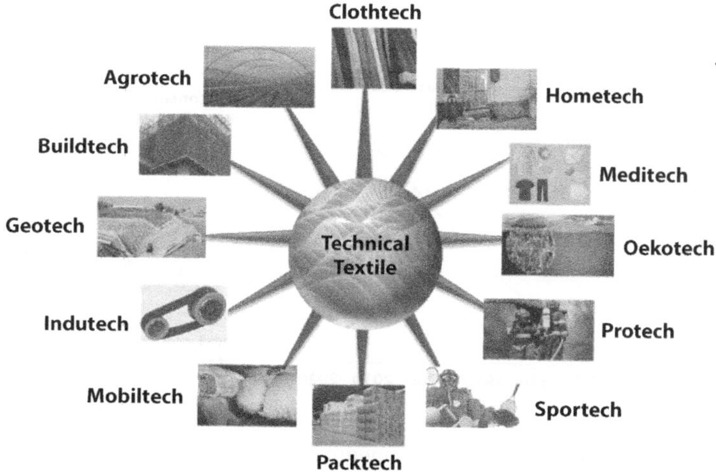

Figure 12.1 Different segments of technical textiles.

causes contribute to textile wastes and create key concerns for the textile industries.

12.3 Textile Waste: Sources and Challenges

Among the vast quantity of clothing produced each year, estimated at 100 billion, a significant proportion of approximately 92 million garments find their ultimate destination in landfills [6]. In the United States, approximately 11.3 million tons of textile waste, which represents about 85% of the total textile waste produced, is disposed of in landfills annually. This equals about 2,150 pieces each second across the country and an average of 81.5 pounds (37 kg) per person yearly. The generation of textile waste within the context of MSW is seeing an upward trend, encompassing the potentially recyclable portion. The quantity of textile waste generated in 2010 was estimated to be over 13.22 million tons, of which only 2.05 million tons were considered suitable for recycling [7]. However, by 2018, textile waste had increased to 17.03 million tons, and the recyclable portion had risen to 2.51 million tons of the total MSW, as illustrated in Figure 12.2a [8]. The annual financial loss due to discarded undergarments and inadequate clothing recycling reaches $500 billion [6].

The textile industry, driven by globalization, has shifted toward rapid and cost-effective clothing production, giving rise to the era of fast fashion. Consequently, the increased consumption of clothing and the resultant surge in textile waste have emerged as significant societal challenges.

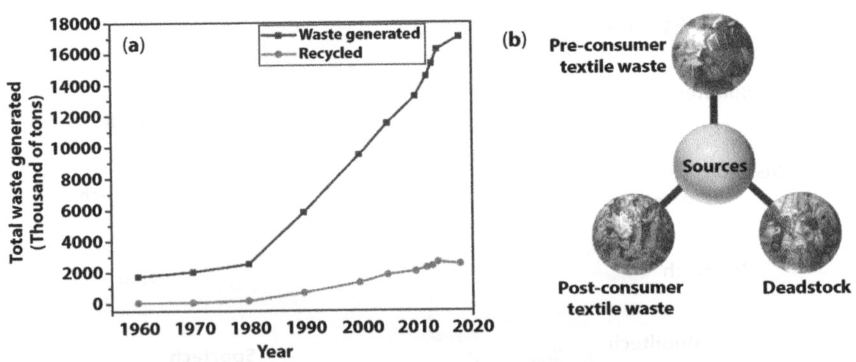

Figure 12.2 (a) The graph illustrates the amount of textile waste produced within the municipal waste stream, with data sourced from the US Environmental Protection Agency [7, 8]. (b) The schematic diagram displays the origins of textile waste generation.

Responding to rising social pressure, it is critical to prioritize recycling for long-term solutions and economic well-being. Textile waste, generated from various sources (as depicted in Figure 12.2b) and characterized by multifaceted complexities, poses a substantial environmental threat in the modern world. Understanding the origins and complexities of textile waste is vital for developing effective waste management and reduction strategies.

12.3.1 Sources of Textile Waste

12.3.1.1 Pre-Consumer Waste

This category covers waste generated within the textile industry before the product reaches consumers. It includes manufacturing byproducts, production offcuts, and rejected or defective textile materials. The waste generation in manufacturing processes contribute significantly to this source of waste [9].

12.3.1.2 Post-Consumer Waste

Post-consumer textile waste arises from discarded textile products no longer used by individuals or households. This category comprises used clothing, linens, towels, and other textile items that have reached the end of their lifecycle. Consumer behaviors, such as fast fashion trends and frequent wardrobe turnover, drive the growth of post-consumer textile waste [10].

12.3.1.3 Deadstock

Deadstock refers to new and unworn garments that remain unsold or are returned, often in the context of retail and fashion industries. Industries often dispose of deadstock through burning or other unsustainable methods, contributing to textile waste [11].

Numerous factors contribute to the challenges associated with textile waste. These challenges include the emission of harmful greenhouse gases during the decomposition of textiles in landfills and water contamination caused by the dyeing and finishing processes in the textile industry [10, 12, 13]. The complex blend of natural and synthetic components in textiles makes recycling and repurposing difficult [14]. Sorting textile waste is laborious and time-consuming due to its diverse composition [15]. Additionally, the prevailing culture of disposable fashion, where garments are frequently discarded after limited use, intensifies the issue of post-consumer textile waste [16]. Addressing these challenges requires a comprehensive approach involving sustainable manufacturing practices, enhanced waste management infrastructure, raising consumer awareness, and adopting innovative recycling technologies.

12.4 Recycling Technologies

Recycling technologies are a broad range of techniques designed to turn waste materials into useful resources as shown in Figure 12.3. These techniques include chemical recycling [17], which breaks down waste into its parts for new products; biological recycling [18], which uses enzymes and microorganisms to break down organic waste; mechanical recycling [19], which transforms materials like plastics and textiles into reusable forms; and upcycling, which creatively turns wasted materials into higher-value products. Pyrolysis and gasification processes break waste into valuable gases, liquids, and solids [20]. These innovative ways minimize waste and align with sustainable practices, making them critical in reusing waste materials in technical textiles and other industries. Several relevant recycling technologies should be considered in this context.

These recycling technologies not only facilitate the transition from waste to value-added products but also align with the broader goals of sustainability and circular economy principles. When applied to technical textiles, they provide chances to fabricate high-performance, eco-friendly materials, which is an important concern in industries where functionality, durability, and sustainability are critical [15]. These technologies continue

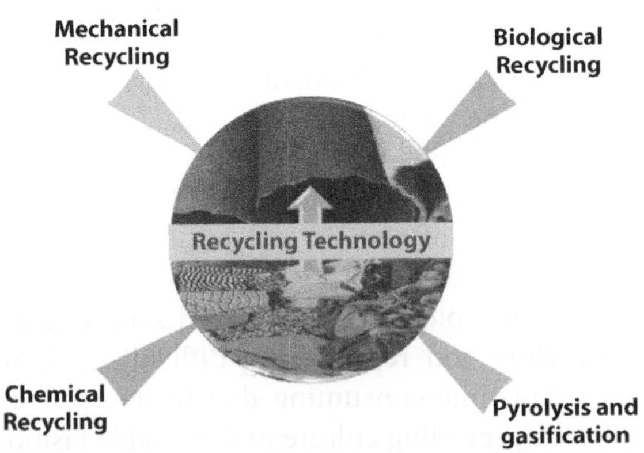

Figure 12.3 Various broad recycling methodologies for textile waste.

to evolve, offering promising prospects for the incorporation of recycled waste into the technical textile sector and contribute to a more environmentally conscious and resource-efficient future.

12.5 Benefits of Using Recycled Waste in Technical Textiles

Using recycled waste in the manufacturing process of technical textiles presents numerous attractive advantages that was shown in Figure 12.4. It makes a considerable contribution to environmental sustainability by lowering the use of raw materials, cutting down on waste, and cutting emissions [21]. This environmentally friendly strategy is also cost-effective, as it reduces material purchase and waste disposal charges. Recycling conserves energy and lowers the environmental effect of raw material extraction. Furthermore, it diversifies raw material sources, improves supply chain resilience, and appeals to an expanding market of environmentally concerned consumers, increasing market competitiveness. The importance of recycled waste materials in technical textiles is further emphasized by job creation, regulatory compliance, and connection with circular economy concepts, making it a sustainable and responsible choice with significant advantages [22]. These benefits highlight the crucial function that recycled waste plays in transforming the technical textile industry into a more resource-efficient and ecologically conscientious sector, in line with the global push toward sustainability.

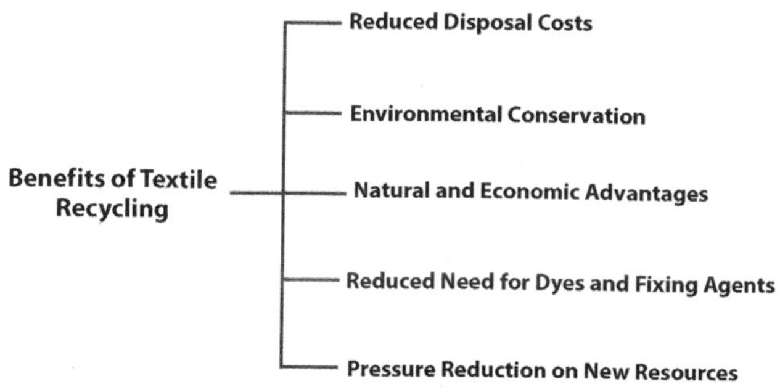

Figure 12.4 Numerous advantages of recycling textiles.

12.6 Application of Recycled Waste in Technical Textiles

The applications of recycled waste in technical textiles are extensive, which is described in Figure 12.5. These materials can be repurposed in apparel, geotextiles, automotive textiles, insulation, filtration systems, and reinforcements for composites. Additionally, recycled waste is finding its way into electronics textiles, medical and healthcare textiles, and environmental protection textiles, further expanding its reach and utility. This discusses the complex life cycle of textile and apparel products, highlighting the diverse stages and global players involved in their production. The life cycle commences of the extraction or cultivation of raw materials from sources like plants, animals, and petroleum, necessitating the participation of the agriculture and chemical industries. Fiber production methods, including techniques like melt spinning, dry spinning, and wet spinning, contribute to the versatile pathways for fabric formation. The subsequent textile conversion stage focuses on the creation of yarns and fabrics with specific properties such as strength, durability, color, pattern, and texture. Further processing is often required to fashion the final garment products. The distribution phase involves intricate logistics, the involvement of wholesalers, and the engagement of retailers. Subsequently, consumers undertake product cleaning and care through domestic laundry during the usage stage [23]. Finally, the disposal stage presents various options,

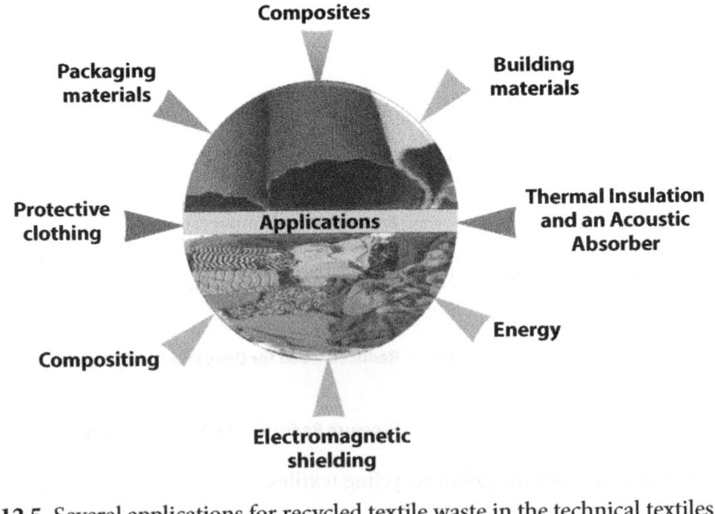

Figure 12.5 Several applications for recycled textile waste in the technical textiles.

including sending used products to landfills or incinerators, as well as promoting recycling or reusing initiatives.

12.6.1 Recycled Fibers in Apparel and Clothing

The utilization of recycled fibers in apparel and clothing is a significant leap forward in the fashion industry's quest for sustainability. Reusing and recycling textiles can help the environment by cutting down on the production of new materials and avoiding certain steps in the life cycle of textile products. This is more eco-friendly than burning or burying textiles. Reusing textiles is seen as even better than recycling, especially if we can use them for a longer time [24]. Recycled fibers, particularly those derived from post-consumer sources like discarded textiles, have opened a world of possibilities in fashion. They offer an eco-conscious solution to the growing issue of textile waste, providing a fresh take on design, production, and consumer preferences. One of the key strengths of incorporating recycled fibers in apparel is their adaptability [25]. These fibers can be used in various garments, from sportswear and outerwear to casual clothing. This versatility is a testament to the industry's creative potential when it comes to reusing materials and reducing the demand for virgin resources [26].

From a consumer perspective, the choice to opt for recycled fiber clothing aligns with the ever-increasing demand for sustainable products. Shoppers are increasingly concerned about the environmental impact of their choices, and recycled fiber clothing provides a way to meet these demands without compromising on style or quality. Another noteworthy aspect is the advancements in recycling technology, which have allowed for improved fiber quality and performance. These recycled fibers can match and, in some cases, even surpass the properties of their virgin counterparts. This means that consumers can enjoy comfortable, durable, and fashionable clothing while reducing their carbon footprint. However, the transition to recycled fiber clothing is not without its challenges [27]. It requires the collaboration of designers, manufacturers, and consumers. It also necessitates overcoming barriers related to consumer education, costs, and supply chain logistics. The review could have delved deeper into these obstacles and potential solutions. Bhuiyan and his colleagues produced chemically bonded non-woven materials by employing recycled cotton and polyethylene terephthalate (PET) and then applied these non-wovens for thermal insulation within building construction as shown in Figure 12.6 [28].

Later in 2021, Sezgin and co-authors engineered composite panels for thermal and acoustic insulation. These panels were created through the

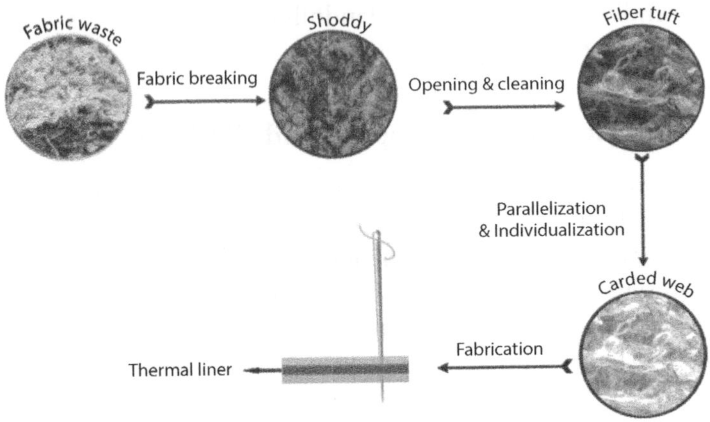

Figure 12.6 Schematics of manufacturing quilt fabric using recycled cotton fiber. The transformative potential of recycled cotton fibers from apparel waste in the creation of superior thermal barrier fabrics [28].

utilization of 100% recycled cotton fibers, which were strengthened with polyethylene and polypropylene [29]. In 2021, Dissanayake and collaborators created a sound insulation material by blending natural rubber with discarded cotton/polyester waste [30]. In 2020, Kamble and Behera showcased the potential use of thermoset composites in the automotive industry. These

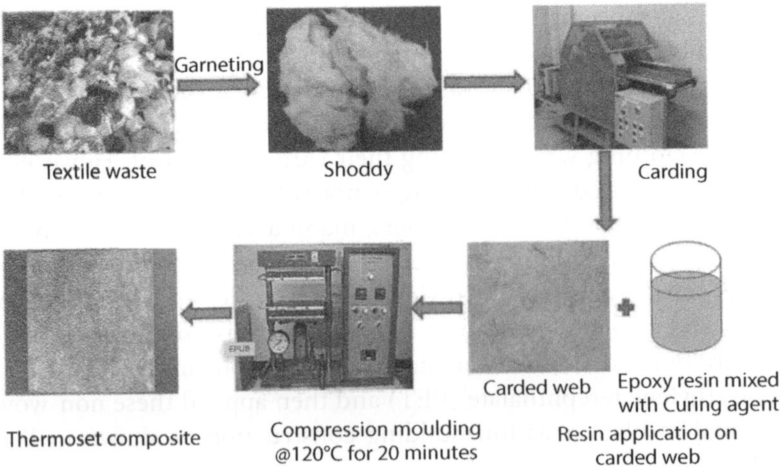

Figure 12.7 The sequential stages involved in the development of composite specimens derived from textile waste [31].

composites were reinforced with recycled cotton fibers obtained from the waste generated during apparel cutting as shown in Figure 12.7 [31].

12.6.2 Hometech

High-performance home textiles, known as Home-Tech, play a crucial role in internal decor, furniture, carpeting, sun shielding, and fire-retardant applications. They are extensively employed in diverse settings such as buildings, ships, caravans, and transportation. The attainment of fire retardancy is typically achieved through the integration of inbuilt fire-retardant materials, such as modacryl fibers, or the application of coatings containing fire retardant components, like phosphorus based Pyrovate [32]. Home-Tech textiles represent a substantial portion, exceeding 35% of the total weight of fibers and textiles, particularly in the realm of non-woven and composite reinforcement materials. Woven fabrics are still employed to a considerable level as furniture and carpet-baking products and curtain header tapes. Hollow fibers, characterized by high-quality insulation properties, find extensive use in beddings and sleep bags. The potential application of recycled waste in technical textiles for Home-Tech emerges as a promising avenue for sustainable material utilization. This entails repurposing waste materials to engineer high-performance textiles, thereby contributing to resource efficiency and

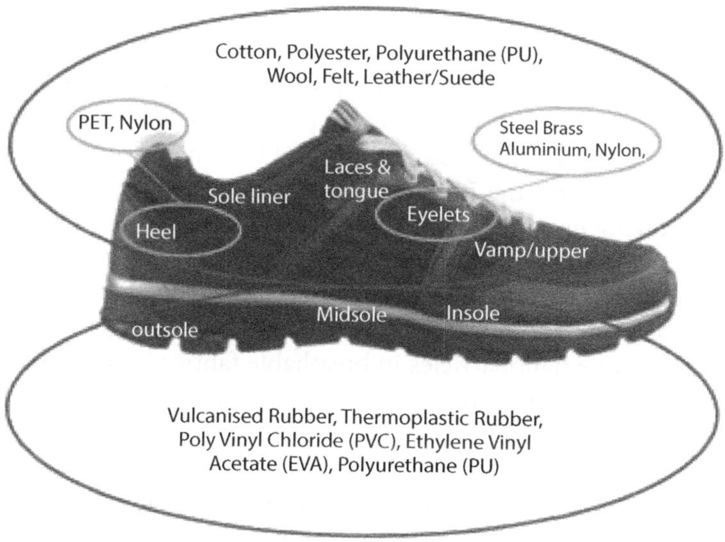

Figure 12.8 A typical sports shoe with main parts and their commonly used materials [34].

addressing environmental concerns in the domain of home textile manufacturing [33]. Jute fibers, traditionally treated with Jute Batching Oil (JBO), face consumer acceptance challenges due to their distinct odor. This study innovatively incorporates aroma microcapsules made from textile waste and imparts a lasting fragrance finish to jute-cotton blended fabric, transforming it into a value-added product while overcoming the typical smell associated with JBO [33]. The key innovation in this academic pursuit was the implementation of air-based separation technologies, effectively distinguishing granulated shoe particles based on size and weight. The study successfully reclaimed four distinct material streams—leathers, textiles, foams, and rubbers as shown in Figure 12.8, with potential applications in surfacing materials, insulation boards, and underlays. This research contributes to the academic discourse on sustainable solutions for managing end-of-life products in the footwear industry [34].

12.6.3 Protech

Technical textiles incorporating high-performance fibers play a pivotal role in providing protection against diverse hazards, ranging from heat and radiation for firefighter clothing, safeguarding welders from molten metals, to ensuring resilience in bulletproof jackets. Protective clothing in the technical field is typically designed to enhance the safety of workers, meeting the criteria outlined by global organizations like the American Society for Testing and Materials (ASTM) and the International Organization for Standardization (ISO) [35].

Many products, predominantly composed of petroleum-based polymers, exhibit prolonged decomposition periods in landfills. However, there are ways to make protective clothing and PPE more environmentally friendly by focusing on the four R's: reduce, reuse, recycle, and recover. There is a growing emphasis on using waste materials in the production of these protective textiles, aligning with sustainability goals. This can be achieved through methods like decontamination using ultraviolet germicidal irradiation, adding a self-detoxifying finish against germs, or incorporating active nanoparticles in breathable fabrics [36]. New use for protective clothing by using recycled materials like Kevlar or strong polyester, known for their toughness. They are using natural substances like chitosan, sodium alginate, and natural rubber as binders for recycled cotton or wool fibers. These materials are then used to build structures like buildings and walls, as well as interior linings, and partitions. This takes advantage of the fibers' abilities to absorb sound, control noise, and provide thermal insulation [37].

The study by Chuang *et al.* focused on recycling high-strength polyester (PET) waste fibers with the aim of creating cost-effective high-performance materials suitable for protective clothing or geotextiles. They used recycled high-strength PET in combination with low melting and woven fabrics through a thermal process to create hybrid fibrous planks. The high-strength was processed in a needle punching machine. The developed materials underwent tests for air permeability, tensile strength, tear resistance, bursting strength, and stab resistance. The results indicated that the recycled PET fabrics maintained higher strength [38].

12.6.4 Pack-Tech

Pack-Tech textiles find applications in packaging for silos, containers, tents, and clothing covers. These textiles are extensively used in creating sacks and bags, traditionally made from materials like jute, flax, and cotton, but more recently from poly(propylene). There is a growing emphasis on incorporating technical textiles made from waste materials in the production of these Pack-Tech textiles. This aligns with sustainability goals, providing an eco-friendly approach to packaging solutions [39]. Biocomposite panels, created from textile waste, showed improved water absorption and thickness swelling compared to wood. The inclusion of wool waste enhanced durability against fungal decay [39].

Hessian a woven fabric typically crafted from natural fibers, predominantly jute. It varies in texture, ranging from stiff and dense to softer variations. Jute fabric can be produced by blending jute fibers with natural or synthetic counterparts. Primarily used for technical or decorative purposes, this natural fabric made from jute yarn is well-suited for applications in gardening, upholstery, bag manufacturing, as well as for packaging and storage [40].

12.6.5 Agrotech

Agro-textiles have been used in the agriculture sector for thousands of years and are an attractive tool for the protection of crops during their entire lifecycle. Currently, the agro-textile market is dominated by polyolefins or petrochemical-based agro-textiles. However, climate change and an increase in greenhouse gas emissions have raised concern about the future oil-based economy, and petroleum-based agro-textiles have become expensive and less desirable in the modern world [41]. The smart way to handle waste is by turning it into something useful. It is like a cycle where we use textile waste to create new things. This approach, called the circular economy, is all about using materials that can be used again, making products last longer, and recycling or reusing waste. Instead of throwing things away or burning them, we are finding ways to take that waste and make better-quality stuff that is good for the environment [42].

In horticulture, farmers commonly use pots or trays made of fossil raw materials like polystyrene, polyethylene, and polypropylene for transplanting seedlings [43]. An alternative is biodegradable pots that allow healthy root growth and can be planted directly into the soil, reducing costs, labor, and environmental pollution [44]. The most commonly used synthetic fibers and regenerated fibers in agrotextiles are shown in Figure 12.9. Bhatt *et al.* produced household items such as lampshades, doormats, wind chimes, coasters, and hanging baskets using flawed and damaged textile material sourced from university research farms. This aligns with the aim of promoting sustainability while minimizing environmental impact and decreasing the generation of waste. As a result, repurposing waste agrotextiles into valuable products serves to reduce the amount of textile waste that would otherwise be discarded [45] Agro-textiles play a vital role in agriculture, aiding in faster crop germination and growth. They also impact plant morphology, leading to taller plants and larger leaf areas. The use of fibers from agro-textiles improves soil conditions and adds organic carbon

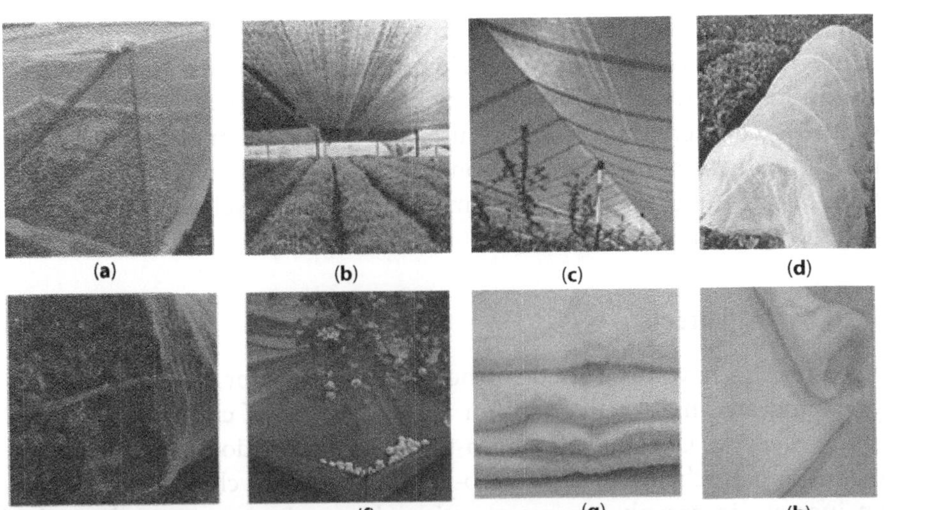

Figure 12.9 Synthetic fibers and regenerated fibers in agrotextiles: (a) insect mesh: nylon and PP; (b) shade net: nylon and PP; (c) anti-hailstone net: PE; (d) frost control fabric: PP; (e) bird protection net: nylon, PE, and PP-17; (f) harvesting net: PE; (g) non-woven mulch: viscose; (h) non-woven mulch: PLA [46].

when degraded. The future of agro-textiles leans toward eco-friendly and biodegradable materials, such as lignocellulosic-based textiles, offering an economical and environmentally friendly option for agriculture.

12.6.6 Automotive Textiles

The use of recycled textile fibers in automotive textiles has gained traction in recent years due to the automotive industry's increasing emphasis on sustainability and environmental responsibility. Incorporating recycled fibers into automotive textiles can have several benefits, including reducing the environmental impact of textile production and contributing to a circular economy. Here are some examples of recycled textile fibers are being used in automotive textiles.

12.6.6.1 Recycled Polyester

Recycled polyester fibers, often sourced from post-consumer plastic bottles or textile waste, are commonly used in automotive textiles. These fibers can be used for seat fabrics, carpets, and headliners. Recycled polyester offers similar performance characteristics to virgin polyester while reducing the consumption of new resources and energy. In one of the studies, the researchers successfully created durable car carpets with recycled PET fibers using a needle-punching process. Results showed that carpets with 85% rPET and 15% bicomponent PET performed almost as well as those with 80% PET and 20% bicomponent PET in terms of looks and durability. Choosing recycled PET not only saves on material costs but also benefits the environment in automotive textiles [47].

12.6.6.2 Recycled Nylon

Recycled nylon fibers are another sustainable option for automotive textiles. These fibers can be used in applications such as carpeting, seat covers, and interior trims. They are typically sourced from discarded nylon products, such as fishing nets and industrial waste [48].

12.6.6.3 Recycled Cotton

Recycled cotton fibers, which come from post-consumer textile waste, can be used in automotive textiles for various applications, including seat upholstery and interior linings. These materials help reduce the need for new cotton cultivation and minimize the waste of existing textiles [49].

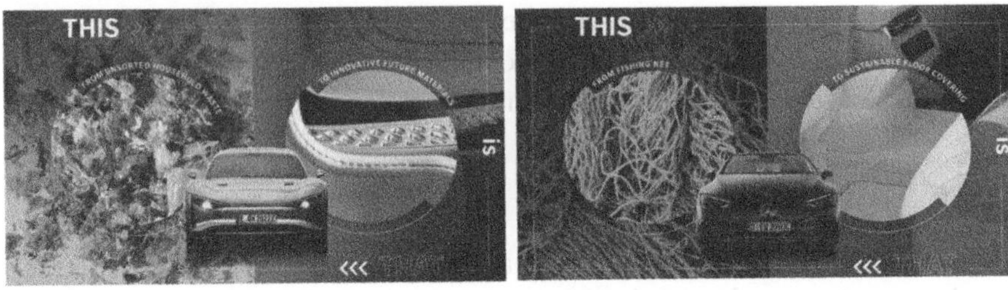

Figure 12.10 Mercedes-Benz incorporates sustainable materials into new vehicles. From recycled textiles in their interiors, we have made a greener narrative, combining luxury with environmental responsibility [51].

12.6.6.4 Recycled Leather Alternatives

To reduce the environmental impact of traditional leather, some automotive manufacturers are using recycled leather alternatives. These materials are made by processing discarded leather and textiles into a new, sustainable material that can be used for interior components like seats as shown in Figure 12.10 and steering wheel covers [50].

12.6.6.5 Recycled Insulation Materials

Recycled textiles can also be used as insulation materials for automotive applications. This can help improve the thermal and acoustic performance of vehicles while reducing the need for new materials. In this study, researchers explored using waste and new wool fibers for car insulation, aiming for an eco-friendly and affordable solution without chemicals [51].

12.6.7 Geotextiles and Construction

The integration of recycled textiles in the production of geotextiles and its applications in the construction industry has gained significant attention due to its potential sustainability and cost-efficiency benefits. Geotextiles made from recycled materials offer a range of advantages, including enhanced environmental performance and performance characteristics. Recycled textiles, which can include materials like discarded clothing, industrial textiles, and even post-consumer plastic bottles, are repurposed to create geotextiles [52]. Geotextiles are permeable fabrics placed in soil, serving various functions in construction and civil engineering. Here are some keyways recycled textiles are being utilized in fabricating geotextiles and in construction.

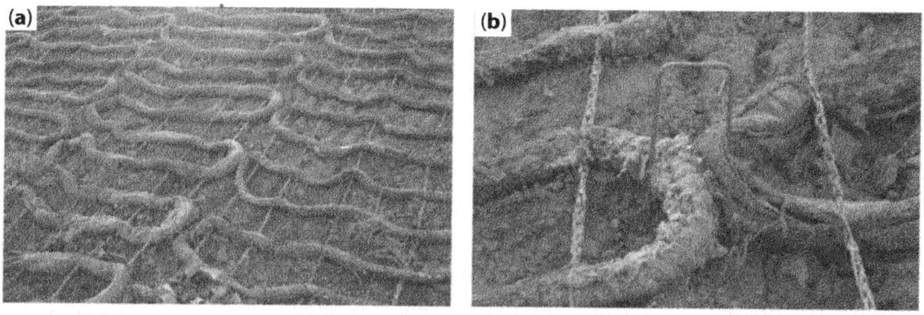

Figure 12.11 Recycled fibers combat slope erosion: Meandrical geotextiles installed with intertwined ropes and a blend of natural and synthetic recycled fibers: (a) the segments spread on the surface of the slope; (b) fastening of the ropes [53].

In a study, recycled fibers and textiles were utilized to protect a steep slope experiencing significant rill erosion as shown in Figure 12.11. Geotextiles made from intertwined thick ropes, incorporating a non-woven fabric composed of a blend of recycled natural and synthetic fibers, were applied to shield one section of the slope. Meanwhile, recycled polyester fibers were mixed with the topsoil to safeguard the other part. The study's results demonstrated that recycled fibers can be effectively and reliably employed to safeguard the slope from erosion and sliding [53].

Another interesting study, a fiber network was created by Zhang *et al.* using discarded fibers (comprising 78% polyester, 20% cotton, and 2% other materials) following a carding process for developing geo-textiles with good ultraviolet-aging properties for heavy metal absorption capabilities, along with its potential to support plant growth as a replacement for riverbank soil or shoreline cover, contributing to ecosystem restoration. The research involved utilizing waste fibers to create a uniform fiber web through carding, followed by reinforcing it using needle-punching technology in multiple layers, ensuring robust mechanical properties for the plant substrate [54].

12.6.8 Medical Textiles

Medical and healthcare textiles are products made from fibers used for medical, surgical, and hygienic purposes. They are specialized and eco-friendly materials. Globally, over 60 million tons of medical waste are produced each year. Reports show that the worldwide market for managing this waste was worth $11.77 billion in 2018 and is expected to reach $17.89 billion by 2026, growing at an annual rate of 5.3%. Every year, people release about 1.5 million tons of tiny plastic particles into the oceans, equivalent to one plastic bag per person each week. These pollutants now make up around 31% of all plastic found in the sea, spreading across marine environments [55].

Creating functional textiles with active materials involves enhancing fibers, treating textiles directly, and constructing 3D textiles with functional features. These textiles exhibit dynamic colors, effective protection, regulated temperature, enhanced antimicrobial capabilities, and various sensing functions. The medical waste from healthcare activities consists of 85% general waste and 15% hazardous or potentially infectious materials, emphasizing the need for efficient management [56]. Innovations in medical textile recycling, including molecular tagging and advanced polymeric implantable devices, aim to address environmental concerns. Despite challenges in recycling, scientific methods are being developed to preserve polymeric bonds, counter bacterial impacts, maintain flexibility, and sustain clinical performance [57].

Prioritizing medical textile recycling and eco-friendly production promotes cost-effectiveness and environmental sustainability, contributing to a healthier global economy. Research on mechanically recycled masks resulted in the development of fabric from reclaimed polypropylene and cotton blends, demonstrating comparable strength to virgin cotton, yet exhibiting some microfiber release during its lifecycle, highlighting the challenges of achieving complete elimination of microfiber release in textiles [58]. Recent studies introduced a method for creating multifunctional non-woven polyester fabrics treated with silver nitrate and aniline monomer, forming nanocomposite layers with silver nanoparticles and polyaniline, showcasing antimicrobial properties, electrical conductivity, and UV-ray blocking capabilities, using recycled PET plastic waste in the fabric production [59]. The study developed a multifunctional non-woven polypropylene fabric by integrating silver nanoparticles and polyaniline, enhancing its antimicrobial, superhydrophobic, UV blocking, and electrical conductivity properties for diverse applications in medical treatments and protective coatings [60]. A sustainable approach repurposes waste polypropylene from face masks and sacks through emulsion-templated 3D printing, resulting in efficient oil-water separation with high adsorption capacity and promising commercial scalability, indicating potential for broader plastic waste management beyond polypropylene [61].

In the coming years, medical textiles will become increasingly important, being used in various body-related devices, external or implanted objects, and healthcare and hygiene products. Medical textile materials will continue to play a crucial role in the production of different medical and surgical items. The global value of the medical textiles market is expected to reach approximately $100 billion in the next 3 to 4 years [57]. It is necessary to prioritize the recycling of medical textiles to ensure a healthier environment. By implementing updated eco-friendly technologies in the production of medical textiles, we can achieve cost-effectiveness and greater environmental friendliness. Initiating a 100% recycling process for medical textiles will mark a significant step toward the future of the global economy and the environment.

12.6.9 Smart Textiles

In response to the challenges posed by the complex nature of textile recycling, the research work primarily focuses on the conversion of second-hand clothing into power-generating electronic textiles through a scalable process. The study emphasizes that the utilized second-hand clothing requires no additional purification before the transformation process [62]. This study introduces a novel recycling method, specifically for materials like acrylic or nylon-6. By utilizing used clothing, an electronic textile is electro-spun and incorporated into a wearable smart mask, showcasing the potential to transform textile waste into a high-performance energy-generating and/or sensing material, a novel approach that repurposes worn-out garments, particularly those composed of blended textiles such as acrylic and nylon as shown in Figure 12.12. By leveraging electrospinning technology, we investigate the feasibility of transforming these discarded materials into energy-generating electronic textiles. This investigation not only addresses the pressing issue of textile waste management

Figure 12.12 A second-hand sweater and raincoat (a) before and (b) after dissolution and electrospinning to create electronic textiles. The blue sweater was acrylic, and the raincoat liner was nylon [63].

but also highlights the potential for sustainable innovation within the realm of functional fabric production [63].

An innovative and environmentally friendly method for producing carbon black by utilizing synthetic textile waste, such as PET and polyester cotton blend, as an alternative to traditional fossil fuels. This study involved the synthesis of carbon black through the utilization of synthetic textile waste, specifically waste polyester fabric (PET) and a blend of polyester cotton (50:50), using a horizontal tube furnace. The synthesis occurred within a temperature range of 700°C~1,000°C, maintaining a consistent heating rate of 5°C/min in a nitrogen environment. Our research addresses the gap in the market for environmentally friendly conductive inks in textile printing. By substituting synthetic textile scraps for fossil fuels in the carbon black manufacturing process, we present an innovative and eco-friendly approach to reduce textile waste in landfills. The resulting conductive inks have diverse applications, including electrical textiles, smart textiles, and wearable technologies [64].

12.7 Environmental and Economic Impact

Huge bulk of clothes and apparels are discarded away every year, causing big problems for the environment. In the USA, approximately 16 million tons of textile waste were generated in 2014, with 10.46 million tons ending up in landfills, leading to environmental pollution through water contamination and greenhouse gas emissions, particularly methane. Repurposing old materials for building insulation offers a sustainable solution to mitigate the adverse impacts of textile waste and landfill issues. Some researchers have looked into using leftover fabrics to make insulation materials. They tried cutting the waste textiles and stitching them together to make stable insulation materials [65]. For example, Trajković and others used old polyester clothes, cutting them up to build the insulation structures. While textile waste can be used as eco-friendly insulation, there are still some issues that need to be dealt with [66]. The aim is to create heat insulation panels for buildings, helping to protect the environment and save energy. This research focuses on making new materials from waste found in the textile industry, like polyester and polyurethane, using a process called thermoforming. The panels had a thermal conductivity ranging from 0.053 to 0.041 W/mK, depending on their density. These values were similar to those of other insulation materials available commercially [67].

To mitigate CO_2 emissions and manage textile waste, a study demonstrates the feasibility of a durable and eco-friendly building material created by blending recycled fabric fibers and industrial by-products with cement, reinforced with Silica Fume (SF) for enhanced mechanical strength and resilience under wet/dry conditions [68]. A novel method converts plastic waste from textiles and construction into a customizable microporous material capable of targeted carbon dioxide (CO_2) absorption, addressing both plastic waste pollution and rising atmospheric CO_2 levels simultaneously [69]. Advancements in waterproof breathable fabrics emphasize recycling, aligning with the growing demand for sustainable products, and are anticipated to contribute significantly to environmental conservation, particularly as the athleisure apparel market continues to embrace eco-friendly options [70]. The integration of waste textiles in geotextile production safeguards slopes against water erosion and fosters environmental conservation, exemplifying a sustainable strategy to combat soil loss and protect vegetation [71]. Integrating recycled poultry feather fiber (PFF) and tamarind kernel cellulose (TKC) in composite fabrication yields eco-friendly composites with enhanced mechanical and thermal properties, offering potential applications in diverse industries while addressing textile recycling and promoting environmental sustainability [72]. Blending ground polypropylene bottle caps with lignin powder enables the production of eco-friendly fibers, offering a sustainable solution to meet textile demand and alleviate concerns regarding resource scarcity [73].

12.8 Conclusion

In conclusion, the integration of recycled textile waste in the industry represents a pivotal step toward a more sustainable and economically viable future. Its environmental impact is profound, reducing the depletion of natural resources, curbing pollution, and lowering greenhouse gas emissions. By diverting textiles from landfills and incineration, it lessens the burden on waste management systems, mitigating environmental harm. Economically, recycling textiles not only leads to cost savings through reduced material acquisition and waste management expenses but also stimulates job creation and fosters economic growth in recycling and manufacturing sectors. Furthermore, the appeal of recycled textiles to eco-conscious consumers enhances market competitiveness, driving demand and creating a positive feedback loop for sustainable practices. This dual impact, both environmental and economic, underscores the importance of embracing recycled textile waste as a cornerstone of responsible industry practices, offering a promising path toward a more sustainable, efficient, and eco-conscious global textile landscape.

References

1. Juanga-Labayen, J.P., Labayen, I.V., Yuan, Q., A Review on Textile Recycling Practices and Challenges. *Textiles*, 2, 1, 174–188, Mar. 2022, doi: 10.3390/textiles2010010.
2. Silpa Kaza, L.Y., Bhada-Tata, P., Van Woerden, F., What a Waste 2.0, 2018, [Online], Available: https://openknowledge.worldbank.org/handle/10986/2174.
3. Bucişcanu, I.-I. and Pruneanu, M., Key Elements of Sustainability in the Field of Technical Textiles, in: *International Symposium "Technical Textiles - Present and Future*, pp. 122–129, Sciendo, Iasi, Romania, 2022, doi: 10.2478/9788366675735-021.
4. ReportLinker, technical-textiles-global-market-report, Globe Newswire, Accessed: Jan. 26, 2024, [Online], Available: https://www.reportlinker.com/p06313445/?utm_source=GNW.
5. Chen, X., Memon, H.A., Wang, Y., Marriam, I., Tebyetekerwa, M., Circular Economy and Sustainability of the Clothing and Textile Industry. *Mater. Circ. Econ.*, 3, 1, 1–9, Dec. 2021, doi: 10.1007/s42824-021-00026-2.
6. 10 Concerning Fast Fashion Waste Statistics, Earth.Org, Accessed: Jan. 29, 2024, [Online], Available: https://earth.org/statistics-about-fast-fashion-waste/.
7. U. Environmental Protection Agency, M. Conservation, R. Branch, R. Conservation, S. Division, E. Office of Resource Conservation, Advancing Sustainable Materials Management: 2014 Tables and Figures, Assessing Trends in Material Generation, Recycling, Composting, Combustion with Energy Recovery and Landfilling in the United States, November 2016, 2014.
8. U. Environmental Protection Agency, O. of Land, E. Management, O. of Resource Conservation, Advancing Sustainable Materials Management: 2018 Fact Sheet Assessing Trends in Materials Generation and Management in the United States, 2020.
9. lan Lau, Y., Reusing pre-consumer textile waste. *Springerplus*, 4, 1–2, Dec. 2015, doi: 10.1186/2193-1801-4-S2-O9.
10. DeVoy, J.E., Congiusta, E., Lundberg, D.J., Findeisen, S., Bhattacharya, S., Post-Consumer textile waste and disposal: Differences by socioeconomic, demographic, and retail factors. *Waste Manage. (Oxford)*, 136, 303–309, Dec. 2021, doi: 10.1016/J.WASMAN.2021.10.009.
11. Napier, E. and Sanguineti, F., Fashion Merchandisers' Slash and Burn Dilemma: A Consequence of Over Production and Excessive Waste?, 2018, [Online], Available: https://ssrn.com/abstract=3289411.
12. Salem, Z., Hamouri, K., Djemaa, R., Allia, K., Evaluation of landfill leachate pollution and treatment. *Desalination*, 220, 1–3, 108–114, Mar. 2008, doi: 10.1016/j.desal.2007.01.026.
13. Mia, R. *et al.*, Review on various types of pollution problem in textile dyeing & printing industries of Bangladesh and recommandation for mitigation. *J. Text. Eng. Fashion Technol.*, 5, 4, 220–226, Aug. 2019, doi: 10.15406/jteft.2019.05.00205.

14. Stefan, D.S., Bosomoiu, M., Stefan, M., Methods for Natural and Synthetic Polymers Recovery from Textile Waste. *Polymers*, 14, 3939, 1–23, Oct. 01, 2022, doi: 10.3390/polym14193939, MDPI.
15. Rahman, S.S., Siddiqua, S., Cherian, C., Sustainable applications of textile waste fiber in the construction and geotechnical industries: A retrospect. *Clean Eng. Technol.*, 6, 100420, Feb. 2022, doi: 10.1016/J.CLET.2022.100420.
16. Yan, R.N., Diddi, S., Bloodhart, B., Predicting clothing disposal: The moderating roles of clothing sustainability knowledge and self-enhancement values. *Cleaner Responsible Consumption*, 3, 100029, Dec. 2021, doi: 10.1016/J.CLRC.2021.100029.
17. Thiounn, T. and Smith, R.C., Advances and approaches for chemical recycling of plastic waste. *J. Polym Sci.*, 58, 10, 1347–1364, May 15, 2020, doi: 10.1002/pol.20190261.
18. Harmsen, P., I. cooperation with Slottje, C., Baggerman, M., Sillekens, E., Biological degradation of textiles And the relevance to textile recycling Paulien Harmsen (WUR) In cooperation with Carolijn Slottje and Michelle Baggerman (ArtEZ) and Ellen Sillekens (Sympany), p. 21, 2021, doi: 10.18174/557073.
19. Egloff, B.A. and Wehrli, D., How Hands-on Experimentation in Mechanical Textile Recycling Influences Existing Waste Management Systems. *J. Text. Des. Res. Pract.*, 9, 2, 210–225, May 2021, doi: 10.1080/20511787.2021.1923200.
20. Dogu, O. *et al.*, The chemistry of chemical recycling of solid plastic waste via pyrolysis and gasification: State-of-the-art, challenges, and future directions. *Prog. Energy Combust. Sci.*, 84, 100901, May 2021, doi: 10.1016/J.PECS.2020.100901.
21. Rani, S. and Jamal, Z., Recycling of textiles waste for environmental protection, 2018, [Online], Available: www.homesciencejournal.com, Accessed: Nov. 13, 2023.
22. Leal Filho, W. *et al.*, A review of the socio-economic advantages of textile recycling. *J. Clean Prod.*, 218, 10–20, May 2019, doi: 10.1016/J.JCLEPRO.2019.01.210.
23. Zhang, L., Leung, M.Y., Boriskina, S., Tao, X., Advancing life cycle sustainability of textiles through technological innovations. *Nat. Sustainability*, 6, 3, 243–253, Dec. 2022, doi: 10.1038/s41893-022-01004-5.
24. Fortuna, L.M. and Diyamandoglu, V., Optimization of greenhouse gas emissions in second-hand consumer product recovery through reuse platforms. *Waste Manage. (Oxford)*, 66, 178–189, Aug. 2017, doi: 10.1016/j.wasman.2017.04.032.
25. Abbate, S., Centobelli, P., Cerchione, R., Nadeem, S.P., Riccio, E., Sustainability trends and gaps in the textile, apparel and fashion industries. *Environ. Dev. Sustain.*, 26, 2837–2864, Feb. 2024, doi: 10.1007/s10668-022-02887-2.
26. Claudio, L., Waste Couture: Environmental Impact of the Clothing Industry. *Environ. Health Perspect.*, 115, 9, A449–A454, Sep. 2007, doi: 10.1289/ehp.115-a449.

27. Mandarić, D., Hunjet, A., Vuković, D., The Impact of Fashion Brand Sustainability on Consumer Purchasing Decisions. *J. Risk Financ. Manage.*, 15, 4, 176, Apr. 2022, doi: 10.3390/jrfm15040176.
28. Bhuiyan, M.A.R., Ali, A., Mohebbullah, M., Hossain, M.F., Khan, A.N., Wang, L., Recycling of cotton apparel waste and its utilization as a thermal insulation layer in high performance clothing. *Fashion Text.*, 10, 1, 1–12, Dec. 2023, doi: 10.1186/S40691-023-00342-Y/FIGURES/6.
29. Sezgin, H., Kucukali-Ozturk, M., Berkalp, O.B., Yalcin-Enis, I., Design of composite insulation panels containing 100% recycled cotton fibers and polyethylene/polypropylene packaging wastes. *J. Clean Prod.*, 304, 127132, 1–9, 2021, doi: 10.1016/j.jclepro.2021.127132.
30. Dissanayake, D.G.K., Weerasinghe, D.U., Thebuwanage, L.M., Bandara, U.A.A.N., An environmentally friendly sound insulation material from post-industrial textile waste and natural rubber. *J. Build. Eng.*, 33, 101606, Jan. 2021, doi: 10.1016/J.JOBE.2020.101606.
31. Kamble, Z. and Behera, B.K., Mechanical properties and water absorption characteristics of composites reinforced with cotton fibres recovered from textile waste. *J. Eng. Fiber Fabr.*, 15, 1–8, 2020, doi: 10.1177/1558925020901530/FORMAT/EPUB.
32. Aldalbahi, A., El-Naggar, M., El-Newehy, M., Rahaman, M., Hatshan, M., Khattab, T., Effects of Technical Textiles and Synthetic Nanofibers on Environmental Pollution. *Polym. (Basel)*, 13, 1, 155, Jan. 2021, doi: 10.3390/polym13010155.
33. Biswas, D., Chakrabarti, S.K., Saha, S.G., Chatterjee, S., Durable fragrance finishing on jute blended home-textiles by microencapsulated aroma oil. *Fibers Polym.*, 16, 9, 1882–1889, Sep. 2015, doi: 10.1007/s12221-015-4829-5.
34. Lee, M.J. and Rahimifard, S., An air-based automated material recycling system for postconsumer footwear products. *Resour. Conserv. Recycl.*, 69, 90–99, Dec. 2012, doi: 10.1016/j.resconrec.2012.09.008.
35. Dolez, P.I., Marsha, S., McQueen, R.H., Fibers and Textiles for Personal Protective Equipment: Review of Recent Progress and Perspectives on Future Developments. *Textiles*, 2, 2, 349–381, Jun. 2022, doi: 10.3390/TEXTILES2020020.
36. Memon, A.H., Peerzada, M.H., Brohi, K.M., Memon, S.A., Mangi, S.A., Performance evaluation of conventional and hybrid woven fabrics for the development of sustainable personal protective clothing. *Environ. Sci. Pollut. Res.*, 27, 27, 33835–33847, Sep. 2020, doi: 10.1007/s11356-020-09610-6.
37. Patti, A., Cicala, G., Acierno, D., Eco-Sustainability of the Textile Production: Waste Recovery and Current Recycling in the Composites World. *Polymers*, 13, 1, 134, 2021, doi: 10.3390/POLYM13010134. Dec. 2020.
38. Chuang, Y.-C., Bao, L., Lin, M.-C., Lou, C.-W., Lin, T., Mechanical and Static Stab Resistant Properties of Hybrid-Fabric Fibrous Planks: Manufacturing Process of Nonwoven Fabrics Made of Recycled Fibers. *Polymers (Basel)*, 11, 7, 1140, Jul. 2019, doi: 10.3390/polym11071140.

39. Mohareb, A.S.O., Hassanin, A.H., Candelier, K., Thévenon, M.F., Candan, Z., Developing Biocomposites Panels from Food Packaging and Textiles Wastes: Physical and Biological Performance. *J. Polym. Environ.*, 25, 2, 126–135, Jun. 2017, doi: 10.1007/s10924-016-0791-6.
40. Todor, M.P., Rackov, M., Kiss, I., Cioata, V.G., Composite solutions with recycled textile wastes. *J. Phys. Conf. Ser.*, 2212, 1, 012030, Feb. 2022, doi: 10.1088/1742-6596/2212/1/012030.
41. Sharma, N., Allardyce, B., Rajkhowa, R., Adholeya, A., Agrawal, R., A Substantial Role of Agro-Textiles in Agricultural Applications. *Front. Plant Sci.*, 13, 895740, 1–11 Jun. 2022, doi: 10.3389/fpls.2022.895740.
42. Todor, M.P., Bulei, C., Kiss, I., Alexa, V., Recycling of textile wastes into textile composites based on natural fibres: the valorisation potential. *IOP Conf. Ser. Mater. Sci. Eng.*, 477, 012004, Feb. 2019, doi: 10.1088/1757-899X/477/1/012004.
43. Sartore, L., Schettini, E., Bignotti, F., Pandini, S., Vox, G., Biodegradable plant nursery containers from leather industry wastes. *Polym. Compos.*, 39, 8, 2743–2750, Aug. 2018, doi: 10.1002/pc.24265.
44. Juanga-Labayen, J.P. and Yuan, Q., Making Biodegradable Seedling Pots from Textile and Paper Waste—Part B: Development and Evaluation of Seedling Pots. *Int. J. Environ. Res. Public Health*, 18, 14, 7609, Jul. 2021, doi: 10.3390/ijerph18147609.
45. Bhatt, P., Rani, A., Gahlot, M., Bhandari, V., Recycling and Reuse of Agrotextiles Used in Agricultural Farms. *Int. J. Curr. Microbiol. Appl. Sci.*, 8, 06, 1566–1576, Jun. 2019, doi: 10.20546/ijcmas.2019.806.188.
46. Dorugade, V. *et al.*, Agrotextiles: Important Characteristics of Fibres and Their Applications–a Review. *J. Nat. Fibers*, 20, 2, 2211290, 1–16 2023, doi: 10.1080/15440478.2023.2211290.
47. Atakan, R., Sezer, S., Karakas, H., Development of nonwoven automotive carpets made of recycled PET fibers with improved abrasion resistance. *J. Ind. Text.*, 49, 7, 835–857, Feb. 2020, doi: 10.1177/1528083718798637/FORMAT/EPUB.
48. Srimahachota, T., Yokota, H., Akira, Y., Recycled Nylon Fiber from Waste Fishing Nets as Reinforcement in Polymer Cement Mortar for the Repair of Corroded RC Beams. *Materials*, 13, 19, 4276, Sep. 2020, doi: 10.3390/MA13194276.
49. Shiferaw, M., Tegegne, A., Asmare, A., Utilization of textile fabric waste as reinforcement for composite materials in car body applications: A review. *Mater. Eng. Res.*, 5, 1, 279–290, Nov. 2023, doi: 10.25082/MER.2023.01.004.
50. Innovative materials for sustainable luxury | Mercedes-Benz Group > Sustainability > Resources, Accessed: Nov. 13, 2023, [Online], Available: https://group.mercedes-benz.com/sustainability/resources/sustainable-materials.html.
51. Cai, Z., Al Faruque, M.A., Kiziltas, A., Mielewski, D., Naebe, M., Sustainable Lightweight Insulation Materials from Textile-Based Waste for the Automobile Industry. *Materials*, 14, 5, 1241, Mar. 2021, doi: 10.3390/MA14051241.

52. Broda, J. *et al.*, Utilisation of textile wastes for the production of geotextiles designed for erosion protection, *J. Text. Inst.*, 110, 3, 435–444, Mar. 2019, doi: 10.1080/00405000.2018.1486684.
53. Broda, J., Grzybowska-Pietras, J., Nguyen, G., Gawlowski, A., Laszczak, R., Przybylo, S., Application of recycled fibres and geotextiles for the stabilisation of steep slopes. *IOP Conf. Ser. Mater. Sci. Eng.*, 254, 19, 192005, Oct. 2017, doi: 10.1088/1757-899X/254/19/192005.
54. Tan, W. *et al.*, The heavy metal adsorption and plant cultivation performance of grafting modified plant medium made with recycled fibers. *J. Clean Prod.*, 329, 129788, 2021, doi: 10.1016/j.jclepro.2021.129788.
55. Karim, N. *et al.*, Sustainable Personal Protective Clothing for Healthcare Applications: A Review. *ACS Nano*, 14, 10, 12313–12340, Oct. 2020, doi: 10.1021/acsnano.0c05537.
56. Zhang, Y. *et al.*, Functional Textiles with Smart Properties: Their Fabrications and Sustainable Applications. *Adv. Funct. Mater.*, 33, 33, 2301607, 1–58, Aug. 2023, doi: 10.1002/ADFM.202301607.
57. Das, S.K., Chinnappan, A., Jayathilaka, W.A.D.M., Gosh, R., Baskar, C., Ramakrishna, S., Challenges and Potential Solutions for 100% Recycling of Medical Textiles. *Mater. Circ. Econ.*, 3, 1, 1–12, Jul. 2021, doi: 10.1007/S42824-021-00023-5.
58. Ramasamy, R. and Subramanian, R.B., Recycling of disposable single-use face masks to mitigate microfiber pollution. *Environ. Sci. Pollut. Res.*, 30, 17, 50938–50951, 2023, doi: 10.1007/s11356-023-25851-7.
59. Alatawi, R.A.S. *et al.*, Production of biologically active non-woven textiles from recycled polyethylene terephthalate. *Luminescence*, 38, 3, 350–359, 2023, doi: 10.1002/bio.4462.
60. Nafady, A., Albaqami, M.D., Alotaibi, A.M., Recycled Polypropylene Waste as Abundant Source for Antimicrobial, Superhydrophobic and Electroconductive Nonwoven Fabrics Comprising Polyaniline/Silver Nanoparticles. *J. Inorg. Organomet. Polym. Mater.*, 33, 5, 1306–1316, May 2023, doi: 10.1007/S10904-023-02562-7/TABLES/5.
61. Rani, S., Kumar, D., Nandan, B., Srivastava, R.K., Emulsion templated porous funnel from polypropylene waste for efficient oil separation and spillage management. *Colloids Surf., A Physicochem. Eng. Asp.*, 650, 129563, Oct. 2022, doi: 10.1016/J.COLSURFA.2022.129563.
62. Damayanti, D., Wulandari, L.A., Bagaskoro, A., Rianjanu, A., Wu, H.S., Possibility Routes for Textile Recycling Technology. *Polymers*, 13, 21, 3834, Nov. 2021, doi: 10.3390/POLYM13213834.
63. Milam-Guerrero, J., Kwon, D.-J., Choi, Y.Y., Al-badani, F., Jiang, J., Schaefer, J., Myung, N.V., From waste to wearable: an alternative waste stream for unusable textiles turned into piezoelectric textiles. *RSC Sustainability*, 1, 2, 326–334, Jan. 2023, https://doi.org/10.1039/d2su00068g

64. Khan, A., Awais, M., Mohsin, M., Recycling textile waste into innovative carbon black and applications to smart textiles: a sustainable approach. *Biomass Convers. Biorefin.*, 1, 1–16, Jun. 2023, doi: 10.1007/S13399-023-04475-4.
65. Jordeva, S., Golomeova-Longurova, S., Kertakova, M., Mojsov, K., Efremov, J., Textile as a sustainable insulating material for buildings. *Tekst. Ind.*, 67, 2, 20–28, 2019, doi: 10.5937/tekstind1902020J.
66. Trajković, D., Jordeva, S., Tomovska, E., Zafirova, K., Polyester apparel cutting waste as insulation material. *J. Text. Inst.*, 108, 7, 1238–1245, Oct. 2016, doi: 10.1080/00405000.2016.1237335.
67. Valverde, I.C., Castilla, L.H., Nuñez, D.F., Rodriguez-Senín, E., De La Mano Ferreira, R., Development of new insulation panels based on textile recycled fibers. *Waste Biomass Valorization*, 4, 1, 139–146, 2013, doi: 10.1007/s12649-012-9124-8.
68. Sadrolodabaee, P., Claramunt, J., Ardanuy, M., de la Fuente, A., Durability of Eco-Friendly Strain-Hardening Cementitious Composite incorporating Recycled Textile Waste Fiber and Silica Fume, in: *fib Symposium*, pp. 321–328, 2022.
69. Savas, P.E. *et al.*, Carbon Dioxide Sorbent from Construction and Textile Plastic Waste. *Adv. Sustain. Syst.*, 7, 6, 2200436, 2023, doi: 10.1002/adsu.202200436.
70. Waterproof breathable fabrics: balancing performance and environmental sustainability, Accessed: Nov. 14, 2023, [Online], Available: https://www.innovationintextiles.com/waterproof-breathable-fabrics-balancing-performance-and-environmental-sustainability/.
71. Grzybowska-Pietras, J., Nguyen, G., Przybyło, S., Rom, M., Broda, J., Properties of meandrical geotextiles designed for the protection of soil against erosion. *E3S Web Conf.*, 00042, 2018, doi: 10.1051/e3sconf/20184900042.
72. Alshahrani, H. and Arun Prakash, V., Development of environmental friendly biocomposites using domestic discarded wastes as potential fibre and filler. *Biomass Convers. Biorefin.*, 1, 1–12, Sep. 2023, doi: 10.1007/S13399-023-04926-Y/FIGURES/12.
73. Soekoco, A.S., Basuki, A., Mardiyati, Eco-friendly Fibre from Recycled Polypropylene of Bottle Cap Waste and Lignin. *IOP Conf. Ser. Mater. Sci. Eng.*, 105, 1, 012001, Jan. 2016, doi: 10.1088/1757-899X/105/1/012001.

13

Nano-Engineered Protective Textiles Using Recycled Wastes

Rupayan Roy[1]* and Pravin P. Chavan[2]

[1]*Department of Fashion Technology, National Institute of Fashion Technology, Kannur, Kerala, India*
[2]*Department of Textile Design, National Institute of Fashion Technology, Daman, India*

Abstract

This book chapter discusses the development and potential applications of protective textiles engineered with nanotechnology and recycled wastes. The chapter begins by outlining the various types of protective textiles currently available, their limitations, and the need for alternative solutions. It then delves into the use of nanotechnology to enhance protective textile properties such as water resistance, UV protection, and antimicrobial properties. The chapter also highlights the importance of recycling waste materials in the production of protective textiles, as it provides a sustainable solution to environmental issues. Various types of recycled waste that can be incorporated into protective textiles are explored, including plastic bottles, textile waste, and agricultural waste. The use of nanotechnology in combination with recycled wastes allows for the creation of protective textiles with enhanced properties such as breathability, durability, and lightweight. Additionally, the chapter explores the potential applications of these nano-engineered protective textiles, such as in the medical, military, and sportswear industries. Overall, this book chapter provides a comprehensive overview of the development and applications of nano-engineered protective textiles using recycled wastes. It highlights the importance of sustainable solutions in the textile industry and showcases the potential for the use of nanotechnology in the creation of protective textiles with enhanced properties.

**Corresponding author*: rupayan.roy@nift.ac.in

Prashansa Sharma and Shilpi Shree Sahay (eds.) Nanotechnology-Assisted Recycling of Textile Waste: Sustainable Tools for Textiles, (379–402) © 2025 Scrivener Publishing LLC

Keywords: Nano-engineering, protective textiles, recycled wastes, nanotechnology, sustainable materials, functional properties, environmental impact

List of Abbreviations

PPE	Personal protective equipment
PET	Polyethylene terephthalate
UV	Ultraviolet

13.1 Introduction

Nano-engineered recycled textile products refer to fabrics that are made from recycled materials and enhanced with nanotechnology for improved properties such as durability, strength, and resistance to stains, water, and bacteria. These textiles are designed to have a smaller environmental impact than traditional textiles, as they reduce waste and minimize the use of resources. Protective textiles, on the other hand, are designed to provide specific functionalities such as fire resistance, chemical resistance, and protection from extreme temperatures. These textiles are used in various applications such as military uniforms, firefighting gear, and medical gowns. They are also being developed for use in smart textiles, which incorporate electronics and sensors for advanced functionality. Both nano-engineered recycled textiles and protective textiles are important areas of research and development in the textile industry.

13.1.1 Overview of Protective Textiles and Their Applications

Protective textiles are specialized fabrics designed to provide protection against various external hazards such as heat, chemicals, radiation, and ballistic threats. These textiles are used in a wide range of applications including personal protective equipment (PPE), military uniforms, industrial clothing, and sportswear. Protective textiles have gained significant attention due to the increasing demand for safer and more reliable protective gear in various industries [1]. Heat-resistant textiles are designed to protect against high temperatures and are used in industries such as firefighting, metalworking, and welding. These textiles are made of materials such as aramid, fiberglass, and carbon fiber, which have high heat resistance properties [2]. Chemical protective textiles are designed to provide protection against hazardous chemicals and are commonly used in industries such as petrochemicals, pharmaceuticals, and agriculture. These textiles are made of materials such as polymeric films and laminates, which provide a barrier against chemicals [3]. Radiation protective textiles are designed to protect against ionizing radiation and are commonly used in industries such as nuclear power, medical radiology, and aerospace. These textiles are made of materials such as lead and tungsten, which have high atomic numbers and can effectively absorb ionizing radiation [4]. Ballistic protective textiles are designed to provide protection against bullets and shrapnel and are commonly used in military and law enforcement applications. These textiles are made of materials such as Kevlar and Dyneema, which have high tensile strength and can absorb the impact energy [5]. In addition to the above applications, protective textiles are also used in sportswear to provide protection against impact and abrasion. Protective clothing for motorcyclists and extreme sports enthusiasts is designed to protect against injury in the event of an accident [6]. The protective textiles play a crucial role in ensuring the safety and well-being of workers, military personnel, and athletes. The development of new and improved protective textiles continues to be an active area of research and development, with the goal of providing even better protection against various hazards.

13.1.2 Challenges and Limitations of Current Protective Textile Materials

Protective textile materials are widely used for safeguarding individuals against environmental hazards such as fire, chemicals, radiation, and biohazards. However, the current protective textile materials have some limitations and challenges that need to be addressed to enhance their protective capabilities. In this section, we will discuss some of the major challenges and limitations of current protective textile materials. Table 13.1 represents the classifications of protective textile materials.

One of the major challenges is the balance between protection and comfort. Protective textiles need to provide adequate protection while still allowing for sufficient airflow to keep the wearer comfortable. This can be a difficult balance to achieve, and current materials often prioritize one over the other [7]. Additionally, the weight of protective textiles can also affect their comfort, making them cumbersome to wear for extended periods [8]. Another challenge is the durability of protective textiles. The repeated use and washing of these materials can cause wear and tear, compromising

Table 13.1 Classification of protective textiles and used materials.

Type of protective textile	Material
Ballistic protective textiles	Aramid fibers and ultrahigh–molecular weight polyethylene fibers
Chemical protective textiles	Polymer coatings, fluoroelastomers, and viton
Fire protective textiles	Aramid fibers, meta-aramid fibers, and para-aramid fibers
UV protective textiles	UV-absorbing agents, UV stabilizers, and titanium dioxide nanoparticles
Antimicrobial protective textiles	Silver nanoparticles, copper nanoparticles, and zinc oxide nanoparticles
Electrically conductive protective textiles	Carbon fibers, graphene, and metallic nanowires
Thermal protective textiles	Ceramic fibers, carbon fibers, and silica fibers

their protective capabilities. Additionally, exposure to environmental factors such as UV radiation and chemicals can also damage the materials [9].

The design of protective textiles is another limitation. Current materials are often designed for specific applications, and it can be challenging to modify them for other uses. This makes it difficult to create multifunctional protective textiles that can be used in a variety of situations [10].

Finally, there is a need for better testing and standardization of protective textiles. While there are established standards for some materials, such as flame-resistant textiles, there is a lack of standards for other types of protective textiles. This can lead to inconsistencies in their protective capabilities and can make it difficult for manufacturers to ensure their products meet safety requirements [11]. While protective textile materials have come a long way in providing safety to individuals, there are still challenges and limitations that need to be addressed to enhance their protective capabilities. These challenges include balancing protection and comfort, durability, design, and testing and standardization.

13.1.3 Introduction to the Concept of Using Recycled Waste Materials for Textile Production

The fashion industry is one of the most polluting industries globally, with significant negative impacts on the environment and human health. The industry's overconsumption of resources, energy, and water, as well as its high levels of waste production, make it a significant contributor to climate change and pollution [12]. In recent years, there has been a growing interest in sustainable fashion and circular economy principles, where waste is viewed as a valuable resource that can be reused and repurposed [13]. One such approach is the use of recycled waste materials for textile production. The use of recycled fibers is illustrated by Figure 13.1.

By avoiding the extraction of fresh resources and lowering waste sent to landfills, recycling waste materials into new textiles helps lessen the environmental effect of the fashion industry [14]. Post-consumer plastic bottles, abandoned fishing nets, and textile trash are a few of the recycled waste products that are frequently employed in the textile industry [15]. Through techniques including mechanical recycling, chemical recycling, and upcycling, these materials are turned into fresh fabrics and textiles [16].

The classification of textile waste and regenerated goods is shown in Table 13.2. By mechanically recycling, waste materials are converted into fibers that are spun into fresh yarns and garments [18]. On the other hand, chemical recycling entails turning the waste material into a chemical solution that can be utilized to produce new fibers [19]. Upcycling is the

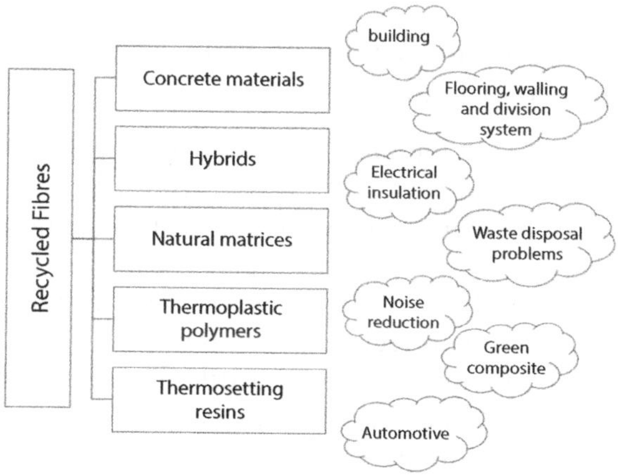

Figure 13.1 Recycled fibers and uses [17].

Table 13.2 Classification of textile waste and the recycled product.

Type of textile waste material	Recycled product
Pre-consumer waste	Re-spun yarn, felt products, and insulation material
Post-consumer waste	Recycled cotton, recycled polyester, and recycled nylon
Industrial waste	Industrial wiping cloths, fiberfill material, and sound insulation material
Scrap fabrics	Paper products, automotive sound insulation, and landscaping fabric
End-of-life textiles	Fiberfill material, non-woven fabrics, and carpet underlay

practice of converting waste materials into new items, such as recycling old denim trousers into a jacket or using leftover leather to create a new bag [20].

The environmental effect of the fashion industry might be considerably reduced by using recycled waste resources for textile manufacture. Additionally, it offers fresh chances for fashion design innovation and originality. But in order for this strategy to be widely used in the sector, issues including the supply and quality of waste materials, the cost of recycling procedures, and the requirement for new supply chains and infrastructure must all be addressed [21].

13.1.4 Advantages and Benefits of Nano-Engineering in Protective Textiles

A key technological development in the area of protective textiles is nano-engineering. Researchers can create textile fibers with improved protective qualities by manipulating materials at the nanoscale. Textiles that provide protection are necessary for a variety of uses, including industrial, military, medical, and sportswear. Enhancing the mechanical strength and durability of protective textiles is one of the main advantages of nano-engineering them. Traditional textile fibers cannot be made to be as strong as nanoscale fibers, which increase the resistance of the latter to tearing, abrasion, and puncture [22]. Additionally, the addition of nanoparticles can lengthen the textile's overall lifespan and durability [23].

Using nanotechnology to design textiles allows for the improvement of thermal regulation, which is another advantage. Researchers can develop fabrics that efficiently transfer heat by regulating the size and distribution of nanopores within the textile fibers [24]. For protective materials worn in harsh settings, such as the equipment worn by firemen, this feature is very advantageous.

Nano-engineering can improve the protective qualities of textiles against a variety of threats in addition to improving mechanical strength and thermal regulation. For instance, adding nanoparticles with anti-microbial properties can make the textile appropriate for use in medical applications by preventing the growth of germs and fungi [25]. Similar to this, UV-absorbing nanoparticles can be applied to fabrics to protect them from dangerous UV rays [26].

Additionally, the creation of smart textiles with cutting-edge features has been made possible by the nanoscale engineering of protective textiles. These fabrics can be designed to sense changes in the environment, such as variations in temperature, moisture, and pressure, and to react to those changes [27]. As an illustration, a smart textile with nano-sensors woven in may identify the presence of hazardous chemicals in the air and warn the wearer of potential health risks. When hazardous compounds have been released into the environment during an emergency or in industrial settings where workers may be exposed to toxic chemicals, this can be especially helpful. The wearer's vital signs, including heart rate, blood pressure, and body temperature, can be tracked by smart textiles. Wireless transmission of this data to medical specialists enables remote patient monitoring and the early identification of health problems. Athletic performance can also be improved by using smart textiles. For instance, a running shirt with motion sensors can monitor the wearer's motions and give feedback on their form, assisting in the development of their running form and reducing the risk of injury. Overall, the incorporation of technology into textiles has the power to completely transform a variety of facets of our life, including healthcare, environmental monitoring, and athletic performance.

13.2 Recycled Waste Materials for Protective Textiles

Individuals can be protected against a variety of environmental, chemical, and physical threats with protective fabrics. The use of recycled waste materials in the creation of textiles has become a trend as people's understanding of sustainability grows. Numerous advantages come with this strategy, including lowered environmental impact, affordability, and distinctive aesthetics. This article presents a summary of the most often utilized recycled waste materials in the creation of protective textiles, highlighting their qualities and advantages.

Plastic is one of the recovered waste products that is most frequently used in the textile industry. Recycling plastic trash allows for the creation of fibers that can be utilized to make protective apparel. Textiles made of plastic have many benefits, such as being lightweight, strong, and resistant to chemicals and the elements [28, 29]. Additionally, recycled plastic fibers can be blended with other fibers, such cotton, to improve their functionality and lower production costs.

Denim is yet another typical textile industry waste material. Waste denim from the textile industry can be recycled and made into new fabrics in large quantities. Protective gear with outstanding tensile strength, abrasion resistance, and thermal insulation capabilities can be made from recycled denim fibers [30–33]. Recycled denim fibers also provide a distinct aesthetic value that is widely sought after in the fashion sector. Another recycled material that is frequently used in the textile industry is cotton waste. Cotton waste produced by the textile industry can be recycled and made into new fibers in large quantities. In essence, protective apparel with strong thermal insulation characteristics, breathability, and comfort can be made from recycled cotton fibers. Additionally, recycled cotton fibers are far more affordable than virgin cotton, making them a superior choice [34–36].

In addition to the aforementioned waste materials, recycled rubber, nylon, and polyester can be utilized to make protective clothing. Recycled polyester fibers are perfect for creating protective apparel for outdoor activities because they are lightweight, strong, and moisture-wicking [37]. Recycled nylon fibers are excellent for constructing protective apparel for harsh settings because they have great tensile strength, abrasion resistance, and UV protection [38]. Rubber from recycled sources may be converted into flexible, strong materials with great impact and puncture resistance, which makes it a perfect material for constructing protective footwear. Numerous advantages come from the utilization of recycled waste materials in the textile industry, such as reduced environmental impact, financial effectiveness, and distinctive aesthetics. The utilization of recycled materials aids in promoting sustainable practices and lowering the carbon footprint of the textile industry by lowering the quantity of waste it produces. Furthermore, the use of recycled materials can lower production costs, making protective textiles more affordable for people from a variety of socioeconomic backgrounds [39, 40].

In conclusion, recycled waste materials present a viable and affordable substitute for virgin resources in the manufacture of protective textiles. Commonly utilized recycled materials include plastic, denim, cotton, polyester, nylon, and rubber. These materials have a variety of qualities and advantages, including light weight, durability, thermal insulation, abrasion resistance, and distinctive aesthetics. The textile industry may lessen its negative effects on the environment and encourage a more sustainable future by adopting sustainable practices and employing recycled materials in textile production.

13.3 Nano-Engineering in Protective Textiles

The textile business has been completely changed by nano-engineering, especially in the creation of protective textiles. By manipulating matter at the nanoscale, nanotechnology offers many advantages, including improved toughness, usefulness, and durability [41]. The different ways that nano-engineering is used to create protective fabrics and how it improves their performance have been covered in this chapter. The various nanotechnology applications for protective fabrics are shown in Table 13.3.

Table 13.3 Different nano-engineering techniques and their application in the protective segment.

Nanotechnology	Protective application
Nanofibers	Filtration, wound dressings, flame retardancy, and UV protection
Nanoparticles	Antimicrobial coatings, UV protection, and flame retardancy
Nanocomposites	Impact resistance, thermal insulation, and UV protection
Nanocoatings	Waterproofing, antimicrobial coatings, and abrasion resistance
Nanosensors	Wearable sensors for monitoring physiological changes
Nano-encapsulation	Controlled release of antimicrobial agents and UV stabilizers
Carbon nanotubes	Electrically conductive textiles and antistatic textiles

13.3.1 Nanofibers in Protective Textiles

Ultrafine fibers with a diameter of less than 100 nm are known as nanofibers. They are perfect for use in protective fabrics because of their good mechanical qualities and high surface area-to-volume ratio. Electrospinning, a method that uses electric fields to form fine fibers from a polymer solution, can be used to create nanofibers. Protective clothing with improved thermal insulation, UV protection, and water resistance can be made using electrospun nanofibers [42]. Additionally, nanofibers' large surface areas can be used to contain useful components like conductive materials and antibacterial agents, making them perfect for creating smart protective fabrics [43].

13.3.2 Nanoparticles in Protective Textiles

Micron-sized particles known as nanoparticles have a diameter of under 100 nm. They are perfect for use in protective textiles due to their distinctive qualities, which include a wide surface area and strong reactivity. To guard against environmental dangers including UV radiation, chemical spills, and biological agents, the nanoparticles can be added to fabrics [44, 45]. For instance, it has been demonstrated that incorporating titanium dioxide and zinc oxide nanoparticles into fabrics results in good UV protection [46]. Furthermore, by strengthening the fiber structure and enhancing abrasion resistance, nanoparticles can be employed to improve the mechanical properties of textiles [47].

13.3.3 Nanocoatings in Protective Textiles

Nanocoatings are thin films that are applied to the surface of textiles to protect them against a variety of risks. They can be used in a variety of ways, including dip coating, plasma deposition, and chemical vapour deposition. The nanocoatings can be utilized to make protective textiles that are more resistant to liquids like water and oil. Additionally, UV and electromagnetic radiation protection can be provided through nanocoatings [48]. Additionally, nanocoatings can be utilized to add useful qualities like self-cleaning, anti-fouling, and anti-bacterial capabilities, making them perfect for usage in protective fabrics for healthcare environments [49, 50].

In conclusion, by improving the performance and functionality of protective fabrics, nano-engineering has transformed the industry. The three most important uses of nanotechnology in the creation of protective textiles are nanofibers, nanoparticles, and nanocoatings. Nanofibers are perfect for usage in smart protective fabrics due to their high surface area and mechanical strength. Nanoparticles can be employed to improve the mechanical qualities of textiles and offer defense against a variety of dangers. Nanocoatings give textiles functional characteristics and improved resistance to liquids. We can increase people's safety and protection in a variety of dangerous locations by incorporating nanotechnology into the manufacture of protective textiles.

13.4 Characterization of Nano-Engineered Protective Textiles Using Recycled Wastes

The utilization of recycled waste in the manufacture of protective textiles has drawn a lot of attention due to the rising demand for eco-friendly textiles. The mechanical, thermal, and protective qualities of such textiles have been further enhanced by the incorporation of nanotechnology [51]. The techniques for evaluating and testing protective textiles' characteristics are covered in this segment, along with examples of studies on the mechanical, thermal, and protective qualities of protective textiles made from recycled materials.

13.4.1 Methods for Analyzing and Testing the Properties of Protective Textiles

Various techniques, including as mechanical, thermal, and protective properties, can be used to examine and test the characteristics of protective textiles. The strength, tear resistance, and abrasion resistance of protective fabrics' mechanical components are crucial for defining their effectiveness and longevity [52]. For instance, tensile strength describes the highest stress that a fabric can endure before breaking or rupturing. A tensile testing machine can be used to test the tensile strength of materials [53]. Contrarily, rip strength describes a fabric's capacity to endure tearing forces. The Elmendorf tear test or the Trapezoidal tear test can be used to measure tear strength. The Martindale abrasion tester can be used to measure a textile's abrasion resistance, which indicates its resistance to surface wear [54].

The ability of protective fabrics to control temperature depends on thermal qualities including thermal conductivity. A thermal conductivity analyzer can be used to measure the thermal conductivity of textiles [55].

A variety of testing techniques can also be used to examine protective qualities such water repellency, chemical resistance, and flame resistance. For instance, the vertical flame test can be used to assess flame resistance, whereas the spray test can be used to assess water repellency [56].

Some studies on the mechanical, thermal, and protective qualities of protective textiles made with nanotechnology and recycled materials.

The characteristics of protective textiles made with nanotechnology and recycled materials have been the subject of several investigations. For instance, Zaghloul *et al.* [57] examined the impact of using graphene oxide and recycled polyester waste in the creation of protective fabrics. According to the study, adding graphene oxide greatly increased the tensile strength and tear strength of the fabrics. The mechanical and thermal characteristics of recycled polyethylene terephthalate (PET) fibers reinforced with multi-walled carbon nanotubes were also examined in a study by Fang *et al.* [58]. Incorporating carbon nanotubes dramatically increased the tensile strength and thermal conductivity of the PET fibers, according to the study.

Wang *et al.* [59] looked at the protective qualities of nonwoven polypropylene textiles that have been reinforced with cellulose nanofibers. According to the study, the addition of cellulose nanofibers considerably increased the materials' ability to resist fire and water. Similar to this, Wang *et al.* [60] examined the mechanical and thermal characteristics of cotton fabrics treated with cellulose nanocrystals. According to the study, the treated fabrics showed better tensile strength and thermal stability.

Due to the rising demand for environmentally friendly textiles, the use of recycled waste in the creation of protective textiles has attracted a lot of attention. The mechanical, thermal, and protective qualities of protective fabrics have been further enhanced by incorporating nanotechnology into their design. The characteristics of protective fabrics can be examined and tested using a variety of techniques, including mechanical, thermal, and protective testing [61]. According to studies, the qualities of protective fabrics can be greatly enhanced by adding recycled garbage and nanoparticles. The potential of recycled waste-based, nano-engineered protective textiles for use in a variety of industries, including the medical, military, and aerospace fields, has to be further investigated. Advanced analytical methods including scanning electron microscopy, X-ray diffraction, and Fourier transform infrared spectroscopy can reveal more information about the composition and characteristics of protective fabrics that have been nano-engineered.

In general, creating environmentally friendly protective fabrics from recycled garbage and nanomaterials is a viable strategy for establishing a circular economy and minimizing negative environmental effects. In addition to offering superior protection, these fabrics can help save natural resources and lessen waste production. To develop more durable and efficient protective fabrics for a brighter future, it is crucial to keep investigating cutting-edge ideas and technology.

13.5 Applications of Nano-Engineered Protective Textiles Using Recycled Wastes

Waste generation has increased as a result of the quick development of industrialization and urbanization, which is turning into a serious environmental issue. Recycling is a successful waste management strategy that not only minimizes trash but also protects natural resources [62]. As a creative response to the issue of waste management, nano-engineered protective fabrics made from recycled garbage have been created recently. These fabrics provide a number of advantages, including greater performance, increased durability, and defense against environmental risks. We shall talk about the uses of protective textiles made from recycled waste in this article.

13.5.1 Recycling of Textile Waste

The management of textile waste has emerged as a key issue for the textile industry because it poses a serious threat to the environment. Recycling textile waste is a sustainable waste management strategy that not only minimizes the impact on the environment but also protects natural resources [63–65]. Waste textiles can be recycled using a variety of techniques, such as mechanical, chemical, and thermal recycling. The method of mechanical recycling entails shredding and combining the used textiles to create new products [66]. Chemical recycling is the process of dissolving used textiles into their individual fibers so they can be turned into new goods. Thermal recycling involves the use of heat to break down textile waste into its constituent fibers, which can subsequently be processed into new goods [67].

13.5.2 Nano-Engineering of Protective Textiles

In nano-engineering, materials are altered at the nanoscale to produce new materials with distinctive features. Textiles that have been nano-engineered to offer improved protection against environmental threats are known as protective textiles [68]. These fabrics are made to be breathable, lightweight, and resistant to fire, chemicals, and UV rays. They also provide protection against these dangers. In order to give protective textiles the appropriate qualities, several nanomaterials, including nanoparticles, nanotubes, and nanofibers, are introduced into the textile matrix [69]. These nanoparticles have a high surface area–to–volume ratio, great strength, and distinctive optical and electrical properties, among other advantages. The performance of the protective textile can be considerably improved by adding these nanoparticles to the textile matrix [70, 71].

13.5.3 Applications of Nano-Engineered Protective Textiles Using Recycled Waste

The creation of protective textiles with nanotechnology utilizing recycled trash has created numerous potentials for increased environmental protection and sustainable waste management. Personal protection equipment, military gear, sportswear, and medicinal textiles are just a few uses for these fabrics [72].

Personal protective equipment (PPE) is equipment that is intended to shield users from harmful substances at work. PPE including gloves, coveralls, and face masks can be made utilizing nano-engineered protective textiles made from recovered trash [73]. These fabrics offer improved defense against dangerous substances like chemicals, radiation, and biological agents. For instance, it is possible to create nanofibers from recovered cotton waste and insert them into textiles to improve their ability to fend against radiation and dangerous substances [74].

Military Equipment: Uniforms, helmets, and body armor can all be produced utilizing nano-engineered protective materials made from recycled trash. These materials can offer improved defense against environmental dangers like heat, chemical exposure, and ballistic impact. For instance, it is possible to create nanotubes from recovered plastic waste and insert them into a textile matrix to provide ballistic protection [75].

Sportswear: Sportswear like cycling jerseys, jogging shorts, and hiking pants can be made utilizing nano-engineered protective materials made from waste. These fabrics can offer improved defense against environmental dangers like UV rays, heat, and dampness. For instance, it is possible to create nanoparticles from recovered polyester waste and insert them into the textile matrix to increase UV protection [83].

Medical Textiles: Medical fabrics including surgical gowns, drapes, and masks can be produced utilizing nano-engineered protective textiles made from recycled waste. These fabrics can offer better defense against biological agents and other dangers that are present in the medical setting. For instance, it is possible to create nanofibers from recovered cotton waste and include them into the textile matrix to increase the effectiveness of the filtration process [83].

Greater environmental protection, sustainable waste management, and improved performance are just a few advantages of employing recycled waste in nano-engineered protective fabrics. These textiles can be utilized in a variety of settings, including as industrial safety gear, military applications, and personal protection equipment. The industry has undergone a revolution thanks to the application of nanotechnology in textile engineering, which has allowed for the creation of materials with increased protective properties and functionality [76, 77].

Environmental friendliness is one of the main benefits of protective fabrics made using nanotechnology. These textiles contribute to waste reduction and a reduction in the textile industry's carbon footprint by using recycled waste materials [78]. Additionally, materials that are strong, light, and easy to wear can be created thanks to the application of nanotechnology. Another important advantage of nano-engineered protective textiles is their ability to improve performance. These textiles can be developed to provide protection against a variety of hazards, such as fire, chemicals, and biological agents. During the manufacturing process, nanoparticles can be introduced to the textile fibers to impart particular qualities including antibacterial, UV-resistant, and water-repellent properties. This makes them perfect for usage in contexts like the military or industry where high-performance textiles are required [79, 80].

Overall, protective fabrics made with nanotechnology provide a practical and efficient way to shield people from a range of risks. These textiles can help minimise waste, lessen the textile industry's carbon impact, and provide protection for those who need it most by utilizing the power of nanotechnology and repurposed waste materials.

13.6 Future Prospects and Challenges

The excellent protection that nano-engineered protective fabrics may offer against physical and environmental threats has attracted a lot of interest in recent years. These textiles can be created from a wide range of materials,

Table 13.4 Future prospects and challenges in nano-engineered protective textiles.

Future prospects	Challenges
Sustainable materials and production methods	High production costs
Enhanced performance and functionality	Regulatory compliance
Integration with wearable technologies	Durability and reliability
Personalized protective textiles	Standardization and certification
Multifunctional protective textiles	Compatibility with existing equipment
Rapid prototyping and customization	Safety concerns related to nanomaterials
Smart and responsive protective textiles	Limited knowledge about long-term health effects of nanomaterials

including recycled trash, offering a greener, more sustainable option to the procedures used to make traditional textiles [81]. Despite the positive potential of these fabrics, there are still issues and restrictions that must be resolved before they can be widely adopted and commercialized. This paper will discuss the possibilities and potential for additional research and development in protective textiles made of nanotechnology and recycled materials, the difficulties and constraints to be overcome for their widespread use and commercialization, and concluding remarks on the outlook for the future [82]. Table 13.4 lists the opportunities for protective fabrics made using nanotechnology as well as forthcoming obstacles.

13.6.1 Opportunities and Potential for Further Research and Development in Nano-Engineered Protective Textiles Using Recycled Waste

Numerous possibilities and possibility for additional research and development exist when recycled wastes are used in the production of protective textiles with nano-engineering. One such chance is the capacity to lessen the environmental impact of textile production procedures [83].

It is well-known that conventional textile manufacturing methods use a lot of energy and produce a lot of waste and pollution. The environmental effect of the manufacturing process for these textiles can be greatly decreased by using recycled wastes.

The capacity to develop distinctive and avant-garde textile materials is another opportunity. The protective qualities of textiles can be improved by adding nanostructures to the matrix of the fabric. For instance, it has been demonstrated that using carbon nanotubes in the production of textiles improves their tensile strength, thermal stability, and electrical conductivity [84]. The production of high-performance textiles can also be done more affordably by using recycled wastes like plastic bottles, rubber tyres, and industrial waste products.

Additionally, the utilization of recycled wastes in protective textiles with nano-engineering can open doors for the creation of textiles with multiple uses. The addition of functional elements like nanoparticles, enzymes, and biomolecules can give textiles extra qualities including antibacterial, self-cleaning, and UV protection [85, 86]. This could result in the creation of textiles that improve comfort and functionality while also offering protection.

13.6.2 Challenges and Limitations to Overcome for Wider Adoption and Commercialization of These Textiles

Although there are possibilities and opportunity for more study and development, there are still obstacles to be solved before commercializing and deploying nano-engineered protective fabrics made from recycled trash. The absence of standardized testing procedures for assessing the performance of these fabrics is one of these issues. To accurately assess these fabrics' protective qualities and to make sure they adhere to industry norms and regulations, it is crucial to develop standardized testing procedures [87–89].

The high production costs for these textiles present another difficulty. The cost of making these textiles can be dramatically raised by adding functional components and nanostructures to the textile matrix. Additionally, the utilization of recycled wastes might necessitate further processing, which would raise the price of production. For this reason, the creation of affordable manufacturing techniques is crucial to the widespread use and commercialization of these fabrics [89–91].

Additionally, both consumers and manufacturers are unaware of and underappreciate the potential advantages of these materials. For raising awareness and promoting the use of these fabrics, the creation of powerful marketing tactics and educational campaigns is crucial [92]. The creation of alliances and partnerships between producers, academic institutions, and regulatory bodies can also speed up the creation and use of these fabrics.

13.6.3 Conclusion

In summary, the use of recycled materials in nano-engineered protective fabrics has enormous potential for improving science and technology. These textiles, which incorporate nanostructures and useful elements, have the potential not only for increased protective properties but also for multifunctionality. However, they face significant obstacles, including the lack of standardized testing procedures, the requirement for the cost-effective incorporation of materials like nanotubes and graphene oxide into production and the challenging scalability barriers. It is crucial to resolve these problems in a productive manner in order to promote wider adoption and guarantee commercial viability. Despite these formidable obstacles, the fascinating potential of nano-engineered protective textiles as a research frontier endures, with the ability to offer enhanced protection and adaptable functionality across a variety of applications, subject to continued advancements in materials science, manufacturing processes, and testing protocols.

References

1. Shahid, M. and Mohammad, F., Perspectives for natural product based agents derived from industrial plants in textile applications–a review. *J. Cleaner Prod.*, *57*, 2–18, 2013.
2. Smith, W.C., An overview of protective clothing-markets, materials, needs, Industrial Textile Associates, 1999.
3. Schreuder-Gibson, H.L., Truong, Q., Walker, J.E., Owens, J.R., Wander, J.D., Jones Jr., W.E., Chemical and biological protection and detection in fabrics for protective clothing. *MRS Bull.*, *28*, 8, 574–578, 2003.
4. Gholamzadeh, L., Asari-Shik, N., Aminian, M.K., Ghasemi-Nejad, M., A study of the shielding performance of fibers coated with high-Z oxides against ionizing radiations. *Nucl. Instrum. Methods Phys. Res., Sect. A*, *973*, 164174, 2020.
5. Roy, R., Laha, A., Awasthi, N., Majumdar, A., Butola, B.S., Multi layered natural rubber coated woven P-aramid and UHMWPE fabric composites for soft body armor application. *Polym. Compos.*, *39*, 10, 3636–3644, 2018.

6. Burgin, S. and Hardiman, N., Extreme sports in natural areas: looming disaster or a catalyst for a paradigm shift in land use planning? *J. Environ. Plann. Manage.*, *55*, 7, 921–940, 2012.
7. Mao, N., High performance textiles for protective clothing, in: *High performance textiles and their applications*, pp. 91–143, Woodhead Publishing, Cambridge, United Kingdom, 2014.
8. Dolez, P.I., Marsha, S., McQueen, R.H., Fibers and Textiles for Personal Protective Equipment: Review of Recent Progress and Perspectives on Future Developments. *Textiles*, *2*, 2, 349–381, 2022.
9. Júnior, H.L.O., Neves, R.M., Monticeli, F.M., Dall Agnol, L., Smart Fabric Textiles: Recent Advances and Challenges. *Textiles*, *2*, 4, 582–605, 2022.
10. Shishoo, R., Recent developments in materials for use in protective clothing. *Int. J. Clothing Sci. Technol.*, *14*, 3/4, 201–215, 2002.
11. Hoffmann, K., Laperre, J., Avermaete, A., Altmeyer, P., Gambichler, T., Defined UV protection by apparel textiles. *Arch. Dermatol.*, *137*, 8, 1089–1094, 2001.
12. Costa, C., Azoia, N., Silva, C., Marques, E., Textile industry in a changing world: Challenges of sustainable development. *U. Porto J. Eng.*, *6*, 2, 86–97, 2020.
13. Kirchherr, J., Reike, D., Hekkert, M., Conceptualizing the circular economy: An analysis of 114 definitions. *Resour. Conserv. Recycl.*, *127*, 221–232, 2017.
14. Singh, N., Hui, D., Singh, R., Ahuja, I.P.S., Feo, L., Fraternali, F., Recycling of plastic solid waste: A state of art review and future applications. *Composites, Part B*, *115*, 409–422, 2017.
15. Choi, T.M., Cai, Y.J., Shen, B., Sustainable fashion supply chain management: A system of systems analysis. *IEEE Trans. Eng. Manage.*, *66*, 4, 730–745, 2018.
16. McNeill, L. and Moore, R., Sustainable fashion consumption and the fast fashion conundrum: fashionable consumers and attitudes to sustainability in clothing choice. *Int. J. Consum. Stud.*, *39*, 3, 212–222, 2015.
17. Patti, A., Cicala, G., Acierno, D., Eco-sustainability of the textile production: Waste recovery and current recycling in the composites world. *Polymers, 13*, 1, 134, 2020.
18. Davidson, M.G., Furlong, R.A., McManus, M.C., Developments in the life cycle assessment of chemical recycling of plastic waste–A review. *J. Cleaner Prod.*, *293*, 126163, 2021.
19. Dogu, O., Pelucchi, M., Van de Vijver, R., Van Steenberge, P.H., D'hooge, D.R., Cuoci, A., Van Geem, K.M., The chemistry of chemical recycling of solid plastic waste via pyrolysis and gasification: State-of-the-art, challenges, and future directions. *Prog. Energy Combust. Sci.*, *84*, 100901, 2021.
20. Janigo, K.A., Wu, J., DeLong, M., Redesigning fashion: An analysis and categorization of women's clothing upcycling behavior. *Fashion Pract.*, *9*, 2, 254–279, 2017.

21. Gwilt, A. and Rissanen, T. (Eds.), *Shaping sustainable fashion: Changing the way we make and use clothes*, Routledge, 2012.
22. Hunley, M.T. and Long, T.E., Electrospinning functional nanoscale fibers: a perspective for the future. *Polym. Int.*, *57*, 3, 385–389, 2008.
23. Xia, Y., He, L., Feng, J., Xu, S., Yao, L., Pan, G., Waterproof and moisture-permeable polyurethane nanofiber membrane with high strength, launderability, and durable antimicrobial properties. *Nanomaterials*, *12*, 11, 1813, 2022.
24. Zhu, F.L. and Feng, Q.Q., Recent advances in textile materials for personal radiative thermal management in indoor and outdoor environments. *Int. J. Therm. Sci.*, *165*, 106899, 2021.
25. Syduzzaman, M.D., Patwary, S.U., Farhana, K., Ahmed, S., Smart textiles and nano-technology: a general overview. *J. Text. Sci. Eng.*, *5*, 1, 1–7, 2015.
26. Yetisen, A.K., Qu, H., Manbachi, A., Butt, H., Dokmeci, M.R., Hinestroza, J.P., Yun, S.H., Nanotechnology in textiles. *ACS Nano*, *10*, 3, 3042–3068, 2016.
27. Kim, B.J., Cha, S.H., Kong, K., Ji, W., Park, H.W., Park, Y.B., Synergistic interfacial reinforcement of carbon fiber/polyamide 6 composites using carbon-nanotube-modified silane coating on ZnO-nanorod-grown carbon fiber. *Compos. Sci. Technol.*, *165*, 362–372, 2018.
28. Karim, N., Afroj, S., Lloyd, K., Oaten, L.C., Andreeva, D.V., Carr, C., Novoselov, K.S., Sustainable personal protective clothing for healthcare applications: a review. *ACS Nano*, *14*, 10, 12313–12340, 2020.
29. Pensupa, N., Leu, S.Y., Hu, Y., Du, C., Liu, H., Jing, H., Lin, C.S.K., Recent trends in sustainable textile waste recycling methods: current situation and future prospects, in: *Chemistry and Chemical Technologies in Waste Valorization*, pp. 189–228, 2018.
30. Soni, A., Das, P.K., Hashmi, A.W., Yusuf, M., Kamyab, H., Chelliapan, S., Challenges and opportunities of utilizing municipal solid waste as alternative building materials for sustainable development goals: A review. *Sustain. Chem. Pharm.*, *27*, 100706, 2022.
31. Zong, B., Sun, B., Cheng, S., Mu, X., Yang, K., Zhao, J., Wu, W., Green production technology of the monomer of nylon-6: caprolactam. *Engineering*, *3*, 3, 379–384, 2017.
32. Okafor, C.C., Madu, C.N., Ajaero, C.C., Ibekwe, J.C., Nzekwe, C.A., Okafor, C.C., Nzekwe, C.A., Sustainable management of textile and clothing. *Clean Technol. Recycl.*, *1*, 70–87, 2021.
33. Kumartasli, S. and Avinc, O., Important step in sustainability: polyethylene terephthalate recycling and the recent developments, in: *Sustainability in the Textile and Apparel Industries: Sourcing Synthetic and Novel Alternative Raw Materials*, pp. 1–19, 2020.
34. Patnaik, A., Mvubu, M., Muniyasamy, S., Botha, A., Anandjiwala, R.D., Thermal and sound insulation materials from waste wool and recycled polyester fibers and their biodegradation studies. *Energy Build.*, *92*, 161–169, 2015.

35. H Cumming, D., A case study engaging design for textile upcycling. *J. Text. Des. Res. Pract.*, *4*, 2, 113–128, 2016.
36. Zeeshan, M., Ali, M., Anjum, A.S., Nawab, Y., Optimization of mechanical/thermal properties of glass/flax/waste cotton hybrid composite. *J. Ind. Text.*, *51*, 5, 768–787, 2021.
37. Ahmad, F., Akhtar, K.S., Anam, W., Mushtaq, B., Rasheed, A., Ahmad, S., Nawab, Y., Recent Developments in Materials and Manufacturing Techniques Used for Sports Textiles. *Int. J. Polym. Sci.*, 2023, https://doi.org/10.1155/2023/2021622.
38. Qin, Y., Zhang, X., Chai, J., Damage performance and compressive behavior of early-age green concrete with recycled nylon fiber fabric under an axial load. *Constr. Build. Mater.*, *209*, 105–114, 2019.
39. Sadeghi, B., Marfavi, Y., AliAkbari, R., Kowsari, E., Borbor Ajdari, F., Ramakrishna, S., Recent studies on recycled PET fibers: production and applications: a review. *Mater. Circ. Econ.*, *3*, 1–18, 2021.
40. Sandvik, I.M., Stubbs, W., Circular fashion supply chain through textile-to-textile recycling. *J. Fashion Mark. Manag.*, 23, 3, 366–381, 2019.
41. Lu, J.J. and Hamouda, H., Current status of fiber waste recycling and its future. *Adv. Mater. Res.*, 878, 122–131, 2014. Trans Tech Publications Ltd.
42. Patra, J.K. and Gouda, S., Application of nanotechnology in textile engineering: An overview. *J. Eng. Technol. Res.*, *5*, 5, 104–111, 2013.
43. Raza, A., Li, Y., Sheng, J., Yu, J., Ding, B., Protective clothing based on electrospun nanofibrous membranes, in: *Electrospun Nanofibers for Energy and Environmental Applications*, pp. 355–369, 2014.
44. Sawhney, A.P.S., Condon, B., Singh, K.V., Pang, S.S., Li, G., Hui, D., Modern applications of nanotechnology in textiles. *Text. Res. J.*, *78*, 8, 731–739, 2008.
45. Roy, R., Chavan, P.P., Rajeev, Y., Praveenraj, T., Kolazhi, P., Sustainable manufacturing practices in textiles and fashion. in: *Sustainable Manufacturing Practices in the Textiles and Fashion Sector*, pp. 1–22, Cham, Switzerland, 2024.
46. Morabito, K., Shapley, N.C., Steeley, K.G., Tripathi, A., Review of sunscreen and the emergence of non-conventional absorbers and their applications in ultraviolet protection. *Int. J. Cosmet. Sci.*, *33*, 5, 385–390, 2011.
47. Asif, A. K. M. A. H. and Hasan, M.Z., Application of nanotechnology in modern textiles: A review. *Int. J. Curr. Eng. Technol.*, *8*, 2, 227–231, 2018.
48. Mahmud, R. and Nabi, F., Application of nanotechnology in the field of textile. *IOSR J. Polym. Text. Eng.*, *4*, 1–6, 2017.
49. Nguyen-Tri, P., Tran, H.N., Plamondon, C.O., Tuduri, L., Vo, D.V.N., Nanda, S., Bajpai, A.K., Recent progress in the preparation, properties and applications of superhydrophobic nano-based coatings and surfaces: A review. *Prog. Org. Coat.*, *132*, 235–256, 2019.
50. Jhaveri, J.H. and Murthy, Z.V.P., A comprehensive review on anti-fouling nanocomposite membranes for pressure driven membrane separation processes. *Desalination*, *379*, 137–154, 2016.

51. Fang, C., Yang, R., Zhang, Z., Zhou, X., Lei, W., Cheng, Y., Wang, D., Effect of multi-walled carbon nanotubes on the physical properties and crystallisation of recycled PET/TPU composites. *RSC Adv.*, *8*, 16, 8920–8928, 2018.
52. Roy, R., Majumdar, A., Butola, B.S., Comparative study of P-aramid based soft and stiff composite panels for protective application. *Fibers Polym.*, 406–412, 2019.
53. Ertekin, M. and Kirtay, E., Tensile properties of some technical core spun yarns developed for protective textiles. *Text. Apparel*, *25*, 2, 104–110, 2015.
54. Witkowska, B. and Frydrych, I., Static tearing, part I: Its significance in the light of european standards. *Text. Res. J.*, *78*, 6, 510–517, 2008.
55. Baxter, S.T., The thermal conductivity of textiles. *Proc. Phys. Soc.*, *58*, 1, 105, 1946.
56. Ahmed, M.T., Morshed, M.N., Farjana, S., An, S.K., Fabrication of new multifunctional cotton–modal–recycled aramid blended protective textiles through deposition of a 3D-polymer coating: High fire retardant, water repellent and antibacterial properties. *New J. Chem.*, *44*, 28, 12122–12133, 2020.
57. Zaghloul, M.Y.M., Zaghloul, M.M.Y., Zaghloul, M.M.Y., Developments in polyester composite materials–An in-depth review on natural fibres and nano fillers. *Compos. Struct.*, *278*, 114698, 2021.
58. Roy, R., Ashmika, R., Chakraborty, A., Polymeric energy harvesting materials: A comprehensive overview. *Emerg. Mater. Technol. Solut. Energy Harvesting*, 33–71, 2024.
59. Wang, D.C., Yu, H.Y., Ouyang, Z., Qi, D., Zhou, Y., Ju, A., Cao, Y., Chain-ring covalently interconnected cellulose nanofiber/MWCNT aerogel for supercapacitors and sensors. *Nanoscale*, *14*, 13, 5163–5173, 2022.
60. Wang, Z., Yao, Z., Zhou, J., He, M., Jiang, Q., Li, A., Zhang, D., Improvement of polylactic acid film properties through the addition of cellulose nanocrystals isolated from waste cotton cloth. *Int. J. Biol. Macromol.*, *129*, 878–886, 2019.
61. Roy, R., Eco-friendly colors: A review of recent advances in natural dyes. *Promot. Sustain. Manag. Technol. Innov.*, 203–215, 2023.
62. Poornima, S., Manikandan, S., Karthik, V., Balachandar, R., Subbaiya, R., Saravanan, M., Nguyen, T.L.C., Pugazhendhi, A., Emerging nanotechnology based advanced techniques for wastewater treatment. *Chemosphere*, 303, 135050, 2022.
63. Tozlu, A., Özahi, E., Abuşoğlu, A., Waste to energy technologies for municipal solid waste management in Gaziantep. *Renew. Sustain. Energy Rev.*, *54*, 809–815, 2016.
64. Holkar, C.R., Jadhav, A.J., Pinjari, D.V., Mahamuni, N.M., Pandit, A.B., A critical review on textile wastewater treatments: possible approaches. *J. Environ. Manage.*, *182*, 351–366, 2016.

65. Yalcin-Enis, I., Kucukali-Ozturk, M., Sezgin, H., Risks and management of textile waste, in: *Nanoscience and biotechnology for environmental applications*, pp. 29–53, 2019.
66. Ribul, M., Lanot, A., Pisapia, C.T., Purnell, P., McQueen-Mason, S.J., Baurley, S., Mechanical, chemical, biological: Moving towards closed-loop bio-based recycling in a circular economy of sustainable textiles. *J. Cleaner Prod., 326*, 129325, 2021.
67. Lee, H. S., Jung, S., Lin, K.-Y.A., Kwon, E.E., Lee, J., Upcycling textile waste using pyrolysis process. *Sci. Total Environ.*, 859, 160393, 2023.
68. Roy, R., and Ashmika, R., Textile products in healthcare: Innovations, applications, and emerging trends. *Emerg. Technol. Health Lit. Med. Pract.*, 288–314, 2024.
69. Jadoun, S., Verma, A., Arif, R., Modification of textiles via nanomaterials and their applications, in: *Frontiers of textile materials: polymers, nanomaterials, enzymes, and advanced modification techniques*, pp. 135–152, 2020.
70. Yin, Z., Lu, H., Gan, L., Zhang, Y., Electronic fibers/textiles for health-monitoring: fabrication and application. *Adv. Mater. Technol.*, *8*, 3, 2200654, 2023.
71. Malik, S., Muhammad, K., Waheed, Y., Nanotechnology: A Revolution in Modern Industry. *Molecules*, *28*, 2, 661, 2023.
72. Peng, X., Meng, X., Yu, B., Chen, H., Liu, Z., Tang, M., Zheng, Y., Sun, Y., Lu, W., Dai, Y., Graphitized and flexible porous textile updated from waste cotton for wearable electromagnetic interference shielding. *Carbon*, 207, 144–153, 2023.
73. Liang, D., Liu, W., Zhong, T., Liu, H., Dhandapani, R., Li, H., Wolcott, M., Nanocellulose reinforced lightweight composites produced from cotton waste via integrated nanofibrillation and compounding. *Sci. Rep.*, *13*, 1, 2144, 2023.
74. Rizal, S., HPS, A.K., Oyekanmi, A.A., Gideon, O.N., Abdullah, C.K., Yahya, E.B., Rahman, A.A., Cotton wastes functionalized biomaterials from micro to nano: a cleaner approach for a sustainable environmental application. *Polymers*, *13*, 7, 1006, 2021.
75. Habibie, S., Review of Nano Technology Development in Textile Industry and the Role of R&D in Indonesia. *Adv. Nanosci. Technol.: Int. J. (ANTJ)*, 5, 1, 1–6, 2019.
76. Wang, Q., Yang, L., Song, J., Preparation, thermal conductivity, and applications of nano–enhanced phase change materials (NEPCMs) in solar heat collection: A review. *J. Energy Storage*, *63*, 107047, 2023.
77. Gong, X., Yin, X., Wang, F., Liu, X., Yu, J., Zhang, S., Ding, B., Electrospun Nanofibrous Membranes: A Versatile Medium for Waterproof and Breathable Application. *Small*, *19*, 2, 2205067, 2023.
78. Harifi, T. and Montazer, M., Application of nanotechnology in sports clothing and flooring for enhanced sport activities, performance, efficiency and comfort: a review. *J. Ind. Text.*, *46*, 5, 1147–1169, 2017.

79. Shaheen, T.I., Nanotechnology for modern textiles: highlights on smart applications. *J. Text. Inst.*, *113*, 10, 2274–2284, 2022.
80. Shah, M.A., Pirzada, B.M., Price, G., Shibiru, A.L., Qurashi, A., Applications of nanotechnology in smart textile industry: A critical review. *J. Adv. Res.*, *38*, 5, 55–75, 2022.
81. Samuel, M.S., Ravikumar, M., John J, A., Selvarajan, E., Patel, H., Chander, P.S., Chandrasekar, N., A review on green synthesis of nanoparticles and their diverse biomedical and environmental applications. *Catalysts*, *12*, 5, 459, 2022.
82. Zia, K.M., Noreen, A., Zuber, M., Tabasum, S., Mujahid, M., Recent developments and future prospects on bio-based polyesters derived from renewable resources: A review. *Int. J. Biol. Macromol.*, *82*, 1028–1040, 2016.
83. Basak, S., Laha, A., Bar, M., Roy, R., Recent advances in protective textile materials. *Adv. Text. Eng. Mater.*, 55–86, 2018.
84. Liu, Y., Wang, X., Qi, K., Xin, J.H., Functionalization of cotton with carbon nanotubes. *J. Mater. Chem.*, *18*, 29, 3454–3460, 2008.
85. Maan, A.M., Hofman, A.H., de Vos, W.M., Kamperman, M., Recent developments and practical feasibility of polymer-based antifouling coatings. *Adv. Funct. Mater.*, *30*, 32, 2000936, 2020.
86. Banerjee, I., Pangule, R.C., Kane, R.S., Antifouling coatings: recent developments in the design of surfaces that prevent fouling by proteins, bacteria, and marine organisms. *Adv. Mater.*, *23*, 6, 690–718, 2011.
87. Qian, H., Kucernak, A.R., Greenhalgh, E.S., Bismarck, A., Shaffer, M.S., Multifunctional structural supercapacitor composites based on carbon aerogel modified high performance carbon fiber fabric. *ACS Appl. Mater. Interfaces*, *5*, 13, 6113–6122, 2013.
88. Memon, H. and Kumari, N., Study of multifunctional nanocoated cold plasma treated polyester cotton blended curtains. *Surf. Rev. Lett.*, *23*, 05, 1650036, 2016.
89. Joshi, M. and Bhattacharyya, A., Nanotechnology–a new route to high-performance functional textiles. *Text. Prog.*, *43*, 3, 155–233, 2011.
90. Holme, I., Innovative technologies for high performance textiles. *Color. Technol.*, *123*, 2, 59–73, 2007.
91. He, T., Wang, H., Wang, J., Tian, X., Wen, F., Shi, Q., Lee, C., Self-sustainable wearable textile nano-energy nano-system (NENS) for next-generation healthcare applications. *Adv. Sci.*, *6*, 24, 1901437, 2019.
92. Peleg Mizrachi, M. and Tal, A., Regulation for promoting sustainable, fair and circular fashion. *Sustainability*, *14*, 1, 502, 2022.

14

Advanced Application of Recycled Textile Generated from Waste in Military Application Using Nanotechnology

Rohit Rai and Prodyut Dhar*

School of Biochemical Engineering, Indian Institute of Technology (BHU), Varanasi, Uttar Pradesh, India

Abstract

Due to high consumption in recent decades, the accumulation of textile waste has increased significantly. The generated waste textiles can be utilized for making value-added products via recycling approaches. Advances in nanotechnology have also shown great potential in textile modification that can be utilized in making smart, robust, and protective textiles integrated with wearable sensing technologies. Cellulose, as the major textile component, has been modified with nanotechnology-based approaches to incorporate flame retardance, mechanical strength, and anti-microbial features, along with stimuli response. The essential requirement of textiles for military applications is self-cleaning and dirt-free textiles; health monitoring; anti-static; anti-allergic; improved wear and tear resistance; capable to minimize injuries, infections, and accidents; security; and comfort. Furthermore, nanotechnology-based modification can equip textiles with sensing technologies for monitoring vital signs; detection of warfare, explosives, and devices; supercapacitors; and batteries for power storage. The utilization of textile wastes in fabricating smart textiles for military applications using nanotechnology is an innovative and upcoming research area. The current chapter discusses textile recycling technologies, nanotechnology modification strategies, and the potential of recycled and modified textiles for military-based applications to improve security, comfort, surveillance, and protection from hazardous situations.

***Keywords*:** Cellulose, self-cleaning, composites, protective materials

**Corresponding author*: prodyut.bce@iitbhu.ac.in

Prashansa Sharma and Shilpi Shree Sahay (eds.) Nanotechnology-Assisted Recycling of Textile Waste: Sustainable Tools for Textiles, (403–434) © 2025 Scrivener Publishing LLC

List of Abbreviations

MT	Million tons
NBC	Nitrated bacterial cellulose
PET	Polyethylene terephthalate
HMDA	Hexamethylene diamine
DMT	Dimethyl terephthalate
CNT	Carbon nanotubes
CNF	Cellulose nanofibers
SWCNT	Single-walled CNT
PVDF	Polyvinylidene fluoride
STF	Shear thickening fluid
MWCNT	Multiwalled CNT
MCC	Microcrystalline cellulose
UPF	UV protection factor
TENG	Triboelectric nanogenerators
PVB	Poly (vinyl butyl)

14.1 Introduction

The textile industry has developed rapidly in recent decades globally due to the rise in demand for modified textiles products with varying designs and quality [1]. Apart from clothes, products, like curtains, carpets, shoes, and home design articles made from synthetic and organic materials, are also a part of the textile industry [2]. Food and Agriculture Organization (FAO) reports that, in 1950, 3.7 kg of textiles per individual was utilized which enhanced to 11.1 kg of fiber per individual by the year 2007. The reason for this enhanced textile consumption is the technical advances and functional textiles that have been developed recently. With the increased demand for textiles, the annual generation rise from 70.6 million tons (MT) in 2007 to 90.8 MT in 2014. It is estimated that the apparel and textile market will exceed 100 MT in 2025 [3]. From an economic perspective, the textile industry has a wide impact as in 2002, the textile industry presented a $400 million business worldwide, accounting for 6% of global trade. The trade of textiles improved to $1.8 trillion dollars globally in 2012 [4]. India and China are the top industries of textiles globally. China, as the world's largest textile exporter, accounts for 54% of the global textiles with a value of $755 in year 2012. India, on the other hand, presents a trade cost of $108 billion dollars, which generates around 4% of Indian GDP. The United

States as the largest consumer and the biggest importer of textile products imported about $151 billion worth of textiles in the year 2012 [5].

The increase in global textile production has also impacted the ecosystem to a great extent as 10% to 20% of products in the textile industry are considered wasted. The annual textile waste in the United States and China is approximated to be 12.4 and 26 MT, respectively, whereas, in Hong Kong, 293 tons of textiles wastes are produced per day [4]. The life cycle of textiles requires large amounts of chemicals, energy, and a high volume of water which generates major pollution footprints on the ecosystem during large-scale textile production [6]. The waste generated from textile industries can be grouped into two categories, i.e., pre-consumer textile waste, generated when textiles are produced, and post-consumer textiles, which is generated after the utilization and discarding of textiles [7]. The pre-consumer waste of textiles also involves raw materials used for textile production i.e., synthetic fiber, peptide fibers, and cellulose. Cellulose fibers are obtained from ramie, hemp, and cotton-based plant fibers, whereas synthetic fibers are made from petroleum-based polymers like acrylic, spandex, nylon, polyester, and polypropylene. Protein fibers are obtained from animal-based sources like silk, cashmere, wood, and angora fibers [8]. Depending on the raw material, the specific types of process and dyeing chemicals are used during the production process of fabrics leading to the generation of different streams of textile waste. Recycling industrial textile waste decreases the requirement for fresh raw materials and chemicals, which reduces the environmental impacts generated. The recycling of textile waste follows the principal of circular economy as the materials can be recycled and returned back to the economy and utilized again [9]. The European Parliament has targeted that about 70% of textile waste must be recycled by the end of 2030, whereas 50% of textile waste is currently recycled in Europe [10]. The utilization of textiles as an insulation material in the construction sector has also encouraged the possibility of recycling the textiles. The insulation textiles come from the processing of old automobiles where synthetic fabrics from upholstery and carpets are recycled. The high cost associated with the management of waste along with less space available for landfilling is making the recycling of waste an important aspect of textile industries [11].

Textiles are ideal substrates and universal interfaces for the integration of electronics, optical devices, and nanomaterials. The integration of advanced technologies and materials makes the prepared textile responsible for magnetic, optical, thermal, chemical, mechanical, and electrical stimuli, which enables the fabrication of wearable devices including data transmission, processing units, and sensors [12]. The high amount of recycled waste generated globally can be modified through nanotechnology-based methods for advanced applications such as the development of robust, stimuli-responsive, and protective textiles [13]. Integration of nanomaterials incorporates multifunctional properties in textiles without altering the comfortability of the substrates. For applications in specified fields such as military applications, the textiles need to satisfy appearance, performance, and weight properties [14]. A major challenge in conventional modification of recycled textiles is the absence of permanent effects and usually loses functionality during cleaning or laundry. However, nanotechnology-based modifications can impart recycled textiles with permanent effects such as conductivity, antistatic, antiwrinkle, hydrophobic, light guidance, scattering, improved mechanical strength, and antibacterial properties. These modifications can be applied in the form of foams, fillers, patterns, and surface coatings without affecting the texture and breathability of the textiles used [12]. The potential for nanotechnology in the modification of recycled textiles is enormous, which can be applied to the development of the next generation of military textiles. Recycled textiles, modified with nanotechnology-based approaches can be improvised for military applications by focusing on the following crucial aspects: (i) Materials for enhanced survival and protection: textiles utilized for military applications must be able to withstand enemy fire during wars, and therefore, materials with lightweight and high mechanical strength are suitable for preparing protective textiles [15]. (ii) Low maintenance and high reliability: materials required for military applications should be able to maintain quality for long spans without active maintenance. Paint chipping during Navy operations is a resource-intensive task that consumes a lot of manpower, is such an example. Along with low maintenance, high reliability of materials is also important to ensure the availability of military assets for military requirements [15]. (iii) Lightweight functional materials: lightweight textiles are a crucial requirement for soldiers as it reduces the overall load carried by personnel, which can improve their efficiency. The ability to sense surroundings, withstanding the enemy fire, increased weapon lethality, communication and response with troops and vehicles, and protection to new threats can be an important aspects of the functional textiles developed from waste [16]. (iv) Low-cost, eco-friendly military textiles: for military acquisitions, the cost is a primary criterion. Recycled textiles with low procurement, processing, maintenance, and disposal costs are, therefore, a favorable choice for advanced military application, as the costs for making new textiles are high. Similarly, military from several nations are focusing on the ecological impacts of military operations.

In line with this agenda, the use of recycled military textiles with low carbon footprints will make the production of military wear sustainable [16]. (v) Stealth materials: for strategical and tactical deployment operations, the military assets are required to remain unobserved or unseen by adversaries, which can be done by applying stealth technologies. Textile as stealth materials is an upcoming research area for incorporating electronic, and multifunctional optical properties that can passively or actively achieve, a state of invisibility. (vi) Self-diagnosing and self-healing textiles: these textiles enhance the survivability to reduce the maintenance of a system [15]. (vii) Electronic and photonic materials for high-speed communications: coordination, control, and communication an important function for military troops; therefore, textiles possessing optical and electronic properties with the capability of signal reception, interpretation, and transmission capacity will be a preferable choice for the development of advanced textiles. Nanotechnology-based textile modification approaches can equip wearables with high-speed communication ability [15]. (viii) Textiles integrated with actuators and sensors: recycled textiles equipped with actuators and sensors can detect or sense biological and chemical agents as well as weapons. This will enable the enhanced efficiency of troops, comfortability, and protection from physical, chemical, and biological hazards [16]. For achieving the goals mentioned above for military textiles, cellulose is a preferable material due to its abundant, renewable, biodegradable, sustainable, and mechanical properties [higher than Kevlar and stainless steel (eight times)] [15]. In recent years, cellulose has been utilized for military applications including self-cleaning cellulose-based textiles, magnetic and electricity responsive materials, energetic nanocomposites for explosives and propellants application, flame-retardant, and high-strength materials. Chen *et al.* [17] prepared nitrated bacterial cellulose (NBC) binder–based energetic materials. The nitramine-based explosives were observed to be dispersed and embedded in the porous matrix of NBC. The composites demonstrate substantial heat release of 3,346 J/g with a peak temperature of 224.8°C, which suggests the possibility of utilization of cellulose-based materials in explosive military applications. Abdulrahim *et al.* [18] made composites using cow horn particulate, banana stem fibers, and polyesters for making military helmets. The helmet prepared was lightweight and shows 174.79% improvement in impact energy and 94.66% increase in tensile strength compared to polyester. Cellulose-based materials, which are the core components of almost all the textiles, can be engineering with nanotechnology-based fabrication for enabling the properties suitable for military applications as discussed in the current chapter.

14.2 Types and Sources of Waste

Generally, the textile wastes are composed of a mixture of synthetic and natural polymers like silk, polyester, wood, nylon, and cotton. Textile products are engineered for improvement of their chemical and heat resistance along with mechanical strength [19]. Wastes that are produced during manufacturing processes of fibrous products are termed industrial textile wastes. The generated textile waste can range from monomers to multimers complex systems based on the manufacturing stage releasing the waste [20]. Post-consumer waste usually comes from fibrous products discarded after the shelf life of the product. The utilized textiles can be used as a replacement for outdated products. One of the most utilized routes for the decomposition of used textiles is by landfilling; however, landfilling of fabrics from biodegradable sources, like silk, cotton, wool, and polylactic acid, produces a large amount of methane during decomposition. To reduce the greenhouse gas emission from landfilling, recycling of textile waste can be performed [19].

14.3 Fiber Waste Recycling Procedures

14.3.1 Fiber Spotting and Sorting

The stream originating from a monomeric waste source is comparatively easy to recycle than one containing a mixture of polymers for the preparation of products with high values and quality. In the textile industry, fibrous waste must be segregated as early as possible in generation as it makes the sorting of polymers inconvenient in a mixed slurry [21]. Figure 14.1 shows an overview of the post-consumed textile waste recycling procedures used currently. The post consumed industrial textiles are sorted first, which can be processed through reuse, recycling, landfilling, and incineration. For recycling, chemical approaches include solvent-based approaches such as deep eutectic solvent and ionic liquids. The other approaches involve melt processing and depolymerization of the used textiles. Customization of the recycling process by categorization is important. For example, the type of fibers that are being recycled are important for depolymerization of nylon and recovery of polymer resins. For carpets, sorting based on the fiber type is done, which can be differentiated using Infrared and Raman spectroscopy techniques. Another approach could be the melting point, which, however, is not able to differentiate polyester and nylon 6,6 [19]. After careful sorting of fibers, the polymers are processed by reducing the size for mechanical and chemical treatments. Recently, Teijin Limited has

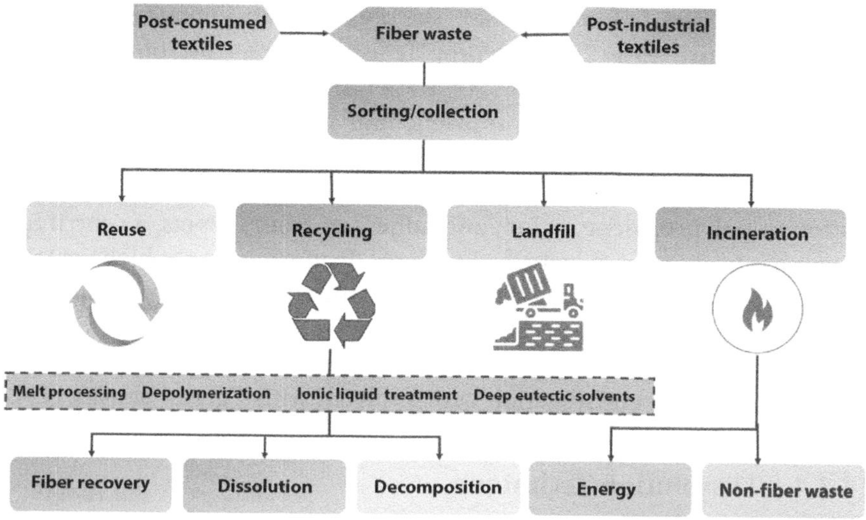

Figure 14.1 Strategies for treatment of textile waste generated post-consumption and their respective fate.

presented the "Eco Circle program" during which a special network was prepared for the recycling of polyester from the sports teams and schools. The collected polyester waste is recycled by depolymerization into virgin fibers and products [22].

14.3.2 Size Reduction

Size reduction of textiles and carpets to smaller sizes is required initially, for any recycling process. Processes such as grinding and shredding are common examples of methods used for size reduction [23, 24]. Generally, the substrate is chopped using a rotary drum equipped with blades against a feeding bed in which the substrate moves against a screen with a specific opening [25, 26]. The pieces that are smaller than the screen drop off, whereas the fiber that does not drop is resented inside the cutting chamber for reprocessing. The shredder has sharp edges for cutting, and a fixed narrow gap between the feeding bed and cutting blades to prevent wrapping of fibers. Low rotational values and high torque are preferred to avoid the melting and heating of polymers [19].

14.3.3 Mechanical Separation

Mechanical methods are used for the separation of carpet components. Du *et al.* recycled nylon 6,6 carpets during which it goes through size reduction and separation processes. The obtained mix contains 15%–20% latex, fillers, dirt, 15%–25% polypropylene, and 50%–70% of nylon. For separation of the shredded fibers, they are immersed in water for density-dependent separation of polypropylene and nylon [27, 28, p. 6]. At the end of the process, 98% pure polypropylene and 98% nylon were obtained which was recycled for the preparation of automotive parts. For the separation of polypropylene, nylon, and adhesives from carpets, a centrifuge system has been prepared which consists of drums rotating at high centrifugal speeds (1,500 g). During the first stage, the adhesive is separated from the polypropylene and nylon fibers using a solvent with 1.15 g/cm^3 density [29]. During the second stage, the separation of polypropylene and nylon was achieved with centrifugation with a solvent with 1 g/cm^3 density [19].

14.3.4 Dissolution Technique

Dissolution has been utilized in recent years for the separation of high-value nylon from waste carpets. For dissolution, hydrochloric acid, alkyl phenols, and aliphatic alcohols are used [19]. Booji *et al.* [30] shredded the carpet wastes into size of 0.5–20 cm^2. The shredded carpet parts were then immersed in methanol as an extraction agent. The solvent-to-carpet weight ratio ranges from 5 to 20, and the optimum dissolution conditions for nylon were temperature of 135°C–140°C and pressure of 0.2–2 MPa for 60 min. Solids from the solution were removed via filtration and cooled down for nylon precipitation. During the extraction, almost no degradation occurred as the nylon extracted and present in the carpet has 90% relative viscosity. One of the drawbacks of using dissolution methods is the requirement of pressure, temperatures, and chemicals involved, and the polymer solubility varies with the change in temperature and pressure [31]. Sikorski *et al.* [32] studied that individual fibers in carpets can be obtained sequentially CO_2 by enhancing pressure (500–1,000 atm) and temperature (170°C–210°C). Another approach to separate the components of the carpet is to utilize the supercritical fluid method in a batch process.

14.4 Approaches in Fiber Recycling

14.4.1 Melt Processing

Melt processing aids in the preparation of resin pellets using thermoplastic polymers by extrusion. The process can also be termed as compounding if more than one type of polymer is included. Fibrous waste like carpets is usually processed through size reduction, followed by densification as the shredded material is bulky [25]. After its melting, the polymer comes through a die for preparation of multiple strands of polymers that are chopped later in the form of pellets. For the preparation of pellets, various types of pelletizers such as strand pelletizer, underwater pelletizer, and water ring pelletizer can be used. The selection of a pelletizer also depends on material properties like melt flow index, melt temperature, pellet quality, and degree of automation required [26]. For example, the water ring method utilizes a rotating knife and water jet to cut the strands as they are extruded. These pelletizers are preferred for polymers like polyethylene because of the low melt flow index. Underwater and strand pelletizers can process a wide spectrum of materials involving polypropylene, polyethylene terephthalate (PET), and nylon. During strand pelletizing, the stands are first cooled in a water bath, followed by drying and chopping the material into pellets. However, a major problem using strand pelletizing is the requirement of constant operator attention so that the strands do not break and the large floor space required for the machine. During underwater pelletizing, the die should be immersed in water and the strands should be chopped by rotating knives before solidification of the polymer. Underwater pelletizing is highly automated and compact and generates pellets having a smooth, consistent, and spherical shape [26]. Melt processing-based polymer recycling is being utilized abundantly by several industries as resins for the generation of new materials that possess almost similar features as virgin polymers.

14.4.2 Polymer Depolymerization

Depolymerization of polymers changes the waste polymers to oligomeric or monomeric forms that can be polymerized and have similar properties to virgin polymers, achieving closed-loop recycling [33]. Chemical recycling utilizes a wide variety of waste belonging to a particular polymer type even if the contamination of multiple polymers is present. Despite the associated advantages of recycling, high cost is associated with depolymerization and purification of products [19]. Polymers like PET and nylon 6 are recycled via depolymerization commercially. Because nylon is a high-value polymer, carpets have been utilized as substrates for the preparation of virgin nylon using depolymerization. For this, carpet waste is collected and processed through shredding prior to depolymerization. Nylon 6 is prepared by polymerization of caprolactam, which is a reversible process. Depolymerization of nylon 6,6 is comparatively more complex than nylon 6 as it is composed of two monomers, i.e., hexamethylene diamine (HMDA) and adipic acid. Depolymerization of nylon 6,6 to obtain HMDA and adipic acid is possible; however, it has not been applied commercially [34]. Another polymer, PET, can be converted back to its base monomers, i.e., terephthalic acid and ethylene glycol on and dimethyl terephthalate (DMT) on methanolysis. The Eco Circle initiative by Teijin Limited has been applied to convert the used textiles to DMT, which is polymerized further into PET for the generation of textile products.

14.4.3 Ionic Liquid (IL) Extraction

Ionic liquids are organic compounds having a low melting point and non-volatile nature at room temperature. It is commonly referred to as a "green solvent" and designer solvent due to its high thermal stability, and reusability. The term designer solvent implies that ILs can be designed using several organic salts with tuneable properties such as lower melting point, extremely high thermal stability, and volatility (non-volatile) [35]. The modular architecture of ILs consists of a cationic core, cationic substituents, anionic core, and anionic substituents [36]. Ionic liquids have been proposed to be synthesized in over a thousand different ways and can be categorized as heterocyclic amines [1-butyl-3-methylimidazolium iodide, 1-butyl-3-methyl pyridinium bis(trifluoromethanesulfonyl)imide, etc.], quaternary cations (tributyl-methyl phosphonium chloride, ethanolammonium nitrate, etc.), and superbase (1,8-diazabicyclo[5.4.0]undec-7-enium acetate, 1,5-diazabicyclo [4.3.0]non-5-enium propionate) [37].

In the year 2002, Swatloski *et al.* used IL (1-butyl-3-methylimidazolium chloride) and microwave heating to dissolve cellulose. The author employed different types of ILs and methods to dissolve cellulose; however. they concluded that 1-butyl-3-methylimidazolium chloride and microwave heating helped in dissolving 25 wt.% of cellulose. Chloride-containing ILs were considered to be the best cellulose solubilizing agent due to the formation of H-bonding between the hydroxyl functional groups and the solvent ions [38]. The cationic and anionic components of IL have been well known to contribute to the disruption of strong cellulose and lignin complexes. According to general trends, the anionic component of ILs preferentially formed H-bonds with the cellulose hydroxyl groups, facilitating the immediate swelling of the biomass [35]. Additional H-bonding between the heavier cations and anion creates repulsive forces in between cellulose chains, creating a solvation enclosure over each monomer unit thus assisting in its disintegration [35, 39]. Interactions between ILs and biomass have also been shown to break the glycosidic linkages between the lignin and the cellulose residues of lignocellulose and hemicellulose fractions, allowing for an incomplete separation of the cellulose [35, 40]. Recent developments regarding the use of ILs in biomass extraction has been focused on developing agro-waste for renewable and sustainable raw materials for various applications such as biofuels, food packaging, and textile reinforcement. Menezes *et al.*, in their work, mixed cellulose-derived orange bagasse into corn starch for the production of cellulose-starch-based composites. The bagasse was first dissolved in an IL of 1-methylimidazolium for the extraction of cellulose fibers. The developed membrane had 0.45MPa tensile strength [41]. Amalini *et al.* developed a cellulose-based film from palm biomass for food packaging applications. The extraction of cellulose was carried out in an IL 1-butyl-3-methylimidazolium chloride. The produced films have opacity in the range of 1.44 to 1.66 and water vapor permeability of ~1.30×10^{-10} g s^{-1} m^{-1} Pa^{-1} [42]. Wei *et al.* recently developed a cellulose-based film for the preservation of pineapple. Cellulose was extracted from the leaves of pineapples using acid (sulphuric acid) and alkali (sodium hydroxide) hydrolysis. Furthermore, the extracted cellulose was dissolved in ILs (1-butyl-3-methylimidazolium chloride), and, then, the produced suspension was sonicated and casted to generate films suggesting the potential of solvent-based approaches for recycling of agricultural biomass waste. The agricultural waste has also been recycled via IL approaches, which dissolves the lignin present in the biomass.

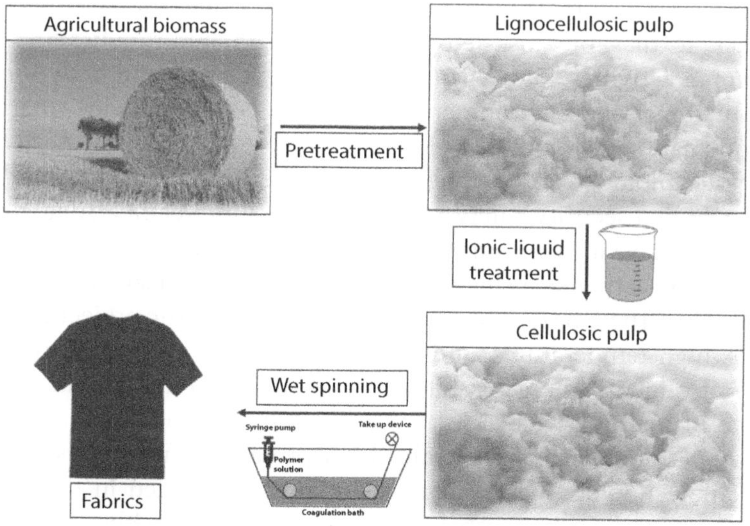

Figure 14.2 Schematic representation of ionic-liquid based recycling of agricultural biomass for textile manufacturing.

The obtained cellulose slurry can be processed with wet spinning for preparation of fabrics as shown in Figure 14.2 [43].

14.5 Textile Modification with Nanotechnology for Military Applications

Military personnel and soldiers encounter countless complex dangers on air, land, and sea that are specifically aimed at harming or maiming them. Therefore, it is compulsory to prepare protective wear like biological and chemical protective clothing, and body armor [44]. Currently, nylon, Nomex, and Kevlar-based fibers are being used for the preparation of protective wear for military applications. Flame retardancy, antiballistic features, and high strength make these fibers optimum for designing textiles for high-technology defense applications. Melt-blown and spun-bonded fabrics are some of the examples of non-woven fabrics that are mainly utilized for manufacturing fire-retardant and barrier protective wear [45]. Protective textiles are advantageous because of their low cost, high strength, comfort, and impermeability to particulate matter. Protective textiles are being modified with nanotechnology, equipping the clothes with (i) lightweight protective clothing; (ii) adaptive suits for switchable camouflage and thermal control; (iii) biological and chemical warfare protection; (iv) flexible antiballistic textiles; (v) self-decontaminating nanofiber fabrics; (vi) wearable or flexible displays on textiles for real-time visual feedback, textiles for exoskeletons, extra strength, and robotics for aiding the human activities; and (vii) microsensors for brain and body sensing, and situational and environmental awareness combined with smart helmets or smart suits [45].

Nanotechnology can be applied for a wide range of textile modifications including security, safety, military, and leisure wear. Personal protective textiles have a role in minimizing the risk of accidents, injuries, and infections by acting as a shield against physical, chemical, and biological hazards [12]. These hazards can be in the form of ballistic projectiles, sharp objects, high temperatures, and fire. Fire-retardant properties in textiles can be of value in emergency situations as well as in the limitation of fire damage in buildings. Fibers can also be made resistant to shrapnel and blade penetration, which can offer high protection from physical damage. Nanotechnology has a huge potential to be applied in the military field for saving lives in combat, with ease of use, and without sacrificing comfort [46]. The application of nanotechnology to textiles will make the soldiers less vulnerable to ecological and enemy-related threats, will reduce war causalities, and will provide special functionalities to the army along with enhanced performance because of the unique properties of textiles obtained at nanoscale dimensions. Properties like flexibility, strength, heat resistance, light weight, and texture comfort can be introduced to the textiles with nanotechnology [47]. Additionally, the combat suits modified with miniaturized electronic nanotechnology-based devices will enable the preparation of multifunctional combat suits for application in areas such as medical monitoring to responsive armor. The fibers used for making suits can be stimuli-responsive, antibacterial, detection of biological weapons, and electronically controlled nanolayers that can make an army invisible virtually [48]. Nanotechnology-based approaches can be utilized for incorporation of "smart" textile properties to the recycled fabrics as shown in

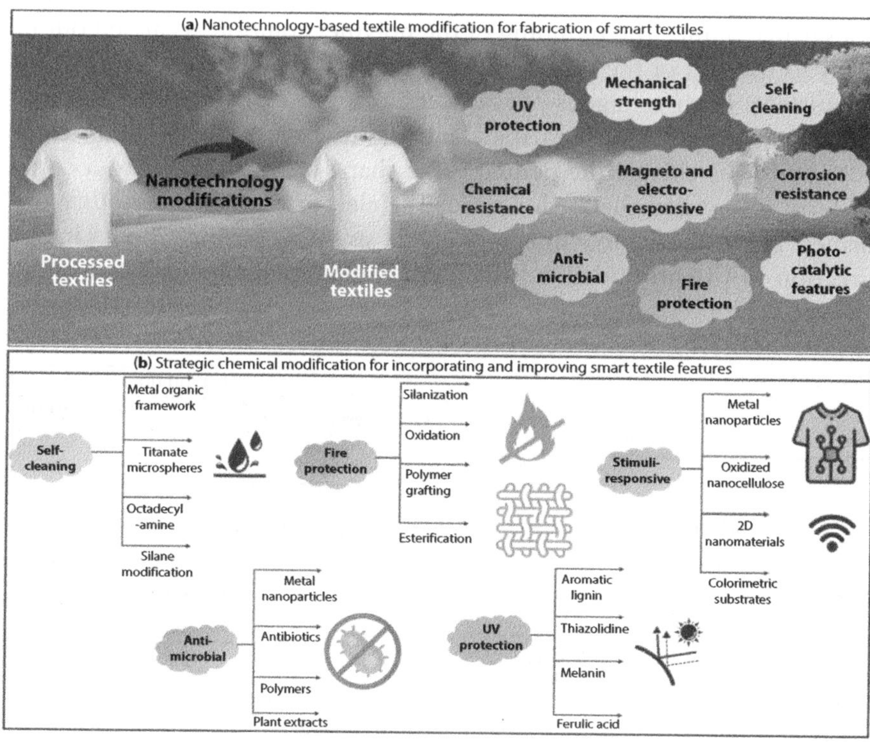

Figure 14.3 (a) Nanotechnology-based textile modifications for making smart textiles; (b) chemical modification and nanomaterial modification strategies for textiles used for incorporation of self-cleaning, anti-microbial, UV-protecting, fire retardant, and stimuli-responsive properties.

Figure 14.3a which can be achieved through physiochemical modification of pristine fabric as shown in Figure 14.3b.

14.5.1 Nanofiber-Based Military Clothing

Nanofiber-based textiles due to their enhanced mechanical strength, tuneability, and ability to be functionalized can be utilized for making high-performance, robust, and stimuli-responsive textiles. Electrospinning is a widely used technique for the production of nanofibers as it is easy to operate and enables low-cost fabrication of nanostructures. The electrospinning procedure typically involves the connection of a direct current power source to produce a high voltage (25–85 kV) in the polymer slurry that is drawn for making nanofibers and collector plates (Figure 14.4). A syringe is used to draw the polymer solution and its storage present at one end. The polymer solution is pumped at a specific flow rate with the aid of a syringe pump. The pumping of the polymer solution is carried out in a way to the end of the tube, resulting in the formation of a hemispherical surface. High voltage is put between the collector plate and polymer suspension, which generates the electrostatic forces. The resultant electrostatic forces change the pendant-like shape of the polymer to a conical form also called the Taylor cone. The electrostatic forces modify the formed Taylor cone into a fine fiber jet via stretching, which moves in the direction of the grounded collector plate and becomes deposited on it.

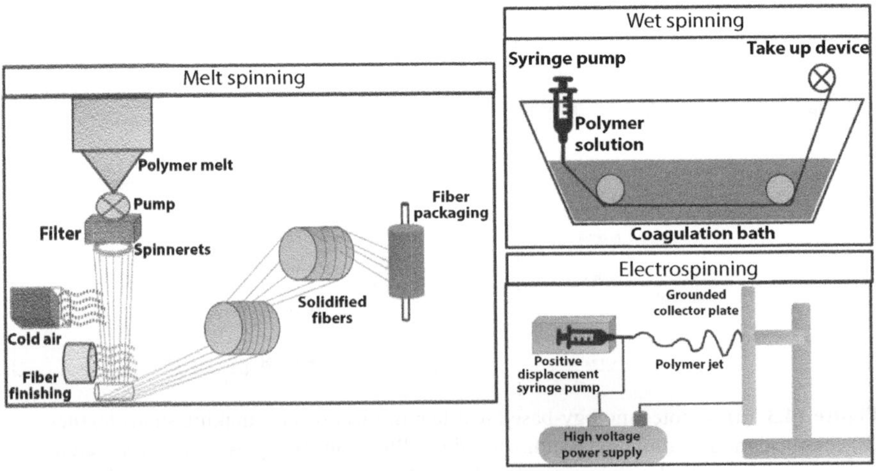

Figure 14.4 Schematic representation of melt spinning, wet spinning, and electrospinning-based processing of textile fibers.

Continuous deposition of fibers on the collector plate forms a non-woven mat with a high surface area–to–volume ratio [45]. The nanofibers prepared with electrospinning have advantages like high strength, surface energy, enables high vapor transmission rate, and softness for behaving as a barrier against fine particulates and microbes [49]. The nanofibers obtained from electrospinning can be processed by spinning into yarn and mats, which can be utilized for multiple applications such as (i) preparation of sensors with high surface area, making polymers sensitive to change and absorption. One such example of nanofiber-based sensors includes the preparation of conductive nanofiber that can be utilized for the preparation of sensors. Qi *et al.* [50] developed piezoresistive carbon nanotubes (CNTs) embedded stretchable and wearable yarns as sensing materials. The nanofiber base sensing yarns were developed by combining polyurethane nanofibers and CNT, wrapping them onto a fiber-shaped stretchable electrode via electrospinning. The nanosensor prepared possesses high-pressure sensitivity of 12.3/N, durability of 10,000 cycles, broad sensing range of 0.001–5 N, large stretch range of 0.1%–220%, and high gauge factors of 114–720. The prepared multimodal, stretchable, and textile-based sensors can be utilized to develop smart military textiles for sensing applications (ii) absorption and nanofiltration-based catalytic breakdown from nanofibers. For example, Shi *et al.* [51] developed cellulose nanofibers (CNFs) and polyethyleneimine-mica nanosheets for salt and dye separation. The two-dimensional nanochannels among the PEI nanosheets with CNF allow the transport of salt and water molecules and restrict dye transfer. The prepared textile shows high hydrophilicity with salt (15.63% of Na_2SO_4, 3.42% for NaCl) and dye retention (98.9% for Congo red, 98.64% for Evans Blue), which indicates high antifouling properties and operation stability. (iii) Insulation properties: for instance, Li *et al.* [52] prepared aerogel-based ultralight textiles from aramid nanofibers for electromagnetic and thermal insulation applications. The aramid nanofibers were modified with CNT via wet spinning followed by freeze-drying. The developed textiles demonstrate thermal insulation in a wide temperature range of −30°C to 200°C, a high specific area of 123–219 m^2/g, and a density of 5–12 mg/cm^3. The prepared aerogel textiles also demonstrate an electromagnetic shielding performance of 30 dB in X-band suggesting the potential of prepared aerogel textiles for electromagnetic shielding and thermal insulation in harsh environments. (iv) Structural applications as reinforcement fabrics for ballistic applications: Yao *et al.* [53] prepared textile materials from meta-aramid nanofibers with diameters ranging from 100 nm to 500 nm with high mechanical and thermal properties, suitable for anti-ballistic applications. (v) Selective gas permeation: for example,

Yang *et al.* [54] developed nanofiber-based Janus epidermal electrode having directional sweat permeability for applications in monitoring of health conditions. The nanofiber-based Janus epidermal electrodes with ability to sweat transport for biopotential monitoring. Janus multilayered membrane prepared is superhydrophillic-hydrolyzed-polyacrylonitrile/polyurethane/silver nanowire with a membrane thickness of 15 μm. The nanofibers can be used in a plethora of applications for developing advanced military-based textiles providing high tuning of features.

14.5.2 Nanotube-Based Fibers and Yarns

Fibers based on nanotubes are made using single-walled CNT (SWCNT), by dispersing the fibers in a surfactant solution, which is performed by delivering these solutions into a rotation bath with polymers like polyvinyl alcohol, and slowly recovering the fibers from the bath allowing the alignment and aggregation of CNTs. The integration of CNTs in the fibers can modify the electrical and mechanical properties [55]. Figure 14.5 describes the types of nanomaterials modification that can be used for making engineered military textiles. For example, Ericson *et al.* [56] generated macroscopic fibers made solely of SWCNTs present in a well-aligned form by fuming sulfuric acid-based wet-spinning. These neat SWCNT fibers present a tensile strength of 116 MPa and Young's modulus of 120 MPa.

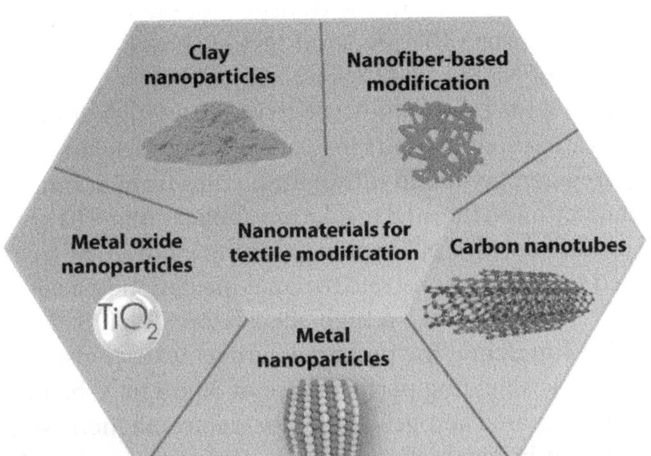

Figure 14.5 Types of nanomaterial modification used for the fabrication of engineered military textiles.

The mechanical properties of the fibers were comparable to laboratory-grade ballistic protection materials possessing mechanical strength of 2.6 GPa and Young's modulus of 138 GPa. The CNT-based nanofibers possess good thermal and electrical conductivity in conjugation to their mechanical properties. Ran *et al.* [57] prepared polyurethane core-sheath nanocomposite with CNT making electro-thermochromic textiles and wearable strain sensors. The prepared textile changed color with temperature as color changed from red (25°C) to yellow (66.4°C) to by applying a voltage of 9V. The prepared composite fibers demonstrate high sensitivity, excellent conductivity, durability (>2,000 cycles), and a wide strain range (0%–310%). The prepared composite textiles can aid in detecting human micro-movements due to large strain range and the color changing ability which can aid in military camouflage. The CNT-based fibers are being used for variety of application including sensors fabrics and actuator for a defense personnels.

14.5.3 Polymer Nanocomposite Fibers and Yarns

Nanocomposites are composed of a matrix, generally made from fiber dispersion of polymer or nanoparticles. Multiple nanoparticles like nano-TiO_2, aluminosilicate, and nano Al_2O_3 have been dispersed in the polymer matrix to generate novel composites with improved thermal, mechanical, and electrical characteristics [45]. For example, polymers having CNTs are an active research field for the preparation of multifunctional materials. Polymers with an easy method of preparation, low density, flexibility, and electric insulation properties can be conjugated with CNTs to improve the thermal, mechanical, and electrical conductivity of the material. These polymers can be applied for chemical sensor preparation, ballistic protection, microwave absorption, and electromagnetic shielding [58]. Polymers, like PET, poly (methyl methacrylate), and poly (vinyl alcohol), are complexed with CNTs to prepare bulk composites, fibers, and films by spinning technologies like solution spinning, electric, and melt spinning. Nanotubes or nanoparticles have also been utilized for the preparation of porous nanocomposites. The traditional porous polymeric foams have a high surface area–to–volume ratio, which enhances the rate of gas and heat release in fire instances. The introduction of nanoclay to foams can significantly reduce the dangers associated with burning. The porous nanocomposites find their applications in making vehicle seats, acoustic absorbents, and shock-absorbing materials [55].

14.6 Defense Applications of Nanotechnology-Based Textiles

14.6.1 Smart Uniforms

The progress in nanotechnology has enabled the fabrication of smart uniforms for soldiers, combined with macro-, micro-, and nanofibers for protection against chemical agents, bioagents, grenade fragments, and bullets. The nanotechnology-modified textiles also provide comfort to the soldier body by offering local cooling, ventilation, and insulation. Development of soldier suits based on nanofiber technology for decontaminating, deactivating, and absorbing capacity for hazards, integrated with biological and chemical sensors for detection of chemical and biological hazards [45]. Asim and Naeem [59] discussed the preparation of denim fabric modified with zinc oxide nanoparticles with self-cleaning properties. The prepared textiles modified with 5% zinc oxide and 7% binder demonstrated self-cleaning properties. Wei *et al.* [60] prepared cooling textiles by aluminum oxide (Al_2O_3) nanoparticles modification of cellulose acetate. The prepared textiles showed solar energy blocking for long-wave infrared radiation at the wavelength range of 8–13 μm. The presence of Al_2O_3 enhanced solar reflectivity and show high thermal conductivity. The combined Al_2O_3-cellulose acetate composites show an improved solar reflectivity from 62.6% to 80.1% and the textile reduces the temperature of the skin model by 2.3°C to 8°C showing cooling performance. The prepared textiles show the potential of nanotechnology in developing smart textiles that can be applied for military applications.

14.6.2 Signature Reduction

A highly classified application of nanotechnology-modified textiles is in the field of developing camouflage materials based on polymer nanocomposites. Camouflage materials that can be tuned dynamically to aid in defense operations are highly valuable as they allowed the equipment and soldiers to achieve a completely concealed presence based on the situation (Figure 14.6). Aloia *et al.* [61] made graphene-based textiles that can absorb wide-range electromagnetic radiation for the reduction of radar signatures in aircraft, ships, and tanks for camouflage and stealth applications. Graphene nanoplatelets filled with polyvinylidene fluoride (PVDF) coatings for generating radar-absorbing coated textiles show minimum bandwidths at −10 dB of 5 GHz and reflection coefficients below −5 dB ranging in frequency 8 GHz to 18 GHz. The produced textile shows a reflection

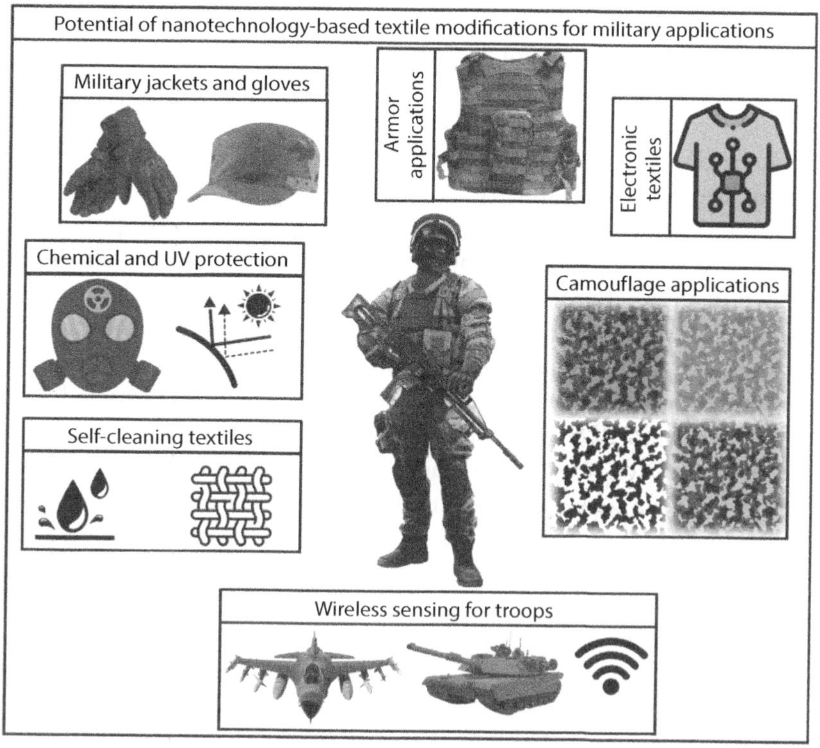

Figure 14.6 Modification of military gears via nanotechnology-based approaches for preparation of next-generation "smart" military wears.

coefficient with a bandwidth at −10 dB upto 77% of the resonant frequency. The nanomodified textiles can be used for military applications in signature reduction and camouflage, for making textiles for advanced military applications. Forsman *et al.* [62] prepared cellulose-based wet foams made from Kraft lignin and *Chlorella vulgaris* microalgae for camouflage applications. The visible light camouflage is crafted with green microalgae, and the brown colored Kraft lignin foams stabilized the foams. The prepared foams demonstrate stealth in the infrared light regions along with high thermal insulation, which suggests its potential for preparing textiles with stealth properties for military camouflage.

14.6.3 Ballistic Protection

For fabrication of vehicle liners and body armor, woven fibers like Nomex and Kevlar are utilized. For instance, Lee *et al.* [63] report the fabrication of nanotechnology-based liquid body armor. For preparation, the conventional body armor based on Kevlar is infiltrated with a fluid with stress-thickening properties made of hard spherical silica nanoparticle–based colloidal dispersion in polyethylene glycol polymer solution. The incorporation of shear-thickening fluid into conventional Kevlar improves the material capacity to absorb energy. Four layers of Kevlar incorporated with shear thickening fluid (STF) show the potential to absorb energy like 10 layers without the STF, suggesting the ballistic performance of the material doubles. The reason for improved ballistic performance is the high shear stresses generated at high volume fractions, which results in hydrocluster formation. The formation of hydroclusters leads to an enhancement in viscosity, which at very high particle loading can lead to a solid-like discontinuous shear thickening. CNT has a high potential for ballistic, high-energy adsorption, and high-strength materials because of its low density, good ductility, high stiffness, tensile strength, and aspect ratio. The effects of various concentrations of multiwalled CNT (MWCNT) on the ballistic resistance performance of epoxy/Twaron composites. MWCNT (1 wt.%) can significantly improve mechanical strength, fracture toughness, and energy absorption. MWCNT has also been combined with epoxy and Kevlar for improved energy absorption and impact strength of composites. With 0.5% of MWCNT addition of composites, the impact strength increased by 45%, whereas energy absorption increased by 27% compared to neat composite. Bessa *et al.* [64] prepared Kevlar-fabric reinforced with polybenzoxazine composites, containing microcrystalline cellulose (MCC) as a filler. During ballistic studies, the maximum energy adsorption of 166.6 J was observed at 1% of MCC filler with an increase of flexural strength and storage modulus of 17% and 180%. The textile modified with nanomaterials shows its potential of ballistic protection application as discussed in the mentioned studies.

14.6.4 Fire-Retardant Fabrics

Reduction of the flammability in fabrics is a desired characteristic from a domestic and military perspective. Conventionally, halogenated hydrocarbons are utilized for the preparation of flame-retardant fabrics; however, these fabrics on combustion generate highly corrosive and toxic gases, which results in multiple ecological problems. Nanometer-sized clays or nanoclay-like montmorillonite are used for the preparation of flame-retardant materials. For this, the finely dispersed nanoclay is dispersed finely in the matrices of the selected polymers as the reinforcement phase, which results in toughness, thermal stability, tensile strength, stiffness, flame-retardant properties, and gas-barrier properties [45]. The incorporation of clay delayed the dripping and melting of fabric along with a reduction of the burning rate by half. The incorporation of clay nanoparticles also results in improved mechanical properties of the fabrics which can be used for reducing the weight and thickness of the prepared fabrics. The common morphologies found in these types of fabrics are as follows: (i) exfoliated: from the silicate layers that are delaminated entirely and evenly dispersed within the polymeric network. Dispersion of silicate layers is difficult and requires the use of a compatible agent containing an organophilic and a hydrophilic group; (ii) the other type of morphology presents an intercalated form, in which monomers or polymers infiltrate the ordered structure of silicate layers [45]. Samanta *et al.* [65] prepared fire-retardant fabrics using jute fabrics modified with nano-ZnO. Bleached jute fabrics were coated with nano-ZnO to impact flame retardance. The modified textiles show a limiting oxygen index (LOI) of 35 with a minimum char length of 1 cm. The ZnO modification also improves the residual char (5%) at the temperature range of 400°C–500°C. Potassium methyl siliconate as a binder (10%) on jute fabrics modified with ZnO (0.01%) can become a potential alternative to commercial flame-retardant materials requiring high pickup for achieving fire retardancy. Jose *et al.* [66] coated the wood fabrics with nanoclay for fire-retardant applications. Modification of wool fabrics with nano-kaolinite enhances flame-retardant properties by 32%. The nanoclay modification of textiles improved the LOI from 25 to 33. The clay incorporation improves the char percent by 5% observed during thermal stability analysis. The incorporation of nanomaterials in textiles has been shown to improve the fire-retardant properties for the development of advanced military-based textiles.

14.6.5 Nanocomposites for Protection From Biological/Chemical Hazards

The prevention of personnel exposure to toxic chemical entities used for warfare is very important for the accomplishment of missions on battlefields. The protection from chemical agents is generally achieved using an activated carbon system, by utilizing impermeable barrier materials and semipermeable materials systems. The activated carbon-based protective coatings offer protection by adsorbing toxic chemicals [67]. The materials used for barrier properties are usually made from rubber and coated with multilayer laminate fabrics present in boots and gloves, which makes the fabrics impermeable and offers protection by working as a physical barrier to chemicals. These materials, however, present limitations like logistics or cost that makes the product undesirable. A low-weight, cheap, low-packing volume, chemical, and biological preventive barrier material would make the process of collective fabrication feasible for biological and chemical protective clothing and shelter systems, improving the safety of personnel during biological and chemical warfare [68]. Filtration or insulating materials utilized for the fabrication of protective suits against chemical warfare agents produce only a physical barrier against toxic chemicals; therefore, the steps for their adequate disposal and decontamination are required. The enhanced need for the development of materials not only provides protection against toxic chemicals but also degrades it to some extent. Nanotechnology-based approaches can be utilized for the preparation of multifunctional textile materials that can decompose organic matter and toxic chemicals. The application of TiO_2 nanoparticles for the preparation of fabrics with photocatalytic degradation of organic matter and toxic chemicals because of their low toxicity, cost, and abundant availability. The standard military textile is modified by TiO_2 nanoparticles by Military Technical Institute for the development of self-decontaminating textiles. The obtained fabrics present photocatalytic activity for Methyl Orange and the ([2-(diisopropylamino)ethyl]-O-ethyl methylphosphonothioate). One of the demerits of using TiO_2 is the reduced mechanical strength of the prepared fabrics, which can be enhanced by incorporating SiO_2 NPs. The photocatalytic activity of the treated dyes was tested with the degradation of organic dyes like Methyl Orange. The photocatalytic degradation follows first-order reaction kinetics calculated by first-order rate constants [45]. Kocic *et al.* [69] prepared knitted fabrics with pure viscose, pure hemp, pure cotton, and pure viscose. The knitted fabrics demonstrated a UV protection factor (UPF) of 118.7, whereas the hemp/cotton blend shows a UPF of 45.1 and hemp-only fabrics demonstrated a UPF of 24.4. Yalcinkaya *et al.* prepared polyvinyl butyrl nanofiber incorporated with copper oxide nanoparticles. The prepared nano-textiles demonstrate excellent antibacterial activity against *Staphylococcus gallinarum* and *Escherichia coli* bacteria. Modification of textiles with nanomaterials can equip the textiles with protective properties against chemical and biological agents that can enable the development of gears resistant to physical, chemical, and biological hazards.

14.6.6 Decontaminating, Healing, and Self-Cleaning Fabrics

The ability to oxidize contaminants and dirt based on the photocatalytic activity of nanoparticles has been confirmed using nanoparticles like

ZnO and TiO_2. The prepared fabric from the nanotechnological modification is favorable for utilization in both civilian and military uses. Textile materials such as wool, polyamide, and polyester, were pretreated with plasma and UV irradiation such that chelating groups for negatively charged TiO_2 molecules like carboxylic acids are introduced. The coating of silver nanoparticles on military garments offers further protection from incoming microbial loads [45]. Rashid *et al.* [70] prepared cost-effective and sustainable textile surfaces doped with Ag and TiO_2 for antibacterial, UV protection, and self-cleaning features. The addition of Ag improves the TiO_2 photocatalytic activity. The textile bonding through the organic-inorganic hybrid system serves as an anchoring site for Ag-doped TiO_2 by the formation of Ti–O–Si bonds. The prepared cotton fabrics demonstrate both passive and active antibacterial activity. The nanocomposite films prepared show a high photocatalytic degradation of upto 80% MB, which comes from the efficient heterostructure by interfacial charge carrier separation in Ag/AgCl/TiO_2. Biswas and Jana [71] prepared textiles by modification with TiO_2, octadecyltrimethoxysilane, and silica nanoparticles for visible light self-cleaning applications. Bukit *et al.* prepared used chitosan and oil palm boiler ash for making antibacterial and self-cleaning composites. The textiles remove the coffee stains by self-cleaning under both sunlight and UV light and show anti-bacterial activity against *E. coli* and *S. aureus* bacteria. Kale *et al.* [72] prepared combined TiO_2 with cellulose on cotton fabric for durable catalytic self-cleaning properties. Orange II dyes were degraded with the self-cleaning textiles prepared, which were stable for upto 10 washing cycles for stiffness and self-cleaning properties. Nanoparticle-based approaches can impart permanent effects to the textiles for self-cleaning abilities and do not result into loss in properties after washing.

14.6.7 Fabrics with Energy Storage and Sensing Properties

Integration of conventional textile materials with nanoparticles converts the normal textiles to sensors. Another example is the incorporation of nanocrystalline piezoceramic particles into fabrics, in which the fabrics can change exerted mechanical force into electrical signals, allowing the monitoring of crucial body activities like pulse and heart rhythm in contact with the skin. The fabrics, made with CNTs, can also be used as energy storage devices integrated with the fabrics [45]. Ye *et al.* [73] prepared a hydrophobic self-repairing power textile for energy-harvesting from water droplets. The prepared triboelectric nanogenerators (TENGs) are composed of poly(vinylidenefluoride-co-hexafluoropropylene)/perfluorodecyltrichlorosilane (PVDFHFP/FDTS), and SiO_2 shows self-repairing, hydrophobic, and energy conversing with good air permeability for energy harvesting from water droplets. The prepared TENG shows an output of 22V and 7.5 nC while performing energy conversion from water and can be applied for modification of textiles for energy harvesting. Baro *et al.* [74] made TENG on a textile platform by modification with ZnO nanoparticles for sweat sensing. The attachment of hydrated Cl ions of saline on the attachment of physiosorbed water molecule of ZnO increases the conduction band. The TENG prepared shows a sensitivity of approximately 0.02 V/μL, with a limit of detection of 4.8 μL. The TENG upon integration into the shoe sole showed a change in sensing on sweat detection, suggesting the potential of the work in real-life sensing applications. Safaee *et al.* [75] prepared optical microfibrous wearable materials for oxidative stress monitoring. Single-walled CNTs were integrated with microfibrous textiles for real-time optical monitoring of hydrogen peroxide in *in vitro* wounds. The individual fibers encapsulate the SWCNT nanosensors for about 21 days without any loss in structural features. The prepared sensors demonstrate a limit of detection for as low as 1 μM and detect hydrogen peroxide upto a concentration of 5,000 μM. The SWCNT has the potential to be integrated into wearables for the detection of hormones, carbohydrates, proteins, and sweat. Cho *et al.* prepared weavable tunicate cellulose/CNT for making high-performance wearable sensors. Nanocellulose obtained from tunicates (TCNF) was homogenously deposited on CNTs for making composites. The fibers are produced with wet spinning for NO_2 sensing properties in parts-per-billion detection with 0.5% and 0.125 ppm. The discussed examples suggest the potential of recent advances in sensing and energy-harvesting technologies that can be integrated in textiles for real-time sensing of metabolites and health parameters for army personnel.

14.7 Utilization of Sustainable Nanotextiles for Military Applications

14.7.1 Nanocellulose Applications for Military Packaging

The goals of commercial packaging are different from military packaging as the latter requires the material preservation for long durations. Military preservation protects the materials during handling, shipment, distribution, and storage. Therefore, materials with excellent barrier properties, mechanical strength, and waterproofing properties are required for packaging in military applications. Packaging materials in the military are used for the protection of beverages, food, electronics, and electrically assembled parts [16]. Nanocellulose can be applied for the preparation of cartons, bags, military fastpacks, cushioning, food packages, wraps, and tapes. It was also observed between the duration of World War I to World War II that containers that were used for shipping military goods were not able to stand up to the rigorous overseas travel. At that time, the major emphasis was only on the conservation of packing materials and visual attractiveness for industrial purposes. The standards for military packaging ensured the protection of equipment and combat readiness and maintained combat performance for the shipments [76]. Nanocellulose has been abundantly used for the preparation of packaging materials. The paper films prepared with nanocellulose have a brittle nature and tend to swell in the presence of water, which limits its application in military packaging. The incorporation of additives like nanoclays, waxes, and surface-active compounds and resins has been done in recent years. The incorporation of stearoyl chloride and TiO_2 into the composites is observed to improve the enhance the hydrophobic properties of pristine nanocellulose to reduce the swelling properties of nanocellulose on exposure of moisture [77]. The addition of nanocellulose is also observed to enhance the mechanical properties of polypropylene, polyethylene, and PET by the addition of 1% to 5% of nanocellulose. Another interesting feature of biopolymer-based packaging materials is their biodegradable features, which remove the burden of properly disposing of water for army personnel. Biopolymers like starch, polyhydroxyalkanoates, and polylactic acid are popular biomaterials used for packaging applications [78]. The complications of using biopolymers are their low thermal stability and mechanical performance, which can be countered using the introduction of nanocellulose to the biopolymers. Nanocellulose due to its high specific surface area and crystallinity has a tunable surface chemistry, which allows the nanocellulose to enhance the water vapor and oxygen barrier properties when utilized as filler or coating in composites or as self-standing thin film [16]. The assessment of nanocellulose toxicity is crucial for the preparation of biocompatible and edible packaging for military applications, which has been studied thoroughly in recent decades.

14.7.2 Application of Nanocellulose for Military Textiles

The textiles required for military applications are desired to possess fire retardant, high strength, hydrophobic, anti-wrinkle, antimicrobial, color-adaptive, self-cleaning, and air- and water-permeable properties. Military textiles are used for making tents, combat uniforms, parachutes, shelters, body armor, explosive ordnance, footwear, camouflage systems, and aircraft materials. Textiles for automatic monitoring of body functions have also been explored in recent years [79]. During military operations in desert areas, crews of armored vehicles, military battle tanks, and soldiers are subjected to high risks of dehydration and heat stroke. Heat stress can induce physiological conditions like seizures, reduced mental alertness, and reduced situational awareness from physical discomfort [80]. On the contrary, troops employed in cold mountains, and arctic operations, have a risk of hypothermia, hypoxia, and UV blindness. Nanocellulose has been observed to have superior mechanical properties than the currently utilized materials for military fabrics like Kevlar. The nanocellulose, due to its thermal insulation can be used for making wound dressings, tampons, and sanitary napkins. The nanocellulose can also serve as an alternative to currently used materials like Kevlar for making body armor [81].

Prevention of troops from glass fragments release during transport collisions also results in significant physical injuries to the soldiers. This presents the possibility of the preparation of goggles and personal protective equipment to reduce the threats occurring from terrorist attacks, natural disasters, and civil unrest. Usually, poly (vinyl butyl) (PVB) is sandwiched between glass sheets [82]. The transparent PVB has the ability to keep the broken glass in place; however, PVB is heavy and expensive and requires thick sheets for protection. PVB usage also affects the transparency of the glass layer along with the visual quality. Nanocellulose has been applied for modifying the drawbacks present by PVB. Maury *et al.* combined nanocellulose with PVB by hot pressing to improve the drawbacks of PVB [83]. The prepared composites showed improved mechanical properties, transparency, and low surface mass. Superhydrophobic nanocellulose-based textiles are also significant for military purposes as it prevents soldiers from getting their uniform dirty and wet. Modifying the water and dirt contaminants from the surface of clothing is of high priority. Several chemical modifications are available such as electro-assisted chemical deposition, dip-coating, wet chemical deposition, sol-gel, chemical etching, spray coating, plasma processing, polymer grafting, and chemical vapor deposition [16]. Another significant nanotechnology-based approach to modifying the materials used in the military is the development of electronic textiles. Electronic textiles can be utilized for locating soldiers, monitoring personnel health, ecological temperature, and communication monitoring.

14.8 Conclusions

Advances in the recycling technologies have enabled the recycling of waste textiles generating and its possible conversion to value added products. Waste textiles can be recycled via melt processing, solvents, and depolymerization-based approaches. The recycled textiles can be modified with nanotechnology-based approaches for development of smart textiles with temperature tolerance, surveillance, scratch resistant, armor-like fabric, and equipped with sensors. The nanotechnology-based modification of waste textile can be developed to prepare smart uniforms, signature reduction, ballistic protection, fire-retardant fabrics, protection from chemical/biological hazards, decontaminated self-cleaning textiles, and sensing textiles for military applications. The recycled textiles modified with nanotechnology show immense potential for application in next-generation smart textiles for application in military applications.

References

1. Singleton, J., *World textile industry*, Routledge, Taylor & Francis, London, 2013.
2. Korhonen, J., Pätäri, S., Toppinen, A., Tuppura, A., The role of environmental regulation in the future competitiveness of the pulp and paper industry: the case of the sulfur emissions directive in Northern Europe. *J. Cleaner Prod.*, 108, 864–872, Dec. 2015, doi: 10.1016/j.jclepro.2015.06.003.
3. I. FAO, A summary of the world apparel fiber consumption survey 2005–2008. Food and Agriculture Organization of the United Nation, International Cotton Advisory Committee, 2011.
4. Pensupa, N. *et al.*, Recent trends in sustainable textile waste recycling methods: current situation and future prospects, in: *Chemistry and Chemical Technologies in Waste Valorization*, pp. 189–228, 2018.
5. Sharma, M. and Dhiman, R., Determinants affecting Indian textile exports: a review. *Biz Bytes J. Manag. Technol.*, 6, 193–199, 2016.
6. Claudio, L., Waste couture: Environmental impact of the clothing industry, National Institute of Environmental Health Sciences, EHP Publishing, North Carolina, 2007.
7. Büyükaslan, E., Jevsnik, S., Kaloğlu, F., A sustainable approach to collect post-consumer textile waste in developing countries. *Marmara J. Pure Appl. Sci.*, 1, 107–111, 2015.
8. Ghaly, A.E., Ananthashankar, R., Alhattab, M., Ramakrishnan, V.V., Production, characterization and treatment of textile effluents: a critical review. *J. Chem. Eng. Process Technol.*, 5, 1, 1–19, 2014.

9. Gordon, S. and Hsieh, Y., *Cotton: Science and technology*, Woodhead Publishing, United Kingdom, 2006.
10. Juanga-Labayen, J.P., Labayen, I.V., Yuan, Q., A review on textile recycling practices and challenges. *Textiles*, 2, 1, 174–188, 2022.
11. Danihelová, A., Němec, M., Gergeľ, T., Gejdoš, M., Gordanová, J., Sčensný, P., Usage of recycled technical textiles as thermal insulation and an acoustic absorber. *Sustainability*, 11, 10, 2968, 2019.
12. Yetisen, A.K. *et al.*, Nanotechnology in textiles. *ACS Nano*, 10, 3, 3042–3068, 2016.
13. Hu, J., Meng, H., Li, G., Ibekwe, S.I., A review of stimuli-responsive polymers for smart textile applications. *Smart Mater. Struct.*, 21, 5, 053001, 2012.
14. Avila, A.G. and Hinestroza, J.P., Tough cotton. *Nat. Nanotechnol.*, 3, 8, 458–459, 2008.
15. N. R. Council, *Materials research to meet 21st-century defense needs*, National Academies Press, United States, 2003.
16. Norrrahim, M.N.F. *et al.*, Nanocellulose: the next super versatile material for the military. *Mater. Adv.*, 2, 5, 1485–1506, 2021, doi: 10.1039/D0MA01011A.
17. Chen, L. *et al.*, Nitrated bacterial cellulose-based energetic nanocomposites as propellants and explosives for military applications. *ACS Appl. Nano Mater.*, 4, 2, 1906–1915, 2021.
18. Abdulrahim, M.Y., Yawas, D.S., Mohammed, R.A., Afolayan, M.O., Hybridization of Polyester/Banana stem Fiber and Cow horn particulate composite for possible production of a military helmet. *Int. J. Sustain. Eng.*, 14, 5, 1170–1180, 2021.
19. Wang, Y., *Recycling in textiles*, Woodhead publishing, United Kingdom, 2006.
20. Braun, M., Levy, A.B., Sifniades, S., Recycling nylon 6 carpet to caprolactam. *Polym. Plast. Technol. Eng.*, 38, 3, 471–484, 1999.
21. Martin, D.L., Wang, Q., Klevisha, D., Challenges of FTNIR for recycling of fibrous textile and carpet waste industries, 2004.
22. Bohnhoff, A. and Petershans, J., Decentralised technology for the sorting of textile floor coverings, 2002.
23. Rowe, R.G., Textile recycling machine. United State Patent, no. 6,061,876, 2000.
24. White, D.W., System and method for decomposing reclaiming and refusing waste carpet materials, Feb. 29, 2000.
25. Hawn, K.L., An overview of commercial recycling technologies and textile applications for the products, Georgia Institute of Technology, 2001.
26. Strzelecki, C., Modern solutions for shredding, grinding and repelletizing post-industrial fiber, nonwovens and carpet scrap, 2004.
27. Herlihy, J., Recycling in the carpet industry. *Carpet Rug Ind.*, 25, 17–25, 1997.
28. Kasserra, H.P., Recycling of Polyamide 66 and 6, in: *Science and Technology of Polymers and Advanced Materials: Emerging Technologies and Business Opportunities*, pp. 629–635, 1998.

29. Roth, J., CENSORTM centrifuge technology for plastic separation, 2002.
30. Booij, M., Hendrix, J.A.J., Frentzen, Y.H., Process for recycling polyamide-containing carpet waste, European Patent, vol. 759, p. 456, 1997.
31. Sarian, A.K., Handermann, A.C., Jones, S., Davis, E.A., Adhya, A., Recovery of polyamides from composite articles, Dec. 15, 1998.
32. Sikorski, M.E., Recycling of polymeric materials from carpets and other multi-component structures by means of supercritical fluid extraction, Aug. 03, 1993.
33. Brown, T., Infinity Nylon–A Never-ending Cycle of Renewal, 2001.
34. Bodrero, S., Canivenc, E., Cansell, F., Chemical recycling of polyamide 6.6 and polyamide 6 through a two step ami-/ammonolysis process, 1999.
35. Rieland, J.M. and Love, B.J., Ionic liquids: A milestone on the pathway to greener recycling of cellulose from biomass. *Resour. Conserv. Recycl.*, 155, 104678, Apr. 2020, doi: 10.1016/j.resconrec.2019.104678.
36. Mai, N.L. and Koo, Y.M., Computer-aided design of ionic liquids for high cellulose dissolution. *ACS Sustain. Chem. Eng.*, 4, 2, 541–547, 2016.
37. Kuzmina, O. *et al.*, Superbase ionic liquids for effective cellulose processing from dissolution to carbonisation. *Green Chem.*, 19, 24, 5949–5957, 2017, doi: 10.1039/C7GC02671D.
38. Swatloski, R.P., Spear, S.K., Holbrey, J.D., Rogers, R.D., Dissolution of Cellose with Ionic Liquids. *J. Am. Chem. Soc.*, 124, 18, 4974–4975, May 2002, doi: 10.1021/ja025790m.
39. Kostag, M., Jedvert, K., Achtel, C., Heinze, T., El Seoud, O.A., Recent Advances in Solvents for the Dissolution, Shaping and Derivatization of Cellulose: Quaternary Ammonium Electrolytes and their Solutions in Water and Molecular Solvents. *Molecules*, 23, 3, 511, Mar. 2018, doi: 10.3390/molecules23030511, Art. no. 3.
40. Wen, J.-L., Yuan, T.-Q., Sun, S.-L., Xu, F., Sun, R.-C., Understanding the chemical transformations of lignin during ionic liquid pretreatment. *Green Chem.*, 16, 1, 181–190, 2014, doi: 10.1039/C3GC41752B.
41. Menezes, D.B. *et al.*, Starch-based biocomposite membrane reinforced by orange bagasse cellulose nanofibers extracted from ionic liquid treatment. *Cellulose*, 28, 7, 4137–4149, May 2021, doi: 10.1007/s10570-021-03814-w.
42. Nor Amalini, A., Noor Haida, M.K., Imran, K., Mohamad Haafiz, M.K., Relationship between dissolution temperature and properties of oil palm biomass based-regenerated cellulose films prepared via ionic liquid. *Mater. Chem. Phys.*, 221, 382–389, Jan. 2019, doi: 10.1016/j.matchemphys.2018.09.028.
43. Elsayed, S. *et al.*, Recycling of superbase-based ionic liquid solvents for the production of textile-grade regenerated cellulose fibers in the lyocell process. *ACS Sustain. Chem. Eng.*, 8, 37, 14217–14227, 2020.
44. Laxminarayana, K. and Jalili, N., Functional nanotube-based textiles: pathway to next generation fabrics with enhanced sensing capabilities. *Text. Res. J.*, 75, 9, 670–680, 2005.

45. Thilagavathi, G., Raja, A.S.M., Kannaian, T., Nanotechnology and protective clothing for defence personnel. *Def. Sci. J.*, 58, 4, 451, 2008.
46. Ngô, C., Van de Voorde, M., Ngô, C., Van de Voorde, M.H., Nanotechnology for the textile industry, in: *Nanotechnology in a Nutshell: From Simple to Complex Systems*, pp. 321–329, 2014.
47. Coyle, S., Wu, Y., Lau, K.-T., De Rossi, D., Wallace, G., Diamond, D., Smart nanotextiles: a review of materials and applications. *MRS Bull.*, 32, 5, 434–442, 2007.
48. Syduzzaman, M.D., Patwary, S.U., Farhana, K., Ahmed, S., Smart textiles and nano-technology: a general overview. *J. Text. Sci. Eng.*, 5, 1, 1–7, 2015.
49. Thandavamoorthy, S., Gopinath, N., Ramkumar, S.S., Self-assembled honeycomb polyurethane nanofibers. *J. Appl. Polym. Sci.*, 101, 5, 3121–3124, 2006.
50. Qi, K. *et al.*, Weavable and stretchable piezoresistive carbon nanotubes-embedded nanofiber sensing yarns for highly sensitive and multimodal wearable textile sensor. *Carbon*, 170, 464–476, 2020.
51. Shi, Q., Zhang, N., Wang, D., Zhang, J., Li, Y., Wang, Z., Polyethylenimine-mica nanosheets/cellulose nanofibers stacked loose nanofiltration membrane for effective dye/salt separation. *Desalination*, 551, 116410, 2023.
52. Li, M. *et al.*, Ultralight aerogel textiles based on aramid nanofibers composites with excellent thermal insulation and electromagnetic shielding properties. *Compos. Commun.*, 35, 101346, 2022.
53. Yao, L., Lee, C., Kim, J., Fabrication of electrospun meta-aramid nanofibers in different solvent systems. *Fibers Polym.*, 11, 1032–1040, 2010.
54. Yang, X. *et al.*, All-Nanofiber-Based Janus Epidermal Electrode with Directional Sweat Permeability for Artifact-Free Biopotential Monitoring. *Small*, 18, 12, 2106477, 2022.
55. Savage, S.J., Defence applications of nanocomposite materials. FOI-Swedesh Defence Research Agency. User Report No. FOI, 2004.
56. Ericson, L.M. *et al.*, Macroscopic, neat, single-walled carbon nanotube fibers. *Science*, 305, 5689, 1447–1450, 2004.
57. Ran, J. *et al.*, Carbon nanotube/polyurethane core–sheath nanocomposite fibers for wearable strain sensors and electro-thermochromic textiles. *Smart Mater. Struct.*, 30, 7, 075022, 2021.
58. Winey, K.I. and Vaia, R.A., Polymer nanocomposites. *MRS Bull.*, 32, 4, 314–322, 2007.
59. Asim, F. and Naeem, F., Investigation of self-cleaning attributes of denim fabric modified with naturally synthesized ZnO nanoparticles. *Pigm. Resin Technol.*, 52, 6, 738–746, 2023.
60. Wei, W. *et al.*, An Al2O3-cellulose acetate-coated textile for human body cooling. *Sol. Energy Mater. Sol. Cells*, 211, 110525, Jul. 2020, doi: 10.1016/j.solmat.2020.110525.
61. D'Aloia, A.G., Bidsorkhi, H.C., De Bellis, G., Sarto, M.S., Graphene Based Wideband Electromagnetic Absorbing Textiles at Microwave Bands. *IEEE Trans. Electromagn. Compat.*, 64, 3, 710–719, 2022.

62. Forsman, N. *et al.*, Microalgae Chlorella vulgaris and kraft lignin stabilized cellulosic wet foams for camouflage. *Soft Matter*, 18, 10, 2060–2071, 2022.
63. Lee, Y.S., Wetzel, E.D., Wagner, N.J., The ballistic impact characteristics of Kevlar® woven fabrics impregnated with a colloidal shear thickening fluid. *J. Mater. Sci.*, 38, 2825–2833, 2003.
64. Bessa, W., Trache, D., Derradji, M., Tarchoun, A.F., Kevlar fabric reinforced polybenzoxazine composites filled with silane treated microcrystalline cellulose in the interlayers: The next generation of multi-layered armor panels. *Def. Technol.*, 18, 11, 2000–2007, Nov. 2022, doi: 10.1016/j.dt.2021.10.005.
65. Samanta, A.K., Bhattacharyya, R., Jose, S., Basu, G., Chowdhury, R., Fire retardant finish of jute fabric with nano zinc oxide. *Cellulose*, 24, 2, 1143–1157, Feb. 2017, doi: 10.1007/s10570-016-1171-z.
66. Jose, S., Shanmugam, N., Das, S., Kumar, A., Pandit, P., Coating of lightweight wool fabric with nano clay for fire retardancy. *J. Text. Inst.*, 110, 5, 764–770, May 2019, doi: 10.1080/00405000.2018.1516529.
67. Rajagopalan, S., Koper, O., Decker, S., Klabunde, K.J., Nanocrystalline metal oxides as destructive adsorbents for organophosphorus compounds at ambient temperatures. *Chemistry–A Eur. J.*, 8, 11, 2602–2607, 2002.
68. Watson, S., Beydoun, D., Scott, J., Amal, R., Preparation of nanosized crystalline TiO 2 particles at low temperature for photocatalysis. *J. Nanopart. Res.*, 6, 193–207, 2004.
69. Kocić, A., Popović, D.M., Stanković, S., Poparić, G.B., Influence of yarn folding on UV protection properties of hemp knitted fabrics. *Hem. Ind.*, 70, 3, 319–327, 2016.
70. Rashid, M.M., Tomšič, B., Simončič, B., Jerman, I., Štular, D., Zorc, M., Sustainable and cost-effective functionalization of textile surfaces with Ag-doped TiO2/polysiloxane hybrid nanocomposite for UV protection, antibacterial and self-cleaning properties. *Appl. Surf. Sci.*, 595, 153521, 2022.
71. Biswas, A. and Jana, N.R., Cotton Modified with Silica Nanoparticles, N, F Codoped TiO2 Nanoparticles, and Octadecyltrimethoxysilane for Textiles with Self-Cleaning and Visible Light-Based Cleaning Properties. *ACS Appl. Nano Mater.*, 4, 1, 877–885, 2020.
72. Kale, B.M. *et al.*, Coating of cellulose-TiO2 nanoparticles on cotton fabric for durable photocatalytic self-cleaning and stiffness. *Carbohydr. Polym.*, 150, 107–113, 2016.
73. Ye, C. *et al.*, A hydrophobic self-repairing power textile for effective water droplet energy harvesting. *ACS Nano*, 15, 11, 18172–18181, 2021.
74. Baro, B., Khimhun, S., Das, U., Bayan, S., ZnO based triboelectric nanogenerator on textile platform for wearable sweat sensing application. *Nano Energy*, 108, 108212, Apr. 2023, doi: 10.1016/j.nanoen.2023.108212.
75. Safaee, M.M., Gravely, M., Roxbury, D., A wearable optical microfibrous biomaterial with encapsulated nanosensors enables wireless monitoring of oxidative stress. *Adv. Funct. Mater.*, 31, 13, 2006254, 2021.
76. Higgs, J., How To Meet Military Packaging Standards & Specifications.

77. Eren, T. and Neren Ökte, A., Polymerization of methacryl and triethoxysilane functionalized stearate ester: Titanium dioxide composite films and their photocatalytic degradations. *J. Appl. Polym. Sci.*, 105, 3, 1426–1436, 2007.
78. Kosior, E., Braganca, R.M., Fowler, P., Lightweight compostable packaging: literature review. *Waste Resour. Action Programme*, 26, 1–48, 2006.
79. Islam, G.N., Ali, A., Collie, S., Textile sensors for wearable applications: A comprehensive review. *Cellulose*, 27, 6103–6131, 2020.
80. Steffens, F., Gralha, S.E., Ferreira, I.L.S., Oliveira, F.R., Military Textiles-An Overview of New Developments. *Key Eng. Mater.*, 812, 120–126, 2019.
81. Owens, J.R., Key elements of protection for military textiles, in: *Functional Textiles for Improved Performance, Protection and Health*, pp. 249–268, Woodhead Publishing, United Kingdom, 2011.
82. Wang, L., *Performance Testing of Textiles: Methods, Technology and Applications*, Woodhead Publishing, United Kingdom, 2016.
83. Maury, C., Crispino, F., Loranger, É, Preliminary study of laminated glass with nanocellulose and poly (vinyl butyral) for safety glazing. *BioResources*, 14, 2, 4157–4170, 2019.

Part IV

QUALITY CONTROL AND REGULATORY ASPECTS OF ADVANCED NANO TEXTILE MATERIAL WITH RESPECT TO INDUSTRIES

15

Global Legislation, Schemes and Standards to Control Environmental Concern for Textile-Generated Waste

Shambhavi Sharma[1,2], Muhammad Jahanzaib[1,2], Jooyeon Lee[1,3] and Duckshin Park[1,2]*

[1]Transportation Environmental Research Division, Korea Railroad Research Institute (KRRI), Chleodobangmulgwan-ro, Uiwang-si, Republic of Korea
[2]Transportation System Engineering, University of Science and Technology (UST), Yuseong-gu, Daejeon, Republic of Korea
[3]Department of Mechanical Engineering, Yonsai University, Seodaemun-gu, Seoul, Republic of Korea

Abstract

The response to address environmental issues associated with textile-generated issues brought on by textile-generated trash includes a diverse strategy of laws, plans, and many international legislative efforts, programs, and standards in place. A sustainable future for our world is stitched together by these complex endeavors. These include laws and regulations, programs and initiatives, standards, and certifications. Extended Producer Responsibility (EPR) laws, circular economy strategies, zero-waste initiatives, textile recycling programs, and green certification programs are among the laws and regulations. The standards and certifications include the Global Organic Textile Standard (GOTS), the International Association for Research and Testing in the Field of Textile Ecology (OEKO-TEX) Standard, and the International Organization for Standardization (ISO Standards). The acceptance and implementation of these efforts might differ by nation and location, and it is crucial to remember. Nations all around the world are weaving legislative frameworks that specify the proper disposal, recycling, and management of textile waste to weave change. These laws provide a safety net, preventing textile waste from degrading into a threat to the environment. Governments all around the world have woven their legislative frameworks with a resolve to

**Corresponding author*: dspark@krri.re.kr

Prashansa Sharma and Shilpi Shree Sahay (eds.) Nanotechnology-Assisted Recycling of Textile Waste: Sustainable Tools for Textiles, (437–460)

counteract the harmful ecological effects of textile waste. For both producers and consumers, these legal strands establish the ground principles for acceptable waste management, recycling, and disposal procedures. These regulations, like guiding threads, work to keep our ecosystem from falling apart and keep trash from becoming a tangled web of pollution. Textile manufacturers and consumers are prompted to reconsider, mend, and recycle by circular economy tactics that unravel waste into threads of rejuvenation. Initiatives to reduce trash stitch together sectors and communities to form a collaborative web that leaves no garbage untangled. New ideas are like spindles that could transform trash into opportunity. Innovative plans include cooperation and innovation into the very fabric of textile waste management as part of the larger sustainability plan. Initiatives in the circular economy are the foundation, turning waste into opportunity by promoting the rethinking, mending, and reusing of textiles. Communities and industries are connected by zero-waste aspirations, which bind them together in the effort to weave collaborative fabrics that produce no trash at all. These plans give the art of waste reduction life; they are not just simple stitches but complex patterns. The sustainability efforts of the textile industry are given shape and purpose by standards, like the warp and weft of a loom. This fabric of transformation is delicately embroidered with certifications like GOTS and OEKO-TEX. While OEKO-TEX weaves assurance that textiles are free of dangerous elements, GOTS, a beacon of purity, makes sure that organic fibers are grown with respect for nature's balance. Similar to an artist's palette, ISO standards add shades of social responsibility and environmental awareness to the weaving, resulting in a well-balanced composition of sustainable practices. The legislation serves as the thread, plans as the weft, and standards as the subtle patterns in this elaborate tapestry of international endeavors. The backbone is made up of legislative threads that provide a well-organized basis for long-term action. A future where garbage serves as a canvas for innovation in the textile industry is something we may imagine thanks to schemes and their imaginative touches. Standards direct the delicate process, creating a tapestry that vibrates with integrity and purpose, just like the expert weaver's hands do. But the canvas is never fully finished, just like with every masterpiece. Innovations, laws, and forms of cooperation are being woven into the fabric of change every day. Together, they lay the groundwork for the textile industry to design a future in which waste is turned into a sustainable thread and woven into the fabric of ecological harmony and prudent stewardship. Innovations, laws, and forms of cooperation are needed to consider with current changing times to build a more sustainable future.

***Keywords*:** Extended producer responsibility (EPR) laws, circular economy, global organic textile standard (GOTS), international association for research and testing in the field of textile ecology (OEKO-TEX) standard, international organization for standardization (ISO Standards)

15.1 Introduction

15.1.1 Background and Significance

The textile industry has been a major driver of economic development, cultural expression, and invention throughout the course of human civilization. The fashion business has continually developed to meet the demands of a changing society, from the delicate textiles of ancient civilizations to current high-speed fashion production. This progress has not, however, been without costs. The same business that has made us comfortable and fashionable has also left behind a path of resource depletion, environmental damage, and moral dilemmas. In an era of increased social and environmental consciousness, it has never been more important than it is now to solve the difficulties facing the textile industry. The industry's connection to climate change and ecological disturbance has been clouded by the large use of water, energy, and raw materials as well as the production of enormous volumes of garbage. The industry's sometimes harmful working conditions have also sparked ethical questions and prompted a call for responsible procedures that put the well-being of employees first. The textile industry is at a crossroads as the world population expands and customers become more aware of the effects of their purchases. A significant adjustment is required to maintain the delicate balance between business, culture, and sustainability. This transition is not only strategically necessary to preserve the survival of an industry that affects every region of the world, but it is also morally necessary.

15.2 Textile Industry and Environmental Impact

15.2.1 Growth of the Textile Industry

Over the past few decades, the textile industry has grown remarkably, becoming one of the greatest industries in the world economy. The sector has experienced quick and significant growth as a result of factors like population increase, urbanization, and shifting customer tastes. The demand for textiles, which are used in apparel, home décor, and industrial uses, is rising, underscoring this increase. This rise has been further accelerated by the availability of fast fashion, global supply chains, and industrial technology improvements, resulting in a proliferation of textile production and consumption on a global scale. Growth in the textile sector has benefited the economy, but it has also harmed the environment.

Important worldwide issues now include things like water contamination, excessive energy use, and the production of textile waste. Because of the severity of these issues, government action is required to guide the sector toward greener practices [1–4, 8].

15.2.2 Environmental Concerns of Textile-Generated Waste

However, some environmental issues have become more prominent as a result of the textile industry's exponential expansion, particularly in light of the enormous volumes of trash that are produced during its existence. The textile industry's supply chain is inherently resource-intensive because it depends heavily on inputs of raw materials, electricity, and water. The significant amount of waste produced at all phases, including manufacturing, processing, and end-of-life disposal, is a major worry. Along with unwanted apparel and textiles, the garbage also comprises a sizable quantity of production waste in the form of trims, offcuts, and flawed goods.

The disposal of textile waste in landfills is one of the most urgent issues. Traditional textile production techniques frequently employ synthetic fibers and chemicals that take a long time to break down, having a long-lasting negative influence on the environment. Furthermore, dumping textiles in landfills increases methane emissions, a powerful greenhouse gas. Furthermore, as untreated wastewater containing dangerous chemicals is frequently discharged into aquatic bodies, the water-intensive dyeing and finishing operations contribute to water contamination [4–8].

15.2.3 Need for Legislative Interventions

There is an urgent need for worldwide legislative reforms given the scope of the environmental concerns posed by textile-generated trash. The linear "take-make-dispose" approach used by the textile industry is unsustainable, necessitating a paradigm change toward more circular and sustainable processes. The regulation and encouragement of these changes can be greatly aided by legislation. Legislative interventions can take many different forms, such as the adoption of Extended Producer Responsibility (EPR) regulations, which require producers and brands to assume liability for a product's whole lifespan. These regulations force manufacturers to create items that are robust, repairable, and recyclable, which cuts waste at the source. Additionally, legislation governing appropriate waste treatment and promoting recycling within the sector might be passed. To reduce the negative environmental effects of improper disposal, import and export limitations can be put in place to stop the movement of hazardous or non-recyclable textiles over international boundaries.

In conclusion, while the expansion of the textile industry has boosted the economy, it has also brought about serious environmental issues relating to waste produced by the business. To reduce the negative effects of this waste and move the sector toward more sustainable practices, quick and effective legislative action is essential. The textile sector may aim to reduce its environmental impact and adopt a circular economy model by putting policies like EPR rules and waste management restrictions into place [5, 6].

15.3 Global Legislation for Textile Waste Management

Global legislation has been created in response to the environmental difficulties raised by textile waste to solve them and promote sustainable practices within the textile industry. The difficulties of managing textile waste on a worldwide scale have been addressed by several significant legislative initiatives and international agreements.

15.3.1 Basel Convention

A significant international convention established to control the transboundary movement of hazardous waste, including certain forms of textile waste, is the Basel Convention on the Control of Transboundary Movements of Hazardous Wastes and Their Disposal. Even while textiles are not often thought of as hazardous trash, some of them may contain chemicals or have been treated with substances that might be detrimental to human health and the environment.

The Basel Convention is an essential tool for preventing the disposal of hazardous waste in developing nations with lax waste management regulations. Prior informed consent and adequate labeling are stressed as prerequisites for trash transportation across international boundaries. The Basel Convention, although not focusing just on textiles, emphasizes the significance of responsible waste management procedures and encourages nations to set up efficient waste disposal processes.

15.3.2 Extended Producer Responsibility (EPR) Laws

Extended Producer Responsibility (EPR) rules require manufacturers and producers to manage the whole lifespan of a product, including final disposal.

EPR legislation forces textile manufacturers to be more proactive in minimizing the environmental effects of their goods throughout their lifecycles.

EPR rules provide businesses with incentives to create items that are recyclable and environmentally friendly. They also promote the creation of take-back programs and recycling campaigns, which raise recycling rates and lower the number of textiles dumped in landfills. EPR rules can encourage the textile sector to adopt a more circular economy strategy, therefore minimizing waste and fostering resource efficiency.

15.3.3 Waste Management and Recycling Acts

Acts for waste management and recycling that specifically target textile waste have been established in several nations. These laws specify procedures for the correct collection, handling, classification, and recycling of textile waste. They support the use of waste management strategies that are ecologically friendly and encourage the diverting of textiles from landfills.

The implementation of systems for the effective management of textile waste is frequently covered by provisions in waste management and recycling legislation that establish recycling goals. By requiring that textiles are handled and processed in ways that minimize harm to the environment and human health, these legal measures help to reduce the environmental effects of textile waste.

15.3.4 Import and Export Restrictions

Some nations have imposed import and export limitations on specific categories of textiles, particularly those containing dangerous materials. These limitations are meant to stop the cross-border movement of textiles that might be hazardous or non-recyclable. These limitations lessen the worldwide environmental effect of textile waste by restricting the international commerce of textiles that do not adhere to certain environmental and safety requirements.

15.3.5 Case Studies: Legislative Approaches in Different Countries

Many nations have established legislative strategies that are adapted to their unique conditions and difficulties with textile waste. Examples of such methods include the following:

Sweden: Sweden passed a law requiring textile manufacturers to collect and recycle a certain portion of the textiles they sell. This law encourages textile collecting and recycling, which lowers waste and preserves resources [1].
Japan: The country has put in place extensive waste management regulations that cover textile waste. These rules support recycling and waste reduction by encouraging efficient garbage sorting and disposal.
United States: To reduce textile waste, numerous states in the U.S. have adopted various legislative strategies. While some jurisdictions have prioritized increasing textile recycling through outreach and education programs, some states have adopted EPR legislation forcing manufacturers to develop take-back programs [2].

These case studies highlight the many legislative strategies that various nations have used to address textile waste and advance more environmentally friendly practices in the textile sector.

Worldwide accords like the Basel Convention as well as national laws like EPR laws, waste management acts, and import/export limitations are just a few of the tactics that make up global legislation for textile waste management. Collectively, these legislative initiatives help to reduce the environmental effects of textile waste, promote ethical waste management, and support a more sustainable future for the textile sector.

15.4 International Schemes and Initiatives

Numerous worldwide programs and efforts have been formed to support sustainability, moral behavior, and efficient resource management in response to the rising environmental problems connected to the textile industry. These programs involve all parties involved in the textile supply chain and seek to develop a more socially and ecologically responsible sector. Here are five noteworthy worldwide programs and projects.

15.4.1 Sustainable Apparel Coalition (SAC)

To lessen the negative effects that clothes and footwear goods have on the environment and society, brands, retailers, manufacturers, NGOs, and academic institutions have come together to join the Sustainable Clothing Coalition. The creation of the Higg Index, a set of tools that enables businesses to evaluate and enhance their sustainability performance across many facets of their operations, from product design to end-of-life considerations, is one of the SAC's most notable accomplishments.

The SAC equips businesses to make educated decisions that result in lower resource consumption, fewer emissions, and more supply chain transparency by using standardized evaluation tools. The coalition's focus on data-driven decision-making aids in the development of a textile sector that is more responsible and sustainable.

15.4.2 Better Cotton Initiative (BCI)

A global initiative called the Better Cotton Initiative (BCI) aims to increase the sustainability of cotton production. Because conventional cotton cultivation uses a lot of pesticides and water, BCI encourages the implementation of more environmentally friendly agricultural methods. BCI-certified cotton abides by standards that lessen cotton farming's impact on the environment while simultaneously enhancing the lives and working circumstances of cotton farmers. The main objective of BCI is to teach and equip farmers with the knowledge and tools that they need to follow best practices, such as conserving water, using less pesticides, and fostering biodiversity. BCI seeks to change cotton production and assure its sustainability over the long term by collaborating with stakeholders along the cotton supply chain.

15.4.3 Zero Discharge of Hazardous Chemicals (ZDHC)

The textile, garment, and footwear sectors have joined forces to launch the Zero Discharge of Hazardous Chemicals (ZDHC) program, which aims to do away with the use of dangerous chemicals in manufacturing. The goal of ZDHC is to guarantee the safe and responsible use of chemicals at every stage of the supply chain, from raw materials to completed goods. ZDHC provides brands and manufacturers with tools and information to help them implement safer chemical management practices. ZDHC helps to lessen the negative effects of hazardous chemicals on ecosystems and human health by promoting openness and traceability in chemical usage.

15.4.4 The Fashion Pact

Three important environmental goals—mitigating climate change, restoring biodiversity, and safeguarding the oceans—are the focus of the Fashion Pact, a commitment-driven project that unifies various fashion firms, brands, and industry leaders. This program highlights the significance of group action in bringing about change and acknowledges that the effect of the fashion industry goes beyond its immediate operations. The Fashion

Pact emphasizes the industry's responsibility in addressing global environmental concerns by pushing member firms to establish ambitious sustainability objectives and collaborate on creative solutions. The stakeholders in the fashion business feel a feeling of shared responsibility thanks to its all-encompassing approach.

15.4.5 Global Recycle Standard (GRS)

The usage of recycled content in goods across several sectors, including textiles, is mandated by the Global Recycle Standard, a voluntary certification scheme. Global Recycle Standard (GRS) authenticates recycled materials, making sure that goods that claim to contain recycled material adhere to strict standards. GRS assesses every step of the supply chain, from sourcing raw materials through manufacturing and labeling. It promotes responsibility and transparency, which makes it simpler for customers to choose sustainable products. The GRS accreditation encourages the use of recycled materials in textile goods, reducing waste and preserving resources. These global programs and activities are crucial in bringing about constructive change in the textile sector. These initiatives assist in reducing the negative environmental and social effects of textile production by encouraging sustainability, ethical sourcing, and responsible methods. These projects' emphasis on collaboration paves the path for a more sustainable and responsible future as the textile industry develops [7–9].

15.5 Standards and Certifications

A variety of standards and certifications have evolved to direct and assess sustainable practices within the textile industry as it battles environmental and social issues. These guidelines offer a framework for ethical concerns, responsible sourcing of resources, and production practices. We look at four well-known standards and accreditations that are influencing the development of sustainable textile practices here.

15.5.1 OEKO-TEX Standard

A well-known certification program, the International Association for Research and Testing in the Field of Textile Ecology (OEKO-TEX) Standard works to assure the security of textile goods and safeguard people, workers, and the environment from dangerous chemicals. It offers a thorough set of standards for textiles at all manufacturing phases, from raw materials to completed goods. The standard evaluates chemicals, dyes, and finishes that might be harmful to human health. For items to receive OEKO-TEX certification, they must pass stringent inspection and testing for dangerous compounds [10]. The accreditation applies to a variety of textile products, including textiles, clothing, and accessories. Manufacturers may show their dedication to creating safe, ecologically friendly, and chemical-free textiles by getting OEKO-TEX certification.

15.5.2 Cradle-to-Cradle (C2C) Certification

A comprehensive system called Cradle-to-Cradle (C2C) Certification assesses items based on their components, production methods, and possibilities for recycling or reuse. The certification promotes the idea of a circular economy, in which goods are made to be environmentally friendly and regenerate themselves. A product's performance is evaluated by C2C Certification according to five criteria: social justice, material health, material reuse, renewable energy and carbon management, and water stewardship. Based on how they perform in these categories, products are given a certification level (Basic, Bronze, Silver, Gold, or Platinum). C2C encourages manufacturers and product designers to think about a product's whole lifespan and promote sustainability and circularity when making design choices.

15.5.3 Global Organic Textile Standard (GOTS)

The most widely recognized certification for organic textiles that considers both social and environmental factors is the Global Organic Textile Standard (GOTS) [9]. GOTS makes ensuring that "organic" textiles meet stringent requirements for sourcing, handling, manufacturing, and labeling raw materials. Products must adhere to strict ecological and social standards to receive GOTS certification, and they must also include a certain number of organic fibers. The GOTS certification assesses several factors, including chemical usage, energy and water use, social responsibility, and supply chain traceability. It makes sure that employees are treated properly and that the production of textiles has a minimal negative impact on the environment. A rising consumer demand for textiles made in an ethical, organic, and sustainable manner is satisfied by GOTS certification.

15.5.4 ISO Standards for Environmental Management

The International Organization for Standardization (ISO) has created several environmental management standards, some of which apply to the textile sector [8]. For example, ISO 14001 offers a framework for businesses to set up and maintain an environmental management system. This approach aids businesses in determining and controlling their environmental effects, particularly those brought on by the manufacture of textiles. Guidelines are provided by ISO standards for actions including waste management, energy conservation, and pollution avoidance. Although not specifically related to textiles, these guidelines provide a useful foundation for businesses looking to enhance their environmental performance and integrate sustainable practices into their daily operations.

To advance sustainable practices within the textile business, standards and certifications are essential. These frameworks offer producers, brands, and consumers clear rules so they may make decisions that support a more ethical and responsible textile industry, whether they are focused on chemical safety, circularity, organic sourcing, or general environmental management. These guidelines support changing behaviors and establishing a more sustainable and socially responsible future as the sector continues to develop.

15.6 Challenges and Limitations

Although adopting sustainable standards, certifications, and initiatives within the textile sector is an essential step in reducing negative effects on the environment and society, several obstacles and restrictions prevent their broad acceptance. These difficulties underscore the difficulties of changing an industry that has historically been resource-intensive into one that is more sustainable. Here, we examine the principal difficulties in establishing environmentally friendly textile practices.

15.6.1 Enforcement and Compliance Challenges

Ensuring the application of sustainable standards and certifications throughout the world's textile supply chain is a huge problem. It is challenging to monitor and confirm adherence to set standards because of the industry's complicated and sometimes fragmented character, with several actors participating in various production phases. Strong monitoring systems, audits, and ongoing follow-up are necessary to ensure that suppliers and manufacturers follow sustainable practices, but these processes may be time- and resource-consuming. Enforcement attempts may be hampered by opaque supplier networks. It becomes difficult to spot non-compliant operations when suppliers" and subcontractors' methods are not visible. It continues to be difficult to strike a balance between promoting voluntary adoption and ensuring compliance.

15.6.2 Lack of Global Consistency

Global inconsistency in sustainable textile techniques is a result of the absence of common norms and laws across nations and regions. Different nations may define "sustainability" or "ethical production" differently. Due to this lack of harmonization, it is difficult for companies with global operations to follow a standardized set of procedures and standards. The inability of customers to make educated decisions is further complicated by the lack of worldwide uniformity. Consumers may find it challenging to recognize truly sustainable items in the absence of uniform labeling or certification systems. This lack of consistency emphasizes the necessity for better coordinated international initiatives.

15.6.3 Technological and Infrastructural Barriers

The implementation of sustainable practices frequently necessitates large expenditures in new infrastructure and technology. Less developed areas might not have the resources that wealthier nations possess to implement cutting-edge technology for waste reduction, energy efficiency, and sustainable materials. The global adoption of sustainable practices may be hampered by this technology gap. Furthermore, the viability of sustainable textile processes may be constrained by inadequate waste management and recycling infrastructure in some locations. It may not always be possible to find the specialist equipment and methods needed for effective recycling and upcycling of textile waste. The textile sector has obstacles in reaching circular economy goals due to a lack of sufficient infrastructure.

15.6.4 Economic Implications for Industry Players

Costs are frequently increased while switching to sustainable practices, especially in the beginning phases of implementation. The financial resources of industry participants, particularly small- and medium-sized firms (SMEs), may be put under financial stress by investments in research, development, and the adoption of new technology. The financial obstacles to sustainability can be a disincentive, particularly in a sector with competitive pricing and money-conscious customers. Additionally, there is a notion that sustainable items are more expensive, which may influence customer choice. Although there are many long-term advantages to sustainability, such as reduced resource consumption and enhanced brand recognition, the early financial challenges can be a barrier for companies looking to adopt sustainable practices.

Several obstacles and problems must be overcome in the quest for sustainable textile methods. Improved monitoring and transparency procedures are necessary to address enforcement and compliance issues. International cooperation and harmonization efforts are required due to the absence of worldwide uniformity. Investments and attempts to increase capability are needed to overcome technological and infrastructure limitations. Finally, addressing the economic ramifications necessitates striking a balance between immediate expenses and long-term rewards. Stakeholders in the textile industry may strive toward a more sustainable and responsible future by addressing these issues as a group.

15.7 Benefits and Impacts of Legislation and Standards

A key factor in changing the textile industry into a more sustainable and responsible sector is the creation and application of laws and regulations. These strategies produce a wide range of advantages and favorable effects in addition to directing industrial operations. The consequences of regulations and standards are felt across the sector and beyond, from lessening environmental impact to promoting innovation and enhancing market dynamics. Here, we examine various regulatory systems' main advantages and effects.

15.7.1 Reduction in Environmental Footprint

The dramatic decrease in the environmental impact of the textile industry is one of the most obvious and important results of laws and norms. Reduced use of hazardous chemicals, better water management, and waste reduction are all required by regulations, which help to lessen the detrimental effects on ecosystems and natural resources. These policies encourage the adoption of cleaner and more sustainable manufacturing techniques by making industry participants responsible for their environmental behaviors. Therefore, the decrease in pollution, energy use, and resource depletion aids in reducing the environmental impact historically connected with the textile manufacturing industry.

15.7.2 Stimulating Innovation and Research

For innovation and research in the textile sector, laws and regulations serve as accelerators. Manufacturers, researchers, and designers are motivated to investigate novel materials, technologies, and processes by the need to adhere to sustainability criteria. This desire for innovation results in the creation of eco-friendly textiles, sophisticated dyeing methods, and fresh approaches to recycling and reusing textile waste. The industry promotes a culture of continual improvement as it works to comply with regulatory requirements, which encourages the development of sustainable solutions with potential uses outside the textile sector.

15.7.3 Economic Advantages and Market Access

Contrary to popular belief, adhering to laws and regulations can have positive economic effects. Businesses are frequently better positioned to maximize resource utilization, reduce waste, and run more smoothly when they implement sustainable practices. Additionally, sustainability-driven programs might draw customers who are prepared to pay more for ecologically responsible goods. For brands that value sustainability, this translates to more market access and perhaps better revenues. Additionally, upholding international standards can promote international business and alliances, boosting a company's standing and competitiveness on a worldwide scale.

15.7.4 Improved Consumer Awareness and Perception

Standards and legislation both help to increase consumer knowledge of sustainable methods used in the textile sector. Consumers now have access to information about the things that they buy as legislation forces businesses to disclose their attempts to comply with ethical and environmental standards. Transparency in sourcing, production methods, and certifications empowers customers to make values-aligned decisions. Because of the increased customer knowledge, there is a greater demand for goods that adhere to sustainable norms, which encourages firms to adopt ethical business practices to accommodate this need. As a result, laws and regulations have a significant impact on how consumers perceive products and how they behave while making purchases.

15.7.5 Social and Ethical Impact

Sustainable laws and regulations also have an impact on social and ethical issues in addition to environmental ones. These policies aid in the development of a more just and equitable industry by encouraging fair labor practices, secure working conditions, and ethical treatment of employees. Consumers respond favorably to improved working conditions and ethical sourcing, which promotes favorable brand perceptions and helps the market become more socially conscious. In the textile business, regulations and standards have broad positive effects on the social, economic, and environmental spheres. These initiatives help the industry move toward a more sustainable and responsible future by stimulating innovation, expanding economic possibilities, and raising customer awareness. As the textile industry accepts these advantages, it paves the path for a more peaceful coexistence of business, society, and the environment.

15.8 Case Studies

Analyzing case studies of sustainable textile practices offers insightful information on the practical effects of governmental regulations, business policies, and corporate efforts [1–7]. These examples show how different tactics are applied, the difficulties encountered, and the results obtained. Here, we examine three noteworthy case studies that provide information on various facets of sustainable textile practices:

15.8.1 Sweden's Textile Recycling Act

The Textile Recycling Act of Sweden is a ground-breaking piece of legislation that addresses textile waste and advances the ideas of the circular economy. The regulation, which goes into effect in 2020, mandates that apparel and textile businesses make sure that at least 20% of the items they put on the market are reused or recycled. This law supports Sweden's greater commitment to sustainability by highlighting the value of decreasing waste and prolonging the lifespan of textiles. The Textile Recycling Act encourages producers to create goods that are long-lasting and recyclable, having an impact on the whole product lifetime. The legislation promotes a culture of resource efficiency and waste reduction by requiring businesses to assume responsibility for the end-of-life stage of their goods. To monitor compliance and measure the actual percentage of textiles that are reused or recycled, the law's implementation has also presented obstacles.

15.8.2 China's Green Fence Policy

China's 2013 Green Fence Policy is an example of a regulatory strategy that affected resource flows in the textile sector and global commerce. Post-consumer textiles were one of the recyclable products whose import limits were tightened under the program. It sought to solve issues with the environment and human health caused by imported garbage, which frequently contained polluted or non-recyclable textiles. The worldwide textile recycling business was significantly impacted by the Green Fence Policy. Due to the resulting waste material flow disturbances, exporters were compelled to raise the caliber of their shipments. The policy's goal was to improve China's environmental position, but it also made clear the necessity for a worldwide effort to manage garbage and recycle it. The instance serves as a reminder of the significance of appreciating how intertwined the global textile supply chain is and the demand for coordinated efforts to solve waste-related issues.

15.8.3 Patagonia's Sustainable Practices

A leading manufacturer of outdoor apparel, Patagonia, is a role model for sustainable operations in the textile sector. The organization's dedication to environmental stewardship is ingrained in its business culture. Initiatives from Patagonia cover a range of topics, including sourcing materials, manufacturing techniques, and end-of-life concerns. Patagonia's strategy stands out for its focus on durability. To increase the lifespan of its clothing, the firm encourages customers to mend and reuse their purchases. Additionally, Patagonia aggressively seeks eco-friendly materials and encourages supply chain transparency. Its "Worn Wear" initiative makes it easier to resell and trade worn Patagonia items, which cuts waste and encourages circularity. Patagonia also makes research and development investments to investigate cutting-edge sustainable materials and technology. Consumer loyalty and trust are enhanced by the company's open disclosure about its methods and issues. This example demonstrates how a comprehensive strategy for sustainability may boost customer involvement, foster brand loyalty, and result in favorable environmental effects.

These case studies highlight the many approaches and difficulties involved in sustainable textile practices. The Textile Recycling Act of Sweden demonstrates the potential of governmental actions to promote waste minimization and circular economy concepts. The worldwide effects of governmental policies on recycling and trash management are highlighted by China's Green Fence Policy. The actions of Patagonia serve as an example of how a company's dedication to sustainability may have an impact on product design, consumer behavior, and industry standards. A more responsible and resilient industry may be created by stakeholders in the textile industry by studying these stories to get knowledge and inspiration for navigating the challenges of sustainable practices.

15.9 Future Prospects and Recommendations

The textile industry's future depends on its capacity to overcome sustainability obstacles and adopt cutting-edge tactics that encourage ethical behavior. To guarantee its long-term sustainability, the textile industry must align with these ideals as people's awareness of environmental and social implications grows. Here, we examine some of the most promising potential outcomes and suggestions for promoting sustainable textile practices:

15.9.1 Strengthening Global Collaboration

Due to its interconnectedness, the textile sector requires international cooperation to efficiently solve its numerous issues. Governments, business leaders, NGOs, and consumers must work together internationally. Harmonization efforts can offer a single framework for sustainability, decreasing ambiguity and improving compliance internationally. Cross-sector collaborations can promote the sharing of innovative ideas, technology, and best practices. Additionally, collaborative platforms can aid in overcoming common challenges like resource conservation and waste management. By cooperating, stakeholders may use their combined skills to bring about significant change on a global level.

15.9.2 Integrating Circular Economy Principles

The transition to a circular economy is crucial for the sustainable transformation of the textile sector. Designing goods with durability, repairability, and recycling in mind is a part of embracing circularity. Manufacturers can use modular designs that make component replacement and disassembly simple. Leasing or subscription-based business models can also encourage product returns and refurbishing. Increasing the lifespan of textiles lessens waste and the need for fresh materials. To promote resource efficiency and reduce environmental impact, brands and policymakers should give priority to incorporating circular economy ideas into their strategy.

15.9.3 Investing in Research and Development

Sustainable growth in the textile sector is driven by innovation. The goal of research and development (R&D) should be to discover new materials, technologies, and processes that have less environmental impact. Alternative fibers, eco-friendly dyeing techniques, water-saving procedures, and cutting-edge recycling technology can all be considered innovations. Resources should be set aside by governments, trade groups, and private businesses to fund R&D activities. Open innovation approaches that support teamwork among various stakeholders can quicken the speed of invention. The sector may advance answers to urgent problems while promoting economic growth by investing in R&D.

15.9.4 Education and Awareness Campaigns

Raising consumer awareness and educating stakeholders along the textile supply chain are crucial to a sustainable shift. Industry participants should place a high priority on teaching staff, partners, and suppliers on sustainable practices and the value of following rules. Campaigns to increase consumer knowledge may influence them to make more sustainable decisions. Brands may use their power to enlighten customers about product lifecycles, sourcing, and materials. Education programs at colleges and universities may develop a future workforce of experts with the knowledge and abilities to advance industrial sustainability.

15.9.5 Policy and Incentive Frameworks

Governments' use of incentive systems and policy initiatives to influence the textile industry's future is vital. Policies may include subsidies for R&D initiatives, tax incentives for businesses adopting sustainable practices, and rules that support responsible sourcing and waste minimization. Frameworks for incentives can assist the development of sustainable materials, stimulate the use of renewable energy in manufacturing, and encourage the use of circular economy models. Governments may help the sector move to more sustainable practices by fostering an enabling environment. The ability of the textile sector to adopt sustainability as a guiding concept will determine its destiny. The industry can overcome its difficulties and steer a sustainable route by enhancing international cooperation, incorporating circular economy ideas, investing in research and development, and conducting education and awareness initiatives. Governments, business leaders, consumers, and other stakeholders working together can create a textile sector that not only prospers economically but also benefits the environment and society.

15.10 Conclusion

An important component of contemporary civilization, the textile industry has long been a part of the fabric of life. It is essential for its role in supplying textiles, apparel, and jobs. However, the industry's substantial environmental and social concerns have come to light precisely because of the boom that gave it worldwide relevance. The massive resource use, pollution, and labor difficulties highlight the pressing need for a change to sustainable methods. We have examined the complicated web of sustainable textile practices throughout this extensive debate, from governmental regulations and international efforts to standards and case studies. As our investigation ends, it becomes clearer how the textile sector might move toward a future that is more accountable, moral, and sustainable. Transformative change is required as a result of the tremendous resource use, waste production, and frequently unsatisfactory working conditions. Legislation, guidelines, initiatives, and case studies that aim to encourage sustainable practices, lessen the environmental impact, and enhance working conditions have begun to develop in the sector as a reaction to these difficulties. It is clear that the path to a more responsible and ethical sector is difficult but hopeful as we approach to the end of our investigation into the complicated terrain of sustainable textile practices.

The system supporting sustainable textile practices is diverse and includes laws, regulations, projects, and partnerships. Legislation has shown to be effective in promoting waste reduction and circular economy ideas, as illustrated by the Swedish Textile Recycling Act. Such laws reset industry norms and encourage the development of items made to be durable and recyclable by requiring producers to be accountable for their products throughout their lives. The capacity of laws to motivate businesses to innovate, cut waste, and embrace resource efficiency is key to its success. Fundamentally, the sustainable development of the textile sector depends on a confluence of governmental actions, international cooperation, business initiatives, and consumer awareness. The Swedish Textile Recycling Act serves as an example of how legislation may reshape industry standards. Legislation like this encourages innovation, resource efficiency, and waste reduction by requiring producers to account for the lifespan of their goods. The act represents a circular economy in which goods are made to last, be reused, and be recycled. The instance of China's Green Fence

Policy also demonstrates the worldwide effects of governmental actions on recycling and trash management. The adoption of the legislation has repercussions on global supply networks, highlighting how interconnected the textile sector is. It stresses the significance of appropriate waste management practices within nations and regions and the necessity of coordinated worldwide efforts to address trash-related concerns.

On the business level, Patagonia's sustainable practices serve as an example of the transformative power of industry leaders dedicated to moral and ethical standards. Patagonia creates a standard for companies who want to incorporate sustainability into their DNA by pushing robust goods, encouraging repair and reuse, and upholding openness. The situation serves as a reminder of the importance of consumer loyalty and understanding in generating demand for goods that uphold sustainable ideals. The practices of industry leaders like Patagonia in the corporate sphere illustrate the transformative impact of deliberate business decisions. Patagonia serves as a model for businesses looking to incorporate sustainability into their operations by emphasizing toughness, reparability, and openness. The example serves as a reminder of the significance of consumer understanding and adherence in generating demand for goods that maintain moral and environmental ideals.

We have discussed how standards and certifications like the OEKO-TEX Standard, C2C Certification, GOTS, and ISO standards for environmental management support sustainable practices throughout our discussion. These standards include recommendations for ethical concerns, responsible material sourcing, and production practices. They enhance the reputation and competitiveness of brands that follow them by directing industry practices as well as customer choices. The GOTS, C2C Certification, and other standards and certifications offer concrete foundations for ethical behavior. These standards have an impact beyond what is seen in the industry; they affect customer decisions and motivate businesses to use moral and environmentally friendly practices. Initiatives like the BCI and the SAC promote innovation, cooperation, and openness, demonstrating the potential of group efforts to address difficulties faced by the whole sector.

Initiatives like the SAC, BCI, ZDHC, The Fashion Pact, and GRS provide venues for collaboration among varied stakeholders. These programs foster the adoption of best practices, openness, and innovation. They highlight the potential for group efforts to address industry-wide problems and create a more ethical and sustainable textile sector.

Prospects and proposals for sustainable textile practices set the foundation for a sectoral transformation, looking ahead. Improving international cooperation is essential to solving problems with no national boundaries.

Resource efficiency, waste minimization, and product longevity are all enhanced by applying the concepts of the circular economy to all aspects of the sector. Research and development funding stimulates innovation and supports finding answers to sustainability problems. Campaigns for education and awareness heighten consumer awareness and encourage changes in purchasing behavior toward more responsible decisions. Looking ahead, the world of sustainable textile techniques offers a canvas full of suggestions and future potential. To overcome barriers and address problems collectively, strengthening global collaboration emerges as a critical method. It becomes essential to include circular economy ideas in product design, production, and waste management. Research and development spending stimulate innovation, which produces answers that close sustainability gaps already present. Campaigns for education and awareness that are aimed at both consumers and industry stakeholders foster a culture of deliberate decision-making.

The textile industry is at a key turning point when we consider the issues around sustainable textile methods. A picture of an industry aiming for a more harmonious connection with the environment, society, and economy is painted by the convergence of laws, standards, initiatives, and case studies. However, obstacles to compliance, a lack of worldwide uniformity, technological constraints, and economic ramifications continue to exist. A thorough and team-based strategy is needed to overcome these challenges.

The threads of legislative efforts, industry partnerships, corporate initiatives, and consumer consciousness interlock to create a fabric that tells a tale of change in the expansive tapestry of the textile industry's sustainable evolution. However, problems still exist, ranging from regulatory issues to financial consequences. All parties involved must have a firm commitment to overcoming these challenges. The textile industry's ambition of a sustainable future is not simply a pipe dream; it is already becoming a reality. The textile business is at a crossroads in a world where environmental issues are becoming more serious and ethical considerations are important. This discourse's ideas and observations provide a road map for how this sector of the economy may move toward change. It is a route that balances economic development with environmental protection, puts the well-being of people first, and weaves sustainability into every facet of its operation.

In conclusion, the textile sector has made great progress and is becoming more cognizant of its social and environmental obligations. Stakeholders in the textile industry are laying the groundwork for a more ethical and responsible sector through new practices, industry initiatives, and regulatory measures. The insights discovered from these debates provide a road map for reaching a more sustainable future, one in which economic progress and environmental preservation coexist and consumers may take pleasure in goods that reflect their beliefs. The textile sector can create a brighter and more sustainable future for future generations through continued dedication, group efforts, and a commitment to change. The ability to alter the textile industry's position on the world stage exists as it moves forward. An industry primed for regeneration is depicted through the lessons acquired from legislative efforts, international partnerships, business undertakings, and pioneering experiences like Patagonia. The textile sector may not only lessen its effect but also weave a fabric that embodies the harmony between mankind, nature, and development via sustainable methods, responsible decisions, and unshakable devotion. The future of the textile industry becomes an endeavor filled with purpose when each participant, from politicians to consumers, accepts their role in this journey. This purpose is to build a world where fashion is not just a statement but a symbol of sustainable, ethical, and responsible choices.

Acknowledgments

The Korea Railroad Research Institute (KRRI) is acknowledged by all of the writers for its assistance.

References

1. Circular Economy Action Plan, European Commission, 2020, https://environment.ec.europa.eu/strategy/circular-economy-action-plan_en.
2. Textile Recycling, California Department of Resources Recycling and Recovery, 2020, https://www.calrecycle.ca.gov/textilerecycling.
3. Textile Waste Reduction, Ministry of Ecological Transition, France, 2021, https://www.ecologie.gouv.fr/404.
4. Waste Textiles Plan, Ministry of Infrastructure and Water Management, Netherlands, 2020, https://www.government.nl/topics/waste/documents/publications/2020/02/19/national-waste-textiles-plan.
5. Sustainable Apparel Coalition, Higg Index, 2021, https://apparelcoalition.org/the-higg-index/.
6. Global Fashion Agenda, 2021, https://globalfashionagenda.org/about-gfa-academy/.
7. Make Fashion Circular, Ellen MacArthur Foundation, 2021, https://ellenmacarthurfoundation.org/topics/fashion/overview.

8. Environmental Management Systems, International Organization for Standardization (ISO), ISO 14001:2015-, 2021, https://www.iso.org/standard/60857.html.
9. Global Organic Textile Standard (GOTS), Standard, 2021, https://global-standard.org/404.
10. Overview of OEKO-TEX Standards, OEKO-TEX, 2021, https://www.oeko-tex.com/en/apply-here/oeko-tex-standard-100.

16

Prevailing Eco-Parameters and Protocols for Nanotechnology in the Textile Industry

Sanduru Sai Keerthana[1], Vivek Dave[1]* and Prashansa Sharma[2]

[1]*Department of Pharmacy, School of Health Sciences, Central University of South Bihar, Gaya, India*
[2]*Department of Home Science, Mahila Mahavidyala, Banaras Hindu University, Varanasi, India*

Abstract
Nanotechnology is widely regarded as a tool for resolving issues and improving the standard of living. With rising customer demand for functional clothing that is produced sustainably, nanomaterials can now be incorporated into textile substrates. Consumption and trash production have increased as a result of the quick fashion cycle in the textile sector. The industry aspires to create safe and environmentally friendly textiles while utilizing the advantages of emerging nanomaterials. To ensure the continued development of nanomaterials in the future, it is imperative to be aware of both their potential benefits and unanticipated risks to the environment and human health. Among all the sectors, apparel industry is one of the most polluting sectors; this could have a severe effect on the environment. The appropriate regulations will allow for the mitigation of problems related to the production of nanotextiles and will support the responsible and safe advancement of nanotechnology. In this chapter, we will discuss about recycling, eco-friendly parameters, protocols, and applications related to nanotechnology.

Keywords: Nanotechnology, recycling, nanomaterials, pollution output, restricted substances, protocols, applications

**Corresponding author*: vivekdave@cusb.ac.in

Prashansa Sharma and Shilpi Shree Sahay (eds.) Nanotechnology-Assisted Recycling of Textile Waste: Sustainable Tools for Textiles, (461–496) © 2025 Scrivener Publishing LLC

List of Abbreviations

VOC	Volatile organic compounds
BOD	Biological oxygen demand
COD	Chemical oxygen demand
SVHC	Substances of very high concern
EU	European Union
PFOA	Perfluorooctanoic acid
ISO	International Organization for Standardization
PPE	Personal protective equipment
REACH	Registration, Evaluation, Authorisation and Restriction of Chemicals
TSCA	Toxic Substances Control Act
OECD	Organization for Economic Cooperation and Development

16.1 Introduction

Nanoparticles, which contain a structural dimension at the nanoscale (1 nm is a billionth of a meter or 10^{-9} of a meter), have received a significant amount of interest in study due to their potential for usage in a variety of scientific and technical disciplines [1]. Despite significant developments in nanotechnology, nothing is known about the possible effects of nanomaterials on human health. Nanoparticles may not be traceable once discharged into the atmosphere without a restoration plan, which could result in a variety of environmental issues. Green nanoscience has been proposed to minimize significant hazards to both the environment and human health from the fabrication and utilization of nanomaterials as well as to hasten the replacement of existing products with new nanoproducts that are more environmentally sustainable [2].

Recently, nanotechnology has been employed to improve the softness, durability, breathability, water resistance, fire resistance, and antibacterial properties of textile fibers and fabrics. The primary uses of nanotechnology in textiles include nanosilver for textiles, nanochemicals for textiles, and nanocoatings for textile materials. Realistic criteria and risk assessments should be developed along the entire supply chain, from manufacturers to customers, to assist the textile industry and the government in identifying applications with inherent risk and implementing the appropriate safety measures.

The rapid expansion and production of the textile industry contribute to economic growth and industrialization. The production process has significant environmental impacts, including waste generation, water and air emissions, and excess waste. Textile wastewater contains pollutants like heavy metals, fluctuating pH levels, dissolved and suspended solids, organic matter, and inorganic matter, all of which negatively affect the environment. Due to the extensive range of pollutants produced by the textile sector, treatment and prevention are highly difficult. Along with waste water and air emissions, other wastes include soft and hard fiber wastes such as yarn spinning wastes, fabrication wastes, residual wastes, and packing wastes. For the growth of the eco-fashion sector, it is the duty of today's designers and producers to employ sustainable materials and methods. The selection of raw materials and the processes used to transform those raw materials into finished goods have a substantial impact and should be properly understood in order to prevent waste, utilize natural resources effectively, and safeguard the environment [3].

16.2 Environmental Implications of Nanotechnology

The nanotechnological techniques can be used to reduce or prevent the damage to our environment. Since nanotechnology-enabled products that pose lower probability of negative consequences for human health and the environment have a marketing advantage, safe designing should ideally be incorporated into the production of nanomaterials. Green technology, as a form of environmental healing, seeks to reduce the harm caused by goods and innovations developed for human convenience. It is expected to minimize environmental damage while enhancing productivity and preserving natural resources. When employed in a variety of processes or applications, green technologies are sustainable and leave no trace. Utilizing natural organic resources and preventing the formation of green gases are goals of green technology [4]. Green technology does not promote any type of environmental devastation. To prevent human intervention, they support the automation of all processes. They are sustainable, enhance the environment, do not promote environmental degradation, improve people's quality of life, and increase human comfort as they do not promote environmental deterioration. Today's main technologies, such as those in aviation, automobiles, biotechnology, computers, telecommunications, education, the internet, renewable energy, atomic and nuclear physics, nanotechnology, and space technology, can all be made eco-friendly by applying the principles of green technology. The innovative technology of the 21st century, nanotechnology will be embraced by all users and play an essential role in resolving both basic and complicated societal problems if it is transformed into a green technology [5–8].

In contrast to conventional approaches, there are numerous green synthesis processes apply green chemistry principles to generate nanoparticles and subsequently nanomaterials. The following are some examples: Using a bottom-up strategy, non-hazardous naturally existing materials should be used as starting materials; magnetic nanomaterials have the ability to be recycled and re-used in nanocatalysis applications. Natural anti-oxidants, such as polyphenols from tea, wine, or agricultural waste, can be used to make metal nanoparticles. Nanoparticles made in the simple synthesis with mild reaction conditions exhibit reduced toxicity and are appropriate for use in environmental clean-up. Plants can synthesize metal nanoparticles with antibacterial properties utilizing a biogenetic reduction process. The reducing agents used include a variety of water-soluble metabolite chemicals. Plant-based biodegradable materials are particularly effective in producing gold and silver nanoparticles which can be applied across various industries. Many crystalline inorganic compounds can be synthesized by using bottom-up, low-temperature approaches such as hydro/solvothermal synthesis, template-assisted procedures, nucleation and growth in solution or suspension, microemulsion, and mini emulsion. Due to its effectiveness and low cost, "green" nanoparticle synthesis refers to the manufacture of nanoparticles by microbes like bacteria, fungi, and viruses, as well as phototrophic eukaryotes such as plants, cyanobacteria, and microalgae, heterotrophic human cell lines, and various other biological materials. Organic synthesis methods with microwave assistance are used to generate metal nanoparticles. To synthesize nanostructures and nanocomposites of metals and metal oxides, such as gold, silver, aluminium, cobalt, platinum, Buckminster fullerene, zinc oxide, and titanium dioxide, bottom-up techniques like the sol-gel method or chemical reduction techniques employ natural sources taken from plants or microbes. The green nanotechnology movement is boosted by these environmentally friendly methods for creating nanomaterials and nanostructures, giving industries the assurance they need to promote nanotechnology-based goods and services [9, 10].

16.2.1 Environmental Influence of Nanomaterials

Nanomaterials can enter the environment through various natural phenomenon (including volcanic activity, deforestation, eutrophication, erosion and deposition, montmorillonite, and sand storms) or deliberate or accidental aspects of human life (such as combustion, excavation and explosion, destruction of reef, and the generation and disposition of nanoparticles). Once released into the atmosphere, these nanomaterials can accumulate in a variety of environmental substrates, including topsoil, groundwater, moisture, and detritus. As nanoparticles are used more

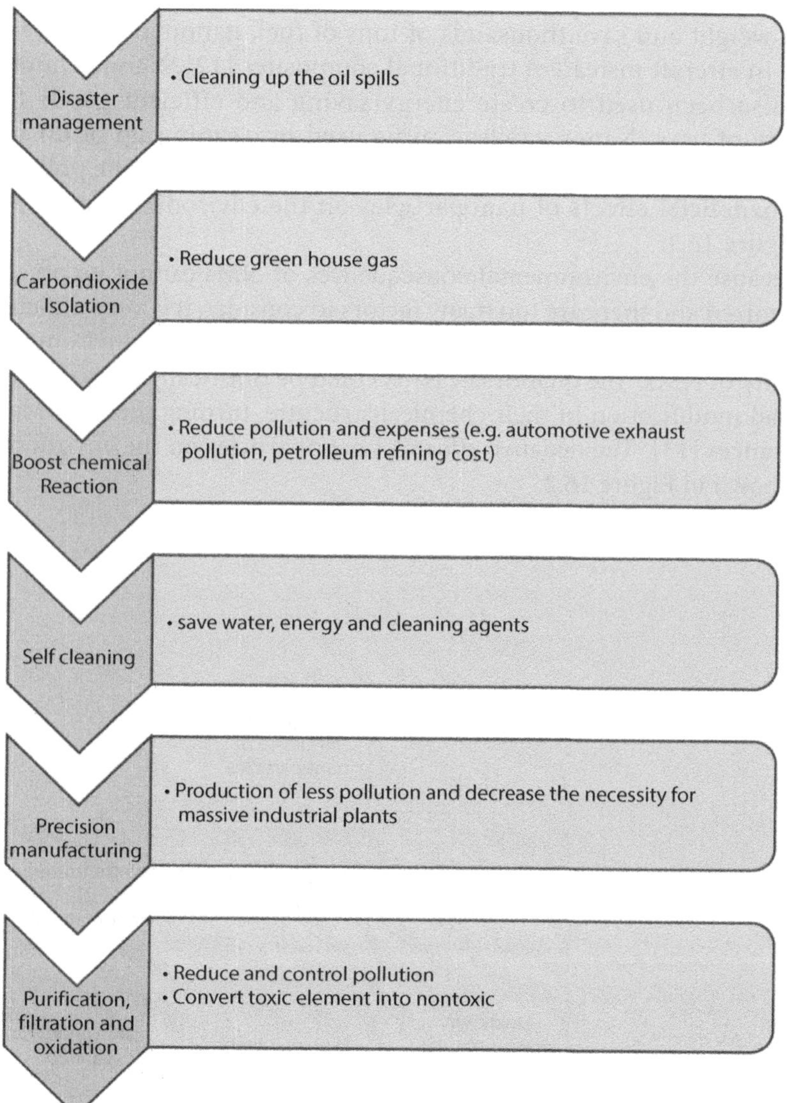

Figure 16.1 Positive influence of nanoparticles on the environment.

frequently for commercial purposes and in a wide range of consumer commodities, there is a greater chance that humans will be exposed to them both directly and indirectly. The environmental repercussions of nanotechnology, both favorable and unfavorable, are significant [11].

Significant social, environmental, and financial advantages are promised by nanotechnology. In the long run, nanotechnology might help reduce the environmental impact of human activity by introducing more innovative energy-saving technologies. For instance, to lower aircraft weight and save thousands of tons of fuel, nanomaterials are utilized in aircraft instead of traditional composites [12]. Nanotechnology has also been used to create energy-saving and efficient goods for a variety of uses. Nanoparticles can be used in cleaning up of oil spills and in carbon dioxide isolation, reducing and controlling the pollution. The beneficial effects of nanoparticles on the environment are shown in Figure 16.1.

Because the environmental consequences of NMs cannot be precisely recognized and there are too many factors to consider, it is very difficult to make conclusions about the ecological repercussions and environmental stability of NMs. The qualities of NMs could be drastically altered by even a small modification in their chemical structure, turning them into lethal substances [13]. The negative effects of nanoparticles on the environment are shown in Figure 16.2.

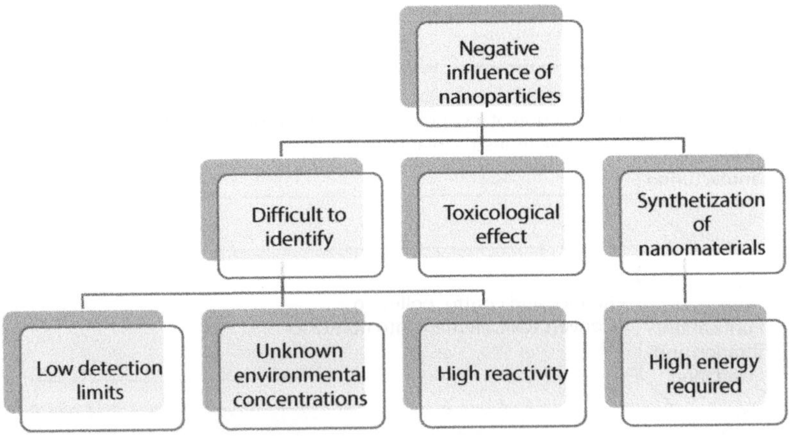

Figure 16.2 Negative influence of nanoparticles on the environment.

16.2.2 Toxicity of Nanomaterials

Nanomaterials have the potential to be toxic to microorganisms and multicellular organisms due to the solubility of hazardous ions or the formation of reactive oxygen species during direct contact [14]. When used in small amounts, they can stimulate plant growth, but, when used in large amounts, they hinder growth. Nanomaterials can alter biochemical pathways or change the expression of genes that are important for cellular functions, which results in phytotoxicity.

Mainly, four representative NMs were examined for their trace concentrations and ecotoxicity. They include titanium dioxide for higher yield and concentrations in the environment; silver for higher yield and environmental toxicity; copper oxide which also contains toxicological properties; and graphene which is increasingly used in nanomaterial production. Although, these four nanoparticles are present in soil and surface waters such as rivers, wetlands, and reservoirs but they are found in higher concentrations in sedimentary deposits, industrial effluents, and sewage water residues [15]. These NMs are ecotoxic in wastewater and sludge, but their ecotoxicity in surface water and soil is less concerning. However, the reliance on other environmental elements may cause laboratory levels of ecotoxicity to be lower under specific instance [16].

16.3 Industrial Use of Nanotechnology in Textiles

Nanotechnology has been introduced to the textile industry; in fact, a brand-new discipline called "nano-finishing" has evolved in the textile finishing sector. Making fabric with nano-sized particles results in several advantageous properties in the fabrics, as compared to previous approaches, which drastically increased mass, complexity, or rigidity. Some of the nano-finishing techniques include ultraviolet blocking, antimicrobial, bacterial resistant, fire resistant, wrinkle-proof, static free, insect or water proofing, and self-cleaning characteristics [17].

The field of textile technology has advanced significantly due to the finishing of fabrics made of natural and synthetic fibers to provide the best feel, surface texture, appearance, and other distinctive aesthetic appeal characteristics. The usage of nanotechnology in novel ways has demonstrated a great deal of potential for significant advancements in fabric finishing. By using the appropriate surface finishing, coating, or modified processes, all of which may be done with the aid of nanotechnology, the surface attributes of a fabric can be adjusted and improved in a variety of ways. Nano-finishing, nano-chemicals for apparel, nanocoating materials and nano-silver particles or smart silver for textiles are the most often used applications of nanotechnology in textiles. Recently, nanotechnology has been used to improve the tenderness, robustness, and permeability of textiles as well as the water resistance, fire resistance, and antimicrobial attributes of fibers, yarns, and fabrics [18].

The creation of stain and water resistance is one of the most popular uses of nanotechnology in the textile sector. To do this, the materials are infused with countless microscopic fibers known as nano whiskers, which enhance the fabric's density and are waterproof. Because the nano whiskers create an air cushion around each fiber, they are able to repel stains [19]. The ability of fabrics, especially synthetic ones, to absorb dyes can be improved by using nanotechnology in the opposite way. Until recently, the majority of polypropylenes were thought to be inappropriate for consumer goods like apparel, window treatments, tablecloths, and floor coverings because they resisted dyeing. A revolutionary approach that is currently being studied involves adding clay nanoparticles that are eco-friendly and acts as natural dye to raw polypropylene stock before it is converted into yarn and fibers. Without harming the fabric, the resulting nanocomposite material might be able to absorb dyes. Nanotex is one of the market-leading fabric technology related company that offers textile improvements based on nanotechnology to the interior design markets for homes, businesses, and fashion. For instance, their product, Aquapel, offers outstanding protection against rain, slush, snow, and spillage; enhances water resistant; and is also eco-friendly. Aquapel improves fabrics at cellular level without changing its natural breathability or texture by chemically glueing hydrophobic "whiskers" to individual fibers using an unique hydrocarbon process. Aquapel is the ideal solution for the environment because it is devoid of PFOA (perfluorooctanoic acid) and fluorocarbons. Silver nanoparticles contain antimicrobial and antibacterial properties help to eliminate odors in clothes. The most popular type of nanotechnology nowadays is nanoparticles of silver [20].

The silver nanoparticles are made smarter through nanotechnology as it is easily incorporated into biopolymers, resins, adhesives, and fibers made of natural and synthetic materials, and it lasts the expected lifespan of the product. Silver's inherent antibacterial properties can be employed to prevent the formation of mould, fungus, and bacteria that produce odors. It has undergone extensive testing and should be environmental friendly and complies with regulatory requirements. It has a history of producing goods for the medical, textile, and industrial sectors. Numerous uses, such as medical apparel, smart clothing, and protective clothing for troops are driving the future growth of textile applications using nanotechnology. Considering the T-shirt as an example, researchers are investigating the possibility of creating a T-shirt with nanotechnology-enhanced fabric that can monitor your heart rate, analyze your sweat, and even keep you cool on a hot summer day [20]. Nanotechnology has the potential to revolutionize the technical textile industry, which would be beneficial to all.

16.4 A Brief Note on Textile Recycling

16.4.1 Textile Recycling

The process of retrieving discarded or damaged clothes for reuse or recycling is known as textile recycling. In the United States, a group representing used clothing and fiber industries, called Secondary Materials and Recycled Textiles, plays a role in this sector. The main processes involved in textile recycling are collection of clothes (either by donation or by picking), sorting of clothes according to their size or material type, and processing of textiles followed by transportation of used apparel and other recovered materials to consumers. The textile marketing is now worth close to one trillion dollar worldwide, including products such as furniture, bedding materials, and apparel [21].

16.4.2 The Importance of Textile Recycling

Recycling of textiles is more important nowadays as there is an increase in production and usage of clothes. Approximately 100 billion apparels are produced annually [22]. According to U.S. Environmental Protection Agency, 5.8% of total municipal solid waste (MSW) was made up of textiles, which is about 17 million tons. The recovery of textiles made from bedsheets and pillow covers is about 15.8% and 13.0% for the textiles manufactured from clothing and shoes [23]. In order to build a world without landfills, textile recycling is a significant problem that needs to be addressed. Natural fibers can decay in landfills over weeks to years, releasing greenhouse gases like CO_2 and methane into the environment. Additionally, because synthetic textiles are made to not disintegrate, they may release harmful compounds into the soil and groundwater nearby, causing pollution of the water and soil [24]. Recycling of textiles has the environmental benefits such as reduction of pollution, decreased

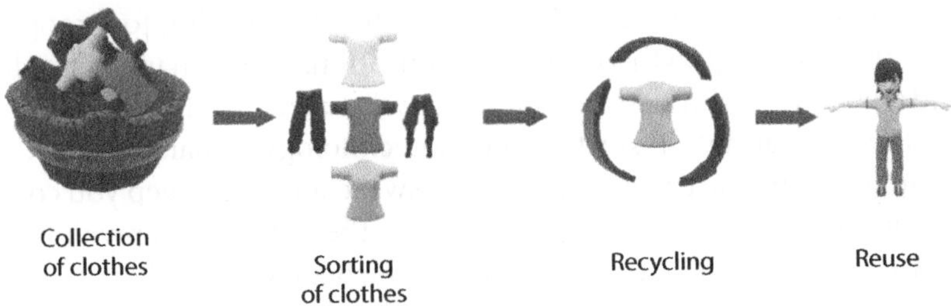

Figure 16.3 Outline image for the process of textile recycling.

demand for dyes, lower consumption of energy and water, and minimize the use of virgin fibers.

16.4.3 Process of Textile Recycling

Textile recycling involves collection of old clothes, and sorting of cloths according to their size and material type. The clothes that are in good condition, are again sent back for reuse, and the remaining textiles can be recycled. Figure 16.3 illustrates the process of textile recycling.

16.5 An Overview of the Industrial Process in Textile Industry

Considering the procedures and their impacts on the environment, which give rise to concepts such as pollutant output, possibilities to prevent pollution and regulatory standards require a comprehensive overview of the industrial technique used in the apparel sector. There are four key operations in the textile industry: spinning, fabric formation, wet processing, and fabrication, which are shown in the Figure 16.4. Fabric formation includes knitting and weaving [25]. Wet processing includes dyeing, printing, finishing, and chemical treatments. Refined products are available during fabrication. Textile manufacturers receive, process, and transform fibers into yarn, thread, or webbing, which is then used to make cloth by knitting or weaving. At various phases of the manufacturing process, these fabrics are colored and finished. Waste of all kinds is produced at each stage of preparation and conversion, and each of these wastes needs to be treated carefully to avoid contaminating the environment [26].

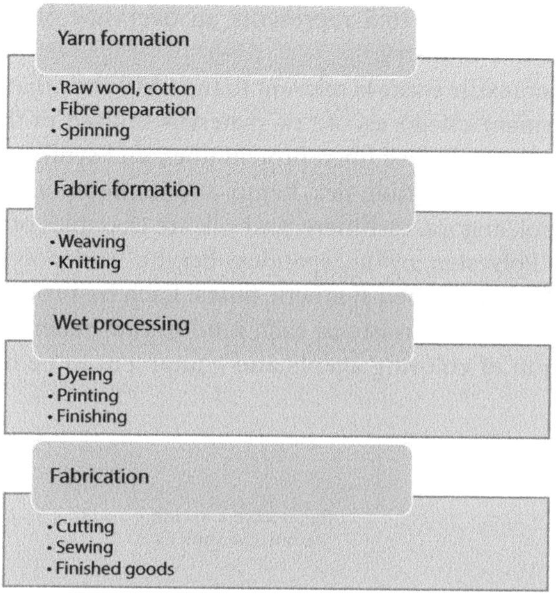

Figure 16.4 Steps of industrial process in textile industry.

16.6 Textile Waste and Its Environmental Problems

The expansion of the textile industry impacts the environment, as an estimated 10%–20% of all textile goods are wasted. According to estimates, the yearly amounts of textile waste in China, the United Kingdom, and the United States are 26.0, 1.0, and 12.4 million tons, respectively. An estimated 293 tons of textile waste are produced everyday in Hong Kong, where the scenario is critical. Every stage of garment's life cycle carries potential environmental risks and concerns due to the pollution left behind by fast fashion. The process of producing textiles also requires substantial energy, water, and chemicals. An increase in textile waste has followed the expansion of the textile industry. Mainly, there are two main types of textile waste: pre-consumer textile waste and post-consumer textile waste [27]. In the former type, the textile waste is produced during the textile manufacturing process. In the later type, the textile waste is produced as a result of consumer use and disposal. The raw materials for the manufacturing of textiles can be taken from renewable or non-renewable resources. These can be processed and utilized in different process of textile manufacturing. Textile waste can be produced after manufacturing can be disposed into the environment. Figure 16.5 represents an overview of environmental impact of the textile industry.

Pre-consumer textile waste is relevant to the textile raw materials because there are three main categories of raw materials utilized in the production of textiles including cellulose fiber, protein fiber, and synthetic fiber. Plant-based resources such as cotton, flex, hemp, and ramie are used to make cellulose fiber. Wool, angora, cashmere, and silk are examples of animal-based protein fibers. Polyester, nylon, spandex, acrylic, and polypropylene are examples of petroleum-based synthetic fibers. Each by-product produces a different source of textile waste as each kind of raw material needs a specific combination of coloring agents and equipment to be manufactured.

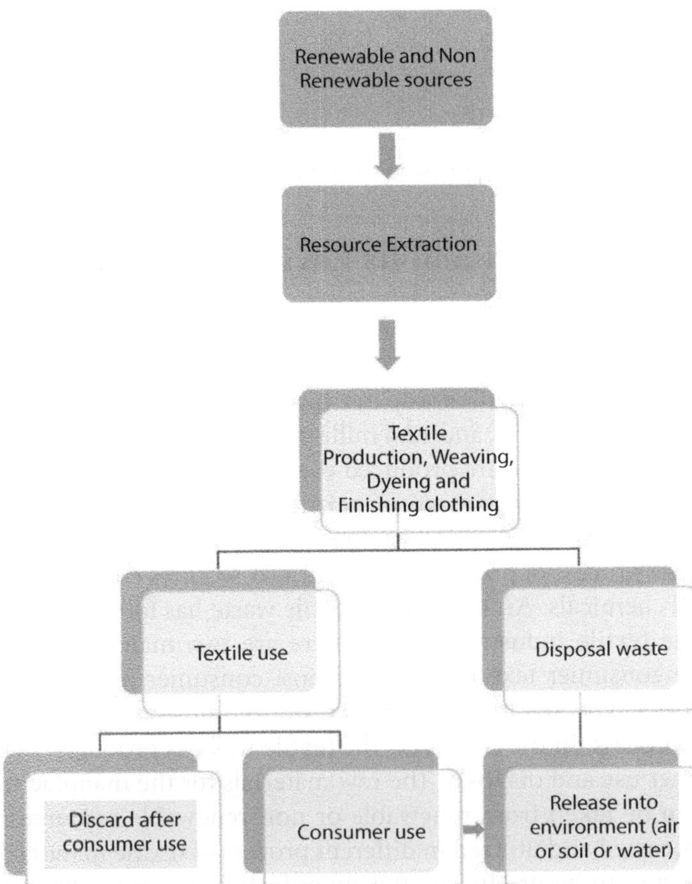

Figure 16.5 Environmental impact of textile industry.

Due to its significant market share in the textile industry, cotton is one of the most important raw materials among all fiber types. China is currently the world's top producer of cotton, but, due to the high demand for cotton in the textile sector, most of the cotton is utilized for domestic purpose. In order to achieve high production, cotton farming requires a high input of water, herbicides, insecticides, and fertilisers, which negatively impacts the environment [28].

Cotton is harvested from cotton plants, and cotton and seeds are separated through the ginning process. The separated cotton is converted into bundles and is sent toward spinning for fabric formation, as shown in Figure 16.6. Cotton fiber is spun into yarn to begin the process of making textiles using cotton as a primary material. The fabric is then created from the yarn through weaving or knitting, colored, and printed. Consequently, the textile waste generated from cotton textile production includes cotton gin waste produced during the manufacturing of yarn and thread, coloring materials, and chemicals that dissolve in waste water and have the potential to be hazardous to human health and the environment [29]. Clothes that have been discarded and contain a variety of components make up the majority of post-consumer textile waste. Due to its inability to decompose, post-consumer textile waste accumulates, occupies space, and poses the risk of spreading infectious diseases, luring pests, and emitting unpleasant odors into the environment. As a result, these may cause environmental problems if waste management is ineffective [30].

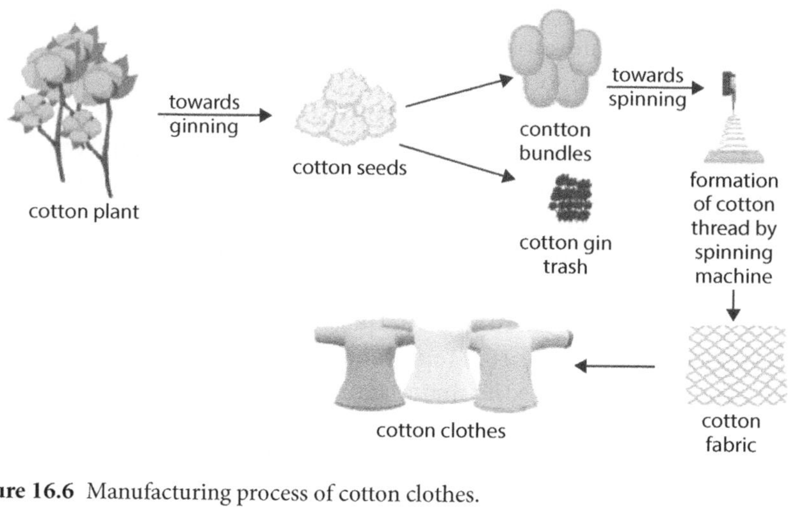

Figure 16.6 Manufacturing process of cotton clothes.

16.7 Pollution Output

Due to the textile industry's rapid expansion and production, the economy and industrialization have improved. The industrial practices implemented in the textile sector have a considerable negative influence on the ecosystem in terms of effluent, industrial pollution, greenhouse gas emission, and surplus waste materials. The water-based textile industry uses a lot of processing water and generates a lot of polluting effluents. If the effluent release is not controlled, then it could seriously harm the ecosystem. The chemicals and treatment processes involved in production play a significant role in wastewater's composition. Among the varieties of wastewater are cleansing water, processing water, storm water, and noncontact cooling water. Wet processing procedures contain a significant amount of waste water from alkali processes, which use a lot of salts, acids or alkali in finishing and dyeing processes. Pollutants in wastewater from the textile industry may include changes in pH and organic and inorganic compounds [31].

Due to the extensive range of air pollutants produced by the textile sector, treatment and prevention are highly difficult. Boilers, resin finishing processes, and drying activities in various processes all produce nitrogen and sulphur oxides. At a temperature of 200°C, the materials are dried and processed, mineral oils released from drying ovens, and other sources leak hydrocarbons. One ton of formaldehyde is thought to be emitted annually from storage tanks and fabric off-gassing at the majority of resin finishing facilities. Chemicals from fabrics evaporating at normal atmospheric pressure are known as "fabric off-gassing" [32]. During the drying process, industrial chemicals such as polycyclic aromatic hydrocarbon (2-methyl naphthalene) or 4-chlorotoluene may seep into the fibers and be released as volatile organic compounds [33].

By producing wastewater, air pollutants, and other wastes as a result of mass manufacturing and due to limited understanding regarding the effects of the procedures used to make textiles and garments, the apparel and the clothing sectors have consistently contributed to environmental deterioration. Consumers, manufacturers, intermediary contractors, and retailers have become more conscious as a result of environmentally friendly strategies and sustainable production techniques, as well as tight quality control and competitiveness. Table 16.1 represents the list of possible releases into the environment during textile manufacturing.

Table 16.1 List of possible releases generated during manufacturing of textiles.

Procedure	Generation of air pollutants	Waste water generation	Generation of leftover waste
Processing of fiber	Generation of air pollutants is very less.	Generation of waste water is very less.	Waste from packaging fibers and scrap from residual waste
Spinning of yarn	Generation of air pollutants is very less.	Generation of waste water is very less.	Discarded packaging, chopped fiber, residual fiber, and debris from cleaning and manufacturing
Slashing and sizing	Generation of volatile organic compounds	Release of biological oxygen demand, COD, and chemicals	Fiber dust, textile waste, packaging waste, and unallocated proportions made of starch
Weaving	Generation of air pollutants is very less.	Generation of waste water is very less.	Lubricant waste, waste of packaging, and fabric scraps
Knitting	Generation of air pollutants is very less.	Generation of waste water is very less.	Lubricant waste, waste of packaging, and fabric scraps
Tufting	Generation of air pollutants is very less.	Generation of waste water is very less.	Lubricant waste, waste of packaging, and fabric scraps
Desizing	Generation of volatile organic compounds from glycol ethers	Release of BOD from lubricants, biocides, abrasion-resistant agents, and chemically synthesized compounds	Pollutants from packaging, fiber fuzz, synthetic fibers, swabs, rags, and diffusers for cleaning, as well as cleaning wastes and irritants

(*Continued*)

Table 16.1 List of possible releases generated during manufacturing of textiles. (*Continued*)

Procedure	Generation of air pollutants	Waste water generation	Generation of leftover waste
Scouring	Generation of volatile organic compounds from glycol ethers	Alkali, surfactants, fats, hydrocarbons, pectin, wax, weaving lubricants, spin finishes, discarded solvents, biocides, and pesticide remnants	Less or no generation of leftover waste
Bleaching	Less or no generation of air pollutants	Generation of hydrogen peroxide and organic preservatives	Less or no generation of residual waste
Singeing	Little amount of exhaust gases released from burners	Less generation of waste water	Less or no generation of unconsumed waste
Mercerizing	Less or no generation of air pollutants	Release of sodium hydroxide	Less or no generation of residual waste
Heat setting	During the fabrication of synthetic fibers, spin finish chemicals were employed.	Less or no generation of waste water	Less or no generation of leftover waste

(*Continued*)

Table 16.1 List of possible releases generated during manufacturing of textiles. (*Continued*)

Procedure	Generation of air pollutants	Waste water generation	Generation of leftover waste
Dyeing	Generation of volatile organic compounds	Generation of minerals, salts, detergents, allergens, chemical processing aids, alkaline compounds, etc., from waste water	Less or no generation of residual waste
Printing	Generation of exhaust gases, lubricants, and acetic anhydride from drying and curing furnace emissions	Generation of solvents, colorants, minerals, suspended solvents, biological oxygen demand, and froth	Less or no generation of consumed waste
Finishing	Pollutants found in chemicals purchased, formaldehyde vapours, hot exhaust, and particulates	Contaminants, suspended chemicals, discarded solvents, BOD, and COD	Scraps of fibers, trimmings, and debris from packaging
Fabrication of product	Less or no generation of air pollutants	Less or no generation of waste water	Scraps of fibers

16.8 Restricted Substances

The European Union Ecolabel for textile products contains a list of the forbidden compounds for textiles, and, depending on the restriction's reach, each restriction must be applied to particular categories of substances.

Achieving Oeko-Tex 100 certification would demonstrate compliance with restricted substance regulations. Products that have earned the Standard 100 by Oeko-Tex certification have undergone safety testing to safeguard your health. This certification guarantees that the product has undergone thorough testing against a list of up to 350 hazardous toxins. This testing includes the fabric, thread, and accessories. Oeko-Tex 100 standard tests may on occasion be supplemented by additional testing. Table 16.2 represents the list of restricted substances [34].

Table 16.2 List of restricted substances with limit values and regulations.

List of substances	Scope of restriction	Limit values	Regulations for candidates
Substances of very high concern (SVHC)	N,N-Dimethylacetamide is one of the substance with a higher limit value.	0.1% w/w	Final product testing is not necessary if raw fiber testing is in compliance with the Oeko-Tex 100 limit value.
	List of derogated substances: For interior textiles Products used for children	0.005% 0.001%	To adhere to restricted substance list verification requirements
Biocides	EU Ecolabel substances with higher restriction Polychlorinated biphenyls Dimethyl fumerate	0.05% w/w 0.1% w/w	When additional testing and verification are required for polychlorinated biphenyls, the summation value can be utilized.
Auxiliaries and surfactants	Oeko-Tex 100 tests and certification shall be accepted for octyl phenol, nonyl phenol, alkyl phenol derivatives, Alkylphenolethoxylates, and their derivatives	50% w/w sum total	

(*Continued*)

Table 16.2 List of restricted substances with limit values and regulations. (*Continued*)

List of substances	Scope of restriction	Limit values	Regulations for candidates
Dyes and carriers	For halogenated carriers, azodyes, CMR (carcinogenic, mutagenic, and reprotoxic) substances, potentially sensitizing dyes and chlorinated benzene, follows and accepts Oeko-Tex 100 tests and certification.	Halogenated dyes- 1.0% w/w Azo dyes: 50% w/w For CMR substances, potentially sensitizing dyes and metal complex dyes. Limits are not applicable.	Use of metal complex dyes, CMR substances and potentially sensitizing dyes must be verified additionally for the EU Ecolabel.
Heavy metals	Heavy metals such as antimony, cadmium, chromium, metal complex dyes, nickel, and mercury follows and accepts Oeko-Tex 100 tests.	Limit values for Children Antimony: 30% w/w Cadmium: 0.1% w/w Metal complex dyes: 1.0% w/w Mercury: 0.02% w/w Limit values for interior textiles Antimony: 30% w/w Cadmium: 0.1% w/w Metal complex dyes: 2.0% w/w Mercury: 0.02% w/w	EU Ecolabel certification required separately for chromium, cobalt, and nickel.

(*Continued*)

Table 16.2 List of restricted substances with limit values and regulations. (*Continued*)

List of substances	Scope of restriction	Limit values	Regulations for candidates
Printing	Printing pastes contains VOC content. Plastisols and restricted phthalates shall not be used.	Less than 5% w/w VOC content 0.1% restricted phthalates	
Easy care	For formaldehyde, Oeko-Tex values are accepted.	For children: 16 ppm Products with skin contact: 75 ppm For interior textiles: 300 ppm	
Antifelting and shrink resistance	Halogenated substances and wool silvers	As applicable	Selected test strategies are required for validation criteria.
Water and oil repellent treatments	Perfluorooctane sulfonate and its derivatives Perfluorocarboxylic acid and its salts include PFOA (perfluorooctanoic acid)	Limits for PFOA is 0.05 mg/kg	EU Ecolabel evaluation for Perfluorocarboxylic acid separately upto their limit values.
Flame retardants	Paraffin, octa bromidiphenyl ether, and polybrominated biphenyls are the flame retardants required to meet ISO and Oeko-Tex 100 standards.	5 mg/kg	
Coatings, laminates, and membranes	Polymers used for coatings should not contain phthalates.	0.1% w/w sum total	Phthalates should be evaluated separately by EU Ecolabel.

(*Continued*)

Table 16.2 List of restricted substances with limit values and regulations. (*Continued*)

List of substances	Scope of restriction	Limit values	Regulations for candidates
Accessories—buttons, rivets, and zips	Phthalates shall not be used in plastic accessories such as dibutylphthalate. Metal accessories such as nickel and chromium are used with in the limit values.	0.1% w/w	Metal accessories incorporating nickel or chromium must undergo additional EU Ecolabel certification.

16.9 Protocols for Industrial Use of Nanotechnology

As represented in Figure 16.7, this section addresses nano-safety concerns of textiles regarding nanotechnology, regulations and control strategies, principles of safe handling of nanomaterials, risk evaluation and assessment, good practices, and test guidelines.

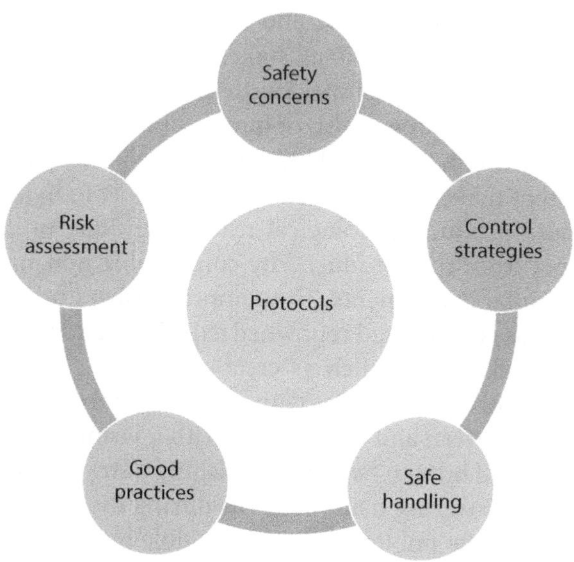

Figure 16.7 Pictorial representation of protocols.

16.10 Nanotechnology-Based Textiles and Nano-Safety Concerns

Nanotechnology is quickly evolving, and it is anticipated to have a significant influence on consumer goods, particularly in the textile sector. With advancements in fiber manufacturing, and fabric design, the scope and applications of nanotechnology in the field of apparel sector are wide. In this way, fabrics can develop peculiar prospects such as UV protection, protection against microbes, auto cleansing, and nano-capsules containing moisturizing elements, deodorizing properties, and repelling properties. This prospective for clothing has been increasing consistently as a result of the commercialization of nanotechnology's global expansion [35].

Nanotextile safety and health implications should take into account the complete life cycle of nanomaterials. safety for all those involved in the trade phases (distribution and retail), and safety for the employees during all stages of manufacturing, particularly in the case of the use of nanostructured materials in fabrics and safety for the customers. Nanoparticles that are discharged from textiles into the air, water, and landfill must also be taken into account because they may have a direct or indirect impact on consumers. Recently, considerations have been expressed about the potential for some nanomaterials to have a harmful influence on the health of humans and the ecosystem, according to a risk assessment of nanotechnology [36]. The sustainable development of nanotechnology for textile finishing must address safety, risk, and regulatory concerns, which have become increasingly important to all stakeholders. Two criteria have been presented by Maynard and Kuempel for identifying nanomaterials that are particularly hazardous to the health of humans [38].

To pose a health risk, the substance must be able to interact with the body in a way that makes its nanostructure accessible to living things. The substance should produce a biological reaction unique to its nanoparticles and distinct from that produced by comparable non-nanoscale substances of the similar components. The comprehensive study of engineered nanostructures has coordinated renowned experts to address the following nanoparticle risk challenges such as creating precise measurements and globally accepted norms for textiles with nanomaterial coatings, determining the exposures pattern and relative risk with a vision toward the effects on the ecosystem and human health, establishing a strategy for the efficient maintenance of engineering nanomaterials during dyeing, and promoting deliberation among the public about the possible benefits and drawbacks of nanotechnology [37].

16.11 Regulation Methods for Nanomaterials

It is crucial that these problems be handled by a strong scientific and regulatory approach that decreases the expense and frequency of testing on animals in order to provide the public and employees who come into contact with these products with enough assurance that they are protected. With potentially dangerous nanomaterials in such production system exposure scenarios, a detailed specific instant risky assessment is necessary, and a foresighted risk-managing strategy must be put in place. The classification of control strategies includes elimination, substitution, engineering controls, and administrative controls, as shown in Figure 16.8.

16.11.1 Elimination

In the hierarchy of controls, elimination is the most ideal strategy. Elimination, as its name suggests, is the process of getting rid of a risk. The design phase, when the material, method, and/or facility are being developed, is typically when eliminating hazards is simplest to do. The removal of a step for incoming nanomaterial inspection during a process step is an example of elimination. The possibility for aerosolization of nanoparticles when conducting an inbound inspection that necessitates opening a package containing them makes the inspector potentially vulnerable. A procedure becomes fundamentally safer by eliminating the inspection phase, which also eliminates the risk [39].

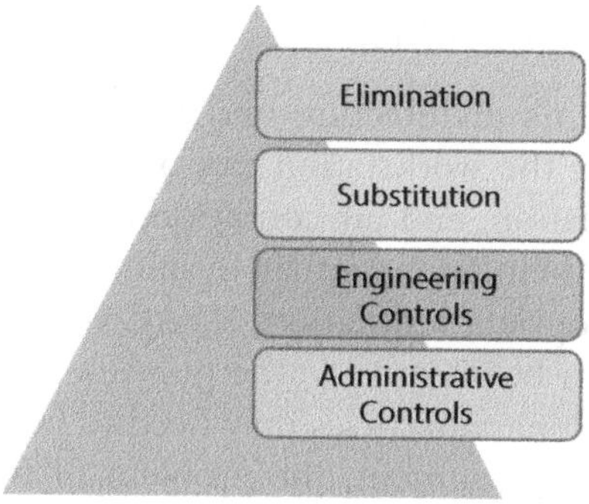

Figure 16.8 Pictorial representation of classification of controls.

16.11.2 Substitution

Substitution serves to substitute one set of conditions with a high risk level with another set of conditions with a lower risk level within the hierarchy of controls. Examples of substitution include switching flammable solvent-based materials for water-based ones, exchanging highly toxic materials for less toxic ones, and modifying harsh operating conditions in a process (e.g., reduced pressure). A nanomaterial may be difficult to replace because it was probably created for its unique features, although it might be feasible in some cases. A level of protection for employees handling the substance will be provided by switching from a dry powder version of the nanomaterial to a slurry.

Elimination and substitution are typically most economically feasible when applied during the design or development phases of a process. Implementation is straightforward and, if started early enough, can yield significant savings over time (e.g., cost of protective equipment, initial investment, and operational expenses for ventilation system). To eliminate a hazard in an existing process, substitution or deletion may necessitate significant adjustments to the machinery and/or operating practices [39].

16.11.3 Engineering Controls

Engineering controls should ideally reduce the difficulty of the process rather than increase it. Making it simpler to perform tasks safely is an appropriate design principle for engineering controls. It is mainly divided into two main categories: ventilation and non-ventilation controls [39].

16.11.3.1 Ventilation Controls

The ventilation system's architecture and the airflow volumes supplied into and vented from the workstation can both affect how efficient it is. Office heating, ventilation, and air conditioning (HVAC); infection control in healthcare; and emission control in industrial operations are a few applications for appropriate ventilation. HVAC ventilation is one example of dilution ventilation but is outside the purview of this article. Because exhaust air containing nanoparticles may need to be cleaned before being discharged into the environment, the problem of filtration has obvious consequences for ventilation.

16.11.3.2 *Non-Ventilation Controls*

A higher level of protection for those who work with nanomaterials can be provided by combining non ventilation controls with ventilation measures. Numerous technologies created for the pharmaceutical sector, such as isolation containment systems and continuous linear systems, are among the other engineering controls besides ventilation. Glove box containment, which can be utilized as an enclosure around small-scale powder processes like mixing and drying, is one of the most prominent flexible isolation methods. The continuous liner system, which encloses the material in a polypropylene bag while enabling the filling of product containers. When powders need to be packed into drums, this approach to product off-loading is frequently used.

16.11.4 Administrative Controls

In existing operations where liabilities are not properly controlled, administrative controls and PPE (personal protective equipment) are typically applied. If engineering control strategies are not practical or do not successfully lower concentrations to a safe level, then this might happen. Administrative controls (which include training, job rotation, work scheduling, and other efforts to prevent limits) and PPE programs may be less expensive to implement but, over the long run, can be highly expensive to manage. Additionally, these measures for worker protection have shown to be less successful than alternative ones and frequently call for a lot of work from the individuals who may be impacted [39].

16.12 Principles for the Safe Handling of Nanomaterials

Nanotechnology has influenced the creation of improved chemical finishes in addition to the creation of diverse fiber composites. In textile finishing, nanoparticles like amphoteric oxides and carbonates of metals and metalloids are also employed to modify surface characteristics and impart textile functionality. Implementing effective control measures based on a thorough knowledge of the nanostructures and the ambiguity surrounding technology is the most efficient strategy to reduce exposure to nanoparticles or complications associated with nanoproducts with nanoparticle coatings on textiles [40].

Numerous governmental organizations and regulatory entities have launched initiatives and programs to address the hazards of nanotechnology for human health and the environment. The International Organization for Standardization and the Organization for Economic Cooperation and Development have established a group of working people and increased their efforts in the area of dealing with the risks of nanotechnological products in the health sector and remaining areas on a transnational level. The development and delivery of a systematic way and disciplined method for identification, management, and decreasing possible safety to environment and health concerns of nanoparticles at all phases of a textile manufacturing process were some different activities that were recognized. Government and industry authorities will lack a stable basis for guaranteeing that nano-products are created and utilized under particular regulations such as REACH and TSCA (Toxic Substances Control Act), which are developed. REACH is a European Regulation and is an acronym for the Registration, Evaluation, Authorisation and Restriction of Chemicals. The TSCA maintain a single list of restricted substances and list of approved substances. The REACH Regulation maintains the Restricted Substances List, which includes substances that are to be phased out of products [41].

16.13 Risk Management and Assurance in Quality for Nano-Coated Textile Products

In order to reduce emissions into the environment and prevent harmful material from endangering the public's health and the ecosystem by leaching into apparel products coated with nanomaterials, analyzing hazardous substances is a crucial first process. Evaluations of the safety of nanoproducts should be conducted, and they should be incorporated into the current guideline for quality, safety, and reliability across the production planning, quality management and product life cycle assessment. It is widely accepted that perilous mitigation should be incorporated into the company's current regulatory and quality assurance standards. The implementation of a sophisticated risk assurance management system is the most effective technique to ensure the security and calibre of nanostructured apparel, such as FERMA (Federation of European Risk Management Association) or AS/NZS 4360:2004 or AS/NZS 4360:1999 (Australia/New Zealand), which determines the required classes of preventive and restraint measures in regard to the use and marketing of nanoproducts that pose a significant hazard to the well-being and safety of consumers. Nano-risk assessment systems, like CENARIOS®, which include criteria for evaluation of risk, risk monitoring, communication, and management of knowledge, were developed in close collaboration with industry requirements because of the rapid expansion of knowledge in occupational health and safety. Extensive familiarity with the nature and risks of nanotextile products will be assumed, and market withdrawal is one potential risk-reduction technique [42].

16.14 Good Practices and Test Guidelines

The control strategies include recommendations for the research that are needed to assess the health and safety hazards and address the environmental risks and exposure. Occupational health surveillance is an integral part, along with an occupational safety and health program for nanomaterial exposure; however, the guideline still needs improvement for use with engineering nanomaterials. Good work practices help reduce exposure to nanoparticles among consumers and workers. Furthermore, the creation of a testing of characteristics, toxicity, and description is necessary for a thorough impact, risk analysis framework, usage, and exposure [43].

The creation of uniform terminology, nomenclature, and standardized methods of measurement and testing in the fields of health, the environment, and safety is the responsibility of the Organization for Economic Cooperation and Development (OECD) and the International Standardization Organization (ISO), in particular. The OECD's chemical committee established the "Working Party on Manufactured Nanomaterials" (WPMN) in September 2006. In order to exchange knowledge and establish an international network, this working group convened sporadic meetings. The working group consists of the following six steering groups:

SG1: Developing a research database
SG2: Research strategy for produced nanomaterials in the areas of environment, health, and safety (EHS).
SG3: Safety testing for a set of manufactured nanomaterials
SG4: Test guidelines for manufactured nanomaterials.
SG5: Collaboration on preventative and voluntary schemes.
SG6: Cooperation in assessing risks and exposure measurement.

The availability of established and validated procedures to measure and test the characteristics of synthetic nanomaterials, particularly their toxicological characteristics, is a necessary prerequisite for the regulation mentioned above. Technology development must incorporate the discussion of opportunities and risks among the public, government, and business sector.

16.15 Applications for Nanotextiles

As represented in Figure 16.9, applications of nanotextiles can be utilized in sports wear, military outfits, electronic textiles, space suit, medical fabrics, and protective clothing. These applications made an easy way for monitoring human health, provide protection for soldiers and astronauts, and can be useful in various other aspects.

16.15.1 Sports Wear

Sports-related research has been heavily influenced by the desire to enhance athletic performance. Examples include wicking fabrics like Coolmax® and Goretex®, which are waterproof and breathable. According to the phase change technology included in Outlasts fabric, excess body heat is absorbed, retained, and then discharged as required, making it even feasible to maintain a stable body temperature [44, 45].

16.15.2 Electronic Textiles

To track the electrolytes in perspiration, Wang *et al.* have printed electrochemical sensors onto fabrics [46]. Additional applications include the use of elastic, flexible electrocardiography patches for cardiac monitoring with wireless signal transmission and the incorporation of biochemical sensors

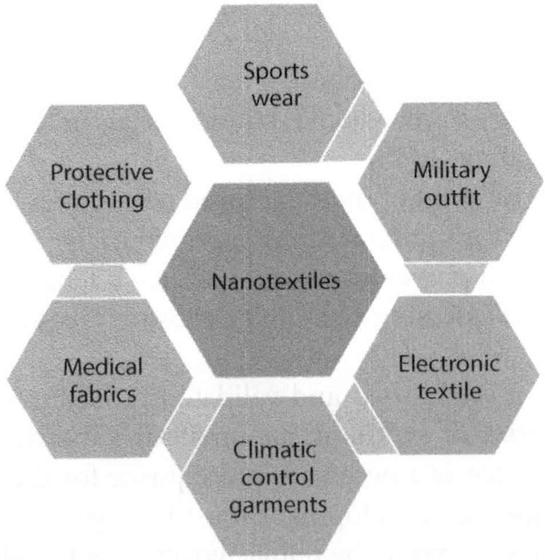

Figure 16.9 Applications of nanotextiles.

into textiles for continuous health monitoring of a person. Fluid management and sampling within the textile structure are required for on-body fluid monitoring [47]. Textile sensors embedded in innovative clothing measure physiological markers like respiration, activity level, and heart rate.

16.15.3 Medical Fabrics

The remote monitoring of patients with chronic illnesses and those undergoing recuperation is significantly aided by smart clothes. It also encourages the concept of preventative healthcare. The medical sector greatly benefit from nano reinforced medical textiles. It is feasible to make wound dressings that are biologically active, offer antimicrobial defense, and promote skin regeneration. These dressings can be applied directly to the affected area or administered through the skin, adjusting to allow air, water, and medication passage as the fabric detects illness or the healing process. Commercially, antibacterial dressings are easily available [48–52].

16.15.4 Military Outfits

To improve protection and survival for all those working in hazardous locations and adverse weather conditions, reliable information technology is required. Innovations are highly beneficial for emergency responders and military personnel. The Smart Shirt by Sensatex™ was initially developed by the Georgia Tech Research Corporation for military use. The T-shirt functions similarly to a computer due to the incorporation of optical and electrical fibers into the apparel. During warfare, the body's vital signs are monitored by a variety of sensors, and optical fibers are employed to identify bullet wounds and specify their exact coordinates [53, 54]. The objective is to develop a force that is more agile, lighter, faster, and aware of its surroundings and potential threats. Medical practitioners can monitor a soldier's health through mobile broadband. The ideal battle suit would be a bulletproof jumpsuit that is a bit thicker than standard spandex, monitors health, repairs wounds, transfers data automatically, and allows doctors to analyze combat casualties remotely so they can react more swiftly and securely [55, 56].

16.15.5 Protective Garments

The body is protected against a variety of situations by protective clothing, which includes medical materials, sportswear, military attire, climatic control garments, space suits, and water-repellent textiles. Nanotextiles are textiles that have been functionalized with nanoparticles to offer advantages like water resistance, odor and moisture removal, increased elasticity and strength, antimicrobials, and thermal resistance, as well as to prevent the permeation of ultraviolet radiation and biomedical clothes, along with many other applications [57, 58]. Due of the unique potential they hold, nanotextile advances are predicted to dominate those in the textile industry. Recently, the apparel sector is one of the sectors that benefits from nanotechnology, included in the moisture management systems used in sports are elements for quick sweat evaporation, luminous clothing, insulation, and other durable materials [59, 60].

16.15.6 Climatic Control Garments

Sweaters, gloves, masks, and other climatic control clothing regulate body temperature. To assist lower the risk of viral infection, it is advisable to wear inexpensive winter clothes with great microbiological protection, especially during pandemics. The pores in the nano-infused nose masks are smaller than those in traditional respiratory masks, which have larger pore diameters [61, 62]. For instance, a nose mask made of a composite of graphene and silver nanoparticles can aid to eliminate 99% of microbiological infiltration, including corona virus. Additionally, conventional masks lack the ability to self-disinfect [62, 63].

Recent years have seen increased attention focused on nanotextile materials. These nanotextiles are multipurpose, specifically produced to order, durable, and very appealing from a business standpoint. One of the main end uses for these fabrics has been in the environmental sector, as well as the medical and health sector, sports apparel, and military apparel. Many facets of our daily life could be impacted by advances in smart nanotextiles, which could also result in clothing that is contextually aware. Passive, non-invasive sensing of users and their surroundings is made possible by new materials that integrate revolutionary technology.

16.16 Conclusion

Nanotechnology has both positive and negative influence on the environment. In this chapter, we discussed various ecological aspects such as green nanotechnology principles, positive and negative influence of nanoparticles on the environment, textile waste and its environmental problems, pollution output, and safety assessment protocols for industries and applications of nanotextiles. The management of the textile industry's industrial activity and associated hazards related to nanotechnology becomes more and more dependent on transparency and risk communication. Recent years have seen regulatory bodies pay much more attention to the hazards associated with engineering nanomaterials potential toxicity. An in-depth understanding of any possible risks to human health and the environment from exposure during the lifecycle of textiles coated with nanoparticles might be acquired by performing a risk evaluation of engineering nanomaterials used in finishing procedures. Nanotextile applications include sportswear, military clothing, medical fabrics, and climatic control garments. Smart nanotextiles have a huge range of possible uses that could change our lifestyles and make things more prevalent.

References

1. OECD Environment Directorate, Ecotoxicology and Environmental Fate of Manufactured Nano-materials: Test Guidelines. Expert Meeting Report. Series on the Safety of Manufactured Nanomaterials No. 40, 2014.
2. Lavicoli, I., Leso, V., Ricciardi, W., Hodson, L.L., Hoover, M.D., Opportunities and challenges of nanotechnology in the green economy. *Environ. Health*, 13, 78–84, 2014.
3. Radhakrishnan, S., Eco-Parameters and Testing of Sustainable Textiles and Apparels, in: *Handbook of Sustainable Apparel Production*, pp. 310–329, 2016.
4. Aithal, P.S. and Aithal, S., Nanotechnology Innovations and Commercialization – Opportunities, Challenges & Reasons for Delay. *Proceedings of National Conference on Changing Perspectives of Management, IT, and Social Sciences in Contemporary Environment*, vol. 14, pp. 1–12, 2016.
5. Shubhrajyotsna, A. and Aithal, P.S., Green and Eco-friendly Nanotechnology-Concepts and Industrial process. *Int. J. Manage. Technol. Social Sci.*, 6, 1, 1–31. https://doi.org/10.47992/IJMTS.2581.6012.0127, 2021.
6. Aithal, P.S. and Rao, P., How Service Industries Can Transform themselves into Green Business Industries. *Int. J. Manage. Sci. Bus. Res. (IJMSBR)*, 5, 4, 150–158, 2016.
7. Aithal, P.S. and Jeevan, P., Strategic Rethinking of Management Education: Green MBA Model. *Int. J. Manage. IT Eng. (IJMIE)*, 6, 1, 55–73, 2016.
8. Rao, P. and Aithal, P.S., Green Education Concepts & Strategies in Higher Education Model. *Int. J. Sci. Res. Mod. Educ. (IJSRME)*, 1, 1, 793–802, 2016.
9. Varma, R.S., Greener approach to nanomaterials and their sustainable applications. *Curr. Opin. Chem. Eng.*, 1, 2, 123–128, 2012.
10. Silva, L.P., Reis, I.G., Bonatto, C.C., Green Synthesis of Metal Nanoparticles by Plants: Current Trends and Challenges, in: *Green Processes for*

Nanotechnology, V. Basiuk and E. Basiuk (Eds.), pp. 259–275, Springer, Cham, 2015.

11. Saleem, H. and Zaidi, S.J., Sustainable Use of Nanomaterials in Textiles and their Environmental Impact. *Materials (Basel, Switzerland)*, 13, 22, 5134, 2020, https://doi.org/10.3390/ma13225134.
12. Kausar, A., Rafique, I., Muhammad, B., Aerospace application of polymer nano-composite with carbon nanotube, graphite, graphene oxide, and nano-clay. *Polym. Plast. Technol. Eng.*, 56, 13, 1–19, 2017.
13. Kabir, E., Kumar, V., Kim, K.-H., Yip, A.C.K., Sohn, J.R., Environmental impacts of nanomaterials. *J. Environ. Manage.*, 225, 261–271, 2018, https://doi.org/10.1016/j.jenvman.2018.07.087.
14. Abbas, Q., Yousaf, B., Ullah, H., Ali, M.U., Ok, Y.S., Rinklebe, J., Environmental transformation and nano-toxicity of engineered nano-particles (ENPs) in aquatic and terrestrial organisms. *Crit. Rev. Environ. Sci. Technol.*, 50, 23, 2523–2559, 2020.
15. Zhao, J., Lin, M., Wang, Z., Cao, X., Xing, B., Engineered nanomaterials in the environment: Are they safe? *Crit. Rev. Environ. Sci. Technol.*, 51, 14, 1443–1478, 2021.
16. Bradford, S.A., Shen, C., Kim, H., Letcher, R.J., Rinklebe, J., Ok, Y.S., Ma, L., Environmental applications and risks of nanomaterials: An introduction to CREST publications during 2018–2021. *Crit. Rev. Environ. Sci. Technol.*, 52, 21, 3753–3762, 2022, https://doi.org/10.1080/10643389.2021.2020425.
17. Yetisen, A.K., Qu, H., Manbachi, A., Butt, H., Dokmeci, M.R., Hinestroza, J.P., Skorobogatiy, M., Khademhosseini, A., Yun, S.H., Nanotechnology in Textiles. *ACS Nano*, 10, 3, 3042–68, 2016.
18. Sawney, A.P.S. *et al.*, Modern Applications of Nanotechnology in Textiles. *Text. Res. J.*, 78, 8, 735, 2008.
19. Yetisen, A.K., Qu, H., Manbachi, A., Butt, H., Dokmeci, M.R., Hinestroza, J.P., Skorobogatiy, M., Khademhosseini, A., Yun, S.H., Nanotechnology in textiles. *ACS Nano*, 10, 3, 3042–3068, 2016, https://doi.org/10.1021/acsnano.5b08176.
20. Marian, R., Nanotechnology in Textile Industry. *Ann. Univ. Oradea Fasc. Textiles Leatherwork*, vol XVI no.2, pp. 83–88, 2015.
21. Horrocks, R., *Recycling Textile and Plastic Waste*, Woodhead Publishers, Cambridge England. ISBN 978-1-85573-306-0, 1996.
22. Pandit, P., Ahmed, S., Singha, K., Shrivastav, S., *Recycling from waste in Fashion and Textiles: A Sustainable and Circular Economic Approach*, Wiley Scrivener, Hoboken, NJ, USA, 2020.
23. U.S. Environmental Protection Agency, Textiles-materials specific data, 2020.
24. Pazoki, M. and Ghasemzadeh, R., *Muncipal Landfill Leachate Management*, 11, 6330, pp. 21–33, Springer International Publishing, Gewerbestrasse Cham, Switzerland Springer, 2020, https://doi.org/10.1007/978-3-030-50212-6.

25. Kozlowski, R.M., *Handbook of natural fibres: processing and applications*, pp. 30–42, Woodhead Publishing, Cambridge, MA, USA, 2012.
26. Ghosh, P. and Gangopadhyay, R., Photofunctionalization of cellulose and lignocellulose fibres using photoactive organic acids. *Eur. Polym. J.*, 3, 625–634, 2000.
27. Pensupa, N., Leu, S.-Y., Hu, Y., Du, C., Liu, H., Jing, H., Wang, H., Lin, C.S.K., Recent Trends in Sustainable Textile Waste Recycling Methods: Current Situation and Future Prospects. *Top Curr. Chem(Z)*, 375, 76, 2017, https://doi.org/10.1007/s41061-017-0165-0.
28. Gordon, S. and Hsieh, Y., *Cotton: Science and technology*, pp. 576–586, Woodhead Publishing, Sawston, 2006.
29. Lee, G.F. and Jones-Lee, A., Environmental impacts of alternative approaches for municipal solid waste management: an overview, Report by G Fred Lee & Associates, El Macero, 1993.
30. Raj, C., Arul, S., Sendilvelan, S., Saravanan, C., Bio gas from textile cotton waste- an alternate fuel for diesel engines. *Open Waste Manage. J.*, 2, 1–5, 2009.
31. United States, Environmental Protection Agency (EPA) 401 M St., SW, Washington, DC; Enforcement and Compliance Assurance (2223A), Profile of the textile industry. EPA/310-R-97-009, September 1997.
32. American Textiles Manufacturers Institute (ATMI), Industrial processes comments on draft of this document, 1997.
33. EPA, *The US Environmental Protection Agency, EPA profile of the textile industry, September 1997*, 1996 U.S.E.P. (n.d.). EPA Office of Compliance Sector Notebook Project: Profile of the Textile Industry https://archive.epa.gov/compliance/resources/publications/assistance/sectors/web/pdf/textilsn-2.pdf.
34. European Commission, EU Ecolabel and Green Public Procurement—Stakeholders, Annexures 3,4,5 restricted substances list, Oeko-Tex 100 read across and risk matrix, February 2013a.
35. Russell, E., Nanotechnologies and the shrinking world of textiles. *Text. Horiz.*, 9/10, 7, 2002.
36. O'Brien, N. and Cummins, E., Recent Developments in Nanotechnology and Risk Assessment Strategies for Addressing Public and Environmental Health Concerns. *Hum. Ecol. Risk Assess.*, 14, 3, 568–592, 2008, https://doi.org/10.1080/10807030802074261.
37. Maynard, A.D. and Kuempel, E.D., Airborne Nanostructured Particles and Occupational Health. *J. Nanopart. Res.*, 7, 6, 587–564, 2005. http://dx.doi.org/10.1007/s11051-005-6770-9.
38. Paik, S.Y., Zalk, D.M., Swuste, P., Application of a Pilot Control Banding Tool for Risk Level Assessment and Control of Nanoparticle Exposures. *Ann. Occup. Hyg.*, 52, 6, 419–428, 2008, https://doi.org/10.1093/annhyg/men041.
39. *Current Strategies for Engineering Controls in Nanomaterial Production and Downstream Handling Processes*, National Institute of Occupational Safety and Health, 2011, Cincinnati, OH: U.S. Department of Health and Human

Services, Centers for Disease Control and Prevention, DHHS (NIOSH) Publication No. 2014–102.
40. Sweet, L. and Strohm, B., Nanotechnology—Life-Cycle Risk Management. *Hum. Ecol. Risk Assess.*, 12, 3, 528–551, 2006, https://doi.org/10.1080/10807030600561691.
41. Chen, J., Yanguang, Sun, C., Safety and Health Assessment of Manufactured Nanoparticles in Nano-coated Textile Products. *Adv. Mater. Res.*, 175–176, 722–728, Trans Tech Publications, Ltd. 2011, https://doi.org/10.4028/www.scientific.net/amr.175-176.722.
42. Rickerby, D.G. and Morrison, Mark., Nanotechnology and the environment: A European perspective. *Sci. Technol. Adv. Mater.*, 8, 1, 19–24, 2007, https://doi.org/10.1016/j.stam.2006.10.002.
43. Siegrist, M., Keller, C., Kastenholz, H. *et al.*, Laypeople's and experts' perception of nanotechnology hazards. *Risk Analysis*, 27, 1, 59–69, 2007. https://doi.org/10.1111/j.1539-6924.2006.00859.x
44. Coyle, S. and Diamond, D., Smart Nanotextiles: Materials and Their Application, *Reference Module in Materials Science and Materials Engineering*. https://doi.org/10.1016/B978-0-12-803581-8.03345-2 2016.
45. Chan, M., EsteVe, D., Fourniols, J.-Y., Escriba, C., Campo, E., Smart wearable systems: current status and future challenges. *Artif. Intell. Med.*, 56, 3, 137–156, 2012, https://doi.org/10.1016/j.artmed.2012.09.003.
46. Yang, Y.L., Chuang, M.C., Lou, S.L., Wang, J., Thick-film textile-based amperometric sensors and biosensors. *Analyst*, 135, 6, 1230–1234, 2010.
47. Coyle, S., Curto, V.F., Benito-Lopez, F., Florea, L., Diamond, D., Wearable bio and chemical sensors, in: *Wearable Sensors, Fundamentals, Implementation and Applications*, E. Sazonov, and M.R. Neuman, (Eds.), Chapter 2.1, pp. 65–83, Academic Press, 2014. https://doi.org/10.1016/B978-0-12-418662-0.00002-7.
48. Jatoi, A.S., Khan, F.S.A., Mazari, S.A., Mubarak, N.M., Abro, R., Ahmed, J., Ahmed, M., Baloch, H., Sabzoi, N., Current applications of smart nanotextiles and future trends. *Nanosensors and Nanodevices for Smart Multifunctional Textiles*, pp. 343–365, 2021. https://doi.org/10.1016/B978-0-12-820777-2.00019-4.
49. Rai, P., Kumar, P.S., Oh, S., Kwon, H., Mathur, G.N., Varadan, V.K. *et al.*, Smart healthcare textile sensor system for unhindered-pervasive health monitoring, in: *Nanosensors, biosensors, and info-tech sensors and systems*, 2012.
50. Massaroni, C., Saccomandi, P., Schena, E., Medical smart textiles based on fiber optic technology: an overview. *J. Funct. Biomater.*, 6, 2, 204–221, 2015. https://doi.org/10.3390/jfb6020204.
51. Juslén, H., Improving Health and Healthcare with Light, in: *Business Briefing: Hospital Engineering and Facilities Management*, 2005.
52. Troester, G., The agenda of wearable healthcare, in: *IMIA Yearbook of Medical Informatics*, R. Haux, and C. Kulikowski, (Eds.) Schattauer, Stuttgart, 2005.

53. Firoozbaksh, B., Jayant, N., Park, S., Jayaraman, S., *Proc. IEEE Int.Conf. Multimedia and Expo*, IEEE Signal Processing Society, Piscataway, NJ, pp. 1253–1256, 2000.
54. Ngo, C. and Van de Voorde, M.H., Nanotechnology for the textile industry, in: *Nanotechnology in a nutshell*, pp. 321–329, Springer, Atlantis Press, Paris, 2014. https://doi.org/10.2991/978-94-6239-012-6_19.
55. Coyle, S., Wu, Y., Lau, K.-T., De Rossi, D., Wallace, G., Diamond, D., Smart Nanotextiles: A review of Materials and Applications. *MRS Bulletin*, 32, 5, 434–442, 2007. doi:10.1557/mrs2007.67.
56. Nesenbergs, K. and Selavo, L., Smart textiles for wearable sensor networks: review and early lessons, in: *2015 IEEE international symposium on medical measurements and applications (MeMeA) proceedings*, 2015.
57. Castellano, J.J., Shafii, S.M., Ko, F., Donate, G., Wright, T.E., Mannari, R.J. *et al.*, Comparative evaluation of silver-containing antimicrobial dressings and drugs. *Int. Wound J.*, 4, 2, 114–122, 2007. https://doi.org/10.1111/j.1742-481X.2007.00316.x.
58. Coyle, S., Curto, V.F., Benito-Lopez, F., Florea, L., Diamond, D., Wearable bio and chemical sensors, in: *Wearable Sensors, Fundamentals, Implementation and Applications*, E. Sazonov, and M.R. Neuman (Eds.), pp. 65–83, Academic Press, 2014. https://doi.org/10.1016/B978-0-12-418662-0.00002-7.
59. Aguda, O.N. and Lateef, A., Recent advances in functionalization of nanotextiles: A strategy to combat harmful microorganisms and emerging pathogens in the 21st century. *Heliyon*, 8, 6, e09761, 2022. https://doi.org/10.1016/j.heliyon.2022.e09761.
60. Joshi, M. and Adak, B., Nanotechnology-Based textiles: a solution for the emerging automotive sector, in: In B. Banerjee (Ed.), *Rubber Nanocomposites and Nanotextiles, Perspectives in Automobile Technologies*, pp. 179–230, De Gruyter, 2019.
61. Hamouda, T., Ibrahim, H.M., Kafafy, H.H., Mashaly, H.M., Mohamed, N.H., Aly, N.M., Preparation of cellulose-based wipes treated with antimicrobial and antiviral silver nanoparticles as novel effective high-performance coronavirus fighter. *Int. J. Biol. Macromol.*, 181, 990–1002, 2021.
62. Rai, M., Bonde, S., Yadav, A., Plekhanova, Y., Reshetilov, A., Gupta, I., Golinska, P., Pandit, R., Ingle, A.P., Nanotechnology-based promising strategies for the management of COVID-19: current development and constraints. *Expert Rev. Anti-Infe. Ther.*, 2020, 20, 10, 1299–1308. https://doi.org/10.1080/14787210.2021.1836961.
63. Sachdev, N.B., {Tech vs. COVID-19: Graphene-Silver Nanotechnology} How Graphene Is Helping in the Fight against COVID-19, 2021.

Index

Printed and bound by CPI Group (UK) Ltd, Croydon, CR0 4YY
19/09/2026
14965102-0003